A. Abraham, L. C. Jain, B. J. van der Zwaag (Eds.)

Innovations in Intelligent Systems

Springer-Verlag Berlin Heidelberg GmbH

Studies in Fuzziness and Soft Computing, Volume 140

Editor-in-chief
Prof. Janusz Kacprzyk
Systems Research Institute
Polish Academy of Sciences
ul. Newelska 6
01-447 Warsaw
Poland
E-mail: kacprzyk@ibspan.waw.pl

Further volumes of this series can be found on our homepage: springeronline.com

Vol. 122. M. Nachtegael, D. Van der Weken, D. Van de Ville and E.E. Kerre (Eds.)
Fuzzy Filters for Image Processing, 2003
ISBN 3-540-00465-3

Vol. 123. V. Torra (Ed.)
Information Fusion in Data Mining, 2003
ISBN 3-540-00676-1

Vol. 124. X. Yu, J. Kacprzyk (Eds.)
Applied Decision Support with Soft Computing, 2003
ISBN 3-540-02491-3

Vol. 125. M. Inuiguchi, S. Hirano and S. Tsumoto (Eds.)
Rough Set Theory and Granular Computing, 2003
ISBN 3-540-00574-9

Vol. 126. J.-L. Verdegay (Ed.)
Fuzzy Sets Based Heuristics for Optimization, 2003
ISBN 3-540-00551-X

Vol 127. L. Reznik, V. Kreinovich (Eds.)
Soft Computing in Measurement and Information Acquisition, 2003
ISBN 3-540-00246-4

Vol 128. J. Casillas, O. Cordón, F. Herrera, L. Magdalena (Eds.)
Interpretability Issues in Fuzzy Modeling, 2003
ISBN 3-540-02932-X

Vol 129. J. Casillas, O. Cordón, F. Herrera, L. Magdalena (Eds.)
Accuracy Improvements in Linguistic Fuzzy Modeling, 2003
ISBN 3-540-02933-8

Vol 130. P.S. Nair
Uncertainty in Multi-Source Databases, 2003
ISBN 3-540-03242-8

Vol 131. J.N. Mordeson, D.S. Malik, N. Kuroki
Fuzzy Semigroups, 2003
ISBN 3-540-03243-6

Vol 132. Y. Xu, D. Ruan, K. Qin, J. Liu
Lattice-Valued Logic, 2003
ISBN 3-540-40175-X

Vol. 133. Z.-Q. Liu, J. Cai, R. Buse
Handwriting Recognition, 2003
ISBN 3-540-40177-6

Vol 134. V.A. Niskanen
Soft Computing Methods in Human Sciences, 2004
ISBN 3-540-00466-1

Vol. 135. J.J. Buckley
Fuzzy Probabilities and Fuzzy Sets for Web Planning, 2004
ISBN 3-540-00473-4

Vol. 136. L. Wang (Ed.)
Soft Computing in Communications, 2004
ISBN 3-540-40575-5

Vol. 137. V. Loia, M. Nikravesh, L.A. Zadeh (Eds.)
Fuzzy Logic and the Internet, 2004
ISBN 3-540-20180-7

Vol. 138. S. Sirmakessis (Ed.)
Text Mining and its Applications, 2004
ISBN 3-540-20238-2

Vol. 139. M. Nikravesh, B. Azvine, I. Yager, L.A. Zadeh (Eds.)
Enhancing the Power of the Internet, 2004
ISBN 3-540-20237-4

Ajith Abraham
Lakhmi C. Jain
Berend J. van der Zwaag (Eds.)

Innovations in Intelligent Systems

Springer

Dr. Ajith Abraham
Computer Science Department
Oklahoma State University
700 N Greenwood Avenue
Tulsa, OK 74106
USA
E-mail: ajith.abraham@ieee.org

Dr. Berend Jan van der Zwaag
Department of Electrical Engineering
University of Twente
P.O. Box 217
7500AE Enschede
The Netherlands
E-mail: b.j.vanderzwaag@utwente.nl

Prof. Dr. Lakhmi Jain
Knowledge-Based Intelligent Engineering
Systems Centre (KES)
University of South Australia
Adelaide, Mawson Lakes
South Australia, 5095
Australia
E-mail: l.jain@unisa.edu.au

ISSN 1434-9922
DOI 10.1007/978-3-540-39615-4

Library of Congress Cataloging-in-Publication-Data

Inovations in intelligent systems / Ajith Abraham, Lakhmi C. Jain, Berend J. van der Zwaag (eds.). p. cm. -- (Studies in fuzziness and soft computing ; v. 140)
Includes bibliographical references and index.
1. Expert systems (Computer science) 2. Artificial intelligence. I. Abraham, Ajith, 1968-
II. Jain, L. C. III. Van der Zwaag, Berend J. IV. Series.
QA76.76.E95I533 2004
006.3'3--dc22

Originally published by Springer-Verlag Berlin Heidelberg in 2004
MyCopy version of the original edition 2004

Cover design: E. Kirchner, Springer-Verlag, Heidelberg
Printed on acid free paper 62/3020/M - 5 4 3 2 1 0
www.springer.com/mycopy

Foreword

An "intelligent" machine embodies artificial intelligence in a general sense, and as a result displays intelligent behavior. The field of artificial intelligence evolved with the objective of developing computers that can think like humans. An intelligent machine relies on computational intelligence in generating its intelligent behavior. This requires a knowledge system in which representation and processing of knowledge are central functions.

Soft computing has effectively complemented conventional AI in the area of machine intelligence, and is widely used in knowledge representation and decision making associated with intelligent machines. It is an important branch of computational intelligence where fuzzy logic, probability theory, neural networks, and genetic algorithms are synergistically used to mimic reasoning and decision making of a human. In this context, fuzzy techniques attempt to approximate human knowledge and the associated reasoning process; neural networks are an artificial representation of the neuron structure of a brain; genetic algorithms follow procedures that are similar to the process of evolution in biological species; and uncertainty and the associated concept of probability are linked to approximation and are useful in representing the randomness in practical systems. Quite effective are the mixed or hybrid techniques, which exploit the advantages of two or more of these areas. In particular, fuzzy logic is useful in representing human knowledge in a specific domain of application and in reasoning with that knowledge to make useful inferences or actions. Artificial neural networks (NN) are massively connected networks of computational "neurons." Their computational power, learning capability, and the ability to approximate nonlinear functions are quite beneficial in intelligent systems. Genetic algorithms (GA) are derivative-free optimization techniques that can evolve analogous to biological evolution, and are applicable in machine intelligence, particularly when optimization is an objective.

As the field of soft computing matures into an engineering discipline, one needs to go from the development and enhancement of a sound analytical foundation to modeling, algorithm development, solution of benchmark problems, computer simulation, implementation in useful prototypes that are sufficiently complex, and rigorous testing and evaluation. The book, *Innovations in Intelligent Systems: Design, Management and Applications*, edited by Ajith Abraham, Lakhmi Jain, and Berend Jan van der Zwaag, makes an important and valuable contribution towards this end. The editors are internationally recognized experts in the field, and this makes the work authoritative. Equally important is the fact that they have carefully chosen a set of contributions that highlight the state of the art of intelligent paradigms.

The book covers such important topics as intelligent multi-agent systems, data mining, case-based reasoning, Bayesian control, universal approximators, and rough sets, all of which are central to intelligent systems and applications. In fact, the investigated techniques like pattern recognition and classification, machine learning, natural language processing, grammar, evolutionary schemes, fuzzy-neural procedures, and intelligent vision are all essential to the development of intelligent machines. The applications given in the book are quite relevant, complementary, and practically useful. They range from medical diagnosis and technical/medical language translation, to power demand forecasting, manufacturing plants, and pedestrian monitoring. In view of the depth and breadth of the coverage and the usefulness of the techniques and applications, *Innovations in Intelligent Systems: Design, Management and Applications* will undoubtedly be a valuable reference for experts and students alike.

Clarence W. de Silva July 2003
The University of British Columbia, Vancouver, Canada
http://www.mech.ubc.ca/facstaff/desilva.shtml

Preface

Machine intelligence refers back to 1936, when Alan M Turing proposed the idea of a universal mathematics machine, a theoretical concept in the mathematical theory of computability. The desire for intelligent machines remained just an elusive dream until the first computer was developed. When the first computer appeared in the early fifties, we admired it as an artificial brain, and we thought that we are successful in creating a low level decision making cognitive machine. Researchers coined the term Artificial Intelligence (AI) and waited for many potential applications to evolve. Research in AI is directed toward building "thinking" machines and improving our understanding of intelligence. As evident, the ultimate achievement in this field would be to mimic or exceed human cognitive capabilities including reasoning, recognition, creativity, emotions, understanding, learning and so on. Even though we are a long way from achieving this, some success has been achieved in mimicking specific areas of human mental activity.

Recent research into AI together with other branches of computer science has resulted in the development of several useful intelligent paradigms, which forms the basis of this volume. This volume is focused on some of the recent theoretical developments and its practical applications in engineering, science, business and commerce. The intelligent paradigms can be roughly divided among knowledge-based systems, computational intelligence and hybrid combinations.

Knowledge-based systems include expert and rule-based systems, intelligent agents and techniques for handling uncertainty (e.g., fuzzy logic). Computational intelligence includes neural networks, fuzzy inference systems, evolutionary computation and other optimization algorithms, rough sets, probabilistic reasoning and so on. The integration of different learning and adaptation techniques, to overcome individual limitations and achieve synergetic effects through hybridization or fusion of these techniques, has in recent years contributed to a large number of new hybrid system designs.

This volume is a rare collection of 18 chapters compiling the latest developments in the state-of-the-art research in intelligent paradigms and some of its practical interesting applications. The chapters are authored by world leading well-established experts in the field. Each chapter focuses on different aspects of intelligent paradigms and is complete by itself.

The volume is divided into two parts: "Theory" (Chapters 1-9) and "Applications" (Chapters 10-18). However, this does not intend to strictly divide the chapters. The theoretic chapters are not limited to theory only; they also illustrate the theory with examples or real-world applications. The chapters in the second part do not only give applications, but treat the underlying theory as well. The division merely indicates the main focus of the chapters contained in the respective parts.

The volume is further organized as follows:

Chapter 1 begins with an introduction to Support Vector Machines and some of the few computationally cheaper alternative formulations that have been developed in recent years. Further, the Multi-category Proximal Support Vector Machine (MPSVM) is presented in detail. The authors use a linear MPSVM formulation in an iterative manner to identify the outliers in the data set and eliminate (reducing) them. A k-nearest neighbor classifier is able to classify points using this reduced data set without significant loss of accuracy. The proposed theoretical frameworks are validated on a few publicly available OCR data sets.

Chapter 2 presents Bayesian Control of Dynamic Systems. Bayesian networks for the static as well as for the dynamic case have gained an enormous interest in the research community of machine learning and pattern recognition. Although the parallels between dynamic Bayesian networks and description of dynamic systems by Kalman filters and difference equations are well known since many years, Bayesian networks have not been applied to problems in the area of adaptive control of dynamic systems. To show how a Bayesian network can control a dynamic system authors exploit the similarities with Kalman Filters to calculate an analytical state space model. The performance of this analytical model is compared with the state space model after training with the EM algorithm and a model whose structure is deduced using difference equations. The experiments show that the analytical model as well as the trained model is suitable for control purposes, which leads to the idea of a Bayesian controller.

Chapter 3 introduces “AppART”: a hybrid neural network based on adaptive resonance theory for universal function approximation. AppART is an Adaptive Resonance Theory (ART) low-parameterized neural model that incrementally approximates continuous-valued multidimensional functions from noisy data using biologically plausible processes. AppART performs a higher order Nadaraya-Watson regression and can be interpreted as a fuzzy logic standard addictive model. Authors present AppART dynamics/training and its theoretical foundations as a function approximation method. Three benchmark problems are solved in order to study AppART from an application point of view and to compare its results with the ones obtained with other models. Finally, two modifications of the original AppART formulation aimed at improving AppART efficiency are proposed and tested.

The authors of Chapter 4 present an algorithmic approach to the main concepts of rough set theory. The rough set theory is a mathematical formalism for representing uncertainty, which can be considered as an extension of the classical set theory. It has been used in many different research areas, including those related to inductive machine learning and reduction of knowledge-based systems. This chapter is focused on the main concepts of rough set theory and presents a family of algorithms for implementing them.

An automated case generation from databases using similarity-based rough approximation is presented in Chapter 5. Knowledge acquisition for a case-based reasoning system from domain experts is a bottleneck in the system development process. It would be useful to derive representative cases automatically from larger, available databases rather than acquiring them from domain experts. Case generation is a branch of data mining that aims at choosing representative cases from large data sets for future case-based reasoning tasks. This Chapter presents two algorithms using similarity based rough set theory to derive cases automatically from available databases. The first algorithm, *SRS1*, requires the user to choose the similarity thresholds for the objects in a database, while the second algorithm, *SRS2*, can automatically select proper similarity thresholds. These algorithms can handle noise and inconsistent data in the database and select a reasonable number of the representative cases from the database. Also these algorithms are easily scalable. The algorithms were implemented and the experimental results showed that their classification accuracy was similar to that of well-known machine learning systems, such as rule induction systems and neural networks.

Chapter 6 introduces a new version of a machine-learning algorithm, FDM, based on a new notion of the fuzzy derivative. The main idea is to describe the influence of the change of one parameter on another. In this algorithm sets of classification rules are generated and a coefficient of significance for every single rule is defined. A new example is classified into a class for which its total degree of membership is maximal. In this way, the effect of a single non-informative rule having occurred by chance is decreased due to the coefficient of significance. The fuzzy derivative method is mainly used to study systems with qualitative features, but it can also be used for systems with quantitative features. The algorithm is applied to classification problems and comparisons made with other techniques.

In Chapter 7, the author explains the model and fixpoint semantics for fuzzy disjunctive programs with weak similarity. In such knowledge representation and commonsense reasoning, we should be able to handle incomplete and uncertain information. In recent years, disjunctive and multivalued, annotated logic programming have been recognized as powerful tools for maintenance of such knowledge's. This chapter presents a declarative model, and fixed-point semantics for fuzzy disjunctive programs with weak similarity – sets of graded strong literal disjunctions. Fuzzy disjunctive programs may contain the binary predicate symbol $\sim$ for weak similarity, which is the fuzzy counterpart of the classical equality. In the end, the mutual coincidence of the proposed semantics will be reached.

Chapter 8 proposes an automated report generation tool for the data-understanding phase. To be able to successfully prepare and model data, the data miner needs to be aware of the properties of the data manifold. The outline of a tool for automatically generating data survey reports for this purpose is described in this chapter. Such report is used as a starting point for data understanding, acts as a documentation of the data, and can be redone if necessary. The main focus is on describing the cluster structure and the contents of the clusters. The described system combines linguistic descriptions (rules) and statistical measures with visualizations. Whereas rules and mathematical measures give quantitative information, the visualizations give qualitative information of the data sets, and help the user to form a mental model of the data based on the suggested rules and other characterizations.

In Chapter 9, the authors propose a framework for a grammar-guided genetic programming system called Tree-Adjunct Grammar Guided Genetic Programming (TAG3P), which uses tree-adjunct grammars along with a context-free grammar to set language bias in genetic programming. The use of tree-adjunct grammars can be seen as a process of building context-free grammar guided programs in the two dimensional space. Authors show some results of TAG3P on the trigonometric identity discovery problems.

The main contribution of Chapter 10 is the development of a framework to determine both agent's behavior and cooperation allowing to express (1) cooperation, (2) adaptability, (3) mobility, and (4) transparency. In a multi-agent environment, each agent could be working at common goals with globally cooperative behaviors. In order to construct a model integrating agent's behavior and cooperation among agents, authors present two approaches for agent collaboration. As for the first approach, a social agency model for constructing a prototype system for guide activities in a laboratory is introduced. The interaction between autonomous agents is then formalized. As for the second approach, an autonomous agent's architecture in social agency aimed at communicating with other agents in knowledge-level is presented.

Chapter 11 presents two frameworks, an action control framework and a safety verification framework for intelligent information systems based on paraconsistent logic program called EVALPSN. Two examples for EVALPSN based intelligent information systems, an intelligent robot action control system and an automated safety verification system for railway interlocking are presented.

Chapter 12 deals with the different neuro-fuzzy paradigms for intelligent energy management. Fusion of Artificial Neural Networks (ANN) and Fuzzy Inference Systems (FIS) have attracted the growing interest of researchers in various scientific and engineering areas due to the growing need of adaptive intelligent systems to solve the real world problems. This chapter presents a fuzzy neural network for developing an accurate short term forecast for hourly power demand and a Mamdani and Takagi Sugeno fuzzy inference system learned using neural network learning technique for controlling the reactive power of a manufacturing plant. Performance of the developed models is compared with neural networks and other connectionist paradigms.

In Chapter 13, the authors use a real-coded genetic algorithm for information space optimization for Inductive Learning. This chapter begins with a presentation of new feature construction methods. The methods are based on the idea that a *smooth* feature space facilitates inductive learning thus it is desirable for data mining. The methods, Category-guided Adaptive Modeling (CAM) and Smoothness-driven Adaptive Modeling (SAM), are originally developed to model human perception of still images, where an image is perceived in a space of index colors. CAM is tested for a classification problem and SAM is tested for a Kansei scale value (the amount of the impression) prediction problem. Both algorithms have been proved to be useful as preprocess steps for inductive learning through the experiments. Authors have also evaluated CAM and SAM using datasets from the UCI repository and the empirical results has been promising.

In Chapter 14, the authors present a hybrid detection and classification system for human motion analysis (moving pedestrians in a video sequence). The technique comprises two sub-systems: an active contour model for detecting and tracking moving objects in the visual field, and an MLP neural network for classifying the moving objects being tracked as 'human' or 'non-human'. The axis crossover vector method is used for translating the active contour into a scale-, location-, resolution-, and rotation-invariant vector suited for input to a neural network according to the most appropriate level of detail for encoding human shape information. Experiments measuring the neural network's accuracy at classifying unseen computer generated and real moving objects are presented, along with potential applications of the technology.

Chapter 15 discusses two applications of the theory of fuzzy sets in investigating and evaluating human learning abilities and cognitive processes. They are an integral part of the Interactivist-Expectative Theory on Agency and Learning (IETAL) and its multiagent expansion known as Multi-Agent Systems Interactive Virtual Environments (MASIVE). In the first application presented, a fuzzy set is defined to ease and automate the process of detection of negative variation in filtered brain waves during the Dynamic Cognitive Negative Variation (CNV) experiment. The automatic detection of brain waveforms that are contingent of negative variation is a crucial part of the experiment that measures individual human learning parameters. By eliminating the direct influence of the human expert, a level of objectivity is being maintained over the duration of the whole experiment. The decision process is significantly shorter, which contributes to more accurate measuring, as is the case in numerous experiments involving human subjects and learning. In the second application, fuzzy sets serve as tools in the process of grading, which are a highly cognitive, but ill-defined problems. The fuzzy evaluation framework that is given is very general, and straightforwardly applicable in any evaluation process when the evaluator is expected to quantize one or several aspects of a given artifact.

In Chapter 16, the authors present a full explanation facility that has been developed for any standard Multi-Layered Perceptron (MLP) network with binary input neurons that performs a classification task. The interpretation of any input case is represented by a non-linear ranked data relationship of key inputs, in both text and graphical forms. The knowledge that the MLP has learned is represented by average ranked class profiles or as a set of rules induced from all training cases. The full explanation facility discovers the MLP knowledge bounds as the hidden layer decision regions containing classified training examples. Novel inputs are detected when the input case is positioned in a decision region outside the knowledge bounds. Results using the facility are presented for a *48*-dimensional real-world MLP that classifies low-back-pain patients. Using the full explanation facility, it is shown that the MLP preserves the continuity of the classifications in separate contiguous threads of decision regions across the *48*-dimensional input space thereby demonstrating the consistency and predictability of the classifications within the knowledge bounds.

Chapter 17 presents a detailed survey of the automatic translation or autocoding systems used in translating unstructured natural language texts or verbatims produced by health care professionals to categories defined by a controlled vocabulary. In the medical domain, over the centuries several controlled vocabularies have emerged with the goal of mapping semantically equivalent terms such as *fever, pyrexia, hyperthermia*, and *febrile* on the same (numerical) value. Translating unstructured natural language texts or verbatims produced by healthcare professionals to categories defined by a controlled vocabulary is a hard problem, mostly solved by employing human coders trained both in medicine and in the details of the classification system. These techniques could also be applied to other problem domains.

The final chapter presents a genetic programming approach for the Induction of a natural language parser. When we try to deal with Natural Language Processing (NLP) we have to start with the grammar of a natural language. But the grammars described in linguistic literature have an informal form and many exceptions. Thus, they are not useful to create final formal models of grammars, which make machine processing of sentences possible. These grammars can be a starting point for the attempts to create basic models of natural language grammar at the most. However, it requires expert knowledge. Machine learning based on a set of sample sentences can be the better way to find the grammar rules. This kind of learning (grammatical inference) allows avoiding the preparation of knowledge about the language for the NLP system. The examples of correct and incorrect sentences allow the NLP systems with the self-evolutionary parser to try to find the right grammar. This self-evolutionary parser can be improved on the basis of new examples. Thus, the knowledge acquired in this way is flexible and easily modifiable. Authors proposed theoretical bases for the use of two classes of evolutionary computation that support-automated inference of fuzzy automaton-driven parser of natural language. This chapter examines the use of edge encoding, a genetic programming approach for induction of parser based on a fuzzy automaton.

Acknowledgments

We are grateful to the authors of this volume and to Greg Huang (Massachusetts Institute of Technology, USA), Pavel Osmera (Brno University of Technology, Czech Republic), Jose Mira (The Universidad Nacional de Educación a Distancia, Spain), Xiao-Zhi Gao (Helsinki University of Technology, Finland), Paulo Jose ad Costa Branco (Institute Superior Technical, DEEC, Portugal), Janos Abonyi, (University of Veszprém, Hungary), José Manuel Benítez (University of Granada, Spain), and Eulalia Schmidt (Polish Academy of Sciences, Poland) for the tremendous service by critically reviewing the chapters within the stipulated deadline. The editors are deeply grateful to Clarence W. de Silva (University of British Columbia) for the encouraging comments on the volume. The editors would like to thank Springer-Verlag, Germany, for the editorial assistance and excellent cooperative collaboration to produce this important scientific work. Last but not the least, we would like to express our gratitude to our colleagues from the Department of Computer Science, Oklahoma State University, USA; Knowledge-Based Intelligent Engineering Systems Centre, University of South Australia, Australia; and the Department of Electrical Engineering, University of Twente, the Netherlands, for supporting us to produce this volume. We hope that the reader will share our excitement to present this volume on "*Innovations in Intelligent Systems: Design, Management and Applications*" and will find this very useful.

Ajith Abraham, Lakhmi Jain, Berend Jan van der Zwaag
April 2003

Contents

Part 1

Theory

Chapter 4. 89

An algorithmic approach to the main concepts of rough set theory
Joaquim Quinteiro Uchôa and Maria do Carmo Nicoletti

Chapter 5. 111

Automated case selection from databases using similarity-based rough approximation
Liqiang Geng and Howard J. Hamilton

Chapter 12. 285

Neuro-fuzzy paradigms for intelligent energy management

Ajith Abraham and Muhammad Riaz Khan

Chapter 13. 315

Information space optimization for inductive learning

Ryohei Orihara, Tomoko Murakami, Naomichi Sueda, and Shigeaki Sakurai

Chapter 14. 343

Detecting, tracking, and classifying human movement using active contour models and neural networks

Ken Tabb, Neil Davey, Rod Adams, and Stella George

Chapter 15. 361

Fuzzy sets in investigation of human cognition processes

Goran Trajkovski

Chapter 16. 381

A full explanation facility for an MLP network that classifies low-back-pain patients and for predicting MLP reliability

M.L. Vaughn, S.J. Cavill, S.J. Taylor, M.A. Foy, and A.J.B. Fogg

Part 1

Theory

Chapter 1

Use of Multi-category Proximal SVM for Data Set Reduction

S.V.N. Vishwanathan and M. Narasimha Murty

Summary. We present a tutorial introduction to Support Vector Machines (SVM) and try to show using intuitive arguments as to why SVMs tend to perform so well on a variety of challenging problems. We then discuss the quadratic optimization problem that arises as a result of the SVM formulation. We talk about a few computationally cheaper alternative formulations that have been developed in recent years. We go on to describe the Multi-category Proximal Support Vector Machines (MPSVM) in more detail. We propose a method for data set reduction by effective use of MPSVM. The linear MPSVM Formulation is used in an iterative manner to identify the outliers in the data set and eliminate them. A k-Nearest Neighbor (k-NN) classifier is able to classify points using this reduced data set without significant loss of accuracy. We also present geometrically motivated arguments to justify our approach. Experiments on a few publicly available OCR data sets validate our claims.

1 Introduction

k-Nearest Neighbor (k-NN) classifiers are one of the most robust and widely used classifiers in the field of Optical Character Recognition [8]. The time required for classification of a test point using a k-NN classifier is linear in the number of points in the training set. One popular method of speeding up the k-NN classifier is to reduce the number of points in the training set by appropriate data selection and *outlier* elimination. It is also well known that the presence of *outliers* tends to decrease the classification accuracy of the k-NN classifier [15].

Support Vector Machines (SVM) have recently gained prominence in the field of machine learning and pattern classification [13, 1]. Classification is achieved by realizing a linear or nonlinear separation surface in the input space. Multi Category Proximal Support Vector Machines (MPSVM), which have been proposed recently, are close in spirit to SVMs but are computationally more attractive [6].

We propose an hybrid classification system, wherein, we first use the MPSVM iteratively to perform data set reduction and outlier elimination. The pre-processed data is now used as the training set for a k-NN classifier. The resultant k-NN classifier is more robust and takes less time to classify test points.

This chapter is organized as follows. In Section 2 we present a brief introduction to VC theory. We also point out a few shortcomings of traditional machine learning algorithms and show how these lead naturally to the development of SVMs. In Section 3 we introduce the linearly separable SVM formulation and discuss its extension to the nonlinear case in Section 4. We try to present a few geometrically motivated arguments to show why SVMs perform very well. We sacrifice some mathematical rigor in order to present more intuition to the reader. While we concentrate our attention entirely on the pattern recognition problem, an excellent tutorial on the use of SVMs for regression can be found in [14].

In the second part of the chapter we propose a new method for data set reduction using the MPSVM. In Section 5 we briefly discuss the MPSVM formulation. We present our algorithm in Section 6. In Section 7 we discuss the experiments carried out on two widely available OCR data sets. Section 8 concludes with a summary of the present work and gives pointers for future research.

2 VC Theory - a Brief Primer

In this section we introduce the notation and formalize the binary learning problem. We then present the traditional approach to learning and point out some of its shortcomings. We go on to give some intuition behind the concept of VC-dimension and show why capacity is an important factor while designing classifiers. More information can be found in [16].

2.1 The Learning Problem

We consider the binary learning problem. Assume that we are given a training set of m labelled patterns

$$X = \{(\mathbf{x_1}, y_1), (\mathbf{x_2}, y_2), \ldots, (\mathbf{x_m}, y_m)\}$$

where $\mathbf{x_i} \in R^n$ are the patterns and $y_i \in \{+1, -1\}$ are the corresponding labels. Further, assume that the samples are all drawn i.i.d (Independent and Identically Distributed) from an unknown probability distribution $P(\mathbf{x}, y)$. The goal of building learning machines is to extract some kind of compact abstraction of the data so that we can predict well on unknown samples drawn from the same distribution $P(\mathbf{x}, y)$. In other words we want to learn the mapping $\mathbf{x_i} \rightarrow y_i$ which accurately models $P(\mathbf{x}, y)$. Such a machine may be parameterized by a set of adjustable parameters denoted by α and the learning function it generates is denoted by $f(\mathbf{x}, \alpha) : R^n \rightarrow \{+1, -1\}$. For example the α's could be the weights on various nodes of a neural network. As is clear different values of α generate different learning functions.

2.2 Traditional Approach to Learning Algorithms

Traditional learning algorithms like the neural networks concentrated their energy on the task of minimizing the empirical error on the training samples [7]. The empirical error is given by

$$E_{emp}(\alpha) = \sum_{i=1}^{m} c(f(\mathbf{x_i}, \alpha), y_i)$$

where $f(\mathbf{x_i}, \alpha)$ is the class label predicted by the algorithm for the ith training sample and $c(.,.)$ is some error function. The hope was that, if the training set was sufficiently representative of the underlying distribution, the algorithm would *learn* the distribution and hence *generalize* to make proper predictions on unknown test samples.

But, researchers soon realized that good training set performance did not always guarantee good test set accuracy. For example consider a naive learning algorithm that *remembers* every training sample presented to it. We call such an algorithm a *memory machine*. The *memory machine* of course has 100% accuracy on the training samples but clearly cannot *generalize* on the test set. In other words, what we are asking for is whether the mean of the empirical error converges to the actual error as the number of training points increases to infinity [16].

2.3 VC Bounds

The empirical risk for a learning machine is just the measured mean error rate on the training set. The $0-1$ loss function incurs a unit loss for every misclassified sample and does not penalize correctly classified samples. Using such a loss function the empirical risk can be written as

$$R_{emp}(\alpha) = \frac{1}{2m} \sum_{i=1}^{m} |f(\mathbf{x_i}, \alpha) - y_i|$$

where the scaling factor 0.5 has been included to simplify later calculations. Given a training set and a value of α, $R_{emp}(\alpha)$ is fixed. The actual risk which is the mean of the error rate on the entire distribution $P(\mathbf{x}, y)$ can be found by integrating over the entire distribution as

$$R_{actual}(\alpha) = \int \frac{1}{2} |f(\mathbf{x_i}, \alpha) - y_i| dP(\mathbf{x}, y)$$

Let, $0 \leq \eta \leq 1$ be a number. Then, Vapnik and Chervonenkis proved that, for the $0-1$ loss function, with probability $1-\eta$, the following bound holds [16]

$$R_{actual}(\alpha) \leq R_{emp}(\alpha) + \phi\left(\frac{h}{m}, \frac{\log(\eta)}{m}\right) \tag{1}$$

where

$$\phi\left(\frac{h}{m}, \frac{\log(\eta)}{m}\right) = m\sqrt{\frac{h(log(2m/h)+1) - log(\eta/4)}{m}} \tag{2}$$

is called the *confidence* term. Here h is defined to be a non-negative integer called the Vapnik Chervonenkis (VC) dimension. The VC-dimension of a machine measures the capacity of the machine to learn complex decision boundaries. In the most abstract sense, a learning machine can be thought of a set of functions that the machine has at its disposal. When we are talking of the VC-dimension of a machine we are talking about the capacity of these functions that the learning machine can implement. In the case of binary classifiers, the VC-dimension is the maximal number of points which can be separated into two classes in all possible 2^h ways by the learning machine.

Consider the *memory machine* that we introduced in Section 2.2. Clearly this machine can drive the empirical risk to zero but still does not generalize well because it has a large capacity. This leads us to the observation that, while minimizing empirical error is important, it is equally important to use a machine with a low capacity. In other words, given two machines with the same empirical risk, we have higher confidence in the machine with the lower VC-dimension.

A word of caution is in order here. It is often very difficult to measure the VC-dimension of a machine practically. As a result it is quite difficult to calculate the VC bounds explicitly. The bounds provided by VC theory are often very loose and may not be of practical use. It must also be borne in mind that only an upper bound on the actual risk is available. This does not mean that a machine with larger capacity will always generalize poorly. What the bound says is that, given the training data, we have more confidence in a machine which has lower capacity. In some sense (1) is a restatement of the principle of *Occam's razor*.

2.4 Structural Risk Minimization

Although the bounds provided by VC theory are not tight, we can exploit them in order to do model selection. A *structure* is a nested class of functions S_i such that

$$S_1 \subseteq S_2 \subseteq \ldots \subseteq S_n \subseteq \ldots$$

and hence their VC-dimensions h_i satisfies

$$h_1 \leq h_2 \leq \ldots \leq h_n \ldots$$

Now, because of the nested structure of the function classes

- The empirical risk R_{emp} decreases as the complexity, and hence the VC-dimension, of the class of functions increases
- The confidence bound (ϕ) increases as h increases.

The curves shown in Figure 1 depict (1) and the above observations pictorially.

These observations suggest a principled way of selecting a class of functions by choosing that class which minimizes the bound on the actual risk over the entire structure. The procedure of selecting the right subset for a given amount of observations is referred to as *capacity control* or *model selection* or *structural risk minimization*.

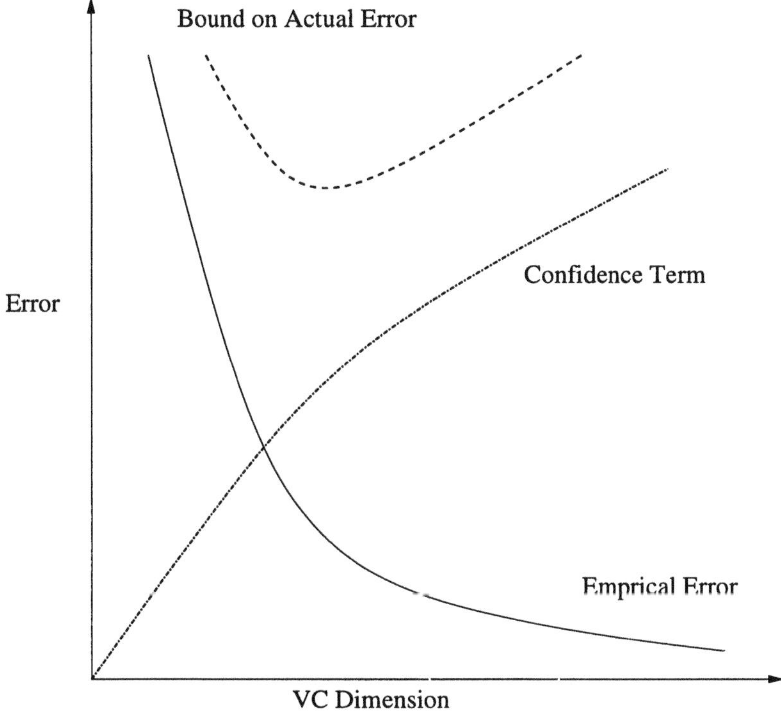

Figure 1. The heavy dotted line represents the bound on the actual error. It is minimized only when empirical error as well as the confidence term are simultaneously minimized.

3 Introduction to Linear SVMs

First, consider a simple linearly separable case where we have two clouds of points belonging to separate classes in 2 dimensions as shown in Figure 2. There are many linear boundaries that can separate these points but the one that is intuitively appealing is the one that maximally separates the points belonging to two different classes. In some sense we are making the best guess given the limited data that is available to us. It turns out that such a separating hyper plane comes with guarantees on its generalization performance. We show in Section 3.3 that selecting such a hyper plane is equivalent to doing structural risk minimization.

3.1 Formulating the Separable Linear SVM Problem

We first formulate the SVM problem for a linearly separable, 2-class problem. As before, $X = \{(\mathbf{x_1}, y_1), (\mathbf{x_2}, y_2), \ldots, (\mathbf{x_m}, y_m)\}$ be a set of labelled data points with

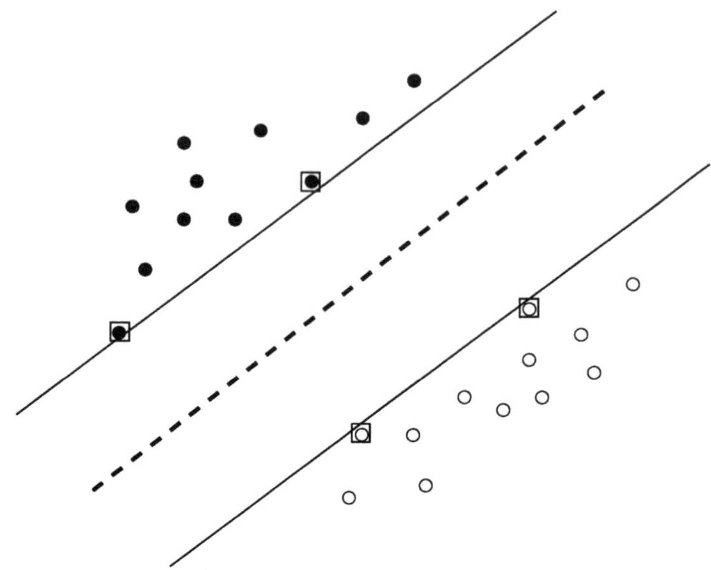

Figure 2. The empty circles and filled circles belong to two different classes. The dashed line represents the maximally separating linear boundary. The boxed points are Support Vectors.

$\mathbf{x_i} \in R^n$ and $y_i \in \{+1, -1\}$. Further, assume that there is a linear hyper plane parameterized by $(\mathbf{w}, b)$ which separates the points belonging to two different classes. We can write the equation for the hyper plane as

$$\mathbf{w}.\mathbf{x} + b = 0$$

where $\mathbf{w} \in R^n$ is the normal to the hyper plane and $|b|/||\mathbf{w}||$ is the perpendicular distance of the hyper plane from the origin ($||\mathbf{w}||$ is the Euclidean norm of $\mathbf{w}$). Let d_+ be the distance of the point closest to the hyper plane with a class label $+1$ and d_- be the corresponding distance for a point with class label -1. We define the margin to be $d_+ + d_-$.

We first show how the problem of maximizing the margin can be expressed as a quadratic optimization problem involving only dot products. Then, we show in Section 4 how we can use the kernel trick to solve this problem in higher dimensions. We require a separating hyper plane that satisfies

$$\mathbf{w}.\mathbf{x_i} + b \geq 1$$

for all points with $y_i = +1$ and

$$\mathbf{w}.\mathbf{x_i} + b \leq -1$$

for all points with $y_i = -1$. This can be expressed compactly as

$$y_i(\mathbf{w}.\mathbf{x_i} + b) \geq 1 \quad \forall i = 1, 2, \ldots m$$

Clearly, points which satisfy the above constraint with equality are the closest points to the hyper plane. If $y_i = +1$ then such points lie on the plane

$$H_1 : \mathbf{w}.\mathbf{x_i} + b = 1$$

which is at a distance of $|1 - b|/||\mathbf{w}||$ from the origin and hence $d_+ = 1/||\mathbf{w}||$. Similarly, if $y_i = -1$ then such points lie on the plane

$$H_2 : \mathbf{w}.\mathbf{x_i} + b = -1$$

which is at a distance of $|-1 - b|/||\mathbf{w}||$ from the origin and hence $d_- = 1/||\mathbf{w}||$. In the simple 2-dimensional case depicted in Figure 2 the boxed points are the closest points to the maximal hyper plane. We can now express our problem of maximizing the margin as

$$\min \frac{1}{2}||\mathbf{w}||^2 \quad st \quad y_i(\mathbf{w}.\mathbf{x_i} + b) \geq 1 \quad \forall i = 1, 2, \ldots m$$

where the factor $\frac{1}{2}$ has been introduced to simplify further calculations. A standard technique for solving such problems is to formulate the Lagrangian and solve for the dual problem. To formulate the Lagrangian for this problem we introduce positive Lagrange multipliers $\alpha_i (\geq 0)$ for $i = 1, 2, \ldots m$ and write

$$L(\alpha, \mathbf{w}, b) = \frac{1}{2}||\mathbf{w}||^2 - \sum_{i=1}^{m} \alpha_i y_i (\mathbf{w}.\mathbf{x_i} + b) + \sum_{i=1}^{m} \alpha_i$$

We want to minimize the above Lagrangian with respect to $\mathbf{w}$ and b which requires that the derivative of $L(\alpha, \mathbf{w}, b)$ with respect to $\mathbf{w}$, and b vanish. Therefore we get

$$\mathbf{w} = \sum_{i=1}^{m} \alpha_i y_i \mathbf{x_i}$$

and

$$\sum_{i=1}^{m} \alpha_i y_i = 0$$

Substituting the above results into the Lagrangian gives us

$$L(\alpha, \mathbf{w}, b) = \sum_{i=1}^{m} \alpha_i - \frac{1}{2} \sum_{i,j=1}^{m} \alpha_i \alpha_j y_i y_j \mathbf{x_i}.\mathbf{x_j}$$

As can be seen, the problem is a convex quadratic optimization problem which involves only the dot products of vectors $\mathbf{x_i}$ and $\mathbf{x_j}$. This is a key observation which will be useful to extend these arguments to the nonlinear case. Another interesting observation is that, α_i's are non-zero for only those data points which lie on either H_1 or H_2. These points are called *Support Vectors* to denote the fact that the solution would have been different if these points had not been present. So in some sense they are *supporting* the current solution. The Support Vectors for our simple 2-dimensional case are shown in Figure 2.

An interesting property of the solution is that it can be expressed completely in terms of the Support Vectors. What it means is that the solution can be specified by only a small fraction of the data points which are closest to the boundary. In some sense the SVMs assign maximum weightage to boundary patterns which intuitively are the most important patterns for discriminating between the two classes.

3.2 Formulating the Non-Separable Linear SVM Problem

Practically observed data is frequently corrupted by noise. It is also well known that the noisy patterns tend to occur near the boundaries [15]. In such a case the data points may not be separable by a linear hyper plane. Furthermore we would like to ignore the noisy points in order to improve generalization performance. Consider for example the scenario depicted in Figure 3. If the outlier is also taken into account then the margin of separation decreases and intuitively the solution does not generalize well.

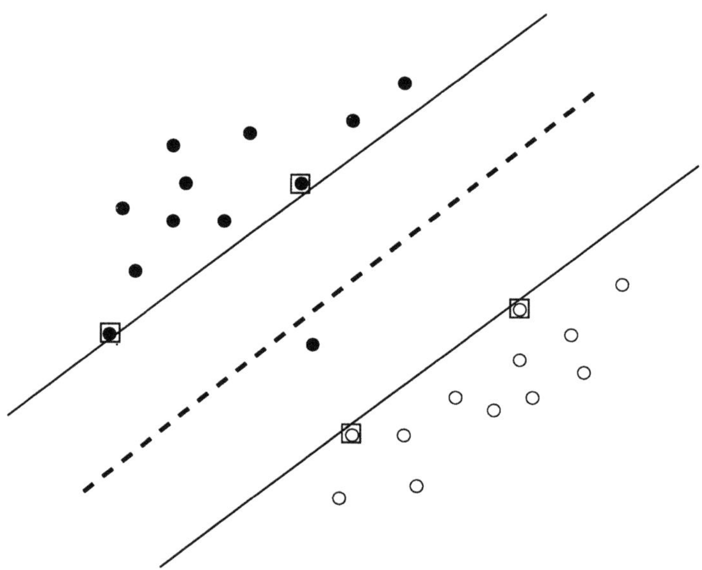

Figure 3. Ignoring the single outlier increases the margin of separation and leads to better generalization.

In this section we use the ideas of the previous section to extend the formulation to the non-separable case. This makes the designed classifier robust to the presence of noisy points. We account for outliers by introducing slack variables $\xi_i \geq 0$ for

$i = 1, 2, \ldots, m$ which penalize the outliers [2]. We then require

$$\mathbf{w}.\mathbf{x_i} + b \geq 1 - \xi_i$$

for all points with $y_i = +1$ and

$$\mathbf{w}.\mathbf{x_i} + b \leq -1 + \xi_i$$

for all points with $y_i = -1$. We now minimize the objective function

$$\min \frac{1}{2}||\mathbf{w}||^2 + C\sum_{i=1}^{m} \xi_i$$

where C is a penalty factor which controls the penalty incurred by each misclassified point in the training set. On solving the above problem using the same methods as outlined in the previous section we get

$$L(\alpha, \mathbf{w}, b) = \sum_{i=1}^{m} \alpha_i - \frac{1}{2} \sum_{i,j=1}^{m} \alpha_i \alpha_j y_i y_j \mathbf{x_i}.\mathbf{x_j} \tag{3}$$

with the additional constraint that

$$0 \leq \alpha_i \leq C \quad \forall i = 1, 2, \ldots m$$

which again is a convex quadratic optimization problem. In this case, if $\alpha_i = C$ then we call such a pattern an *error vector*.

Another interesting formulation penalizes the error points quadratically rather than linearly [4]. In other words we minimize the objective function

$$\min \frac{1}{2}||\mathbf{w}||^2 + C\sum_{i=1}^{m} \xi_i^2$$

This formulation has been shown to be equivalent to the separable linear formulation in a space that has more dimensions than the kernel space [9].

3.3 SVM and SRM

We now try to provide some intuition on how SVMs implement the SRM principle discussed in Section 2.4. A ball of radius R around a point $\mathbf{x} \in R^n$ is defined as

$$B_R(\mathbf{x}) = \{\mathbf{y} \in R^n : ||\mathbf{y} - \mathbf{x}|| < R\}.$$

The following lemma from [16] relates the margin of separation to the VC-dimension of a class of hyper planes.

Lemma 1. *Let, R be the radius of the smallest ball containing all the training samples $\mathbf{x}_1, \mathbf{x}_2, \ldots, \mathbf{x}_m$. The decision function defined by a hyper plane with parameters $\mathbf{w}$ and b can be written as*

$$f_{\mathbf{w},b} = \text{sgn}(\mathbf{w}.\mathbf{x} + b)$$

The set $\{f_{\mathbf{w},b} : ||\mathbf{w}|| \leq A, A \in R\}$ has VC-dimension h satisfying

$$h < R^2A^2 + 1$$

Recall that the margin of separation between hyper planes is given by $2/||\mathbf{w}||$. A large margin implies a small value for $||\mathbf{w}||$ and hence a small value of A, thus ensuring that the VC-dimension of the class $f_{\mathbf{w},b}$ is small. This can be understood geometrically as follows. As the margin increases the number of planes, with the given margin, which can separate the points into two classes decreases and thus the capacity of the class decreases. On the other hand a smaller margin implies a richer class of admissible hyper planes which translates to higher capacity. SVMs also penalize misclassified points in order to ensure that R_{emp} is minimized. Thus, SVMs select the optimal hyper plane which minimizes the bound on the actual risk R_{actual}. This also justifies our intuition that the hyper plane with the largest margin of separation is in some sense the optimal hyper plane.

4 The Kernel Trick for Nonlinear Extensions

As we noted in Section 3.1, the final quadratic optimization problem is expressed in terms of dot products between individual samples. Assume that we have a nonlinear mapping

$$\phi : R^n \to R^d$$

for some $d \gg n$ such that

$$k(\mathbf{x_i}, \mathbf{x_j}) = \langle \phi(\mathbf{x_i}), \phi(\mathbf{x_j}) \rangle \tag{4}$$

and

$$k(\mathbf{x_i}, \mathbf{x_j}) = k(\mathbf{x_j}, \mathbf{x_i}). \tag{5}$$

Then we can write (3) as

$$L(\alpha, \mathbf{w}, b) = \sum_{i=1}^{m} \alpha_i - \frac{1}{2} \sum_{i,j=1}^{m} \alpha_i \alpha_j y_i y_j k(\mathbf{x_i}, \mathbf{x_j}) \tag{6}$$

which allows us to work in a d-dimensional space ($d \gg n$) where the data points are possibly linearly separable. The function k is called as a kernel function.

Now consider the case the data points are not vectors from R^n but are drawn from an arbitrary input domain χ. Suppose, we are able to find a mapping

$$\phi : \chi \to \mathcal{H}$$

where $\mathcal{H}$ is a Hilbert space. Furthermore, suppose the mapping ϕ satisfies (4) and (5). We can apply our SVM algorithm for classifying such data also. Thus the advantage of using a kernel function is that we can work with non-vectorial data as long as the corresponding kernel function is known.

The natural question to ask is whether we can always find a Hilbert space $\mathcal{H}$ and a mapping ϕ such that we can express dot products in $\mathcal{H}$ in the form of (4). It turns out that certain class of functions which satisfy the Mercer's conditions are admissible as kernels. The rather technical condition is expressed as the following two lemmas [3].

Lemma 2. *If k is a continuous symmetric kernel of a positive integral operator K of the form*

$$(Kf)(\mathbf{y}) = \int_C k(\mathbf{x}, \mathbf{y}) f(\mathbf{x})\, d\mathbf{x}$$

with

$$\int_{C \times C} k(\mathbf{x}, \mathbf{y}) f(\mathbf{x}) f(\mathbf{y})\, d\mathbf{x}\, d\mathbf{y} \geq 0$$

for all $f \in L^2(C)$ where C is a compact subset of R^n, it can be expanded in a uniformly convergent series (on $C \times C$) in terms of Eigenfunction ψ_j and positive Eigenvalues λ_j

$$k(\mathbf{x}, \mathbf{y}) = \sum_{j=1}^{N_F} \lambda_j \psi_j(\mathbf{x}) \psi_j(\mathbf{y}),$$

where $N_F \leq \infty$.

Lemma 3. *If k is a continuous kernel of a positive integral operator, one can construct a mapping ϕ into a space where k acts as a dot product,*

$$\langle \phi(\mathbf{x}), \phi(\mathbf{y}) \rangle = k(\mathbf{x}, \mathbf{y})$$

We refer the reader to Chapter 2 of [13] for an excellent technical discussion on Mercer's conditions and related topics.

A few widely used kernel functions are the following:

- Gaussian Kernels of the form

$$k(\mathbf{x}, \mathbf{y}) = \exp\left(-\frac{||\mathbf{x} - \mathbf{y}||^2}{2\sigma^2}\right)$$

  with $\sigma > 0$ and
- Polynomial Kernels

$$k(\mathbf{x}, \mathbf{y}) = (\langle \mathbf{x}.\mathbf{y} \rangle)^m$$

  where m is a positive integer.
- Sigmoid Kernels

$$k(\mathbf{x}, \mathbf{y}) = \tanh(\kappa(\langle \mathbf{x}.\mathbf{y} \rangle) + \Theta)$$

  where $\kappa > 0$ and $\Theta < 0$.

Thus, by using kernel functions we can map data points onto a higher dimensional space where they are linearly separable. Note that this mapping is implicit and at no point do we actually need to calculate the mapping function ϕ. As a result all calculations are carried out in the space in which the data points reside. In fact any algorithm which uses similarity between points can be *kernelized* to work in higher dimensional space.

5 Multi Category Support Vector Machines (MPSVM)

Training a SVM involves solving a quadratic optimization problem which requires the use of optimization routines from numerical libraries. This step is computationally intensive, can be subject to stability problems and is non-trivial to implement [11]. Attractive iterative algorithms like the SMO, NPA and SSVM have been proposed to overcome this problem [11, 9, 19]. These algorithms solve the quadratic optimization problem using various heuristics, some of which are geometrically motivated. Most of these algorithms scale as quadratic or cubic in the input size.

The MPSVM algorithm is motivated by SVMs but is computationally more attractive since it solves a linear optimization problem instead of a quadratic problem. We briefly discuss the MPSVM formulation in this section. Our discussion follows [5, 6].

Let, $X = \{x_1, x_2, \ldots x_m\}$ be a set of m points in n-dimensional real space R^n represented by a $m \times n$ matrix A. We consider the problem of classifying these points according to the membership of each point in the class $A+$ or $A-$ as specified by a given $m \times m$ diagonal matrix D which contains $+1$ or -1 along the diagonal. For this problem, the Proximal Support Vector Machine (PSVM) formulation for a linear kernel with $\nu > 0$ is given by [5]

$$\begin{aligned} \min_{(w,\gamma,y)\in R^{n+1+m}} \quad & \frac{\nu}{2}\|Ny\|^2 + \frac{1}{2}\left\| \begin{bmatrix} w \\ \gamma \end{bmatrix} \right\|^2 \\ s.t \quad & D(Aw - e\gamma) + y = e \end{aligned} \tag{7}$$

Here e is a vector of ones and N is a diagonal normalization matrix used to account for the difference in the number of points in the two classes. If there are p samples in class $A+$ and $m-p$ samples in $A-$ then, N contains $1/p$ on rows corresponding to entries of class $A+$ and $1/(m-p)$ on rows corresponding to entries of class $A-$. y is an error variable and corresponds to the deviation of a point from the proximal plane.

Applying the KKT conditions and solving the above equations yields

$$w = \nu A^T DN[I - H(\frac{I}{\nu} + H^T NH)^{-1} H^T N] \tag{8}$$

and

$$\gamma = -\nu e^T DN[I - H(\frac{I}{\nu} + H^T NH)^{-1} H^T N] \tag{9}$$

Mangasarian et al. have proposed a nonlinear extension to the PSVM formulation using the *kernel trick* [5]. We do not use the nonlinear formulation in our experiments.

The MPSVM is a straight forward extension to PSVM where multiple classes are handled by using the one against the rest scheme, i.e., samples of one class are considered to constitute the class $A+$ and the rest of the samples are considered to belong to class $A-$, this is repeated for every class in the data set [6].

6 Data Set Reduction

Our algorithm for data set reduction is conceptually simple to understand. We reduce the size of the data set in two ways

- Boundary patterns which are most likely to cause confusion while classifying a point are pruned away from the training set.
- Very *typical* patterns of the class which are far removed from the boundary are often not useful for classification and can be safely ignored [12].

For each of the classes, we use the MPSVM algorithm to obtain separating planes. For the next iteration we retain only those points which are enclosed by these separating planes. These are points which satisfy

$$x^T w - \gamma \in (-1, 1) \tag{10}$$

This reduced set is used iteratively as input for the MPSVM algorithm. Once the number of data points enclosed by these separating planes becomes less than a preset threshold (T) we treat them as boundary points and prune them from the training set. We state this in Algorithm 1.

Algorithm 1 EliminateBoundary(A, D, ν, T, M)

```
Reduced := A
for classLabel=1:No Of Classes do
  ToDelete := Reduced
  for i=1:M do
    [W, γ] = MPSVM(ToDelete, D, ν)
    ToDelete := {x ∈ Reduced : x^T W − γ ∈ (−1, 1)}
    if size(ToDelete) < T then
      break
    end if
  end for
  Reduced := Reduced \ ToDelete
end for
return Reduced
```

During the first iteration, all points of a given class which lie on the other side of the separating plane are classified as far removed from the boundary and are removed. Assuming that samples of one class belong to $A+$ and the rest of the samples are classified as $A-$, these are points which satisfy

$$x^T w - \gamma > 1 \tag{11}$$

We state this procedure in Algorithm 2.

In both the algorithms, A is the full training set, D is a diagonal matrix indicating the class label of the training samples, T is the threshold on the number of samples to discard, M is the maximum number of iterations to perform per class.

Algorithm 2 EliminateTypical(A, D, ν)

```
Reduced := A
for classLabel=1:No Of Classes do
    ToDelete := Reduced
    [W, γ] = MPSVM(ToDelete, D, ν)
    ToDelete = {x ∈ Reduced : x^T W − γ > 1}
    Reduced := Reduced \ ToDelete
end for
return Reduced
```

7 Experimental Results

Our experiments were performed using two well known publicly available data sets used widely by us and other researchers [18, 12, 10].

The first training set consists of 6670 pre-processed hand written digits ($0-9$) with roughly 667 samples per class. The test set consists of 3333 samples with roughly 333 samples per class. All the samples had a dimensionality of 192 after initial preprocessing and there were no missing values.The second dataset that we used was the well known USPS dataset. It consists of a training set with 7291 images and a test set with 2007 images. Each image has a dimensionality of 256 and there are no missing values.

7.1 *k*-Nearest Neighbor Classifier Using Full Data Set

We performed *k*-Nearest Neighbor classification of the test set using all the samples in the training set. We experimented with values of *k* between 1 and 10 and the best classification accuracy for both the datasets is reported in Table 1.

Table 1. Accuracy of *k*-Nearest Neighbor using complete training set.

Dataset	% Accuracy	k-value
OCR	92.5	5
USPS	94.469	5

7.2 *k*-Nearest Neighbor Classifier Using Reduced Data Set

We used the Linear MPSVM to reduce the training set size as described in Section 6. A few outliers from the OCR data set which were eliminated by our algorithm are shown in Figure 4. We experimented with various values of threshold and ν and report the best results in Table 2 and 3. As the number of training points decreases the density of data points also decreases. We observed that the best classification accuracies are found for lower values of *k*.

Table 2. Accuracy of k-Nearest Neighbor on the OCR dataset using a reduced training set.

Threshold	ν	No. Eliminated	k-value	% Accuracy
50	1	2177	7	91.749175
100	1	2900	3	90.579058
150	1	3556	1	89.558956
200	1	3796	1	88.898890
250	1	4516	1	86.858686
300	1	4979	1	84.878488
50	5	2229	5	91.929193
100	5	2909	3	90.369037
150	5	3439	1	89.318932
200	5	3875	1	88.748875
250	5	4500	1	87.548755
300	5	4978	1	86.558656

Table 3. Accuracy of k-Nearest Neighbor on the USPS dataset using a reduced training set.

Threshold	ν	No. Eliminated	k-value	% Accuracy
50	1	3103	3	94.120578
100	1	3831	1	93.771799
150	1	3980	1	93.423019
200	1	4710	1	92.575984
250	1	4886	1	92.426507
300	1	5126	1	92.326856
50	5	3083	1	94.020927
100	5	3802	5	93.373194
150	5	4112	1	93.124066
200	5	4424	1	92.924763
250	5	4851	1	92.526158
300	5	5198	1	91.778774

7.3 Discussion of the Results

As can be seen, the best accuracy is obtained for the k-Nearest Neighbor classifier which uses the full training set for classification. But, even after elimination of a substantial number of training samples the classification accuracy does not drop significantly. In fact, more than half the samples can be throw away with out significantly affecting test set accuracy. Saradhi et al. report a accuracy of 86.32% on the OCR dataset after elimination of 2390 *atypical* patterns and 75.61% after elimination of 3972 *atypical* patterns [12]. Thus, it can be clearly seen that our method is effective in reducing the size of the data set without losing the patterns that are important for classification.

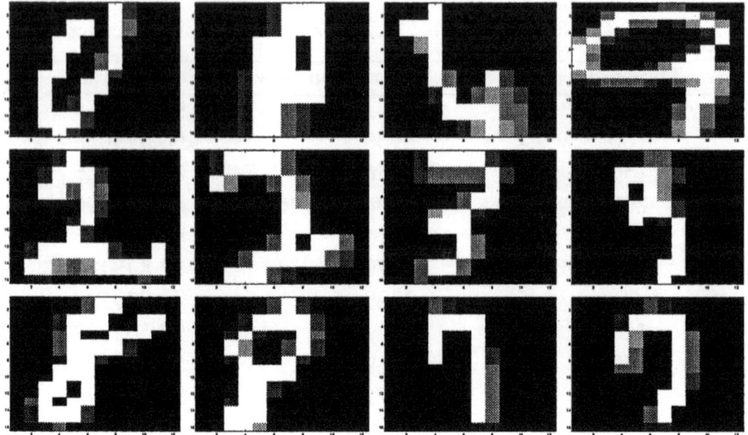

Figure 4. Examples of outliers detected by our algorithm. Actual class labels (L to R) are 0, 0, 4, 9 (Row1) 2, 2, 3, 3 (Row 2) 8, 8, 7, 7 (Row 3).

Another interesting feature is that as the data set size reduces the best classification accuracies are found for smaller k values. This can be explained by the fact that the density of training points in a given volume is decreasing because of dataset pruning. As a result the number of neighbors of a point in a given volume is also decreasing and thus results in better classification accuracies for lower values of k.

8 Conclusion

We have proposed a conceptually simple algorithm for data set reduction and outlier detection. In some sense the algorithm comes closer to methods like bootstrapping which work by increasing the separation between the classes. Our algorithm increases separation between classes by eliminating the noisy patterns at the class boundaries and thus leads to better generalization. It also identifies the typical patterns in the dataset and prunes them away leading to a smaller data set.

One immediately apparent approach to extend our algorithm is to use the Nonlinear MPSVM formulation which may lead to better separation in higher dimensional kernel space. At the time of writing results of this approach are not available. The main difficulty in implementing this approach seems that even a rectangular kernel of modest size requires a large amount of memory because of the large dimensionality of the problem. Use of some dimensionality reduction algorithm like the one proposed by us may be explored to overcome this limitation [17].

Saradhi et al. report good results by applying data set reduction techniques after bootstrapping the data points [12]. This is an area of further investigation.

References

1. C.J.C. Burges. A tutorial on support vector machines for pattern recognition. *Data Mining and Knowledge Discovery*, 2(2):121–167, 1998.
2. C. Cortes and V. Vapnik. Support vector networks. *Mahine Learning*, 20:273–297, 1995.
3. R. Courant and D. Hilbert. *Methods of Mathematical Physics*, volume 1. Interscience, New York, 1953.
4. T. Friess, N. Cristianini, and C. Campbell. The kernel adatron algorithm: a fast and simple learning procedure for support vector machine. In *Proceedings of 15th International Conference on Machine Learning*. Morgan Kaufman, 1998.
5. G. Fung and O.L. Mangasarian. Proximal support vector machine classifiers. In D. Lee, F. Provost, and R. Srikant, editors, *Proceedings KDD2001: Knowledge Discovery and Data Mining*, pages 64 – 70, New York, August 26-29 2001.
6. Glenn Fung and O.L. Mangasarian. Multicategory proximal support vector classifiers. Technical Report 01-06, Data Mining Institute, July 2001.
7. Simon Haykin. *Neural Networks: A Comprehensive Foundation*. Prentice Hall, second edition, July 1998.
8. L. Holmstrom, P. Koistinen, J. Laaksonen, and E. Oja. Neural and statistical classifier - taxonomy and two case studies. *IEEE Transactions on Neural Networks*, 8(1):5 – 17, January 1997.
9. S.S. Keerthi, S.K. Shevade, C. Bhattacharyya, and K.R.K. Murthy. A fast iterative nearest point algorithm for support vector machine classifier design. *IEEE Transactions on Neural Networks*, 11(1):124, 2000.
10. Y. LeCun P. Simard and J. Denker. Efficient pattern recognition using a new transformation distance. In S.J. Hanson, J.D. Cowan, and C.L. Giles, editors, *Advances in Neural Information Processing Systems*, volume 5, pages 50 – 58, San Mateo, CA, 1993. Morgan Kaufmann.
11. J.C. Platt. Fast training of support vector machines using sequential minimal optimization. In B. Schölkopf, C. Burges, and A. Smola, editors, *Advances in Kernel Methods: Support Vector Machines*. MIT Press, Cambridge, MA, December 1998.
12. V. Vijaya Saradhi. Pattern representation and prototype selection for handwritten digit recognition. Master's thesis, Indian Institute of Science, Bangalore, India, June 1999.
13. Bernhard Schölkopf and Alexander J. Smola. *Learning with kernels : Support vector machines, regularization, optimization and beyond*. Adaptive Computation and Machine Learning. The MIT press, Cambridge, MA, first edition, 2002.
14. Alex Smola and Bernhard Schölkopf. A tutorial on support vector regression. Technical Report NC2-TR-1998-030, NeuroCOLT, 1998.
15. D. Stork, R.O. Duda, and P.E. Hart. *Pattern Classification and Scene Analysis*. Wiley-Interscience, 2nd edition, 2000.
16. V.N. Vapnik. *The Nature of Statistical Learning Theory*. Springer, New York, 2nd edition, 2000.
17. S.V.N. Vishwanathan and M. Narasimha Murty. Use of Kohonen map for di-

mensionality reduction. Technical Report IISC-CSA-1999-8, Indian Institute Of Science, Bangalore, India, December 1999.

18. S.V.N. Vishwanathan and M. Narasimha Murty. Kohonen's SOM with cache. *Pattern Recognition*, 33:1927 – 1929, 2000.
19. S.V.N. Vishwanathan and M. Narasimha Murty. SSVM : A simple SVM algorithm. In *Proceedings of the IJCNN'02*, Honolulu, Hawaii, May 2002. IEEE Computer Society.

Chapter 2

Bayesian Control of Dynamic Systems

Rainer Deventer, Joachim Denzler, and Heinrich Niemann

Summary. Bayesian networks for the static as well as for the dynamic case have gained an enormous interest in the research community of machine learning and pattern recognition. Although the parallels between dynamic Bayesian networks and description of dynamic systems by Kalman filters and difference equations are well known since many years, Bayesian networks have not been applied to problems in the area of adaptive control of dynamic systems.

To show how a dynamic system can be controlled by a Bayesian network we exploit the similarities to Kalman Filters to calculate an analytical state space model. The performance of this analytical model is compared to a state space model trained with the EM algorithm and to a model whose structure is deduced using difference equations. The experiments show that the analytical model as well as the trained model are suitable for control purposes, which leads to the idea of a Bayesian controller.

Keywords: dynamic Bayesian networks, Kalman filter, dynamic system, controller

1 Introduction

Bayesian networks (BN) for the static as well as for the dynamic case have gained an enormous interest in the research community of artificial intelligence, machine learning and pattern recognition. Recently, BN have been applied also to static problems in production, since production processes become more and more complex so that analytical modeling and manual design are too expensive. One example for the successful application of BN to a *static system* in production are the quality evaluation and process parameter selection in order to reach an acceptable quality level [6].

Although the parallels between BNs on the one side and Kalman filters respectively the description by difference equations are well-known since many years, BN have not been applied to problems in the area of adaptive control of *dynamic systems*. Adaptive control of dynamic systems is one major problem in production processes. Compared to classical control methods BN have the advantage that the model (of the static or dynamic system) can be trained from examples if the model is not available in analytical form. During training missing information can be handled [25] [7], which makes BN superior to other self adaptive systems like artificial neural networks. Finally, BN can also calculate the most suitable input which leads to a desired output given the information that is entered as evidence in the BN.

In this chapter it is shown that BN can also act as controller. We exploit the well-known similarities between BN and Kalman filters on the one side and the description of dynamic systems by difference equations on the other side to model and control linear dynamic systems using dynamic Bayesian networks (DBN). We show, how the model is used to calculate appropriate input signals for the dynamic system in order to achieve a required output signal. The desired value is entered as evidence. Then, marginalization results in the most likely values of the input nodes.

The performance of the controller is evaluated using the steady state error, the time until the desired value is reached, the integral of the squared error and overshoot. This is done for both the reference and the disturbance reaction of the control loop.

The performance of a controller, inferred from a mathematical model, is compared with two different controllers, the first one is based on a state-space description, the second on difference equations. Both are trained with impulse and step responses. For the evaluation of the difference equations it is necessary to use not only the signals from the time step before, but to handle data longer ago. Models like that are denoted as higher-order Markov Models.

The training is done by the well-known EM-algorithm and simplified by normal forms which are used to reduce the number of parameters of the model and thus time complexity and search space during the training.

All three models show a good performance; the desired value is reached with low overshoot and the deviation of the actual value from the desired value is below 3% for both trained models.

Even if we focus in this chapter on the modeling of stationary, linear dynamic systems of second order, the extension to control nonlinear systems is straight forward using hybrid BN. A hybrid BN uses both discrete and continuous nodes. Discrete nodes are nodes whose values are taken from a finite set, for example the results when tossing a coin may be either heads or tails. In control theory continuous nodes with real numbers or vectors as domain are from greater importance. Using a hybrid Bayesian network means that the Gaussian distribution of the continuous nodes are replaced by a mixture of Gaussians [30].

An additional advantage of our approach is the possibility to learn not only the parameters of a given structure, but both the structure [9] and probabilities [1] of the Bayesian network from examples.

This chapter is structured as follows. In Section 2 the term control system is defined and the analytical descriptions applied in control theory are introduced. Section 3 deals with Bayesian networks and it is shown how the parameters of the DBN are obtained using the similarities to Kalman filters, a special type of DBN. The usage of BNs to generate control signals is explained in Section 4. The results obtained with this new type of controller are presented in Section 5. The chapter finishes with a short conclusion.

2 Dynamic Systems and Control Theory

2.1 The Aim of a Controller

Humans are regularly encountered with controlling tasks. Typical examples are driving a car. Here the driver has to take care of his speed to avoid being stopped by the police. Additionally the temperature of the heating in the car has to be controlled. A lot of controllers are also found in industry. Generally spoken it is the task of the controller to keep a value $\vec{q}$, for example the speed of a car, as close as possible to a desired value $\vec{w}$, for example the maximal speed allowed. If the environment is changing the controller should change the input so that the difference between the desired value and the current value is reduced as fast as possible.

This section is structured as follows. First the main elements of control loops are introduced together with a block diagram which shows the controller and the dynamic system to be controlled. Afterwards a mathematical description of the dynamic system is given. The description is taken from control theory, to ensure a broad field of application. Additionally this description is employed to deduce the structure and parameters of a Bayesian network used for control purposes. Using normal forms described in Section 2.3 results in a reduction of the number of parameters to be learned.

The main elements of the driver example discussed at the beginning can be found in Figure 1, where all variables are one dimensional. The desired value w is compared with the output q of the system, which means the desired speed is compared with the current speed. The difference e between desired and current value is used as input for the controller which calculates the input for the actuator. The changed input value of the dynamic system results in a new output, where the manipulation reaction describes how the system responds to new inputs. A crucial feature is the feedback loop which compares the output with the desired value.

Additionally the controller has to deal with influences from the environment. The reaction of the controlled system to the disturbing value z' is modeled by the disturbance reaction. As the disturbance value z' can usually not be measured it is easier to deal with the system's reaction z to the disturbance variable directly. From now on z is added to the output y of the system

$$q(t) = y(t) + z(t) \tag{1}$$

and thus regarded as disturbance variable instead.

As the block diagram shows, conventional controller uses the error as input. In contrary a model based controller is able to predict the system's reaction and calculate the new input accordingly. It will be shown how a Bayesian model can act as such a controller for a dynamic process and which steps can be taken to simplify the training of it. Simplified training means that some parameters are not changed during training which leads to a more stable controller.

To judge the quality of a controller a measure is needed. In control theory a lot of measures are known and the selection depends on the application. A metric that is

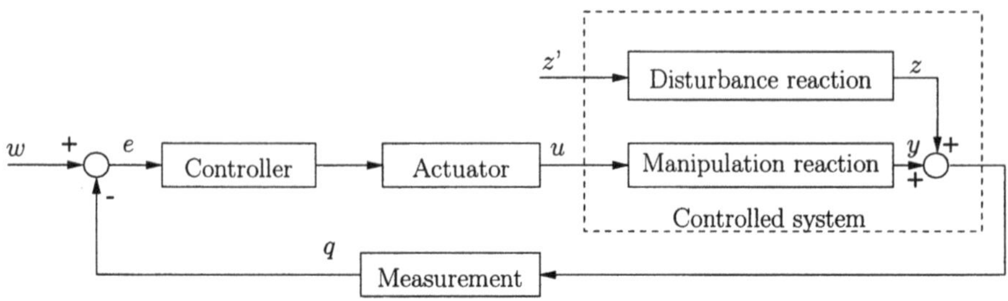

Figure 1. Block diagram of a controlled system.

frequently used integrates the squared error

$$Q = \int_0^\infty (e(t) - e_\infty)^2 dt \tag{2}$$

with $e(t)$ as the difference between the desired value $w(t)$ and the actual value $q(t)$. The steady state error e_∞, the error which remains at $t = \infty$, is subtracted to guarantee that the integral is limited even if a small error remains.

This measure has to be adapted in two ways. First we have only limited time. So the upper limit is changed from infinity to t_{max}, where t_{max} is reached when the system has converged to its final value. We assume convergence if the output has not changed for the last 20 time slices.

Additionally our measurements are only done at discrete time steps. So the integral is replaced by a sum

$$Q_{\text{d}} = \sum_{t=0}^{t_{\text{max}}} \Delta T (e_t - e_\infty)^2 \tag{3}$$

from $t = 0$ to t_{max}. To distinguish between discrete and continuous time we will use t in parentheses for continuous time, and as an index, for example e_t in discrete time systems. The time between two measurements is denoted by ΔT.

We use the following example scenario for testing. First the desired value is changed from 0 to 10 and after convergence the disturbance variable is changed from 0 to 1. Thus the regarded disturbance is 10% of the desired value which is enough to test the performance of the controller. For both cases Q_{d} and e_∞ are measured. Additionally the overshoot of the system, the difference between the maximal output q_{max} and the desired value w, is measured when the desired value is changed. To get an impression how much time is needed until the system reaches its new desired value the settling time, the time between changing and reaching the desired value w is given.

The next section deals with two different mathematical descriptions of the manipulation reaction (confer Figure 1). The comparison of the state space model with a Kalman filter results in an identification of a suitable Bayesian network for modeling the behavior of such a system. The description by a difference equation results in a different structure of the Bayesian network, which has the advantage that there are less unobservable variables. The drawback is that the Markov assumption does no longer hold, thus most of the standard tools can no longer be used.

2.2 Controlled Systems

A dynamical system may be regarded as a black box with several input and output signals $\vec{u}$ and $\vec{y}$ respectively, where the output does not depend solely on the input signal, but additionally on an internal state $\vec{x}$. Linear, time invariant systems with one dimensional in- and output, are regularly described by differential equations

$$\sum_{i=0}^{n} a_i \frac{d^i y(t)}{dt^i} = \sum_{j=0}^{m} b_j \frac{d^j u(t)}{dt^j} \tag{4}$$

whereas only systems with $m <= n$ are physically realizable. It is always possible to transform a differential equation of n-th order to n coupled differential equations of first order,

$$\frac{d\vec{x}(t)}{dt} = \vec{A}\vec{x}(t) + \vec{B}\vec{u}(t) \tag{5}$$

$$\vec{y}(t) = \vec{C}\vec{x}(t) + \vec{E}\vec{u}(t) \tag{6}$$

called the state-space description, which is also used for multivariate systems, that is $\vec{u} \in \mathbb{R}^i$ and $\vec{y} \in \mathbb{R}^o$. The transition matrix $\vec{A}$ describes the transition from one state $\vec{x}$ to the next, $\vec{B}$ the influence of the input $\vec{u}$ on the state. The output $\vec{y}$ depends on the state, as described by $\vec{C}$ and on the input $\vec{u}$, depicted by $\vec{E}$. Unless $n = m$ in (4), matrix $\vec{E} = 0$ and changes of the input have no immediate effect on the output.

To get a good impression of a dynamic system it is helpful to regard the step response of such a system, that is the system's output signal $\vec{y}(t)$, when $\vec{u}(t)$ is changed from 0 to 1 at $t = 0$. Figure 2 displays the step response of a system described by $q + 0.1\frac{dq}{dt} + 0.01\frac{d^2q}{dt^2} = 2u(t)$, which is one of our test systems.

The reader will notice that a second order system may overshoot and needs a long time to converge to a new value. In this chapter we will show how to calculate an input signal, so that overshooting is avoided and the output settles quickly to its new value. This method is based on Kalman filters as described in Section 3.3.

As a simple example a car with mass M, accelerated by a force F and slowed down by friction and a spring, is discussed in the following. The friction is proportional to the product of a constant b and the velocity $v = \frac{dy}{dt}$, the excursion y of the spring causes a force ky. Thus, the following equation holds:

$$M\frac{d^2y(t)}{dt^2} + b\frac{dy}{dt} + ky = F(t). \tag{7}$$

By substitution $x_1 = \frac{dy}{dt}$ and $x_2 = y$ the example can be transformed to a system of differential equations of first order

$$\begin{bmatrix} \frac{dx_1(t)}{dt} \\ \frac{dx_2(t)}{dt} \end{bmatrix} = \begin{bmatrix} -\frac{b}{M} & -\frac{k}{M} \\ 1 & 0 \end{bmatrix} \begin{bmatrix} x_1(t) \\ x_2(t) \end{bmatrix} + \begin{bmatrix} \frac{1}{M} \\ 0 \end{bmatrix} F(t) \tag{8}$$

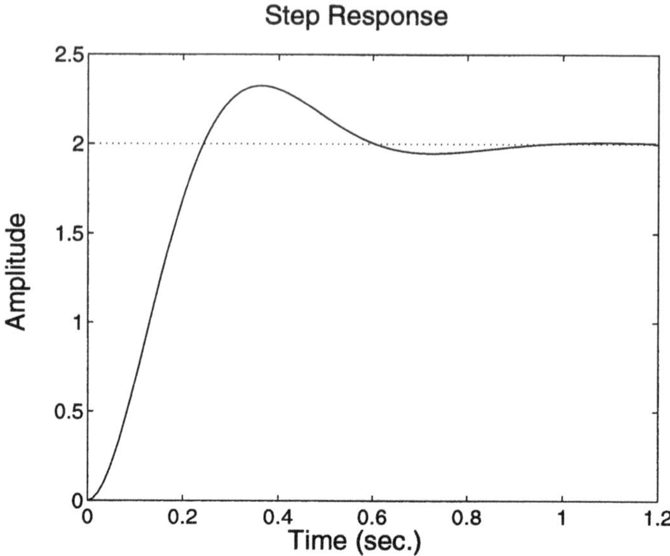

Figure 2. Step response of a second order system.

$$y(t) = x_2(t), \tag{9}$$

called the state space representation. In (5) and (6) $\vec{A}$ is an $n \times n$ matrix, where n denotes the order of the differential equation. For single input/single output systems, that is both y and u are scalars instead of vectors, $\vec{B}$ and $\vec{C}$ are vectors of length n and $\vec{E}$ is a scalar. Thus there are $n^2 + 2n + 1$ parameters in the state space description. Comparing the number of parameters in the state space description with the maximal number of parameters in differential equation (4) leads to the consideration that there are many possible state space descriptions for the same differential equation. Reducing the number of parameters would result in a smaller search space of possible state space descriptions which means a more robust and effective learning process. That is exactly what is done by normal forms.

2.3 Normal Forms

As mentioned in Section 2.2 there is no unique state space description for a given differential equation. As a simple example imagine a dynamic system with gain K. To model such a system the fortification can either take place between the input $\vec{u}$ and the state $\vec{x}$ or between the state $\vec{x}$ and the output $\vec{y}$.

Control theory distinguishes different types of normal forms. As this chapter is no introduction to control theory only the observable canonical form is discussed. The reason to use the observable canonical form is of practical nature and explained later.

In the observable canonical form the four matrices get a special form. The first thing to do is to normalize the regarded differential equation (4), so that the parameter $a_n = 1$. Then the state space description is changed, so that the matrices $\vec{A}$, $\vec{B}$, $\vec{C}$ and $\vec{E}$ are equal to

$$\vec{A} = \begin{bmatrix} 0 & 0 & 0 & \cdots & 0 & 0 & -a_0 \\ 1 & 0 & 0 & \cdots & 0 & 0 & -a_1 \\ 0 & 1 & 0 & \cdots & 0 & 0 & -a_2 \\ \cdots & & & & & & \cdots \\ 0 & 0 & & \cdots & 1 & 0 & -a_{n-2} \\ 0 & 0 & & \cdots & 0 & 1 & -a_{n-1} \end{bmatrix} \quad (10)$$

$$\vec{B} = \begin{bmatrix} b_0 \\ b_1 \\ \vdots \\ b_{n-1} \end{bmatrix} \quad (11)$$

$$\vec{C} = [0\,0\,0 \cdots 1] \quad (12)$$

$$\vec{E} = b_n \ . \quad (13)$$

The important aspect is that the introduction of normal forms is neither a restriction of the dynamic system, nor is the input/output-behavior of the system changed. Only the internal state of the system is concerned. Using the observable canonical form the number of free parameters is reduced to $2n + 1$ which is equal to the number of the parameter in the differential equation if $n = m$ and a_n is normalized to 1. That means that no redundancy is left. More details about linear control theory are found in almost any introduction to control theory, for example [24] is useful for understanding normal-forms. The transformation to normal forms is described in [33].

2.4 Description of Dynamic Systems by Difference Equation

In the state-space description (5) and (6) the state $\vec{x}$ usually can not be observed. Even if the parameters of a Bayesian network can be trained despite the occurrence of unobserved variables, it is more cumbersome. According to our experience the number of iterations and examples needed is higher in the presence of unobserved variables. Thus it is desirable to start modeling with a description without hidden variables.

The starting point is again the description by the differential equation (4). The derivations of a function f can be approximated by the differences

$$\left.\frac{df}{dt}\right|_{t=k\Delta T} \approx \frac{f(k\Delta T) - f([k-1]\Delta T)}{\Delta T} \quad (14)$$

$$\left.\frac{d^2 f}{dt^2}\right|_{t=k\Delta T} \approx \frac{f(k\Delta T) - 2f([k-1]\Delta T) + f([k-2]\Delta T)}{(\Delta T)^2} \quad (15)$$

$$\left.\frac{d^3 f}{dt^3}\right|_{t=k\Delta T} \approx \frac{f(k\Delta T) - 3f([k-1]\Delta T) + 3f([k-2]\Delta T) - f([k-3]\Delta T)}{(\Delta T)^3} . \quad (16)$$

Derivations of higher order can be approximated in a similar manner. Thus it is possible to rewrite (4), using expressions (14) to (16) instead of the derivations. This procedure results in a difference equation, which can be solved for $\vec{y_t}$

$$y_t = -\sum_{i=1}^{n} \alpha_i y_{t-i} + \sum_{i=1}^{n} \beta_i u_{t-i} \; . \tag{17}$$

Please note that the coefficients in (4) and (17) are different. Thus it is not possible to rewrite the differential equation to a difference equation without adapting the coefficients. The adaptation can be calculated using the z-Transformation, for other methods see [31]. But as only the structure to be deduced from (17) is used a description of the transformation algorithm is omitted.

Equation (17) has the important property that there are no unobservable state nodes left. Instead of using state nodes former in- and outputs are used to predict the next output. The next section gives a short introduction to Bayesian networks. The equations used to calculate the joint distribution are compared with the description given in this section to infer a Bayesian network structure for modeling of dynamic systems. This can be done both for the state-space description and for the representation by a difference equation. Later it will be shown how these models can act as controller.

3 Bayesian Networks

Particularly at the beginning of the development of Bayesian networks mostly discrete nodes were used [22, 10]. More recent approaches discuss hybrid Bayesian networks which uses discrete and continuous nodes at the same time [3, 2]. As we deal with the modeling of technical processes it is assumed that all nodes are continuous ones.

Modeling with Bayesian Networks is equivalent to learning a probability distribution $p(X_1, X_2, \cdots, X_n)$ which represents the data as well as possible.

Assuming independencies between the variables, the joint distribution simplifies to

$$p(x_1, x_2, \cdots, x_n) = p(x_1) \cdot \prod_{i=2}^{n} p(x_i | pa(i)) \tag{18}$$

with $pa(i)$ being the instantiation of $Pa(X_i)$. This means that the distribution of a node X_i depends only on its parents $Pa(X_i)$. Instantiation of a node or a set X denotes an observation x within the domain of X. For example imagine X is used to represent the velocity of a car. When the car is driving in a city the domain of X is equal to the typical speed between 0 and 60 km/h. An instantiation for X is the observation that the current speed of the car is 54.2 km/h.

To guarantee the evaluation of the Bayesian network all the continuous nodes are Gaussian ones. The parameters can either be determined by an expert or trained by examples, for the latter see Section 3.1.

Usually the dependency of variables is displayed graphically in a directed acyclic graph. In this graph a link $X_1 \rightarrow X_2$ from a node X_1 to a node X_2 means that X_2 is influenced by X_1. As a simple example let us have a look at the two different equations representing a dynamic systems. In the state-space description the state $\vec{x}_{t+1}$ depends on the former state $\vec{x}_t$ and on the input $\vec{u}_t$. Additionally the output $\vec{y}_t$ depends on the state $\vec{x}_t$ and on the input $\vec{u}_t$. The latter connection is only necessary if $n = m$ in (4). Thus this connection is depicted as a dashed line, because it can be omitted in many cases. This results in the Bayesian network depicted in Figure 3.

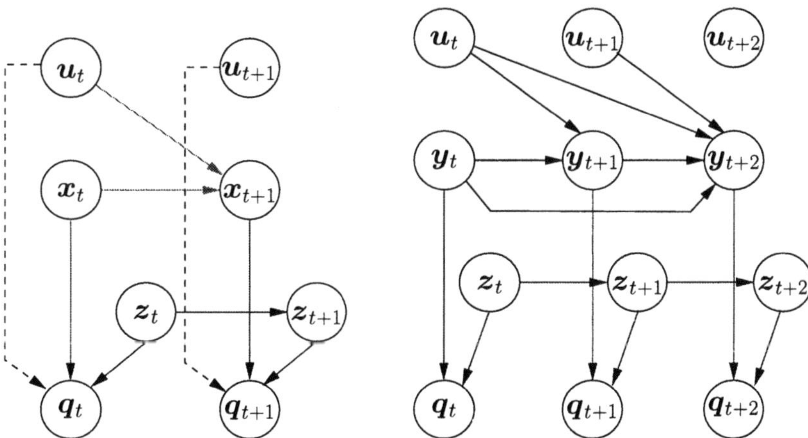

Figure 3. Bayesian network for the state-space representation of (5) and (6).

Figure 4. Bayesian network for the representation by difference equation.

In a similar way the difference equation (17) is mapped to a Bayesian network. In Figure 4 the mapping is shown for a second order system. Please note that the calculation of $\vec{y}_t$ depends not only on $\vec{y}_{t-1}$ but also on $\vec{y}_{t-2}$.

There are several types of BNs, distinguishable by the type of nodes used. We restrict ourselves to normally distributed, continuous nodes, that is

$$p(\vec{x}|\vec{y}) = \mathcal{N}(\vec{\mu}_{X_0} + \vec{W}_X \vec{y}, \vec{\Sigma}_X), \tag{19}$$

where $\vec{y}$ denotes observations for $\vec{Y}$, the parent nodes of X. The vector $\mu_{\vec{X}_0}$ is the mean when no parent exists or all parent have zero values. The weight matrix $\vec{W}_X$ is used to characterize the influence of $\vec{Y}$ on $\vec{X}$. The matrix $\vec{\Sigma}_X$ denotes the covariance of the normal distribution.

The restriction to normally distributed nodes enables us to use the inference algorithms described in [19], avoiding time consuming sampling procedures. The main tasks of the inference algorithm, including the junction-tree algorithm described in [19], are the calculation of marginal distributions (see equation 20) and to determine how the distribution changes, when observations are made. Imagine once again a random variable modeling the speed of a car driving in a city. When nothing else is known a good estimation for the speed might by 52 km/h. When the observation is made, that the car is driving towards a red light in 10m distance the estimation of the speed has to be changed. In this case it would be the task of the inference algorithm to keep track of obeservations and to change the estimations for the other variables accordingly. When acting as a controller the inference algorithm keeps track of the distribution of the input variable given the desired value and the input and output values of the past.

The usage of the junction tree algorithm has the additional advantage, that there is no need to bother about convergence problems. This is important as a controller has to react in real-time.

One of the most important operations on BNs is the calculation of marginal distributions. Given a full distribution $p(X)$ with $X = \{X_1, \cdots, X_n\}$ an arbitrary distribution $p(X \backslash C)$ with $C \subset X$ can be calculated by integration over all variables in C:

$$p(X \backslash C) = \int_C p(X) dC \ . \tag{20}$$

The parameters of a Bayesian network might be inferred, for example by comparison to a Kalman filter, or a training algorithm may be used. There are several training algorithms available, see [26] or [25]. The most popular is the so called EM algorithm [28, 7] which is able to deal with missing data, and both discrete and continuous variables at the same time. Convergence is guaranteed, but only to a local extremum. Thus it is of advantage to find a good initialization of the parameters, which might be found by comparison to the used normal form.

A more detailed description of the algorithms used for BNs is given in [19, 20] or [12].

3.1 Training of Bayesian Networks

As mentioned in the introduction to Bayesian networks in Section 3 a frequently used training algorithm for Bayesian networks is the EM algorithm. It was first introduced by [5] and is used not only for Gaussian networks, that is Bayesian networks with only continuous nodes. It is also applied for the adaptation of Hidden Markov Models, Bayesian networks with discrete nodes, (confer [4]) hybrid Bayesian networks (see [28, 27]), or even for structure learning [8].

The training method from [27] describes the general case of training hybrid Bayesian networks, using both discrete and continuous nodes. The formulas given here are simplified to match the pure Gaussian case, discussed in this chapter.

During the training of Gaussian networks the means, variances and weights of all nodes have to be adapted. It is possible to adapt the parameter of each node separately, so it is sufficient to concentrate on a simple network displayed in Figure 5.

For the training imagine an additional node with constant evidence of 1. This allows to include the calculation of the mean in the calculation of the weight vector. For example in Figure 5 the mean of node X is calculated by multiplication of the weight vector with the instantiation of the parent nodes y_1, y_2 and 1 as the last component

$$\vec{\mu} = \left[w_{13}\ w_{23}\ \mu_{X_0} \right] \begin{bmatrix} y_1 \\ y_2 \\ 1 \end{bmatrix} , \tag{21}$$

with w_{ij} being the weight between node i and j. When $\vec{H}$ denotes the weight vector $\vec{W}$ together with the mean μ_{X_0}

$$\vec{H} = [\vec{W}\ \mu_{X_0}] \tag{22}$$

the distribution of $\vec{x}$

$$p(\vec{x}|\vec{y}) = c|\vec{\Sigma}|^{-1/2} \exp(-\frac{1}{2}(\vec{x} - \vec{H}\vec{y})^{\mathrm{T}}\vec{\Sigma}(\vec{x} - \vec{H}\vec{y}) \tag{23}$$

depends on the distance between $\vec{x}$ and the mean $\vec{H}\vec{y}$. The matrix $\vec{\Sigma}$ denotes the covariance, $|\vec{\Sigma}|$ is the determinant, and $\vec{\Sigma}^{-1}$ the inverse of the covariance. The transpose of a matrix is denoted by T. The constant c guarantees that the integral over the complete domain is 1.

The log-likelihood $\log \prod_{l=1}^{N} p(x_l|y_l, \vec{\Theta})$ of the N examples depends on the parameters $\vec{\Theta}$ to be trained. When some of the nodes are not observed the distribution of the unobserved nodes has to be taken into account. Thus the conditioned expectation E of the log-likelihood

$$L = -\frac{1}{2}\sum_{l=1}^{N} \mathrm{E}[\log \vec{\Sigma} + (\vec{x}_l - \vec{H}\vec{y}_l)^{\mathrm{T}}\vec{\Sigma}^{-1}(\vec{x}_l - \vec{H}\vec{y}_l)|e_l, \vec{\Theta}] + c_{\mathrm{L}} \tag{24}$$

is used. It depends also on the evidence e_l of the remaining nodes of the Bayesian network. For example observations made for possible parents of Y_1 and Y_2 also have an influence on p, but are regarded as constants during the optimization of the parameters of node X. Parameters not relevant for the maximization of the log-likelihood for a special node are combined in the constant c_{L}.

During the maximization the statistics on the left hand side of Table 2 are used, known as essential sufficient statistic.

The new parameters $\hat{\vec{H}}$ for the weights and the mean

$$\hat{\vec{H}} = (\sum_l \vec{x}_l\vec{y}_l^T)(\sum_l \vec{y}_l\vec{y}_l^T) \tag{25}$$

can now be calculated by the essential sufficient statistics. If the node whose parameters are updated has no parents its mean is calculated by

$$\hat{\vec{\mu}}_0 = \frac{\sum_l x_l}{N} . \tag{26}$$

The second parameter to be learnt is the covariance matrix

$$\hat{\vec{\Sigma}} = \frac{\sum \vec{x}_l\vec{x}_l^{\mathrm{T}}}{N} - \hat{\vec{H}}\frac{\sum_l \vec{y}_l\vec{x}_l^{\mathrm{T}}}{N} . \tag{27}$$

Table 1. Training data used for update of the network of Figure 5.

Y_1	Y_2	X
17	2	5
14.8	2.2	5.2
18.2	2.1	5.3
15.1	1.9	4.9
16.3	1.7	5

Table 2. Essential sufficient statistics for parameter update.

	Value for example
$\sum_l \vec{x}_l \vec{x}_l^T$	129.14
$\sum_l \vec{y}_l \vec{y}_l^T$	$\begin{bmatrix} 1333.0 & 161.2 & 81.4 \\ 161.2 & 19.8 & 9.9 \\ 81.4 & 9.9 & 5 \end{bmatrix}$
$\sum_l \vec{x}_l \vec{y}_l^T$	$\begin{bmatrix} 413.91 & 50.38 & 25.4 \end{bmatrix}$
$\sum_l \vec{y}_l \vec{x}_l^T$	$\begin{bmatrix} 413.91 \\ 50.38 \\ 25.4 \end{bmatrix}$

Imagine the training data in Table 1 were given. The first factor of the denominator results in

$$\sum_l \vec{x}_l \vec{y}_l^T = 5\,[17\ 2\ 1] + 5.2\,[14.8\ 2.2\ 1] + \cdots = [413.91\ 50.38\ 25.4]\ . \tag{28}$$

The results of calculating the values of the essential sufficient statistics is given in Table 2. Updating the parameters of node X leads to $\hat{\vec{H}} = [0.0505\ 0.5919\ 3.086]$ respectively $\hat{\vec{\Sigma}} = 0.00716$.

For the discussed example all values are observed, but sometimes, for example when training a Bayesian network for the state space description, the training has to be executed with missing or unobserved values. For this case there is no closed solution for the maximization of the log-likelihood. The reason is that the calculation of the most probable values for the hidden nodes depends on the parameters to be updated and vice versa.

A frequently used approximation algorithm is the so called EM algorithm. Its main idea is to improve the parameters iteratively. Each iteration consists of two steps. In its first step the (E)xpectation of the unobserved nodes is calculated, based on the parameters of the previous iteration and used for the update of the essential sufficient statistic instead of the observed values. The iteration starts with a set of arbitrary parameters. The expected values are calculated using one of the inference algorithms for Bayesian networks.

In the next step the log-likelihood is (M)aximized. Of course this does not lead immediately to an optimal parameter set, but it is guaranteed that the EM-algorithm converges to a local maximum. To approximate an optimal solution as close as possible it is from advantage to start with initial parameters as close as possible to the optimal ones. This is done by using the a-priori knowledge from the modeled domain, for purpose of control the knowledge about normal forms is used to improve training results.

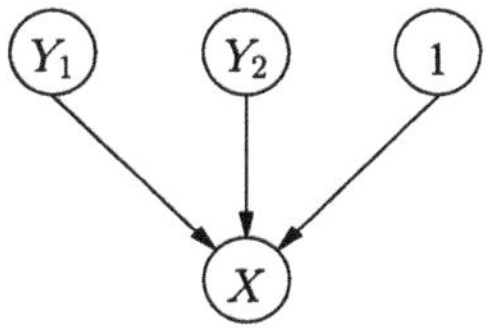

Figure 5. A simple Bayesian network.

3.2 Dynamic Bayesian Networks

For many purposes a static description is sufficient, confer for example [6]. But there are a lot of applications when time is an important factor, that is the distribution of a variable $X(t)$ depends not only on other variables, but also on its own value at previous time steps. Examples are systems described by (5) and (6). For such cases dynamic Bayesian networks (DBN) are developed, which are able to monitor a set of variables at arbitrary, but fixed points in time, that is time is no longer a continuous variable.

For each modeled point in time a static Bayesian network is used. These time slices are linked to represent the state of a variable at different points in time. Regarding the state-space description of (5) the state $\vec{x}_{t+1}$ depends on the input $\vec{u}_t$ and the state $\vec{x}_t$.

In a DBN the states are regarded as normally distributed, that is

$$p(\vec{x}_{t+1} \mid pa(\vec{x}_{t+1})) = p(\vec{x}_{t+1} \mid \vec{x}_t, \vec{u}_t) = \mathcal{N}(\vec{A}_{BN}\vec{x}_t + \vec{B}_{BN}\vec{u}_t, \vec{\Sigma}) \ . \tag{29}$$

For the evaluation a DBN can be interpreted as a static BN with equal parameters for all time slices respectively between the time slices. A deeper introduction is found in [15] and [16]. Well-known DBNs are Hidden Markov Models and Kalman filter which are mostly used in control theory for tracking and prediction of linear dynamic systems.

3.3 Kalman Filter

Our aim is to develop a controller which uses a DBN as model to generate the control signals. As a first step the model used for systems described by (5) and (6) will be developed. As a result we will get the structure, the weight matrices and the mean values of a DBN. In Section 4 this DBN is used to calculate the necessary input signals via marginalization.

In control theory Kalman filters are a well-known method for tracking and prediction of stationary, linear systems as mentioned in Section 3.2. Furthermore they are a special case of DBNs, so the results obtained for Kalman filters, for example in [11], may be used without any changes.

The state $\vec{x}(t)$ of a homogeneous systems, that is $\vec{u}(t) = 0$, is calculated as follows:

$$\vec{x}(t) = \vec{x}(t_0)\vec{\Phi}(t, t_0) \tag{30}$$

$$\vec{\Phi}(t, t_0) = \sum_{i=0}^{\infty} \vec{A}^i \frac{(t - t_0)^i}{i!} \ . \tag{31}$$

As DBNs represent a discrete time system $\vec{\Phi}$ cannot be used directly as weight matrix in a DBN. Discrete time systems are described by difference equations

$$\vec{x}_{k+1} = \vec{A}_{BN}\vec{x}_k + \vec{B}_{BN}\vec{u}_k \tag{32}$$

that are solved by

$$\vec{A}_{BN} = \vec{\Phi}(t_{k+1}, t_k) = \sum_{i=0}^{\infty} \vec{A}^i \frac{(t_{k+1} - t_k)^i}{i!} \tag{33}$$

$$\vec{B}_{BN}\vec{u}_k = \int_{t_k}^{t_{k+1}} \vec{\Phi}(t_{k+1}, \tau)\vec{B}\vec{u}(\tau)d\tau \ . \tag{34}$$

If we restrict ourselves to systems with a constant

$$\Delta T = t_{k+1} - t_k \tag{35}$$

and assuming that the input remains constant during a timeslice, then

$$\vec{\Phi}(\Delta T) = \vec{\Phi}(t_{k+1}, t_k) \tag{36}$$

stays constant for all k and (34) simplifies to

$$\vec{B}_{BN} = \Delta T \sum_{i=0}^{\infty} \frac{\vec{A}^i \Delta T^i}{(i+1)!} \vec{B} \ . \tag{37}$$

To build a DBN, which incorporates these equations $\vec{B}_{BN}$ is used as weight matrix between the input nodes and the state nodes. The matrix $\vec{\Phi}(\Delta T)$ describes the transition from one state to the next and is therefore used as weight matrix for the inter slice connection between two states in neighboring time slices. This means that the state at time $t+1$ is calculated by

$$\vec{x}_{t+1} = \begin{bmatrix} \Phi(\vec{\Delta T}) \ \vec{B}_{BN} \end{bmatrix} \begin{bmatrix} \vec{x}_t \\ \vec{u}_t \end{bmatrix} . \tag{38}$$

In a BN the mean $\vec{\mu}$ is equal to $\vec{\mu} = \vec{\mu}_0 + \vec{W}\vec{y}$. Thus $\vec{\mu}_0$ has to be set to zero and $W = \begin{bmatrix} \Phi(\vec{\Delta T}) \ \vec{B}_{BN} \end{bmatrix}$. The output depends linearly on the state and is not time dependent, thus the matrix $\vec{C}$ and $\vec{E}$ may be used unchanged also in a discrete time system. In Figure 6 two time slices of the second order system used to model the example of Section 2 are shown. Please note that the rectangles in this picture do not represent any random variable, but the weight matrices and how they are used to calculate the means of the nodes, represented by circles.

As a further consequence the dimension of the hidden state nodes respectively the number of the state nodes is equal to the order of the differential equation describing the system.

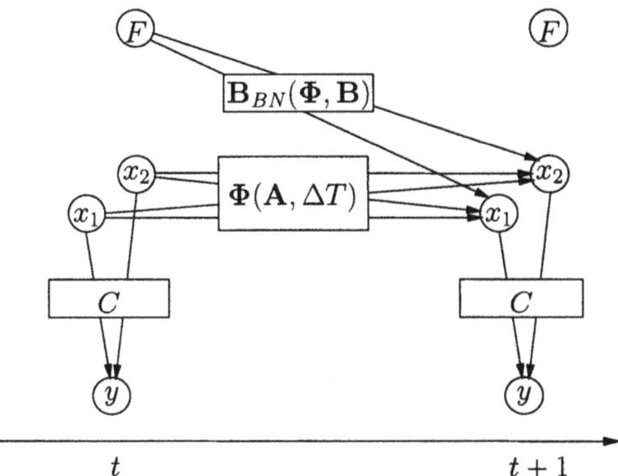

Figure 6. Weight matrices in a Kalman filter.

3.4 State-Space Model

The structure of the Bayesian network modeling the manipulation reaction is already discussed in Section 3, confer Figure 3. Assuming that the system to be modeled is time invariant, or at least the changes are slow in comparison to the sampling frequency, the weights are the same for all time slices. These weights can be either calculated, as discussed in Section 3.3 or trained. For the training the knowledge about normal forms should be used to shrink the search space which results in better training results. But when (33) and (37) are applied to (10) and (11) to calculate an adaptation to time discrete systems the typical form of the normal form is destroyed. But control theory shows that also for discrete time systems normal form exists. For the observable canonical form only the last column has to be adapted, that is only the parameters a_i and b_i changes.

Now each element in a matrix being zero can be interpreted as a missing link, each time a position in $\vec{A}$ is equal to one a weight can be clamped. In Figure 6 the connection $x_1 \rightarrow y$ can be removed, because the weight of this connection in matrix $\vec{C} = [0\ 1]$ is zero. Additionally the link $x_{1,t} \rightarrow x_{1,t+1}$ is unnecessary (confer matrix $\vec{A}$). Thus the usage of normal forms has two advantages. First the number of input nodes connected to the output is reduced which results in higher speed. Second the weights of all nodes being connected to the output node is known. Thus it can be clamped and is not changed during training. Indeed this has been the reason to use the observable canonical form.

The remaining knowledge about normal forms is used to find good initializations for $\vec{A}$ and $\vec{B}$, which is very important when using the EM algorithm which converges to a local extremum, because the log-likelihood is monotonic increasing in each iteration.

Until now only the control transfer function of the system is modeled, that is the reaction when the manipulated variable u is changed. Experiments, not described in this chapter, show that a system where only input, output and state variables are modeled, is sufficient for building a controller as long as only the manipulation reaction of the system is concerned. If disturbance variables occurs an additional slice for the disturbance variables has to be used. According to (1) the disturbance value $\vec{z}$ has to be added to $\vec{y}$ to come to the measured output $\vec{q}$. The weight of these links are set to one. Additionally there should be a possibility to estimate the disturbance variable z_{t+1} using the estimation of the past. Thus a link $z_t \rightarrow z_{+1}$ is necessary. It is assumed that environmental changes are relatively slow, so that the statistical properties of z_t are approximately equal to z_{t+1}. Thus a weight of one is taken.This results in Figure 3.

The additional node z has two different functions. When a perfect model is used the only task of z is to model the disturbance variable. In our test scenario a reference value $w = 10$ and a disturbance variable of $z = 1$ is used. In this case the state x must take on a value, so that $y = q - z = 9$. When trained models are used z has the additional function to make up the differences between the model and the reality.

3.5 Structure of the Higher-Order Markov Model

The model discussed in the last section has unobservable state nodes. Particularly for systems of higher order this might result in problems during the training. To reduce the number of so called hidden nodes the representation of a dynamic system by difference equation (17) was discussed in Section 3. As the calculation in this model does not depend solely on the direct predecessor the Markov assumption does no longer hold for this model. Unfortunately most of the standard Bayesian network tools are therefore unable to deal with such a model. For our experiments we decided to map the model in Figure 4 to a model which meets the Markov assumptions. This is done by introducing redundant nodes, that is nodes situated in different time slices which should have the same value. One method to guarantee this assumption is by assignment of the same value to two different nodes, for example the former input u is assigned to two different values.

The second possibility is the generation of a link with a fixed weight of one. This is done for the undisturbed output y. The usage of these links is depicted in Figure 7.

The modeling of the disturbance variable is the same as in the state-space model, two links $z_t \rightarrow q_t$ and $y_t \rightarrow q_t$ with a fixed weight of one are used to model (1). Additionally a link $z_t \rightarrow z_{t+1}$ allows prediction of the disturbance variable.

4 Calculation of Control Signals

In Section 3.3 we have shown how to set the weight matrices and mean values of the DBN. The questions about the time-difference ΔT and the covariance matrices are still open. We first discuss the generation of the input signal before we deal with the remaining parameters. For the generation of the input variable $\vec{u}$ a DBN with a fixed number of time slices as depicted in Figure 8 for the state-space model or in Figure 7 for the higher-order Markov model is used.

To generate the manipulated value $\vec{u}_{t+1}$ the first part of the input nodes is used to

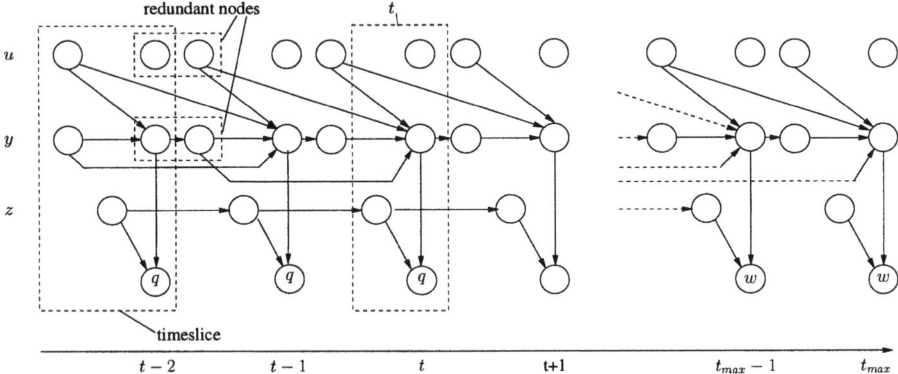

Figure 7. Higher-order Markov model with redundant nodes.

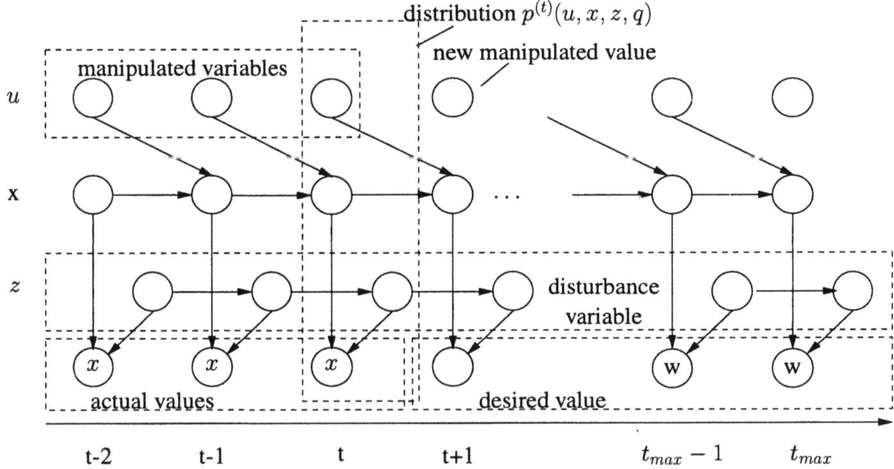

Figure 8. Principle structure of a BN used for control purposes.

enter the history. In Figures 7 and 8 these are the nodes $\vec{u}_{t-2}$ up to $\vec{u}_t$, for our experiments we used 10 nodes for the representation of the past. Moreover the observed output values are stored and entered as evidence using the nodes $\vec{q}_{t_0}$, the oldest stored output value, till $\vec{q}_t$. The state cannot be observed, so usually no evidence is given for the random variable $\vec{x}$. Now it is the task of the DBN to calculate a signal that can be used to change the system's output to the desired signal and to keep that output constant. To tell the system to do so the desired value $\vec{w}$ is also entered as evidence. This means the desired future values for the output nodes are treated as they were already observed and entered as evidence for all the nodes $\vec{q}_{t+2}$ till $\vec{q}_{t_{\max}}$. No evidence is given for $\vec{q}_{t+1}$ as this value is determined by an already calculated input. To control the plant it is necessary to calculate the value $\vec{u}_{t+1}$ which leads to the desired value $\vec{w}$. This can be done by marginalization. Our controller, implemented using the BN toolbox [29] is tested with a dynamic system, simulated by Simulink. So the input is passed to the simulation of the dynamic system and the resulting output is calculated. Then a complete cycle is finished. The used input and the resulting output are added to the history and the next input is calculated. To ensure that the calculation of the input signal is not limited to a certain amount of time the evidence is shifted to the left after each time step, that is the oldest input and output values are deleted. To come to a stable state estimation at the end of each cycle the state x_{t_0+1} is estimated and entered after shifting to the left as evidence for x_{t_0}.

Then the current signal is entered at time t. The future values may remain unchanged if the desired value is not changed. This works well for slow systems, for systems with the ability to oscillate this would result in oscillating input signals. Thus it is necessary to damp the input. To do so a weighted sum of u_t to u_{t+k} is used. For our experiments, described in Section 5, a weighted sum of 4 nodes are used, whereas the highest weight is used for u_{t+1} and the lowest weight for u_{t+4}.

It remains the question, what ΔT is appropriate for the dynamic system. According to control theory a system $K\vec{u}(t) = \vec{q}(t) + T_1 \frac{d\vec{q}}{dt} + T_2^2 \frac{d^2\vec{q}}{dt^2}$ of second order has a natural angular frequency of $\omega_0 = \frac{1}{T_2}$. The minimal sampling rate is twice the frequency to be measured.

Our first experiments are done with very small covariances, because we used an accurate model based on an analytical description and are not interested in loosing information due to great covariances. Please note that zero covariances are not possible, due to matrix inversion during evaluation. As a consequence we got an accurate model that was unable to calculate appropriate control signals. The reason is that small covariances at the input nodes, together with zero mean values results in a high probability for an input close to zero which can not be used for control purposes. Therefore we changed the covariance of the input node to a maximum to tell the system, that there is no a-priori information about the correct value of the input signal. The other covariances remain unchanged to keep the accurate modeling behavior.

5 Experiments

In the last section the idea of the Bayesian controller is presented. This section presents first the test systems together with the criteria used for comparison. In Section 5.2 the inferred model is applied to the test-systems to get an impression about the performance of the new type of controller under ideal conditions.

Afterwards the trained state-space and higher-order Markov model are applied to the same test systems and compared with the results of the inferred controller. It will be demonstrated that the inferred as well as the trained controllers show a good performance, but the effort for training is less for the higher-order Markov model.

5.1 Test Systems

Our experiments were done with three different systems of second order which are simulated by Simulink. These systems are described by differential equations

$$K\vec{u}(t) = \vec{q}(t) + T_1\frac{d\vec{q}}{dt} + T_2^2\frac{d^2\vec{q}}{dt^2} \tag{39}$$

and their behavior depends mainly on the two parameters T_1 and T_2. For the time constants of the test systems see Table 3. If the damping

$$D = \frac{T_1}{2T_2} \tag{40}$$

is greater than one the system has no tendency to overshoot which means these systems are easy to control. Systems with $0 < D < 1$ have the ability to oscillate and overshooting can be regarded when the input signal is changed.

At the beginning of each test the desired value is set to 0 and shortly afterwards it is changed to 10. This results in the peak of the input signal and a steep raise of the output signals which can be seen in Figures 9, 10, and 11. After convergence of the output signal, the disturbance input z is changed to 1. Thus at the first moment the output changes to a value close to 11. Confer for example Figure 9 at $t = 4.1$ s. Then the controller changes the input, so that the output reaches the desired value once again. As a dynamic system with gain 2 is used for the tests the new input signal converges to a value $u = 4.5$. The signals are stored, so that the quality measures can be calculated.

Table 3. Description of test systems.

Number	K	T_1	T_2	Description
1	2	1	0.1	Damped system with gain two which has no tendency to overshoot
2	2	0.1	0.1	System with $D < 1$ which means there is a tendency to overshoot
3	10	0.05	0.1	System with high gain and a large tendency to overshoot

5.2 Experiments with Calculated Models

First experiments were done with models whose parameters were retrieved analytically. The reason for these experiments are to show that Bayesian networks used in the manner described in Section 4 can be used for control purposes and on the other hand to get comparative values. The used quality measures are described in Table 4, the results of our experiments in Table 5. The signals of system 2 can be seen in Figure 9.

At the beginning a steep raise of both the input and the output signal can be observed. The output signal reaches its maximum of 10.56 and after 0.5s the error is below the 1% level. At $t = 4.1$s the disturbance variable is changed to $z = 1$ and half a second later the effect of the disturbance disturbance is nearly vanished.

Table 4. Used quality measures.

Quality measure	Description
$Q_d(z = d)$	Squared error sum as defined in (3). Tests done with a disturbance variable $z = d$.
Overshoot	The difference between the maximal output value q_{max} and w.
e_∞	The remaining error $e = q - w$ after convergence took place.
$t_{\max}(z = 0, c\%)$	Raise time until the output has changed from $q = 0$ to $q = w$ with a deviation of less then $c\%$ of the desired value. Measurements are made with disturbing value $z = 0$.
$t_{\max}(z = 1, c\%)$	Settling time until the error is smaller than $c\%$ starting with the occurrence of the disturbance input.

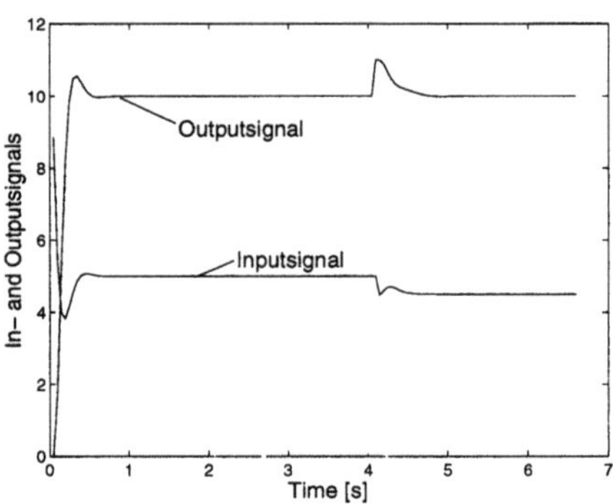

Figure 9. Signals for system 2 controlled by a calculated controller.

In all three cases the system shows a very good performance, the desired value is reached fast and with nearly no deviation. Also when the disturbance value is changed from 0 to 1 the system reacts as intended. The input is changed so that the desired value is reached once again. Thus a Bayesian model is a suitable mean to build a controller. The next chapter deals with the results obtained with a trained controller. Of course one has to take into account that a trained model is less accurate then an inferred one.

5.3 Comparison of Trained Models

The results described in the last section are based on calculated Bayesian networks. As our aim is to use Bayesian models as self adaptive controller we have tested our approach also with trained Bayesian networks. First the results with the state-space model will be discussed. Afterwards the outcome of the experiments with the higher-order Markov models are presented and compared with the state-space model.

The training material is calculated by simulating the system's response to the following input signals, which are used for both the state-space and the higher-order Markov model.

Table 5. Results of experiments with calculated models.

Test System	1	2	3
$Q_d(z=0)$	8.3956	9.6172	10.2806
$e_\infty(z=0)$	-0.0151	-0.0061	-0.0003
Overshoot	-0.0151	0.5616	0.9181
$Q_d(z=1)$	0.1490	0.1868	0.2267
$e_\infty(z=1)$	0.0015	-0.0049	-0.0148
$te_{\max}(z=0,1\%)$	0.4500	0.4500	0.7000
$te_{\max}(z=0,3\%)$	0.3500	0.4000	0.4500
$te_{\max}(z=1,1\%)$	0.4500	0.5500	0.6000
$te_{\max}(z=1,3\%)$	0.3000	0.3500	0.4500

- Step response with different input signals.
- Pulse response.

Particularly the first one is used in control theory for system identification in simple cases. In-depth analysis to find out which training signals would be best are still under investigation. First experiences show that the used signals provide results above the average.

The BNs are trained with 20 iteration. In each iteration 40 different time series are used. After five iterations a new training set is generated, to avoid convergence to an insufficient local extremum.

In our experiments 20 iterations are usually sufficient for convergence. The experiments are repeated 10 times, to control the robustness of the results. Best results are obtained for system 2, an example is shown in Figure 11, the complete results are given in Table 6. The values given in the table are rounded to an accuracy of two positions after decimal point, thus e_∞ = -0.00 signifies that $0 > e_\infty > -0.005$. The mean of the steady state error e_∞ is based on absolute values. This is done to avoid that positive and negative values are averaged to zero. Table 6 shows that also in the trained state-space model the desired value w is reached with nearly no deviation. Additionally the squared error sum Q_d is slightly better than for the inferred model. The price for the better squared error sum is a higher overshoot of 0.88 in comparison to 0.56 of the inferred model. The reason for the better squared error sum Q_d might be that the variations in the mathematical models are clamped to a fixed value which is sufficient for control purposes but is not proven to be optimal.

The settling time for the inferred model is better for the inferred model, but still acceptable for the trained model. Concerning the disturbance reaction similar observations are made. For both systems there is nearly no steady state error after settling, the time until the desired value w is reached is once again better for the inferred model.

Regarding Table 7 leads to the same conclusions. The squared error sum is better for the trained system due to a greater overshoot and the time until the effect of the disturbance is eliminated is better for the inferred system. Additionally one problem

Table 6. Results of experiments with state-space model, system 2.

Experiment number	1	2	3	4	5	6	7	8	9	10	mean
$Q_d(z=0)$	9.45	9.46	9.47	9.50	9.50	9.40	9.43	9.51	9.52	9.44	9.47
$e_\infty(z=0)$	−0.01	−0.00	−0.01	−0.01	−0.00	−0.00	−0.01	−0.01	−0.01	0.00	0.00
Overshoot	0.72	1.03	0.79	0.68	1.11	1.29	0.78	0.76	0.67	0.93	0.88
$Q_d(z=1)$	0.26	0.27	0.27	0.29	0.27	0.26	0.27	0.28	0.29	0.28	0.27
$e_\infty(z=1)$	0.01	0.02	0.00	0.01	0.01	0.01	0.01	0.01	0.01	0.02	0.01
$te_{max}(z=0,1\%)$	0.55	0.85	0.65	0.55	0.85	0.70	0.60	0.80	0.75	0.85	0.72
$te_{max}(z=0,3\%)$	0.45	0.70	0.45	0.45	0.70	0.55	0.45	0.50	0.45	0.55	0.53
$te_{max}(z=1,1\%)$	0.70	0.75	0.75	0.85	0.70	0.75	0.70	0.80	0.90	0.85	0.78
$te_{max}(z=1,3\%)$	0.60	0.60	0.60	0.65	0.60	0.60	0.60	0.60	0.65	0.65	0.62

Table 7. Results of experiments with state-space model, system 1.

Experiment number	1	2	3	4	5	6	7	8	9	10	mean
$Q_d(z=0)$	7.46	7.43	7.45	7.61	7.37	7.69	7.67	7.63	7.51	noConv.	7.53
$e_\infty(z=0)$	0.01	−0.01	0.01	0.00	−0.01	0.01	0.02	0.01	0.00	noConv.	0.01
Overshoot	0.53	0.31	0.52	0.45	0.38	0.55	0.49	0.50	0.53	noConv.	0.48
$Q_d(z=1)$	0.15	0.15	0.16	0.16	0.15	0.18	0.17	0.17	0.16	noConv.	0.16
$e_\infty(z=1)$	0.03	0.02	0.03	0.04	0.02	0.04	0.05	0.04	0.03	noConv.	0.03
$te_{max}(z=0,1\%)$	0.70	0.60	0.70	0.70	0.50	0.75	0.75	0.75	0.70	noConv.	0.68
$te_{max}(z=0,3\%)$	0.50	0.35	0.45	0.50	0.40	0.55	0.55	0.55	0.50	noConv.	0.48
$te_{max}(z=1,1\%)$	0.55	0.45	0.50	0.55	0.45	0.55	0.60	0.55	0.55	noConv.	0.53
$te_{max}(z=1,3\%)$	0.30	0.30	0.30	0.35	0.30	0.35	0.35	0.35	0.35	noConv.	0.33

is revealed by the tenth experiment. Here no convergence took place (the mean values are calculated based on the first 9 experiments only), thus a more stable system is required. Here the higher-order Markov model will show its superiority.

The results of the tests with the third system, presented in Table 8 shows the greatest difference between the inferred and the trained model. Regarding the third and fourth experiment, the steady state error $e_\infty(z=0)$ in the reference reaction is greater than 3% of the desired value. Thus the times $te_{max}(z=0,c\%)$ would be infinite according to the definition, the values given in the table are the times until convergence is reached. Also in the disturbance reaction the remaining steady state error is greater than 1% of the desired value for three cases. From interest is also that $Q_d(z=1)=0.02$. This values seems to be the best error sum in the table. Only together with the steady state error one is able to interpret this value correctly. The steady state error for the undisturbed case $e_\infty(z=0)=-0.51$, the steady state error regarding the disturbance reaction $e_\infty = 0.41$. Together with the fact that $z = 1$ it is possible to conclude that the controller has scarcely reacted in that case. Thus in this case it is better to regard Q_d as a measure until convergence has happened than as an measure for the error. When there is only a small steady state error it is possible to look at Q_d as an error.

Regarding only the successful cases 1,2,5 - 9 demonstrates that also for the third system a Bayesian controller works satisfactorily, provided a successful training. The mean of Q_{d} for these subset is 9.92, and the steady state error $e_\infty(z = 0) = 0.01$, and when an disturbance value occurs $e_\infty = 0.05$. The results are improved when smaller ΔT is used for training, but also in this case the inferred model is clearly better.

Table 8. Results of experiments with state-space model, system 3.

Experiment number	1	2	3	4	5	6	7	8	9	10	mean
$Q_{\text{d}}(z = 0)$	10.03	9.82	10.01	25.37	10.06	9.84	9.87	10.00	9.86	11.39	11.63
$e_\infty(z = 0)$	0.03	0.01	−0.51	1.95	0.02	0.00	0.01	0.02	0.00	0.08	0.26
Overshoot	1.15	0.86	1.75	7.27	1.24	0.91	1.04	1.26	1.04	0.88	1.74
$Q_{\text{d}}(z = 1)$	0.52	0.45	0.02	0.15	0.52	0.39	0.49	0.43	0.36	0.92	0.43
$e_\infty(z = 1)$	0.09	0.06	0.41	2.54	0.04	0.03	0.04	0.05	0.02	0.19	0.35
$te_{\max}(z = 0, 1\%)$	1.00	0.60	4.70	18.00	1.00	0.45	0.50	0.90	0.50	1.80	2.95
$te_{\max}(z = 0, 3\%)$	0.50	0.40	4.70	18.00	0.55	0.45	0.45	0.55	0.45	0.65	2.67
$te_{\max}(z = 1, 1\%)$	1.85	1.40	3.70	14.05	1.30	0.95	1.25	1.00	0.85	5.50	3.19
$te_{\max}(z = 1, 3\%)$	1.00	0.95	3.70	14.05	0.95	0.75	0.90	0.80	0.70	1.80	2.56

The three examples point out that a Bayesian controller is a good mean for controlling linear systems. When based on a mathematical description the desired value is reached, so that nearly no overshoot is observed. Also when working with trained models a high accuracy is shown, provided that a successful training took place. It is well-known that the used EM-algorithm may converge to a local extremum, so that good initializations are essential for good training results. We used the knowledge about normal forms to improve the training, but in seldom cases the parameters found by training are not sufficient for control. In the rest of this section we will show that the higher-order Markov model provides better results. The first improvement is that less training iterations are necessary. As we assumed that the disturbing value $z = 0$ during training the observed output $q = y$. Thus there are no hidden units left, the training of the system is stopped after 5 iterations. Thus a lot of training time is saved which is important for online training. The second advantage is, that the training result is stable, in all cases a steady state error below 1% of the desired value is reached. Convergence to the desired value is observed in all the cases, so that using the higher-order Markov model is a great improvement in comparison to the state-space model.

In Table 9 the test results with system 2 are described. The main result is, that the training result is stable, that is in all of the cases nearly the same results are obtained with very low dispersion which seems to be one of the most important points for practical use. The steady state error for system 2 is slightly greater than for the inferred and the trained state space model. But as the deviation from the desired value w is below the 1% level of w this should not be regarded as a great disadvantage.

The smaller overshoot of 0.45 in comparison to 0.56 of the inferred state space model respectively 0.88 of the trained state space model results in a greater squared error sum of 9.82 in comparison to 9.62 (analytical state-space) respectively 9.47 (trained state-space). In contrary the values for the disturbance reaction are better

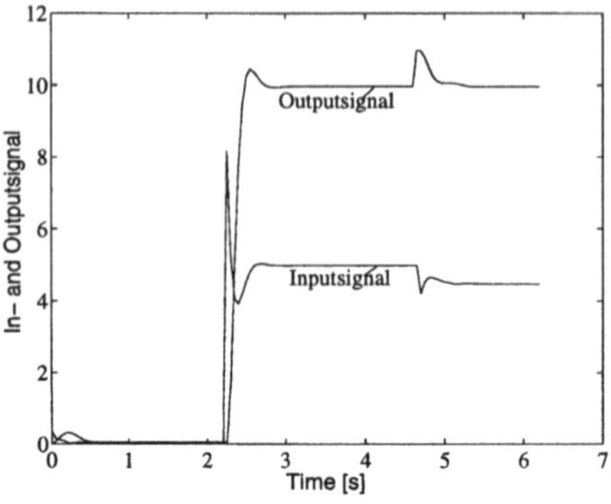

Figure 10. Signals for system 2 controlled by a trained higher-order Markov-model controller.

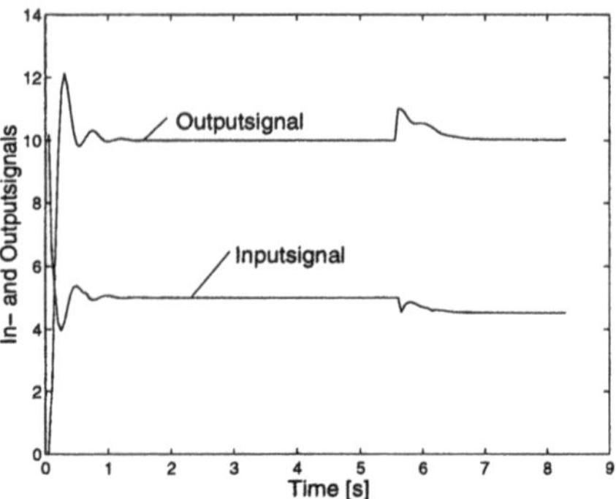

Figure 11. Signals for system 2 controlled by a trained state-space controller.

in comparison to the state space model. A smaller settling time can be observed together with a smaller error sum.

Table 9. Results of experiments with higher-order Markov model, system 2.

Experiment number	1	2	3	4	5	6	7	8	9	10	mean
$Q_d(z=0)$	9.82	9.83	9.82	9.82	9.82	9.82	9.82	9.83	9.83	9.82	9.82
$e_\infty(z=0)$	0.03	−0.03	−0.05	−0.04	−0.03	−0.05	−0.04	−0.03	−0.04	−0.04	0.04
Overshoot	0.45	0.45	0.43	0.44	0.45	0.43	0.44	0.46	0.44	0.44	0.45
$Q_d(z=1)$	0.16	0.16	0.16	0.16	0.16	0.16	0.16	0.16	0.16	0.16	0.16
$e_\infty(z=1)$	−0.04	−0.04	−0.06	−0.05	−0.04	−0.05	−0.05	−0.04	−0.05	−0.05	−0.05
$te_{\max}(z=0,1\%)$	0.45	0.45	0.45	0.45	0.45	0.45	0.45	0.45	0.45	0.45	0.45
$te_{\max}(z=0,3\%)$	0.40	0.40	0.40	0.40	0.40	0.40	0.40	0.40	0.40	0.40	0.40
$te_{\max}(z=1,1\%)$	0.35	0.35	0.35	0.35	0.35	0.35	0.35	0.35	0.35	0.35	0.35
$te_{\max}(z=1,3\%)$	0.25	0.25	0.25	0.25	0.25	0.25	0.25	0.25	0.25	0.25	0.25

The results for system 1 are of great interest, because a lot of frequently occurring systems have a damping greater than 1, so that they are comparable with system number 1. On the other side the state-space model failed in one case which must not happen for a controller in practical use.

But the higher-order Markov model converged in all ten cases, so the results are much better than for the trained state-space system. In that case it is necessary to keep in mind that the results for the trained state-space system in Table 7 are calculated upon the 9 cases with successful training. The measured squared error sum of the experiments with system 1 is better for the state-space model, whereas for the disturbance reaction the higher-order Markov model provides better results both regarding the squared error sum and the settling time $te_{\max}$. As for the analytical state-space model there is nearly no overshoot which is the reason for the smaller error sum of the trained state-space model.

In the third system the training of the state-space model was insufficient in 3

Table 10. Results of experiments with higher-order Markov model, system 1.

Experiment number	1	2	3	4	5	6	7	8	9	10	mean
$Q_d(z=0)$	8.51	8.51	8.51	8.51	8.51	8.51	8.51	8.51	8.51	8.51	8.51
$e_\infty(z=0)$	−0.09	−0.09	−0.05	−0.07	−0.05	−0.10	−0.08	−0.07	−0.06	−0.06	0.07
Overshoot	−0.09	−0.09	−0.05	−0.07	−0.05	−0.10	−0.08	−0.07	−0.06	−0.06	−0.07
$Q_d(z=1)$	0.11	0.11	0.11	0.11	0.11	0.11	0.11	0.11	0.11	0.11	0.11
$e_\infty(z=1)$	0.02	0.02	0.05	0.04	0.05	0.00	0.03	0.04	0.04	0.05	0.03
$te_{max}(z=0,1\%)$	0.75	0.75	0.60	0.60	0.60	2.40	0.65	0.65	0.60	0.60	0.82
$te_{max}(z=0,3\%)$	0.45	0.45	0.40	0.40	0.40	0.45	0.40	0.40	0.40	0.40	0.42
$te_{max}(z=1,1\%)$	0.30	0.60	0.30	0.30	0.30	0.60	0.30	0.30	0.30	0.30	0.36
$te_{max}(z=1,3\%)$	0.25	0.25	0.25	0.25	0.25	0.25	0.25	0.25	0.25	0.25	0.25

Table 11. Results of experiments with higher-order Markov model, system 3.

Experiment number	1	2	3	4	5	6	7	8	9	10	mean
$Q_d(z=0)$	9.90	9.89	9.89	10.10	9.89	9.89	9.90	9.89	9.91	9.89	9.92
$e_\infty(z=0)$	0.01	0.01	0.01	0.03	0.01	0.01	0.01	0.01	0.01	0.01	0.01
Overshoot	1.15	1.15	1.15	1.15	1.15	1.15	1.15	1.15	1.15	1.15	1.15
$Q_d(z=1)$	0.19	0.19	0.19	0.18	0.19	0.19	0.19	0.19	0.19	0.19	0.19
$e_\infty(z=1)$	−0.04	−0.04	−0.04	0.00	−0.04	−0.04	−0.04	−0.04	−0.04	−0.04	−0.03
$te_{max}(z=0,1\%)$	0.70	0.65	0.65	1.70	0.65	0.65	0.65	0.65	0.75	0.65	0.77
$te_{max}(z=0,3\%)$	0.45	0.45	0.45	1.05	0.45	0.45	0.45	0.45	0.45	0.45	0.51
$te_{max}(z=1,1\%)$	0.60	0.60	0.60	0.55	0.60	0.60	0.60	0.60	0.60	0.60	0.60
$te_{max}(z=1,3\%)$	0.25	0.25	0.25	0.30	0.25	0.25	0.25	0.25	0.30	0.25	0.26

of 10 cases. Also here the higher-order Markov model shows its superiority. Nearly no deviation from the desired value is observed, and the squared error sum for the reference reaction ($z = 0$) is better than for the state-space model. Only the settling time te_{max} of the inferred state space model is better. Best values are obtained for the disturbance reaction. The squared error Q_d and the settling time te_{max} is better than for the state-space model.

Considering the results for all three systems the higher-order Markov model clearly outperforms the state-space model. Due to less hidden variables the training can be performed with less iterations and training examples.

Additionally stable training results are obtained, the resulting steady state-error is in all cases below 1% of the desired value. Also the settling time is in most of the cases (The exception is $te_{max}(z = 0, 3\%)$ for system 1) better than for the trained state space.

To ensure applicability in real world domains the most important point to be improved is reaction in real-time. In the simulations described here the time for calculating the new input signal and the system's response is approximately 1.5s. The first step to be done is to use compiled programs instead of interpreted ones to get an better impression of the performance which can be reached. Later on approximative algorithms may be used to further improve the performance. Additional opportunities to save time are to decrease the number of time slices used or to use the fact that the evidence for some of the nodes does not change from one time step to the next.

5.4 Systems of Higher Order

The test systems in the previous section were all single input single output (SISO) systems of second order. To deal with higher-order systems there are five different possibilities which will be discussed here, mostly from the theoretical point of view. But most of the solutions are restricted to SISO systems.

- Increasing the dimension or the number of state nodes.
- Splitting of the system in a series of systems of first and second order.
- Splitting of the system in parallel subsystems.
- Increasing the number of past nodes used to predict the output.
- Approximation of a higher-order system by a system of second order.

Theoretically it is possible to model a system of higher order by using state nodes with the same dimension as the system's order. Of course that is possible also for systems with vector valued in- and output. But experiments, which uses the inference algorithms described in [19], show that numerical problems occur when calculating the inverse of a potential's covariance matrix. As each clique in a junction-tree (see [19] or [13]) contains the node itself and the node's parents the potential which contains the state nodes has at least twice the dimension of the state node. Maybe using the improved inference algorithm introduced in [21] might improve the numeric stability.

To cope with that problem a dynamic system of higher order can be split into several controlled systems of second order being connected in series.

Usually engineers do not work directly with the differential equation (4), but with the Laplace-transform

$$F(s) = \int_0^\infty f(t)\exp(-st)dt \tag{41}$$

so that (4) is transformed to

$$Y(s)\sum_{i=0}^{n} a_i s^i = U(s)\sum_{j=0}^{m} b_j s^j \ . \tag{42}$$

A linear, time invariant system without dead time is then described by its transfer function

$$G(s) = \frac{Y(s)}{U(s)} = \frac{\sum\limits_{j=0}^{m} b_j s^j}{\sum\limits_{i=0}^{n} a_i s^i} \tag{43}$$

being the quotient of two polynomials. For systems with more dimensional in- and output, the usage of transfer functions is still possible, but in this case $G(s)$ is a polynomial matrix [18].

Rewriting the transfer function using the fact that each polynomial is a product of polynomials of first and second order results in

$$G(s) = \frac{Y_1(s)}{U_1(s)}\frac{Y_2(s)}{U_2(s)}\cdots\frac{Y_{n/2}}{U_{n/2}}, \tag{44}$$

where $Y_i(s)$ and $U_i(s)$ are polynomials of maximal second order. As a multiplication of two Laplace transformed is equal to the serial connection of two subsystems, SISO systems of higher order can be modeled by a series of first or second order systems. In the Bayesian network this results in state nodes of dimension two and thus the matrix inversion is numerically stable.

In special cases it is also possible to map a system of higher order in parallel subsystems. To get these subsystems it is necessary to rewrite $G(s)$ to a sum of different terms, as a Laplace transformed $G_1(s) + G_2(s)$ is mapped to two parallel systems whose output is added to get the output of the complete system.

Assuming that all poles of $U(s)$ are real $G(s)$ can be rewritten by a sum of terms

$$G(s) = \sum_{k=1}^{p} \left(\frac{c_{k,1}}{(s - s_k)} + \frac{c_{k,2}}{(s - s_k)^2} + \cdots + \frac{c_{k,r_k}}{(s - s_k)^{r_k}} \right) \tag{45}$$

where p is the number of different poles, s_k is a pole of $U(s)$ and r_k denotes how often $(s - s_k)$ occurs in the decomposition of $U(s)$. Thus a system of n-th order with n different real poles (That is $s_k = 1, 1 <= k <= n$) can be divided in n parallel subsystems. But as some parameters of this sum, for example p and s_k are not known before modeling the structure of the overall system is not known. Thus the splitting into independent parallel subsystem leads to no predefined structure.

Starting with the difference equation (17) systems of higher order can be modeled be increasing the number of past signals used to calculate the predicted output. The trick with the redundant nodes is in principle also possible for higher orders, but dealing with those models get more and more cumbersome, as the number of redundant nodes per slice is $2 * (n - 1)$ with n as the order of the system, as for each slice $u_t, u_{t-1}, \cdots u_{t-n-1}, y_t, y_{t-1}, \cdots, y_{t-n-1}$ needs to be represented. Thus for practical use it seems better to enhance the used tool so that also dynamic higher-order Markov models can be handled.

For some systems of higher order (for the restrictions see [32]) also an approximation with a system of second order

$$G(s) = \frac{K}{(1 + T_1 s)(1 + T_2 s)} \tag{46}$$

might be fruitful.

6 Conclusions

Starting from an analytical description the structure of a dynamic Bayesian network was developed and an explanation was given how to calculate the parameters of a Bayesian network. These networks are used as controller by using the desired value as evidence and using the marginal distribution of the input nodes as input signal for the controlled system.

The performance of a Bayesian controller based on trained and analytically retrieved models is compared regarding both the reference and the disturbance reaction of the controlled system. Even if the analytical model performs better, the trained model shows a very good performance. The steady state error is below 1% in most of the cases. For the future there will be several points of interest. To get a better impression of the practical applicability the performance of a Bayesian controller has to be compared with PID controller which are widely used in industry and it has to be examined how the Bayesian controller can be optimized with respect to special constraints, for example no overshoot or limited input. Other important aspects are the ability to react in real time and the modeling of non-linearities. An approach to deal with non-linearities are hybrid Bayesian networks and Particle filters. Using hybrid Bayesian networks might lead to problems with the real time requirement, as U. Lerner [23] has proven that inference in hybrid Bayesian networks in NP-hard, unless P = NP.

Using Bayesian networks for control purposes is a relatively new approach, so there is nearly no literature about controllers based on BNs. Welch [34] proves that BNs might be used in real time for control purposes, but he is restricted to static BNs and the choice between two possible actions. Additionally D. Koller [17] deals with the modeling of a physical process, particularly with the fault detection using particle filters, but the model is not used for control purposes.

Acknowledgments

This work was funded by the "Deutsche Forschungsgemeinschaft" (DFG) under grant number SFB 396, project-part C1; only the authors are responsible for the content of this chapter.

The experiments were done with Matlab, Simulink to simulate the dynamic systems and the BN-Toolbox [29], an expansion of Matlab which is freely available at http://www.ai.mit.edu/~murphyk/Software/BNT/bnt.html.

References

1. X. Boyen and D. Koller. Approximate learning of dynamic models. In *Proc. of the 11th Annual Conference on Neural Information Processing Systems, Denver, Colorado*, pages 396–402, December 1998.
2. R. Cowell. Advanced inference in bayesian networks. In Jordan [14], pages 27–49.
3. R. Cowell. Introduction to inference for bayesian networks. In Jordan [14], pages 9–26.
4. Robert G. Cowell, A. Philip Dawid, Steffen L. Lauritzen, and David J. Spiegelhalter. *Probabilistic Networks and Expert Systems*. Statistics for Engineering and Information Science. Springer, New York, 1999.
5. A. P. Dempster, N. M. Laird, and D. B. Rubin. Maximum likelihood from incomplete data via the EM algorithm. *Journal of the Royal Statistical Society, B*, 39:1–38, 1977.

6. R. Deventer, J. Denzler, and H. Niemann. Non-linear modeling of a production process by hybrid Bayesian Networks. In Werner Horn, editor, *ECAI 2000 (Berlin)*, pages 576–580. IOS Press, August 2000.
7. N. Friedman. Learning belief networks in the presence of missing values. In *Fourteenth Inter. Conf. on Machine Learning (ICML97)*, 1997.
8. N. Friedman. The Bayesian structural EM algorithm. In *Fourteenth Conf. on Uncertainty in Artificial Intelligence (UAI)*, pages 129–138, 1998.
9. N. Friedman, K. Murphy, and S. Russel. Learning the structure of dynamic probabilistic networks. In *Twelfth Annual Conference on Uncertainty in Artificial Intelligence*, pages 139–147, 1998.
10. Dan Geiger and David Heckerman. Learning gaussian networks. In *Proceedings of the Tenth Conference on Uncertainty in Artificial Intelligence*, pages 235–243, 1994.
11. A. Gelb, editor. *Applied optimal estimation.* MIT Press, Cambridge, Massacusetts, USA, 1994.
12. F. V. Jensen. *An introduction to Bayesian networks.* UCL Press, 1996.
13. Finn. V. Jensen and Frank Jensen. Optimal junction trees. In R.L Mantaras and D. Poole, editors, *Proceedings of the Tenth Conference on Uncertainty in Artificial Intelligence*, pages 360–366. Morgan Kaufmann, July 1994.
14. M. I. Jordan, editor. *Learning in Graphical Models.* MIT Press, Cambridge, Massachusetts, 1999.
15. U. Kjærulff. A computational schema for reasoning in dynamic probabilistic networks. In *Proceedings of the Eighth Conference of Uncertainty in Artificial Intelligence*, pages 121–129. Morgan Kaufmann Publishers, San Mateo, California, 1992.
16. Uffe Kjærulff. dhugin: A computational system for dynamic time-sliced bayesian networks. *International Journal of Forecasting, Special Issue on Probability Forecasting*, pages 89–111, 1995.
17. D. Koller and U. Lerner. Sampling in factored dynamic systems. In A. Doucet, J.F.G. de Freitas, and N. Gordon, editors, *Sequential Monte Carlo Methods in Practice*, chapter 21, pages 445–464. Springer, 2000.
18. Vladimír Kuěra. *Analysis and Design of Discrete Linear Control Systems.* Prentice Hall, New York, London, Toronto, Sydney, Tokyo, Singapore, 1991.
19. S. L. Lauritzen. Propagation of probabilities, means, and variances in mixed graphical association models. *Journal of the American Statistical Association*, Vol. 87(420):1098–1108, December 1992.
20. S. L. Lauritzen. *Graphical Models.* Oxford University Press, 1996.
21. S. L. Lauritzen and F. Jensen. Stable Local Computation with Conditional Gaussian Distributions. Technical Report R-99-2014, Aalborg University, Department of Mathematical Sciences, September 1999.
22. S. L. Lauritzen and D. J. Spiegelhalter. Local computation with probabilities on graphical structure and their application to expert systems. *Journal of the Royal Statistical Society*, pages 157–224, 1988.
23. Uri Lerner and R. Parr. Inference in hybrid networks: Theoretical limits and practical algorithms. In *Proceedings of the 17th Annual Conference on Uncer-*

tainty in Artificial Intelligence, pages 310–318, 2001.
24. David C. Luenberger and David G. Luenberger. *Introduction to Dynamic Systems: Theory, Models, and Application.* John Wiley, 1979.
25. D W McMichael, Lin Liu, and Heping Pan. Estimating the parameters of mixed bayesian networks from incomplete data. In Robin Evans, Lang White, Daniel McMichael, and Len Sciacca, editors, *Proceedings of Information Decision and Control 99*, pages 591–596, Adelaide, Australia, February 1999. Institute of Electrical and Electronic Engineers, Inc.
26. Stefano Monti and Gregory F. Cooper. Learning hybrid bayesian networks from data. In Michael I. Jordan, editor, *Learning in Graphical Models*, pages 521–540. Kluwer Academic Publishers, 1998.
27. Kevin P. Murphy. Fitting a Conditional Gaussian Distribution. Technical report, University of California, Computer Science Division (EECS), October 1998.
28. Kevin P. Murphy. Inference and Learning in Hybrid Bayesian Networks. Technical Report CSD-98-990, University of California, Computer Science Division (EECS), January 1998.
29. Kevin P. Murphy. The Bayes Net Toolbox for Matlab. *Computing Science and Statistics*, 33, 2001.
30. K. G. Olesen. Causal probabilistic networks with both discrete and continuous variables. *IEEE Transaction on Pattern Analysis and Machine Intelligence*, 3(15):275–279, 1993.
31. Gerd Schulz. *Regelungstechnik, Mehrgrößenregelung – Digitale Regelungstechnik – Fuzzy Regelung*. Oldenburg Verlag, München, Wien, 2002.
32. Heinz Unbehauen. *Regelungstechnik I.* Vieweg, 1997.
33. Heinz Unbehauen. *Regelungstechnik II.* Vieweg, 1997.
34. R. L. Welch and C. Smith. Bayesian control for concentrating mixed nuclear waste. In *Proceddings of the 15th conference of uncertainty in artificial intelligence*, 1999.

Chapter 3

AppART: a Hybrid Neural Network Based on Adaptive Resonance Theory for Universal Function Approximation

Luis Martí, Alberto Policriti, and Luciano García

Summary. AppART is an adaptive resonance theory low parameterized neural model that incrementally approximates continuous–valued multidimensional functions from noisy data using biologically plausible processes. AppART performs a higher–order Nadaraya–Watson regression and can be interpreted as a fuzzy logic standard additive model. In this chapter we describe AppART dynamics and training. We discuss the approach it makes to hybrid neural systems and deal with its theoretical foundations as a function approximation method. Two modifications to AppART, aimed at improving AppART efficiency, are proposed and tested. We also discuss the combination AppART with growing neural gas networks. Finally, four benchmark problems are solved in order to study AppART from a practical point of view and to compare its results with those obtained from other models.

Keywords: hybrid neural systems, adaptive resonance theory, RBF networks, neuro–fuzzy systems

1 Introduction

Neural networks have been successfully applied to a number of problems, in particular to problems where there is not a deep knowledge of the phenomena and other methods tend to fail. The approximation of functions from samples is a quite common real–life class of problem. There are many neural models that efficiently solve either function approximation problems in general terms or some particular problems like classification, pattern recognition, clustering and time–series prediction. This success is due to these models main characteristics, in particular:

- inductive learning, generation of a model of the training data;
- generalization when presented with previously unseen data;
- ability to handle incoherent, incomplete, or noisy data;
- compact and distributed knowledge representation.

However, in recent years there has been a steadily increasing amount of attention paid to what have been called *hybrid neural systems* [98]. These approaches are aimed to create knowledge systems that preserve the aforementioned features of neural systems and incorporate some characteristics of symbolic artificial intelligence.

The features of such systems should, for example:

- allow the insertion of a priori problem domain knowledge (either in the form of rules, predicates, automata, language grammars, etc.) to a given network;
- allow the extraction of a formal interpretation of a network;
- provide a justification or explanation of a given network response;
- incorporate into the network dynamics other kinds of learning, i.e., deductive [57], causal [38] and abductive [49].

From a symbolic artificial intelligence point of view these hybrid approaches should allow the system to encode highly non–linear relations, which are difficult to express symbolically; should improve these models' sensitivity to noise, and their typical lack of generalization power and robustness, when dealing with partially corrupted or incomplete information.

There have been strong arguments regarding the significance of archiving a working model that successfully combines both computation paradigms (cf. [67]). Nevertheless, there has been a lot of discussion on whether it is actually possible to create a system that unifies the characteristics of biologically plausible (neural networks) and psychologically plausible (symbolic artificial intelligence) paradigms in a unique artificial knowledge system.

It has been argued [32] that neural networks lack of two paradigmatic features of symbolic systems: they do not represent mental states with combinatorial syntax and semantics and their codings do not hold the structure sensitivity of higher mental processes. Evidence has been provided regarding the inability of current neural models of capturing the representational complexity of natural languages [77]. These arguments cast a shadow of doubt on the realization of an hybrid system that shares the features of both computational models. However, these two paradigms rely on the same working hypotheses, that is that cognition can be modeled by computation [64]. Therefore, at least in theory, such system can exist [46].

In recent years a number of hybrid connectionist models have been proposed. Some of them provide, at some degree, a realization of relevant symbolic information processes, such as variable binding [11, 86], predicates unification [96, 97], etc.. This set of results, although not yet fulfills the expectations created by hybrid systems, provide a feasible foundation for the significant further development of such systems.

The formulation of such systems has been repeatedly addressed with diverse neural networks approaches (see [12, 65] for reviews). However, neural models commonly used as a foundation for those approaches are the ones used as universal function approximators, such as the multi–layer perceptron (MLP) [81] and the radial basis function networks (RBF) [9, 68, 78]. These models have to deal with what has been called *high parameterization*. High parameterization is given by the existence of hard–to–guess parameters, such as the number of layers of a network, the number of nodes in a layer, the form of the activation function of a node, etc.. This inconvenient hinders the achievement of adequate results in problems that do not have a recognizable model.

A low parameterized model also simplifies the application of optimization methods, like genetic algorithms [42], simulated annealing [50] or tabu search [41], in order to obtain optimal networks when solving some particular problem. These optimization methods have shown discrete results when applied to models like the MLP and RBF. This kind of results are partially caused by the exhaustive search and computationally intense processes involved in these methods.

The general regression neural network (GRNN) [82, 84], is a low parameterized extension of RBF networks that has successfully solved many function approximation problems. However, this model critically suffers from the curse of dimensionality [3] as its training set grows. There have been a recent attempt [90] to improve the model by introducing some simplifications and providing an enhanced training algorithm. Nevertheless, the problem of the curse of dimensionality remains not properly handled.

GRNN and RBF networks share the localized nature of input prototype layer nodes. This property makes these networks suitable for representing symbolic knowledge, as it was shown by [31, 83, 87]. From a neural networks perspective this feature allows the dynamic local modification of the network topology and nodes weights as training proceeds.

The issue of finding the right network topology and set of weights for a given problem have been repeatedly addressed by using various heuristic training approaches. However only few approaches [8, 30, 66] have focused on the resolution of this issue in general terms. Instead most of these approaches have focused on solving specific formulations. In particular, they are either oriented to find the optimum weights for a predetermined network architecture [4, 10, 68, 72, 102, 103] or to generate a classification or pattern recognition network architecture and its corresponding weights [34, 35, 37, 47, 48, 63, 69].

Adaptive resonance theory (ART) [44] is a theory of cognition that has been recognized as one of the modern artificial neural network theory achievements. ART based networks have some features, such as match–based stable learning and intrinsic self–organization, that are appealing for constructing a hybrid system [2]. Moreover, the search process involved in the production of a network output in ART networks is also of interest since it can be interpreted as a sort of hypothesis testing mechanism [19]. Some of these features can also sensibly improve the high parameterization problem associated to MLP and RBF networks. However, most of the efforts dedicated to ART models have not addressed the problem of creating a hybrid system or the problem of universal function approximation. Instead, they have focused on the development of unsupervised learning clustering models such as ART1 [17], ART2 [16], ART3 [18], fuzzy ART [23], Gaussian ART [99], etc. and supervised learning models capable of learning multidimensional maps such as ARTMAP [22], fuzzy ARTMAP [21], Gaussian ARTMAP [99], RAM [101], ART-EMAP [26], distributed ARTMAP [25], ARTMAP-IC [24], fusion ARTMAP [1], boosted ARTMAP [94], etc.. Recently there have been some works that deal with ART hybrid systems, as cascade ARTMAP [88], and ART function approximation like PROBART [58, 85], FasArt [13, 15] and FasBack [14, 15].

This chapter describes a smooth function approximation ART network (AppART) [59]. AppART is an ART low parameterized neural network that incrementally approximates continuous–valued multidimensional functions from noisy data using biologically plausible processes. AppART performs a higher–order Nadaraya–Watson regression [70, 95] and can be interpreted as a fuzzy system. AppART allows the on–line insertion of fuzzy if–then rules as well as transformation of the network to a set of fuzzy rules. It provides a straight through way of justifying a network response as rules are encoded explicitly by the network. As part of this chapter we discuss AppART relations with other function approximation models and AppART's theoretical power as a function approximation method.

We also propose two modifications to the original formulation: the use of asymmetric Gaussian receptive fields in AppART input classification layer and the on–line estimation of the initialization value of the widths of the Gaussian activated input layer nodes. These two changes are aimed at rendering the resulting model more efficient. In the same key, we discuss the fusion of AppART and growing neural gas networks. The resulting model maintains AppART properties while provides a more efficacious training algorithm.

In the rest of this chapter we first describe AppART's general dynamics as a neural network. Then we deal with its interpretation as a hybrid neural system and its functional approximation properties. After this, we deal with the three aforementioned modifications. Finally, four benchmark problems are solved in order to study AppART from a practical point of view and to compare its results with the ones obtained from other models. The practical impact of the proposed modifications to the model are also analyzed.

2 The Dynamics of AppART

AppART creates classes of similar inputs that have similar outputs. Each class has a desired output value assigned. The output of the network to a given input is the weighted mean of the degree of membership of the input to the classes stored and the desired output values assigned to each class. A match tracking mechanism induces the creation of more specific classes when the prediction of the network differs from the expected output at some degree.

AppART (see Figure 1) has a layer of afferent or input nodes, *F1*, a classification layer, *F2*, a prediction layer, *P*, and an output layer *O*.

The F2 layer stores classes of inputs. Its activation is a combined measure of the similarity of the input and the prototype of each class, and the size of the given class. This layer dynamics is based on Gaussian ART. Other ART models, like fuzzy ART or ART2 could be used for this task but they have some inconveniences. Fuzzy ART inputs features are restricted to the interval $[0, 1]$. Usually inputs that have a different range are normalized or scaled. This two processes can eventually destroy the information contained in a given input sample [29]. Moreover, in some cases, especially in on–line learning situations, the exact range of the inputs can not be guessed from a given set of samples. Models based on Gaussian ART or ART2 are

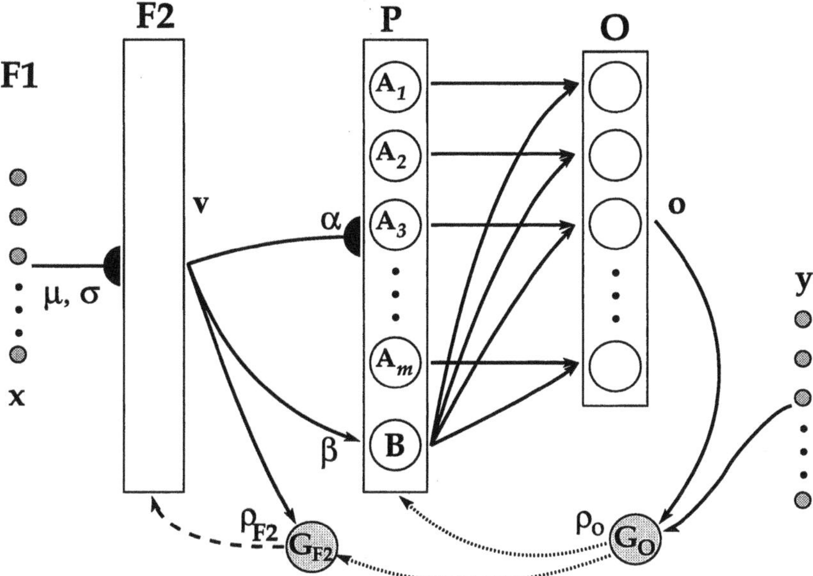

Figure 1. AppART architecture. Input $\vec{x}$ is presented to F1 and then propagated to F2. F2 activation $\vec{v}$ measures the degree of membership of $\vec{x}$ to F2–defined categories. If $\vec{v}$ does not meet the G_{F2} vigilance a node is committed in F2. Prediction $\vec{o}$ is calculated by P and O. If $\vec{o}$ does not meet the G_O error criterion a match tracking algorithm takes place.

useful in overcoming these inconveniences because they do not have restrictions on the inputs as fuzzy ART does. However, ART2 classifies inputs of the form $n\vec{x}$, with $n \in \mathbb{N}$, under the same category [33]. This property, although desirable under many circumstances, is not suitable for function approximation applications.

The network's output is obtained by propagating the output of the F2 layer through the P and O layers using normalized RBF scheme [39] similar to the GRNN summation and output layers, respectively.

2.1 Equations

When an input $\vec{x} \in \mathbb{R}^n$ is presented to the input layer it is propagated to the F2 layer. F2 has N^* nodes, with N of them committed. Each committed node models a local density of the input space using Gaussian receptive fields with mean $\vec{\mu}_j$ and standard deviation $\vec{\sigma}_j$. A node is activated if it satisfies the match criterion. That is, the match function,

$$G_j = \exp\left(-\frac{1}{2}\sum_{i=1}^{n}\left(\frac{x_i - \mu_{ji}}{\sigma_{ji}}\right)^2\right), \; j = 1, \ldots, N, \tag{1}$$

must be greater than the F2 vigilance parameter, ρ_{F2}; according to this, the input strength of a node is computed as

$$g_j = \begin{cases} \frac{\eta_j}{\prod_{i=1}^{n} \sigma_{ji}} G_j, \text{ if } G_j > \rho_{F2} \\ 0 \qquad \text{otherwise} \end{cases} , \tag{2}$$

$$\rho_{F2} > 0 ,$$

where η_j is a measure of the node a priori activation probability.

The activation of each node is then calculated normalizing the node's input strength,

$$v_j = \frac{g_j}{\sum_{l=1}^{N} g_l} . \tag{3}$$

This activation can be interpreted as the conditional probability $p(j|\vec{x})$ of being the node j the one that best encodes input $\vec{x}$.

The prediction and output layers conjointly calculate the prediction of the network. The P layer contains two types of nodes: *A* and *B*. There are as many A nodes as features are in the output vector $\vec{y} \in \mathbb{R}^m$ and only one B node. A and B nodes calculate its activations, a_k and b respectively, as weighted sums of the F2 activation vector

$$a_k = \sum_{j=1}^{N} \alpha_{kj} v_j , \; k = 1, \ldots, m , \tag{4}$$

$$b = \sum_{j=1}^{N} \beta_j v_j . \tag{5}$$

Each α_{kj} represents the sum of values of the output feature k learned when F2 node j was active. The β weight vector is a measure of how much each F2 node has learned.

Nodes on the output layer are connected with one–to–one connections to a corresponding A node and to the B node. Their outputs,

$$o_k = \begin{cases} \frac{a_k}{b} \text{ if } b > 0 \\ 0 \text{ otherwise} \end{cases} , \tag{6}$$

represent the most probable output value, $E(\vec{o}|\,\vec{x})$, expected after an input $\vec{x}$ has been presented. When performing a simulation some steps should be taken to handle correctly the difference of the semantics of the zero as no activation and as a prediction.

One of the principal characteristics of ART neural models is its biological plausibility. AppART has been devised with this in mind. The major arguable point lies in the Gaussian receptive fields of the F2 nodes. However, it has been shown by Poggio and Girosi [78] how they can be implemented in a biologically plausible way.

2.2 Error Detection and Match Tracking

After the presentation of an input, if no F2 node is active, then an uncommitted node must be committed. The task of detecting when an input is not sufficiently coded in F2 is accomplished by the F2 gain control, G_{F2}, that fires if no committed nodes are active. The signal

$$\Gamma_{F2} = \begin{cases} 1 \text{ if } \max_{j=1,\ldots,N} v_j = 0 \\ 0 \text{ otherwise} \end{cases} \tag{7}$$

is used to commit an uncommitted node. It can also be used to offer an "I don't know" answer during the non–adaptive use of a network.

Incorrect predictions are detected by the output gain control, G_O. $\mathrm{G_O}$ compares the similarity between the prediction of the network and the desired network output using a relative measure of the error

$$\Gamma_O = \begin{cases} 1 \text{ if } \frac{|\vec{o}-\vec{y}|}{|\vec{y}|} > \rho_o \\ 0 \text{ otherwise} \end{cases} , \tag{8}$$

with $0 \leq \rho_O \leq 1$, the output vigilance parameter, or an absolute error approach

$$\Gamma_O = \begin{cases} 1 \text{ if } |\vec{o}-\vec{y}| > \rho_o \\ 0 \text{ otherwise} \end{cases} , \tag{9}$$

with $\rho_o > 0$.

The choice of the method largely depends on the characteristics of the problem being solved and the results expected.

If $\mathrm{G_O}$ remains inactive learning takes place in F2 and P. If $\mathrm{G_O}$ fires then a match tracking mechanism takes care of rising the F2 vigilance from its base value $\overline{\rho_{F2}}$, that is the minimum vigilance accepted. The purpose is to reset currently active categories that might be interfering in the calculation of an accurate prediction.

A straight through approach could be setting the vigilance value as the minimum activation,

$$\rho_{F2} = \min_{j=1,\ldots,N} v_j , \tag{10}$$

therefore deactivating the less active F2 node.

Another solution is a one-shot match tracking algorithm, similar to the one used by Grossberg and Williamson [45],

$$\rho_{F2} = \exp\left(-\frac{1}{2} \sum_{j=1}^{N} v_j \sum_{i=1}^{n} \left(\frac{x_i - \mu_{ji}}{\sigma_{ji}} \right)^2 \right) . \tag{11}$$

The efficacy of both approaches will be compared in a subsequent section.

2.3 Learning

As other ART networks, AppART is an on–line learning neural network. Therefore, all adaptation processes have local rules.

In F2, μ_j and σ_j are updated using a learning rule based on the gated steepest descent learning rule [43]. The gated steepest descent,

$$\epsilon \frac{dw_{ji}}{dt} = y_j^* \left[f\left(x_i\right) + w_{ji} \right] , \tag{12}$$

is a learning law for an adaptive weight w_{ji}. The postsynaptic activity y_j^* modulates the rate at which w_{ji} tracks the presynaptic signal $f\left(\vec{x}_i\right)$.

Equation (12) has the discrete–time formulation

$$w_{ji}\left(t+1\right) = \left(1 - \epsilon^{-1} y_j^*\right) w_{ji}\left(t\right) + \epsilon^{-1} y_j^* f\left(x_i\right) . \tag{13}$$

Modifying (13) we can obtain the F2 learning equations which are similar to Gaussian ART's. The constant change rate ϵ is replaced by η_j, which represent the cumulative category activation,

$$\eta_j\left(t+1\right) = \eta_j\left(t\right) + v_j , \tag{14}$$

and, therefore, the amount of training that has taken place in the jth node.

The use of η_j equally weights inputs over time with the intention to measure their sample statistics.

The presynaptic signal $f\left(x_i\right)$ is substituted by x_i and x_i^2, respectively, for learning the first and second moments of the input,

$$\mu_{ji}\left(t+1\right) = \left(1 - \eta_j^{-1} v_j\right) \mu_{ji}\left(t\right) + \eta_j^{-1} v_j x_i , \tag{15}$$

$$\lambda_{ji}\left(t+1\right) = \left(1 - \eta_j^{-1} v_j\right) \lambda_{ji}\left(t\right) + \eta_j^{-1} v_j x_i^2 . \tag{16}$$

The standard deviation,

$$\sigma_{ji}\left(t+1\right) = \sqrt{\lambda_{ji}\left(t+1\right) - \mu_{ji}\left(t+1\right)^2} , \tag{17}$$

is calculated using (15) and (16) as in [100].

In the P layer α_{kj} is adapted to represent the corresponding cumulative expected output learned by each A node. The differential equation formulation of this process,

$$\varepsilon \frac{d\alpha_{kj}}{dt} = v_j y_k , \tag{18}$$

can be transformed to the discrete–time formulation

$$\alpha_{kj}\left(t+1\right) = \alpha_{kj}\left(t\right) + \varepsilon^{-1} v_j y_k , \tag{19}$$

where $\varepsilon > 0$, is a small constant.

The weights of the B node, β_j, are updated in a similar way but tracking the amount of learning that have taken place in each F2 node. Its differential equation is very similar to (18),

$$\varepsilon \frac{d\beta_j}{dt} = v_j . \tag{20}$$

The discrete–time version of (20) is

$$\beta_j(t+1) = \beta_j(t) + \varepsilon^{-1} v_j \,. \tag{21}$$

Note that in (19) and in (21) the same ε is used.

AppART is initialized with all categories uncommitted ($N = 0$). When a category is committed N is incremented. The new category is indexed by N and initialized with $v_N = 1$, $\eta_N = 0$. Learning will proceed as usual but a constant γ_i^2 will be added to each λ_{Ni} to set $\sigma_{Ni} = \gamma_i$. The value of γ_i has a direct impact on the quality of learning. A larger γ_i slows down learning in its corresponding input feature but warranties a more robust convergence. If all input features have approximately the same standard deviation a common $\gamma_i = \gamma_{common}$ can be used.

If an expected output, $\vec{y}$, is present then $\alpha_{kN} = y_k$, for $k = 1, \ldots, m$, and $\beta_N = 1$. Otherwise, $\alpha_{kN} = 0$, for $k = 1, \ldots, m$, and $\beta_N = 0$.

Learning in AppART can be summarized in five possible scenarios:

- *An input is presented but no F2 node becomes active.* F2 activation is calculated by (1)–(3). G_{F2} fires, as no F2 node is active causing the commitment of a new F2 node.
- *An input is presented and some F2 nodes become active.* The F2 activation vector is propagated to P and then to O by (4) and (6), generating a prediction. Active nodes in F2 learn following (15)–(17).
- *An input and an expected output are presented in the same learning interval, no F2 node becomes active.* A new node is committed as in case 1. The P layer weights of the newly committed node are set to correctly predict the expected output.
- *An input and an expected output are presented in the same learning interval, activating some F2 nodes and no match tracking is needed.* The prediction of the network, generated by (1)–(6), does not fires G_O, allowing learning to take place in F2 and P.
- *An input and an expected output are presented in the same learning interval, activating some F2 nodes and match tracking is needed.* As the prediction does not match the vigilance criterion of G_O the F2 vigilance is risen deactivating some F2 nodes. The process of prediction accuracy testing and F2 vigilance raising is repeated until the prediction sufficiently matches the expected output, after which the network behaves as in case 4, or until no F2 nodes are active, when the network acts as in case 3.

As in other ART models AppART prediction accuracy depends on the training patterns presentation order. Here a voting strategy [21] similar to the one used in other supervised mapping ART models would be useful. However, as AppART performs a functional approximation, the prior existing voting strategy can not be directly transferred to it as is. The voting committee is kept but the way the final output is chosen changes. Instead of selecting the most popular output value it is calculated as the average of their outputs

$$\hat{o}(\vec{x}) = \frac{\sum_{h=1}^{H} o_h(\vec{x})}{H} \,, \tag{22}$$

with H, the number of committee members, $o_h(\vec{x})$, the output of the h committee member and $\hat{o}(\vec{x})$ the final prediction.

The standard deviation

$$\hat{s}(\vec{x}) = \sqrt{\frac{\sum_{h=1}^{H} (\hat{o}(\vec{x}) - o_h(\vec{x}))^2}{H}} \tag{23}$$

could be used as a measure of the quality of the prediction.

3 Symbolic Knowledge Representation with AppART

During the last two decades neural networks have evolved into an standard tool for solving numerous complex real–life problems. However, facts as that neural networks did not use a priori domain knowledge, could not give a justification for a given response, and that the stored knowledge could not be easily translated into a human–readable format have limited the application of these technologies in core problems.

These limitations have led to the creation of knowledge systems that attempt to combine neural networks and symbolic artificial intelligence systems. These hybrid neural systems could be an attractive alternative to traditional neural and symbolic systems, as it was mentioned in the introduction.

There have been many approaches to this subject. These approaches could be classified in there broad groups [65]: (i) unified architectures, (ii) transformational architectures and (iii) modular architectures. The first group consists of systems that implements all processing dynamics in a connectionist way. The second group includes the systems that either perform transformations from symbolic to neural representations, from neural to symbolic representations, or both. The last group comprises systems that have neural or symbolic modules that interact with each other with a certain degree of coupling.

Neural models used in most hybrid approaches are based on either MLP, RBF, or Kohonen [51] networks [12]. ART based systems have been recently started to be applied in this subject [20, 27, 88, 89]. The cascade ARTMAP [88, 89] neural network is probably the most relevant result in this field. This neural network is capable of representing chained symbolic if–then rules of the form

(1) if A and B then C
(2) if C and D then E

and thus capable of performing a multi–level inference. It also includes a method for inserting and extracting symbolic rules.

AppART also addresses the hybrid neural system issue as it can encode fuzzy if–then rules in a similar way of fuzzy logic's [104] standard additive model (SAM) [52].

A SAM is a fuzzy function approximator $\widetilde{F} : \mathbb{R}^n \rightarrow \mathbb{R}^m$ that stores N fuzzy if–then rules of the form "if $\vec{x} \widetilde{\in} \mathbf{A}_j$ then $\vec{y} \widetilde{\in} \mathbf{B}_j$", with $\mathbf{A}_k \subset \mathbb{R}^n$ and $\mathbf{B}_k \subset \mathbb{R}^m$.

Assuming that all rules are weighted by w_j , $\widetilde{F}$ can be formulated as

$$\widetilde{F_h}(\vec{x}) = \frac{\sum_{j=1}^{N} c_j w_j V_j a_j(\vec{x})}{\sum_{j=1}^{N} w_j V_j a_j(\vec{x})}, \quad h = 1, \ldots, m, \tag{24}$$

$$= \sum_{j=1}^{N} c_j p_j(\vec{x}) \, . \tag{25}$$

Each $a_j : \mathbb{R}^n \to [0,1]$ is the fuzzy membership for its corresponding fuzzy set $\mathbf{A}_j$. The membership to each $\mathbf{B}_j$ set is computed using the centroids, c_j, and the volumes, V_j. Reformulating $\widetilde{F}$ by introducing the convex weights $p_j(\vec{x})$ it becomes a convex sum of the then–part centroids.

AppART extends the SAM model. From equations (1)–(6) we can formulate AppART's fuzzy function approximator, $\breve{F}$, which takes the form

$$\breve{F}_h(\vec{x}) = \frac{\sum_{j=1}^{N} \alpha_{hj} v_j(\vec{x})}{\sum_{j=1}^{N} \beta_j v_j(\vec{x})} \, . \tag{26}$$

Each F2 node defines a fuzzy set $\mathbf{A}_j^{F2}$ and its activation, v_j, can be interpreted as a membership function of $\mathbf{A}_j^{F2}$. However, the F2 nodes activations are not plain membership functions. Each v_j is a composite measure of the membership degree, the volume of its corresponding $\mathbf{A}_j^{F2}$ and the rule's weight or relevance. The membership degree is computed as the Gaussian match function (1). The volume is computed in (2) by dividing by the multiplication of the deviations (see appendix A for details). By doing this AppART primes active nodes that represent smaller fuzzy sets, working under the assumption that the event of an input belonging to a smaller, more particular class carries a larger amount of information than if the input belongs to a more general or broader class. Finally, the cumulative category activation η_j is used as a measure of the rule's importance.

The A and B nodes are jointly used to empirically approximate the centroid and volume of the membership function of the consequent of the rule.

3.1 Rule Insertion

Inserting a given fuzzy rule to a network has the restriction that the rule's antecedent membership function must be approximated by a set of Gaussian functions. This is equivalent to say that each rule

if $\vec{x}\widetilde{\in}\mathbf{A}_j$ then $\vec{y}\widetilde{\in}\mathbf{B}_j$

should be decomposable into the disjunctive form

if $(\vec{x}\widetilde{\in}\mathbf{A}_{j1}) \vee \ldots \vee (\vec{x}\widetilde{\in}\mathbf{A}_{jL})$ then $\vec{y}\widetilde{\in}\mathbf{B}_j$

where, for each fuzzy set $\mathbf{A}_{jl}$ its corresponding membership function $a_{jl}(\vec{x})$ has a Gaussian formulation with mean $\vec{\mu}^{(jl)}$ and deviation $\vec{\sigma}^{(jl)}$.

From the disjunctive form the insertion of the rule into the network is an straight through process and can be performed at any moment of the network training.

For each $\mathbf{A}_{jl}$ a F2 node, with index N, is committed in the same way as it was explained before (see Section 2.3) and its mean and deviation are made equal to the $a_{jl}(\vec{x})$ ones

$$\vec{\mu}_N = \vec{\mu}^{(jl)} , \tag{27}$$

$$\vec{\sigma}_N = \vec{\sigma}^{(jl)} . \tag{28}$$

Then, the connections from the F2 node N to the A and B layer nodes have to be set by assigning them with the value of the centroid of $\mathbf{B}_j$

$$a_{kN} = c_j v_j , \tag{29}$$

$$b_N = v_j . \tag{30}$$

It is not specified in the SAM definition on what range are defined the rules weights w_j. Therefore, there is not a general method for translation from a rule's w_j to a F2 node η_j. One solution is to normalize all w_j, i.e., by allowing $w_j^* = \frac{w_j}{\sum_h w_h}$, and then set $\eta_j = \kappa w_j$, with κ as the maximum initial cumulative node activation.

After a rule is inserted it can be subjected to the same adaptation process that takes place in the rest of the network during the training process. However, this adaptation can be inhibited if rule's weight, represented by the cumulative node activation η_j, is set to a relatively large value. This action will also induce the network to pay more attention to the rule. This is something that makes sense, since, if it is desired that a rule is not changed by the adaptation process, it is probably because there is substantial evidence of it correctness.

3.2 Rule Extraction and Results Interpretation

As we showed in the above section there is an explicit two–way relationship between AppART's architecture (i.e., F2 nodes and their corresponding A and B nodes weights) and the rules encoded by it. A straight method for converting AppART into a set fuzzy rules consists in, for each F2 node create a fuzzy rule. The antecedent of this rule will be defined by the fuzzy set associated to the node's match function.

The consequent centroid and volume should be constructed from the values of the weights of the connections from the F2 node to the A and B nodes. In particular,

$$c_{kj} = \frac{\alpha_{kj}}{\beta_j} \tag{31}$$

and

$$V_j = \beta_j . \tag{32}$$

The weight of the rule is calculated from the F2 node cumulative activation and the volume of the antecedent's fuzzy set. The total F2 nodes weights

$$\omega_j = \frac{\eta_j}{\prod_{l=1}^{n} \sigma_{jl}} \tag{33}$$

are converted into a normalized form

$$w_j = \frac{\omega_j}{\sum_{h=1}^{N} \omega_h} \tag{34}$$

in order to simplify later calculations.

The fuzzy rules generated from an AppART network can be used to understand or justify a given network response. As AppART stores the information in a localized way by looking at what parts of the network are active we can determine what rules are being used in the production of a response.

3.3 AppART and Other Hybrid Neural Systems

Thanks to the ART based learning we can state that AppART is an autonomous learning SAM. That is, that AppART dynamics performs as SAM and self-adapts its weights and topology to fit the complexity of the problem being solved. As we have shown before, this self-modification process can be directly interpreted as the incorporation of rules to a knowledge system.

AppART also allows the addition of a priori knowledge in the form of fuzzy rules and the extraction of rules of the system.

This characteristics make AppART stand in between in the unified and transformational classes of hybrid systems.

If we compare AppART with other ART hybrid systems, in particular with cascade ARTMAP some interesting points can be noted. The most notorious is the fact that it can only encode discrete if–then associations. This is derived from the winner-take-all activation used in cascade ARTMAP. Cascade ARTMAP also inherits the inefficient coding of fuzzy categories [99] related to fuzzy ARTMAP networks. On the other hand, AppART does not supports the encoding of chained rules. This inconvenience could be overcame by introducing feed-back connections from the output to the input nodes. However, this solution must be investigated with more depth.

There are some other ART neural models that have to do with hybrid neural systems. Two of them, PROBART and FasArt will be discussed in the next section since they are more related to function approximation.

4 AppART as a Function Approximation Method

AppART's hybrid design provides ground for multiple interpretations. In this section we will deal with AppART as a function approximation method, we will comment on this theoretical capabilities and its relations with other methods.

The problem of function approximation can be formulated as the inverse problem:

Definition 1. (Inverse problem) Given the set of pairs $\Psi = \{\langle \vec{x}^{(1)}, \vec{y}^{(1)} \rangle, \ldots, \langle \vec{x}^{(L)}, \vec{y}^{(L)} \rangle\}$, with $\vec{x}^{(l)} \in \mathbb{R}^n$ and $\vec{y}^{(l)} \in \mathbb{R}^m$, find the function $\vec{y} = F(\vec{x})$ that satisfies the conditions:

$$F\left(\vec{x}^{(l)}\right) = \vec{y}^{(l)}\,,\ l = 1, \ldots, L. \tag{35}$$

The problem of function approximation can be reformulated as a regression problem, where F is approximated as the conditional mean $E\left(\vec{y}\middle|\vec{x} = \vec{X}\right)$. This conditional mean can be formulated as

$$y_h = E\left(y_h|\vec{x} = \vec{X}\right) = \frac{\int_{-\infty}^{+\infty} y_h f\left(\vec{x} = \vec{X}, y\right) dy_h}{\int_{-\infty}^{+\infty} f\left(\vec{x} = \vec{X}, y\right) dy_h}\,, \tag{36}$$
$$h = 1, \ldots, m\,,$$

where $f\left(\vec{x} = \vec{X}, y\right)$ is a conditional probabilistic density function. If f is unknown it most be guessed from observations.

The Nadaraya–Watson regression [70, 95] does not assumes any form of f, as it is induced from observations, assuming that f is continuous and have a smooth derivative:

$$y_h = \hat{F}_h\left(\vec{x}\right) = \frac{\sum_{l=1}^{L} y_j^{(l)} \exp\left(-\frac{\left\|\vec{x}-\vec{x}^{(l)}\right\|}{2\sigma^2}\right)}{\sum_{l=1}^{L} \exp\left(-\frac{\left\|\vec{x}-\vec{x}^{(l)}\right\|}{2\sigma^2}\right)}\,. \tag{37}$$

A GRNN is a neural implementation of (37) [6].

From (26) we can state that AppART extends a GRNN as it stores a standard deviation for each input feature and generates only the needed amount of inputs prototypes instead of using the whole training set to build the network. AppART generates just the needed amount of Gaussian basis (F2 nodes) to meet the statistics of the training set. Because of that AppART must generate a more compact code than a similar GRNN and adapt the statistics it has generated to a changing environment.

Following that AppART can be interpreted as a generalization of the GRNN. Even more, AppART with certain parameters values behaves as a GRNN.

Theorem 1. *AppART with* $\rho_{F2} = 0$, $\rho_O = 0$ *and* $\gamma_i = \gamma_{common}$, $i = 1, \ldots, n$ *behaves as a GRNN.*

Proof: If $\rho_O = 0$ then for each $\left\langle \vec{x}^{(l)}, \vec{y}^{(l)} \right\rangle$ not currently stored in the network a new category will be committed as the match tracking algorithm will not find any acceptable F2 activation. The network built following this will have a F2 node for all non-duplicate training samples and no adaptation will take place. As $\overline{\rho_{F2}}$ is set to 0 then it will be no restriction on the activation of F2 nodes. As no learning will take place the values of the standard deviation will remain constant. This changes in the dynamics mimic the GRNN as the network is constructed using the whole training set, all category nodes bias the output of the network and all categories have the same standard deviation [28].

The above theorem allows us express that AppART is a stable learning higher order neural implementation of the Nadaraya–Watson regression. □

The GRNN can be viewed as a normalized radial basis function expansion [6]. This allows us to transitively apply to AppART two important properties of RBF networks: the *universal approximation* [53, 76] and the *best approximation* [40] properties. The first one ensures AppART's capability of approximating any function with any desired degree of accuracy. The second guarantees the existence of a configuration of AppART's parameters that approximates any given function with any desired degree of accuracy. Regarding this last property, it should be noted that MLPs do not share it [40], and that, to our knowledge, it have not been shown to occur in any other ART network.

The fact that AppART possesses these two properties provides a theoretical safety net when solving practical problems.

4.1 Connections with Other ART Models

If we compare AppART with other ART function approximation network, in particular PROBART, FasArt and FasBack some interesting differences can be spotted. As the discussion of the details of each model is out of the scope of this chapter, we will concentrate on commenting these differences.

PROBART [58] is a network based on fuzzy ARTMAP. Fuzzy ARTMAP has two fuzzy ART modules, one generates classes of network's inputs and the other creates classes of the network's outputs. A map field binds classes of inputs and outputs as many–to–one discrete associations. A match tracking process is used to correct wrong predictions by generating more refined classes.

PROBART modifies fuzzy ARTMAP as changes the formulation of the map field by collecting probabilistic information regarding the association between an given class of inputs and a given class of outputs, and by omitting the match tracking process. PROBART in its original formulation used a winner–take–all activation on the fuzzy ART modules. A later modification by [85] allowed distributed activations.

FasArt [13, 15] is also based on fuzzy ARTMAP. It changes the way fuzzy ART computes the category choice by introducing a triangular–shape fuzzy membership function and a weight, that is used in centroid calculation operation. FasBack [14, 15] introduced a backpropagation algorithm aimed at finding the right values of some network parameters and minimizing the network complexity.

The PROBART with distributed activations offers a solution similar to AppART's, but by giving up the match tracking mechanism it looses one of the main features of ARTMAP networks: the capacity to create smaller, more particular, classes of inputs that encode more specific associations.

FasArt addresses, as AppART, the formulation of an ART fuzzy logic system. FasArt keeps the ARTMAP winner–take–all and inter–ART map field formulation. This inhibits FasArt of from exploiting the generalization power that lies in a distributed activations scheme. The use of a backpropagation algorithm in FasBack to determine some network parameters does not allows the network to be trained in an on–line learning fashion and introduces some error–based learning deficiencies as catastrophic forgetting and convergence problems.

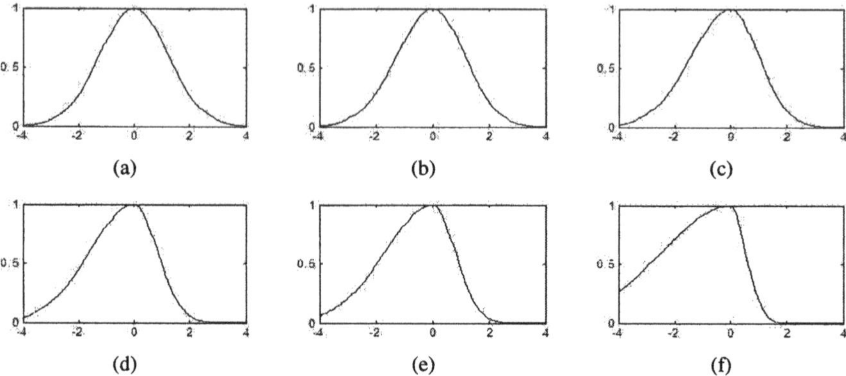

Figure 2. Evolution of the asymmetric widths of a Gaussian receptive field of an AppART network while solving the approximation of the Mackey–Glass equation (see Section 7). It is notable that as training proceeds the left side of Gaussian grows steadily while the right side shrinks itself.

5 Improvements on the AppART Model

In this section we propose two modifications to the original AppART model. These changes may improve AppART overall performance by obtaining a better prediction accuracy, a more compact input coding and shorter training times.

The modifications consist in the use of asymmetric Gaussian receptive fields and an improved selection of the initial widths of the Gaussian receptive fields.

5.1 Asymmetric Gaussian Receptive Fields

The use of asymmetric Gaussian receptive fields in the input classification layer, that is, the use of different widths at both sides of the Gaussian bells, could lead to a faster and more efficient description of the input space as well as to the enrichment of the semantic capacity of the fuzzy rules that can be encoded by the network.

The introduction of this modification implies that some equations must be rewritten. In particular, equation (1) must be reformulated as

$$G_j = \exp\left(-\frac{1}{2}\sum_{i=1}^{n}\left(\frac{x_i-\mu_{ji}}{s_{ji}}\right)^2\right), \tag{38}$$

$$s_{ji} = \begin{cases} \sigma_{ji}^{[\geq]} & \text{if } x_i-\mu_{ji}\geq 0 \\ \sigma_{ji}^{[<]} & \text{if } x_i-\mu_{ji}<0 \end{cases}, \tag{39}$$

in order to use the different values of the widths, $\sigma_{ji}^{[\geq]}$ and $\sigma_{ji}^{[<]}$, at both sides of the exponential.

Learning equations (16) and (17) also need to be modified. They now must be able to cope with fitting both $\sigma_{ji}^{[\geq]}$ and $\sigma_{ji}^{[<]}$. With this in mind, we introduce an *asymmetric learning rate*, $\varsigma \in [0, 1]$, that will modulate the amount of change of the second moment of the input on the side of the Gaussian bell that does not contains the input at a given time. The resulting equations are as follows:

If $x_i - \mu_{ji} \geq 0$, then

$$\lambda_{ji}^{[\geq]}(t+1) = \left(1 - \eta_j^{-1} v_j\right) \lambda_{ji}^{[\geq]}(t) + \eta_j^{-1} v_j x_i^2 , \tag{40}$$

$$\lambda_{ji}^{[<]}(t+1) = \left(1 - \varsigma\eta_j^{-1} v_j\right) \lambda_{ji}^{[<]}(t) + \varsigma\eta_j^{-1} v_j x_i^2 ; \tag{41}$$

in other case,

$$\lambda_{ji}^{[<]}(t+1) = \left(1 - \eta_j^{-1} v_j\right) \lambda_{ji}^{[<]}(t) + \eta_j^{-1} v_j x_i^2 , \tag{42}$$

$$\lambda_{ji}^{[\geq]}(t+1) = \left(1 - \varsigma\eta_j^{-1} v_j\right) \lambda_{ji}^{[\geq]}(t) + \varsigma\eta_j^{-1} v_j x_i^2 . \tag{43}$$

Here $\lambda_{ji}^{[\geq]}$ and $\lambda_{ji}^{[<]}$ represent the second moment of the inputs at both sides of the Gaussian curve.

Finally, the widths are updated using

$$\sigma_{ji}^{[\geq]}(t+1) = \sqrt{\lambda_{ji}^{[\geq]}(t+1) - \mu_{ji}(t+1)^2} , \tag{44}$$

and

$$\sigma_{ji}^{[<]}(t+1) = \sqrt{\lambda_{ji}^{[<]}(t+1) - \mu_{ji}(t+1)^2} . \tag{45}$$

When committing a new node both $\lambda_{Ni}^{[\geq]}$ and $\lambda_{Ni}^{[<]}$ could be set to the same initialization value and, as training proceeds, adapt them separately.

The choice of the value for the asymmetric learning rate, ς, has direct implications on the learning process. As ς approaches 1 learning in the widths get less asymmetric, until $\varsigma = 1$, when the modification reduces itself to the symmetric one. On the other hand, as ς moves toward 0, inputs that lie on one side of the curve have a lesser impact on the width of the other side. This could lead to potentially dangerous situations. For example, if there are many training samples that lie on one side of the curve the other side could remain relatively untrained and do not represent correctly the nature of the input class being modeled. Situations like this should clearly be avoided.

In Figure 2 it can be observed how a receptive field modifies its widths as training proceeds. In the next section we will show how the choice of different values of ς influences prediction accuracy.

5.2 Optimal Initialization of the Widths of the Gaussian Receptive Fields

One of the most obscure matters when solving a particular problem with AppART lies in the initialization of the widths of the radial basis of its Gaussian activated nodes. It is caused by the necessity of understand the problem statistics before hand, at least at some degree.

Although AppART adapts the values of the widths, a rather large or small initialization value could lead either to the excessive overlapping of the input classes or the creation of "holes" in the input domain that remain uncovered by the classes generated.

This problem is closely related to the optimization of the widths of the hidden layer units in an RBF network, which have been addressed before a number of times. There are two fundamental streams [39]. The first fits the widths using a supervised training method based on the backpropagation algorithm [78] or on the generalized cross-validation [73]. The other attempts to provide a method for local adaptation of the widths (i.e., [34, 68]).

As the second class of methods is the one that suits best AppART conception our approach adheres to it. The foundation of this kind of methods was laid out by Moody and Darken [68]. They, for a given node, applied a k–nearest neighbor algorithm to discover its nearest nodes. Then the widths of the node are set to the root of the mean of the square distances. This approach was modified by Fritzke [34] by substituting the k–nearest neighbor algorithm with the topological order used in its supervised growing neural gas networks [36].

In order to apply such methods to AppART they must be adapted to suit the needs of its ART related training characteristics. Our approach, instead of a k–nearest neighbor algorithm, that introduces an unnatural computational overload, builds the set of neighboring nodes using the F2 match function (1) as a measure of distance. That means that only the nodes that are active at a given time are the ones that take part in the estimation of the initial widths of the node about to be created. Active nodes are supposed to be relatively near to the center of the new node forming a sort of topological neighborhood. Each node influences the computation of the width based on the value of its match function.

The formulation of this approach is the following: for a given new node N, set

$$\gamma_{Ni} = \min\left(\gamma_{max}, \sqrt{\frac{\sum_{G_j > \rho_{F2}} \left(x_i - \mu_{ji}\right)^2 G_j}{\sum_{G_j > \rho_{F2}} G_j}}\right) . \tag{46}$$

If there is no $G_j > \rho_{F2}$ then $\gamma_{Ni} = \gamma_{max}$. The maximum width, γ_{max}, prevents that F2 nodes committed earlier during training get large widths initial values. An excessively wide F2 node might not be able to encode the details of the input space and might become a source of noise during computation.

6 Combining AppART and Growing Neural Gas

We now describe a modification to AppART that introduces some growing neural gas [36] based training characteristics. The resulting network, which we called GasART [60], keeps the general input propagation dynamics of AppART, while modifies the way new classes are inserted after performing the match tracking process and the laws that control the way these classes are modified to fit the complexity of the training data. GasART keeps the theoretical properties and hybrid neuro–fuzzy system interpretation of AppART while provides a more efficient mechanism for creating input classes.

Growing neural gas (GNG) networks are intrinsic self–organizing neural networks based on the neural gas [63] model. This model relies in a competitive Hebbian learning rule [62]. It creates an ordered topology of inputs classes and associates a cumulative error to each. The topology and the cumulative errors are conjointly used to determine how new classes should be inserted. Using these heuristics the model can fit the network dimension to the complexity of the problem being solved. Although GNG was originally meant for solving unsupervised learning problems (i.e., clustering and vector quantization); it was extended to supervised RBF networks [34]. Later, the model have been further expanded encompass to the incremental generation of neuro–fuzzy systems [37].

A GNG–RBF network is a RBF network in which the hidden layer RBF nodes are linked together defining a topological order. Each link has assigned an age. When training starts the network is has two randomly initialized nodes. These two nodes are bound with each other. The training algorithm that fits the network could be separated in three interrelated concurrent processes: adaptation, node insertion and node deletion. These processes could be summarized as follows:

- *Adaptation.* For a given training input/output pair:
    1. Adaptation takes place in the *best–matching node* (BMN), which is the node that is closest to the input. Nodes bound to the BMN will adapt their weights with a lesser rate.
    2. A second best–matching node is selected. If there is no link between the BMN and the 2ndBMN the link is created with age equal to zero. If the link already exists its age is reset to zero. The age of the rest of the nodes connected to the BMN is incremented.
    3. The output layer is updated using the Delta Rule [81].
    4. The error accumulator associated to the BMN is incremented by the prediction error.
- *Node insertion.* A new node is inserted to the network after a predefined number of loops of training steps or after observing that there is no improvement on the network performance. The new node is inserted somewhere between the node with higher cumulative error and one its topological neighbors. The selection of second node remains a heuristic decision, i.e., the neighbor with higher cumulative error, the more distant neighbor, etc. [34].
  The cumulative errors of the two nodes involved are reduced by a prefixed amount. The link between those nodes disappears and two new links from the two old nodes to the new one are created with their age set to zero.
- *Node and link deletion.* A link is eliminated if its age grows beyond a certain maximum age. If a node becomes isolated (not bound to any other node) it is removed.

The combined use of these three processes renders GNG training Hebbian in spirit [61].

The GNG training scheme was adapted to fit AppART characteristics. The resulting network, GasART, combines ART and GNG networks. It modifies the learning laws of AppART to improve its accuracy and performance while keeping its main characteristics.

6.1 GasART Modifications of AppART Dynamics

The overall AppART's input propagation dynamics remains the same in GasART. However, some modifications are introduced to handle the cumulative errors of GNG networks. In GasART each F2 node has an associated cumulative error, e_j. This cumulative error is computed during training to represent the a priori error probability of the node given its activation.

In order to compute the cumulative errors the G_O gain control must be split. A node *E* is introduced. This node computes the prediction error, ξ, measuring it either as an relative error,

$$\xi = \frac{|\vec{o} - \vec{y}|}{|\vec{y}|} , \tag{47}$$

or an absolute one,

$$\xi = |\vec{o} - \vec{y}| ; \tag{48}$$

each related with the relative (8) and absolute (9) output gain controls of AppART, respectively.

In correspondence with these changes the GasART output gain control, G_O, detects incorrect predictions by comparing ξ with the output error vigilance parameter, ρ_o,

$$\Gamma_O = \begin{cases} 1 \text{ if } \xi > \rho_o \\ 0 \text{ otherwise} \end{cases} , \; \rho_o > 0 . \tag{49}$$

After these processes the G_{F2} gain control and match tracking mechanism are performed as specified for AppART.

GasART is initialized with two nodes committed ($N = 2$). The centers of these nodes could be either initialized to the first two inputs in the training set or to random values. Each node is bounded with each other with a link that indicates its topological vicinity. This link has an age, κ, that is initialized to $\kappa = 0$ when a new bind is set up. A link between two nodes is removed if its age, κ, is bigger than a certain κ_{max}.

If the signal Γ_{F2} fires, then an uncommitted node is committed and N is incremented. The newly committed node is indexed by N and initialized with $\upsilon_N = 1$, $\eta_N = 0$ and $e_N = 0$. The two units with larger accumulated error are bound to the node N. Subsequently, learning will proceed as normal.

6.2 Learning in GasART

First, the prediction errors obtained in this input presentation are added to the cumulative error of the nodes that had interfered in the production of the prediction. The amount of error accumulated is modulated by the node activation, v_j,

$$e_j(t+1) = e_j(t) + \xi v_j \,. \tag{50}$$

The cumulative category activation, η_j, is updated in the same way as in AppART,

$$\eta_j(t+1) = \eta_j(t) + v_j \,. \tag{51}$$

The core difference between GasART and AppART lies in the way F2 nodes are involved in the learning process. GasART does not grants all actives nodes to freely learn in a proportional amount to each node activation value. Instead a *best–matching node* (BMN) is determined by selecting the node with maximum activation, v_j. A *second best–matching node* (SBMN) is also selected using the same criteria. If the BMN and the SBMN are not connected then a link between them is established with age, κ set to zero. If the link already exists its age is reset back to zero.

The center of the BMN receptive field is updated by allowing

$$\mu_{\mathrm{BMN}i}(t+1) = \left(1 - \eta_{\mathrm{BMN}}^{-1} v_{\mathrm{BMN}}\right)\mu_{\mathrm{BMN}i}(t) + \eta_{\mathrm{BMN}}^{-1} v_{\mathrm{BMN}} x_i \,, \tag{52}$$

All nodes bound to the BMN, represented as the set Δ_{BMN} also modify their centers

$$\mu_{ji}(t+1) = \left(1 - \upsilon\eta_j^{-1} v_j\right)\mu_{ji}(t) + \upsilon\eta_j^{-1} v_j x_i,\; j \in \Delta_{\mathrm{BMN}} \,. \tag{53}$$

but with a rate, $0 \le \upsilon \le 1$, that controls the amount of change. After this, the age of the links emanating from the BMN is incremented.

Either when a node is committed or after a training iteration, the standard deviations, $\vec{\sigma}_j$, should be computed.

An AppART–flavored approach is to also keep track of the second moment of the inputs,

$$\lambda_{\mathrm{BMN}i}(t+1) = \left(1 - \eta_{\mathrm{BMN}}^{-1} v_{\mathrm{BMN}}\right)\lambda_{\mathrm{BMN}i}(t) + \eta_{\mathrm{BMN}}^{-1} v_{\mathrm{BMN}} x_i^2 \,, \tag{54}$$

$$\lambda_{ji}(t+1) = \left(1 - \eta_j^{-1} v_j\right)\lambda_{ji}(t) + \eta_j^{-1} v_j x_i^2 \,,\; j \in \Delta_{\mathrm{BMN}} \,, \tag{55}$$

and use it to estimate the standard deviations,

$$\sigma_{\mathrm{BMN}i}(t+1) = \sqrt{\lambda_{\mathrm{BMN}i}(t+1) - \mu_{\mathrm{BMN}i}(t+1)^2} \,; \tag{56}$$

$$\sigma_{ji}(t+1) = \sqrt{\lambda_{ji}(t+1) - \mu_{ji}(t+1)^2} \,,\; j \in \Delta_{\mathrm{BMN}} \,. \tag{57}$$

This approach adheres to the on–line local adaptation dynamics philosophy embodied in the conception of AppART.

Another approach emerges from the RBF and GNG related works (see Section 5.2 for a brief discussion). In this case the standard deviations are set to the mean distance between the center of the node and the centers of the nodes that are bound to it [34]:

$$\vec{\sigma}_j = \frac{1}{[\Delta_j]} \sum_{l \in \Delta j} \left|\vec{\mu}_j - \vec{\mu}_l\right| \,, \tag{58}$$

with $[\Delta_j]$ the cardinality of Δ_j. This formulation rescinds of the use one standard deviation value of each feature of the input vector and of the local learning philosophy. However it has been experimentally proved to provide better results in problems with low–dimensional inputs.

Learning in the P and O layers remain as specified for AppART in Section 2.3.

7 Simulations

We now focus on solving four benchmark problems: the fifth–order chirp function approximation, Mackey–Glass equation approximation, the predictions of the dynamics of a Puma 560 robotic arm and the DNA promoter recognition.

The first two problems are function approximation problems. They are meant for studying AppART performance as a function approximating algorithm and comparing it with other neural models like MLP, RBF, GRNN, fuzzy ARTMAP (FAM), Gaussian ARTMAP (GAM), PROBART and FasBack. In all cases where AppART, FAM, GAM, PROBART and FasBack are applied a voting strategy is used. The number of voting runs was set to a 10 per cent of the size of the training set. In both tests the AppART output gain control uses the absolute error measure (9).

In the case of MLP a backpropagation algorithm was used for fitting the network. For RBF networks a hybrid training method (unsupervised in the hidden layer and supervised in the output layer) [68] was used. Both of these training techniques are broadly discussed elsewhere (cf. [6]).

The mean squared error (MSE)

$$E = \frac{1}{mL} \sum_{l=1}^{L} \sum_{k=1}^{m} \left(y_k^{(l)} - o_k^{(l)} \right)^2 , \tag{59}$$

with $\vec{y}^{(l)}$ and $\vec{o}^{(l)}$, the expected output and the network prediction, respectively, associated with pattern l,and L, the size of the test set; is used for comparing results.

After assessing the models performance in the first two problems we will proceed to a more complex function approximation problem: the prediction of the dynamics of the Puma 560 robotic arm. With this problem we intend to evaluate the results produced by the networks with the ones obtained from other advanced neural models (i.e., [74]).

The fourth and final problem is meant for testing AppART and GasART as hybrid neural systems. Their results are compared with the ones obtained using cascade ARTMAP, KBANN [91, 93], NofM rule extraction algorithm [92], fuzzy ARTMAP, MLP and other machine learning algorithms like ID–3 [79], k–nearest neighbor (KNN) [29] and consensus pattern analysis [71].

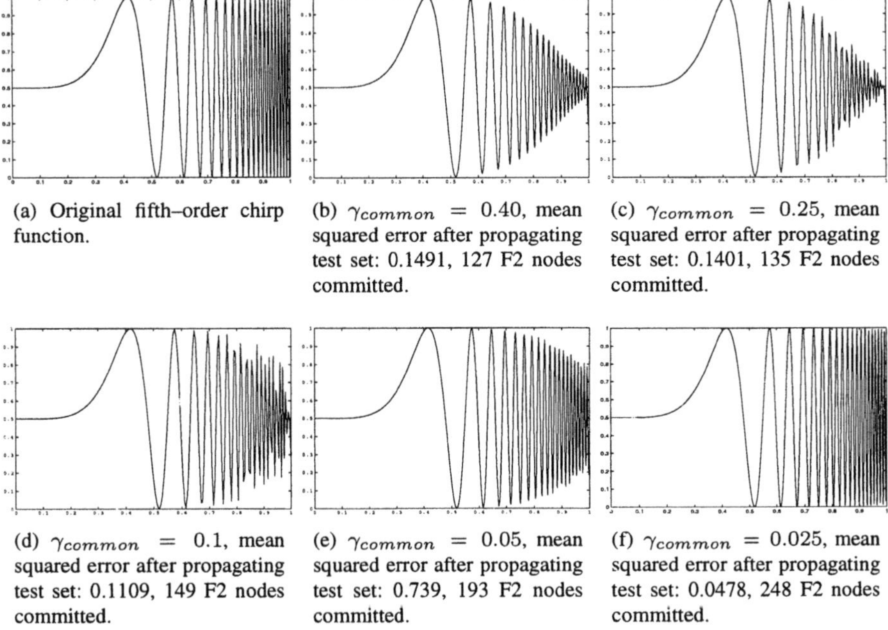

(a) Original fifth–order chirp function.

(b) $\gamma_{common} = 0.40$, mean squared error after propagating test set: 0.1491, 127 F2 nodes committed.

(c) $\gamma_{common} = 0.25$, mean squared error after propagating test set: 0.1401, 135 F2 nodes committed.

(d) $\gamma_{common} = 0.1$, mean squared error after propagating test set: 0.1109, 149 F2 nodes committed.

(e) $\gamma_{common} = 0.05$, mean squared error after propagating test set: 0.739, 193 F2 nodes committed.

(f) $\gamma_{common} = 0.025$, mean squared error after propagating test set: 0.0478, 248 F2 nodes committed.

Figure 3. AppART approximation of the fifth–order chirp function with different values of the initial standard deviation γ_{common}. As no notable difference was obtained using either minimum activation or one–shot match tracking mechanisms the one that offered best results in any case is shown.

7.1 Fifth–Order Chirp Function Approximation

This problem consists on the estimation of the fifth–order chirp function (see Figure 3(a)) formulated as

$$f(x) = \frac{1}{2} + \frac{1}{2} \sin\left(40\pi x^5\right) . \tag{60}$$

A set of 10000 samples was generated having $x \in [0, 1]$. 70% of the samples were randomly extracted and used as the training set and the rest were used as a test set.

Figure 3 shows the result of propagating the test set after training AppART with different sets of training parameters. It is clearly noticeable that as γ_{common} reduces the quality of the prediction increases, as well as the amount of nodes created in the F2 layer. No substantial difference in the accuracy of the prediction was noticed when using one of the two match tracking mechanisms described. However, when using the one-shot match tracking mechanism the speed of the training process was reduced at least by one fourth. The results obtained with other models are summarized in Table 1. Here AppART performs best. FasBack also obtained good results but with a large amount of training epochs.

Table 1. Mean squared errors after propagating the test set in the fifth–order chirp function approximation problem.

Model	MSE obtained	Training epochs
MLP	0.4362	30000+
RBF	0.2701	10000
GRNN	0.1540	150
FAM	0.1802	140
GAM	0.1521	45
PROBART	0.1435	50
FasBack	0.0915	10000
AppART	0.0803	30

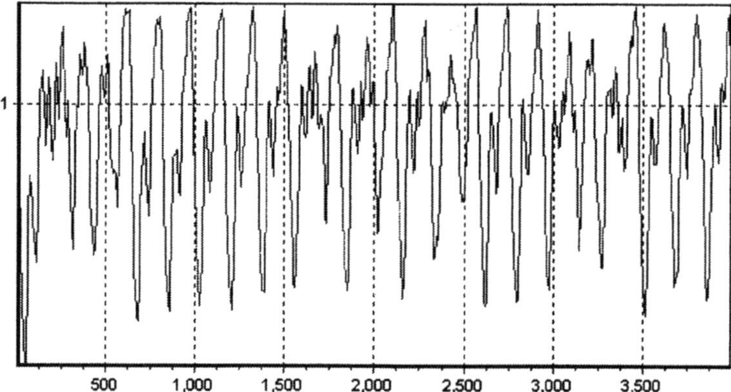

Figure 4. The Mackey–Glass equation.

7.2 Mackey–Glass Equation

The Mackey–Glass equation (see Figure 4),

$$\frac{dx}{dt} = \frac{ax\,(t-\tau)}{[1 + x^c\,(t-\tau)]} - bx\,(t)\ , \tag{61}$$

is a time–delay differential equation that has been proposed as a model of white blood cell production [56]. The constant values are commonly set to $a = 0.2$, $b = 0.1$ and $c = 10$. The delay parameter τ determines the behavior of the system. For $\tau > 16.8$ the system produces a chaotic attractor.

In our simulations we have chosen $\tau = 30$ and an input window of 6 time steps elements as in [54]. From a generated 4000 patterns set, two sets were extracted, one with 3200 pattern used for network training and one with 800 patterns used as test set. As all input features are equally distributed since they are equally spaced time

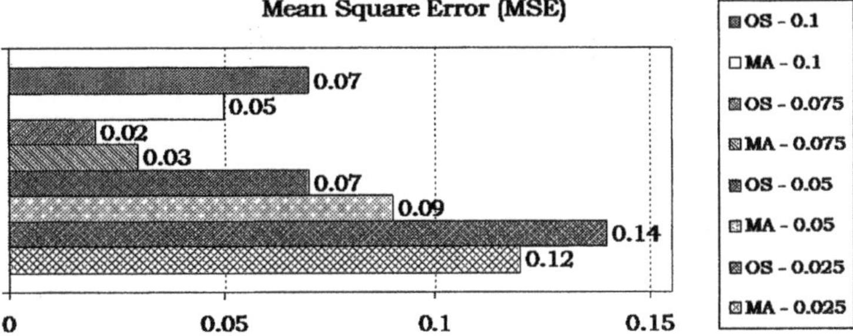

(a) AppART test set results using both types of match tracking algorithms and different values of γ_{common}

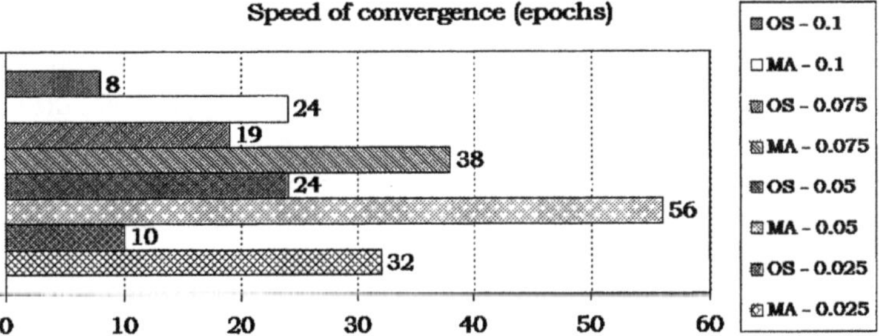

(b) AppART speed of training using different training parameters and match tracking algorithms.

Figure 5. AppART test set results approximating the Mackey–Glass equation using one–shot (OS) and minimum activation (MA) match tracking algorithms. The numbers on the right are the values of the initial standard deviation γ_{common}.

steps elements the initial standard deviations of each input feature $\gamma_1, \ldots, \gamma_n$ were set to a common value γ_{common}.

Figure 5(a) shows the AppART prediction errors. Here it is clear the effect of the initial standard deviation on the accuracy of AppART, but there is no evidence of which match tracking algorithms exhibits a better performance.

However the difference of convergence speed when using one of the match tracking methods is obvious in Figure 5(b). The one–shot match tracking performs generally twice as fast as the minimum activation.

When applying other neural models (see Table 2), the best configuration of AppART outperforms all of them. AppART also generates a more compact coding with respect to the rest of the ART networks, since AppART created 793 and 698 input prototypes using minimum activation and one-shot match tracking, respectively, while FAM generated 2107, GAM 2358, PROBART 1824 and FasBack 1092.

7.3 Assessing the Modifications to AppART

We now return to the problem of the Mackey–Glass equation approximation. We intend to compare the results obtained from the original AppART formulation and the modifications introduced in Sections 5 and 6.

For these tests we used the same data sets and training methodology of the previous section. Some results obtained there are also reproduced here for comparison reasons.

The first experiment has to do with the use of asymmetric Gaussian receptive fields, as described before in Section 5.1. In Table 3 it is shown the results obtained after training AppART with different values of the asymmetric learning rate, ς. For each value of ς we searched for the optimal set of training parameters. Therefore the results displayed are the best ones obtained for a given value of ς.

By analyzing these results some comments emerge. It is noticeable that as the value of ς approaches 1 the value of the prediction accuracy does not differs significantly from the best obtained. However, the number of nodes needed to archive this accuracy is larger that the ones created with smaller values of ς. This is also true for the amount of iterations required by the network to meet its minimum error.

On the other hand, as ς approaches 0 the resulting network performs rather poorly, much worse than the original (symmetric) AppART formulation.

The results yielded by AppART and GasART before and after introducing the estimation of the initial widths and the asymmetric Gaussian receptive fields are shown in Table 4. Networks trained with the initial value of the widths optimized had a smaller test set mean square error and a more compact coding of the input space. It is notable that the combined use of both modifications described in Section 5 offered even more accurate predictions and more compact codings.

Table 2. Test set mean square errors approximating the Mackey–Glass equation with different neural models.

Neural Model	MSE obtained
MLP	0.2406
RBF	0.2173
GRNN	0.1453
GNG-RBF	0.0301
FAM	0.0972
GAM	0.1647
PROBART	0.0721
FasBack	0.0598
AppART-OS	0.0262
AppART-MA	0.0319

Table 3. Test set mean square errors (MSE), nodes created and training epochs needed by AppART solving the Mackey–Glass equation problem with asymmetric Gaussian receptive fields with different values of the asymmetric learning rate, ς.

Asymmetric learning rate	MSE	Nodes used	Training epochs
$\varsigma = 0$	0.1840	96	46
$\varsigma = 0.20$	0.1151	108	51
$\varsigma = 0.40$	0.0972	92	37
$\varsigma = 0.60$	0.0618	69	16
$\varsigma = 0.80$	0.0249	549	21
$\varsigma = 1$ (Original AppART)	0.0262	698	19

The results yielded by GasART deserves special attention. The results displayed in Table 4 show that GasART not only makes a more accurate prediction by two orders of magnitude than AppART but also commits fewer nodes thus creating a more compact input space representation. On the other hand, GasART does not noticeably improves its performance with the introductions of the changes proposed. This might be caused by the nature of the problem that does not allows to further improve the accuracy of the prediction.

7.4 Prediction of the Dynamics of a Puma 560 Robotic Arm

We now turn the attention to the problem of predicting the angular acceleration of the links of the Puma 560 robotic arm given a set of joint angles, velocities and torques. This problem is provided within the DELVE standardized test environment [80].

Table 4. Test set mean square errors (MSE) approximating the Mackey–Glass equation using AppART and GasART with the modifications suggested: optimization of the widths initialization (OWI) and asymmetric Gaussian receptive fields (AGRF).

Neural Model	MSE obtained	Nodes used
Original AppART	0.0262	698
AppART+OWI	0.0225	523
AppART+AGRF	0.0249	549
AppART+AGRF+OWI	0.0195	491
GasART	0.0005	527
GasART+OWI	0.0005	484
GasART+AGRF	0.0004	493
GasART+AGRF+OWI	0.0004	482

Table 5. Normalized standardized mean error when solving pumadyn32nh with different training set sizes using multi–layer perceptron (MLP), radial basis function networks (RBF), RBF with regularized forward selection (RFS), RBF with regression trees (RT), RBF with growing neural gas (GNG), AppART and GasART.

Set size	MLP	RBF	RFS	RT	GNG	AppART	GasART
64	0.9706	0.9638	0.8403	0.8524	0.8103	0.8158	0.8003
128	0.9845	0.9305	0.8267	0.8164	0.7998	0.7583	0.7904
256	0.9572	0.9274	0.8363	0.6395	0.5410	0.5120	0.4285
512	0.9467	0.9085	0.8243	0.5132	0.5173	0.4743	0.4156
1024	0.9403	0.8957	0.8270	0.3920	0.4023	0.4102	0.3823

Using this dataset has the added advantage of allowing us to compare AppART and GasART performance with other state-of-the-art neural networks training strategies (i.e., [74]).

DELVE contains different instances of this problem each with either 8 or 32 inputs, high or low non-linearity and high or low noise. Because of the briefness of this communication the results shown here use the instance called pumadyn32nh, that has 32 inputs, high non-linearity and high noise. This instance, according to the simulations performed, is the more complex of all. Different training set sizes were used in order to study the way the models responds to this variation.

The results obtained are shown in Table 5. In this case we have included in the analysis two models formerly used in [74]. One implements a regularized forward selection of candidate nodes to build a RBF network [5, 55, 73] and the other infers the centers and widths of the RBF from the nodes of a regression tree [75]. For the sake of uniformity and reproducibility we have DELVE's default error measure: the normalized standardized mean error. It should also be mentioned that the models used in the previous experiment were also tested but their results were not significant.

Some conclusions arise from these results. It is clearly visible that as the training set grows the performance of the models networks increases. Its also notable that GasART slightly outperforms the other models. This is very important since this fact can be used as a sort of empirical validation of GasART.

7.5 DNA Promoter Recognition

Promoter are short DNA sequences that precede the beginning of genes. A method for the recognition of promoter DNA sequences allows the identification of the location of genes in large and uncharacterized sequences of DNA. A promoter sequence can be experimentally detected because it is located where a protein called RNA polymerase binds to the DNA sequence.

This area of problems have been addressed using different approaches, as diverse as, test inductive theory–refinement, inductive logic–programming, dynamic

Table 6. Results of different knowledge systems when solving the DNA promoter recognition problem using or not a priori knowledge.

Knowledge System	Rules or nodes used	Error per cent
ID–3	—	17.9
KNN (K=3)	105	12.3
Consensus Sequences	—	11.3
MLP	16	7.5
Fuzzy ARTMAP	20.6	6.5
KBANN	16	2.9
Cascade ARTMAP	28.9	2.0
NofM rules	12	3.8
Cascade ARTMAP rules	19.5	3.0
AppART without a priori knowledge	18.3	2.5
AppART with a priori knowledge	12	1.8
AppART w/o a priori knowledge rules	18.3	2.6
AppART with a priori knowledge rules	12	1.8

programming, neural networks, etc.. This repeated use have turned this problem into a standard problem for model performance comparison.

In this particular case we will address the prokaryotic promoter recognition using a data set [7] that consists of 106 patterns. Each DNA pattern consists of a 57 position window, and each position takes one of the nucleotide values: A, G, T and C, after adenine, guanine, thymine and cytosine, respectively. An imperfect domain theory [71] comes as a companion of the data set. This domain theory, if applied directly only classifies correctly half of the data set.

Although bigger data sets exist this one was chosen because it has been applied in some other related works [88, 89, 91, 93] and thus allows a direct comparison of the results.

Table 6 summarizes the results obtained after applying different knowledge systems. For training AppART we used the same methodology of [88] in order to render the results comparable. Different values of its AppART training parameters were tested. Here, the best results are shown. The results that do not deal with AppART are taken from [88].

Some comments emerge from these results. First, it is notable the improvement of the predictions by first inserting the a priori domain theory into the network, since AppART with a priori knowledge outperforms the rest of the models tested while keeping a compact set of rules. This is probably a consequence of AppART's distributed activation of F2 nodes that enable AppART to have a greater generalization power. It could also be noted that the SAM built using the rules extracted from AppART network offers similar, if not equal, results to the ones obtained with the original network.

Although this problem was chosen mainly because of its advantages as a test problem, since it have been addressed with many diverse approaches, we should remark its importance from a practical point of view. This field of bioinformatics, that is, the field biology related computational problems, is one of the current core problems of science. The application of AppART in further studies related to this subject might reveal some other interesting results.

8 Concluding Remarks

We have dealt with a novel adaptive resonance theory neural network called AppART. AppART incrementally approximates any function with any degree of accuracy from noisy training samples. It incorporates features of ART models, such as match-based on–line learning and deep self–organization, in order to build an autonomous learning neuro–fuzzy system that dynamically fits its topology and parameters (weights) to fit the complexity of the problem being solved.

We have shown how AppART extends the fuzzy logic's standard additive model and therefore performs a sort of fuzzy inferencing. Straight through methods allow the the on–line insertion and extraction of fuzzy rules and the generation of an explanation of a given network response.

Also as part of this chapter we have shown that AppART has the universal and best approximation properties and have discussed AppART relations with similar neural models.

We have also described two modifications to the original model: the use of asymmetric Gaussian receptive fields and the estimation of the optimal initialization value for the widths of the Gaussian receptive fields. These two modifications have been shown to improve AppART prediction accuracy and input space coding compactness.

Finally, we have addressed the formulation of a model that blends AppART with the supervised growing neural gas model. The resulting network, which we called GasART, yielded more accurate results than the original AppART.

In the benchmark tests carried out AppART outperformed all neural models tested, generating compact knowledge representations.

Acknowledgments

The authors wish to thank the Dipartimento di Matematica e Informatica of the Università degli Studi di Udine for its support on the conception and elaboration of this work. Dr. Luis Martí, Sr. provided some insightful comments on the drafts of this work.

Appendix

Representing the Fuzzy Set Volume in the F2 Nodes Activation

The volume or area of a fuzzy set $\mathbf{A}_j$ is represented as

$$V_j = \int a_j\left(\vec{x}\right) d\vec{x} \tag{A.1}$$

$$= \int \ldots \int a_j\left(x_1, \ldots, x_n\right) dx_1 \ldots dx_n \,, \tag{A.2}$$

with $a_j\left(\vec{x}\right)$ the membership function of $\mathbf{A}_j$.

Substituting in (A.2) the membership function used in the F2 nodes we obtain

$$V_j^{F2} = \int \ldots \int \exp\left(-\sum_{i=1}^{n}\left(\frac{\mu_{ji} - x_i}{\sigma_{ji}}\right)^2\right) dx_1 \ldots dx_n \,. \tag{A.3}$$

Transforming (A.3) we get

$$V_j^{F2} = \prod_{i=1}^{n} \int \exp\left(-\left(\frac{\mu_{ji} - x_i}{\sigma_{ji}}\right)^2\right) dx_i \,. \tag{A.4}$$

As

$$\int \exp\left(-\left(\frac{\mu_{ji} - x_i}{\sigma_{ji}}\right)^2\right) dx_i = \sqrt{K\sigma_{ji}} \tag{A.5}$$

we can reformulate (A.4) and obtain that

$$V_j^{F2} = K^{\frac{n}{2}} \prod_{i=1}^{n} \sigma_{ji} \,. \tag{A.6}$$

Therefore we can assert that the multiplication of the standard deviations σ_{ji} is a measure of the volume of the fuzzy sets defined by the F2 nodes.

References

1. Y. R. Asfour, G. A. Carpenter, S. Grossberg, and G. W. Lesher. Fusion ARTMAP: A neural network architecture for multi-channel data fusion and classification. In *Proceedings of the World Congress on Neural Networks (WCNN'93)*, volume 2, pages 210–215, Hillsdale, NJ, 1993. Lawrence Erlbaum Associates.
2. G. Bartfai. An Adaptive Resonance Theory-based neural network capable of learning via Representational Redescription. In *Proceedings of IEEE International Joint Conference on Neural Networks (IJCNN'98)*, pages 1137–1142, Anchorage, 1998.
3. R. Bellman. *Adaptive control processes: A guided tour.* Princeton University Press, Princeton, 1961.
4. M. Bianchini, P. Fasconi, and M. Gori. Learning without local minima in radial basis function networks. *IEEE Transactions on Neural Networks*, 6:749–756, 1995.
5. C. M. Bishop. Improving the generalisation properties of radial basis function neural networks. *Neural Computation*, 3(4):579–588, 1991.
6. C. M. Bishop. *Neural Networks for Pattern Recognition.* Clarendon Press, Oxford, 1995.
7. C. L. Blake and C. J. Merz. UCI repository of machine learning databases. http://www.ics.uci.edu/~mlearn/MLRepository.html, 1998.
8. E. Blanzieri, P. Katenkamp, and A. Giordana. Growing radial basis function networks. In *Proceedings of the Fourth Workshop on Learning Robots*, Karlsruhe, Germany, 1995.
9. D. S. Broomhead and D. Lowe. Multivariate functional interpolation and adaptive networks. *Complex Systems*, 2:321–355, 1988.
10. D. S. Broomhead and D. Lowe. Multivariable functional interpolation and adaptive networks. *Complex Systems*, 2:321–329, 1998.
11. A. Browne and R. Sun. Connectionist variable binding. *Expert Systems: The International Journal of Knowledge Engineering and Neural Networks*, 16(3): 189–207, 1999.
12. A. Browne and R. Sun. Connectionist inference models. *Neural Networks*, 14: 1331–1355, 2002.
13. J. M. Cano, Y. A. Dimitriadis, M. J. Araúzo, and J. López. FasArt: A new neuro–fuzzy architecture for incremental learning in systems identification. In *Proceedings of the 13th World Congress of IFAC*, volume F, pages 133–138, San Francisco, 1996.
14. J. M. Cano, Y. A. Dimitriadis, M. J. Araúzo, and J. López. FasBack: Matching error based learning for automatic generation of fuzzy logic systems. In *Proceedings of the Sixth IEEE International Conference on Fuzzy Systems*, volume 3, pages 1561–1566, Barcelona, 1997.
15. J. M. Cano, Y. A. Dimitriadis, E. Gómez, and J. López. Learning from noisy information in FasArt and FasBack neuro–fuzzy systems. *Neural Networks*, 14:407–425, 2001.

16. G. A. Carpenter and S. Grossberg. ART2: Stable self–organization of pattern recognition codes for analog inputs patterns. *Applied Optics*, 26:4919–4930, 1987.
17. G. A. Carpenter and S. Grossberg. A massively parallel architecture for a self–organizing neural pattern recognition machine. *Computer Vision, Graphics and Image Processing*, 37:54–115, 1987.
18. G. A. Carpenter and S. Grossberg. ART3: Hierarchical search using chemical transmitters in self–organizing pattern recognition architectures. *Neural Networks*, 3:129–152, 1990.
19. G. A. Carpenter and S. Grossberg. A self–organizing neural network for supervised learning, recognition and prediction. *IEEE Comunications Magazine*, 30:38–49, 1992.
20. G. A. Carpenter and S. Grossberg. Integrating symbolic and neural processing in a self–organizing architecture for pattern recognition and prediction. In V. Honavar and L. Uhr, editors, *Artificial Intelligence and Neural Networks: Steps Toward Principled Integration*, pages 387–421. Academic Press, San Diego, CA, 1994.
21. G. A. Carpenter, S. Grossberg, N. Markuzon, J. H. Reynolds, and D. Rosen. Fuzzy ARTMAP: A neural network architecture for incremental supervised learning of analog multidimensional maps. *IEEE Transactions on Neural Networks*, 3:698–713, 1992.
22. G. A. Carpenter, S. Grossberg, and J. H. Reynolds. ARTMAP: Supervised real–time learning and classification of non–stationary data by a self–organazing neural network. *Neural Networks*, 4:565–588, 1991.
23. G. A. Carpenter, S. Grossberg, and D. B. Rosen. Fuzzy ART: Fast stable learning and categorization of analog patterns by an adaptive resonance system. *Neural Networks*, 4:759–771, 1991.
24. G. A. Carpenter and N. Markuzon. ARTMAP–IC and medical diagnosis: Instance counting and inconsistent cases. *Neural Networks*, 11(2):323–336, 1998.
25. G. A. Carpenter, B. L. Milenova, and B. W. Noeske. Distributed ARTMAP: A neural network for fast distributed supervised learning. *Neural Networks*, 11 (5):793–813, 1998.
26. G. A. Carpenter and W. D. Ross. ART–EMAP: A neural network architecture for object recognition by evidence accumulation. *IEEE Transactions on Neural Networks*, 6(4):805–818, 1995.
27. G. A. Carpenter and A.-H. Tan. Rule extraction: From neural architecture to symbolic representation. *Connection Science*, 7(1):3–27, 1995.
28. M. Caudill. GRNN and bear it. *AI Expert*, 8:28–33, 1993.
29. R. O. Duda and P. E. Hart. *Pattern Classification and Scene Analysis.* John Wiley, New York, 1973.
30. A. Esposito, M. Marinaro, D. Oricchio, and S. Scarpetta. Approximation of continuous and discontinuous mappings by a growing neural RBF–based algorithm. *Neural Networks*, 13:651–665, 2000.
31. J. A. Feldman, G. Lakoff, D. R. Bailey, S. Narayanan, T. Regier, and A. Stol-

cke. Lo — the first five years of an automated language acquisition project. *Artificial Intelligence Review*, 10(1–2):103–129, 1996.
32. J. A. Fodor and Z. W. Pylyshyn. Connectionism and cognitive architecture: A critical analysis. In S. Pinker and J. Mehler, editors, *Connections and Symbols*, pages 3–71. MIT Press, Cambridge, MA, 1988.
33. J. A. Freeman and D. M. Skapura. *Neural networks: Algorithms, applications and programming techniques*. Addison–Wesley, Reading, 1991.
34. B. Fritzke. Fast learning with incremental RBF networks. *Neural Processing Letters*, 1:2–5, 1994.
35. B. Fritzke. Growing cell structures — A self-organizing network for unsupervised and supervised learning. *Neural Networks*, 7:1441–1460, 1994.
36. B. Fritzke. A growing neural gas network learns topologies. In G. Tesauro, D. S. Touretzky, and T. K. Leen, editors, *Advances in Neural Information Processing Systems*, volume 7, pages 625–632. MIT Press, Cambridge, MA, 1995.
37. B. Fritzke. Incremental neuro–fuzzy systems. In *Application of Soft Computing, SPIE International Symposium on Optical Science, Engineering and Instrumentation*, San Diego, CA, 1997.
38. H. Geffner. *Default reasoning: Causal and conditional theories*. MIT Press, Cambridge, MA, 1992.
39. J. Ghosh and A. Nag. An overview of radial basis function networks. In R. J. Howlett and L. C. Jain, editors, *Radial Basis Function Neural Network Theory and Applications*. Physica–Verlag, Heidelberg, 2000.
40. F. Girosi and T. Poggio. Networks and the best approximation property. *Biological Cybernetics*, 63:169–176, 1990.
41. F. Glover and M. Laguna. Tabu search. In C. R. Reeves, editor, *Modern heuristic techniques for combinatorial problems*, pages 70–150. Blackwell, Oxford, 1993.
42. D. E. Goldberg. *Genetic Algorithms in Search, Optimization and Machine Learning*. Addison–Wesley, Reading, 1989.
43. S. Grossberg. How does the brain build a cognitive code? *Psycologial Review*, 87:1–51, 1980.
44. S. Grossberg. *Studies of Mind and Brain: Neural Principles of Learning, Perception, Development, Cognition, and Motor Control*. Reidel, Boston, 1982.
45. S. Grossberg and J. R. Williamson. A self–organizing system for classifying complex images: Natural texture and synthetic aperture radar. Technical Report CAS/CNS-TR-96–002, Boston University, Boston, MA, 1996.
46. V. Honavar. Symbolic artificial intelligence and numeric artificial neural networks: Toward a resolution of the dichotomy. In R. Sun and L. Bookman, editors, *Computational Architectures Integrating Symbolic and Neural Processes*, pages 351–388. Kluwer, New York, 1994.
47. T. Hoya and A. G. Constantinides. An heuristic pattern correction scheme for GRNNs and its application to speech recognition. In *Proceedings of the IEEE Workshop on Neural Networks for Signal Processing*, pages 351–359, Cambridge, U.K., 1998.
48. Y. S. Hwang and S. Y. Bang. An efficient method to construct a radial basis

function neural network classifier. *Neural Networks*, 8:1495–1503, 1997.
49. A. Kakas, R. Kowalski, and F. Toni. Abductive logic programming. *Journal of Logic and Computation*, 6(2):719–770, 1993.
50. S. Kirkpatrick, C. D. Gelatt, and M. P. Vecchi. Optimization by simulated annealing. *Science*, 220:671–680, 1983.
51. T. Kohonen. Self–organized formation of topologically correct feature maps. *Biological Cybernetics*, 43:59–69, 1982.
52. B. Kosko. *Fuzzy Engineering*. Prentice Hall, New York, 1997.
53. J. Kowalski, E. Hartman, and J. Keeler. Layered neural networks with gaussian hidden units as universal approximators. *Neural Computation*, 2:210–215, 1990.
54. S. Lawrence, A. C. Tsoi, and A. D. Black. Function approximation with neural networks and local methods: Bias, variance and smoothness. In P. Bartlett, A. Burkitt, and R. Williamson, editors, *Australian Conference on Neural Networks (ACNN'96)*, pages 16–21. Australian National University, 1996.
55. D. J. C. MacKay. Bayesian interpolation. *Neural Computation*, 4(3):415–4478, 1992.
56. M. C. Mackey and L. Glass. Oscillation and chaos in physiological control systems. *Science*, 197:287–289, 1977.
57. Z. Manna and R. Waldinger. *The Logical Basis for Computer Programming*, volume 1: Deductive Reasoning. Addison–Wesley, Reading, 1985.
58. S. Marriott and R. F. Harrison. A modified fuzzy ARTMAP architecture for the approximation of noisy mappings. *Neural Networks*, 8:619–641, 1995.
59. L. Martí, A. Policriti, and L. García. AppART: An ART hybrid stable learning neural network for universal function approximation. In A. Abraham and M. Koeppen, editors, *Hybrid Information Systems*, pages 93–120, Heidelberg, 2002. Physica–Verlag.
60. L. Martí, A. Policriti, L. García, and R. Lazo. AppART + growing neural gas = high performance hybrid neural network for function approximation. In M. Bhattacharya and A. Abraham, editors, *Proceedings of the Sixth International Conference on Knowledge–Based Intelligent Information and Engineering Systems*, Amsterdam, 2002. IOS Press.
61. T. Martinetz. Competitive hebbian learning rule forms perfectly topology preserving maps. In *International Conference on Artificial Neural Networks (ICANN'93)*, pages 427–434, Heildelberg, 1993. Springer–Verlag.
62. T. M. Martinetz. Competitive Hebbian learning rule forms perfectly topology preserving maps. In *International Conference on Artificial Neural Networks (ICANN'93)*, pages 427–434, Amsterdam, 1993. Springer–Verlag.
63. T. M. Martinetz, S. G. Berkovich, and K. J. Shulten. Neural–gas network for vector quantization and its application to time–series prediction. *IEEE Transactions on Neural Networks*, 4:558–560, 1993.
64. W. S. McCulloch and W. Pitts. A logical calculus of ideas immanent in nervous activity. *Bulletin of Mathematical Biophysics*, 5:115–133, 1943.
65. K. McGarry, S. Wermter, and J. MacIntyre. Hybrid neural systems: From single coupling to fully integrated neural networks. *Neural Computing Surveys*,

2:62–93, 1999.
66. J. R. Millán. Learning efficient reactive behavioral sequences from basic reflexes in a goal–oriented autonomous robot. In *Proceedings of the Third International Conference on Simulation of Adaptive Behavior: From Animals to Animats 3*, pages 266–274, Cambridge, MA, 1994. MIT Press.
67. M. Minsky. Logical versus analogical or symbolic versus connectionist or neat versus scruffy. *AI Magazine*, 12(2):34–51, 1991.
68. J. Moody and C. Darken. Fast learning in networks of locally–tuned processing units. *Neural Computation*, 1:281–294, 1989.
69. M. Musavi, W. Ahmed, K. Chan, K. Faris, and D. Hummels. On the training of radial basis function classifiers. *Neural Networks*, 5:595–603, 1992.
70. E. A. Nadaraya. On estimating regression. *Theory of Probability and Its Application*, 10:186–190, 1964.
71. M. C. O'Neill. Escherichia coli promoters: I. Consensus as it relates to spacing class, specificity, repeat substructure and three dimensional organization. *Journal of Biological Chemistry*, 264:5522–5530, 1989.
72. M. J. L. Orr. Regularization in the selection of radial basis function centers. *Neural Computation*, 7:606–620, 1995.
73. M. J. L. Orr. Optimising the widths of radial basis functions. In A. P. Braga and T. B. Ludermir, editors, *Fifth Brazilian Symposium on Neural Networks*, Belo Horizonte, Brazil, 1998.
74. M. J. L. Orr, J. Hallam, A. Murray, and T. Leonard. Assessing RBF networks using DELVE. *International Journal of Neural Systems*, 10(5):397–415, 2000.
75. M. J. L. Orr, J. Hallam, K. Takezawa, A. Murray, S. Ninomiya, M. Oide, and T. Leonard. Combining regression trees and radial basis function networks. *International Journal of Neural Systems*, 10(6):453–465, 2000.
76. J. Park and I. W. Sandberg. Universal approximation using radial basis functions. *Neural Computation*, 3:246–257, 1991.
77. S. Pinker and A. Prince. On language and connectionism: Analysis of a parallel distributed processing model of language acquisition. In S. Pinker and J. Mehler, editors, *Connections and Symbols*, pages 73–193. MIT Press, Cambridge, MA, 1988.
78. T. Poggio and F. Girosi. Networks for approximation and learning. *Proceedings of the IEEE*, 78:1481–1496, 1990.
79. J. R. Quinlan. Induction of decision trees. *Machine Learning*, 1:81–106, 1986.
80. C. E. Rasmussen, R. M. Neal, G. E. Hinton, D. van Camp, M. Revow, Z. Ghahramani, R. Kustra, and R. Tibshirani. *The DELVE manual, (version 1.1)*. University of Toronto, Toronto, ON, 1996.
81. D. E. Rumelhart, G. E. Hinton, and R. J. Williams. Learning internal representations by error propagation. In *Parallel Distributed Processing: Explorations in the Microstructure of Cognition*. MIT Press, Boston, 1986.
82. H. Schiøler and U. Hartman. Mapping neural network derived from the Parzen window estimator. *Neural Networks*, 5:903–909, 1992.
83. P. Smolensky. On the proper treatment of connectionism. *Behavioral and Brain Sciences*, 11:1–74, 1988.

84. D. Spetch. A general regression neural network. *IEEE Transactions on Neural Networks*, 2:568–578, 1990.
85. N. Srinivasa. Learning and generalization of noisy mappings using a modified PROBART neural network. *IEEE Transactions on Signal Processing*, 45(10): 2533–2550, 1997.
86. R. Sun. On variable binding in connectionist networks. *Connection Science*, 4:93–124, 1992.
87. R. Sun. *Integrating Rules and Connectionism for Robust Commonsense Reasoning*. Wiley, New York, 1994.
88. A.-H. Tan. Cascade ARTMAP: Integration neural computation and symbolic knowledge processing. *IEEE Transactions on Neural Networks*, 8(2):237–250, 1997.
89. A.-H. Tan. Supervised adaptive resonance theory and rules. In J. C. Jain, B. Lazzerini, and U. Halici, editors, *Innovation in ART Neural Networks*, pages 55–86. Physica–Verlag, 2000.
90. D. Tomandl and A. Schober. A modified general regression neural network (MGRNN) with new, efficient training algorithms as a robust 'back box'–tool for data analysis. *Neural Networks*, 14:1023–1034, 2001.
91. G. G. Towell and J. W. Shavlik. Directed propagation of training signals through knowledge–based neural networks. Technical Report CS TR 1990 989, University of Wisconsin, Computer Sciences Department, Madison, WI, 1990.
92. G. G. Towell and J. W. Shavlik. Extracting rules from knowledge–based neural networks. *Machine Learning*, 13(1):71–101, 1993.
93. G. G. Towell and J. W. Shavlik. Knowledge–based artificial neural networks,. *Artificial Intelligence*, 70:119–165, 1994.
94. S. J. Verzi, G. L. Heileman, M. Georgiopoulos, and M. J. Healy. Boosted ARTMAP. In *Proceedings of the International Joint Conference on Neural Networks (IJCNN'98)*, volume 1, pages 396–401, Alaska, 1998.
95. G. S. Watson. Smooth regression analysis. *Sankhya: The Indian Journal of Statistics*, 26:359–372, 1964.
96. V. Weber. Connectionist unifying Prolog. In R. F. Albrecht, C. R. Reeves, and N. C. Steele, editors, *Proceedings of the International Conference on Artificial Neural Nets and Genetic Algorithms*, pages 213–20, Heidelberg, 1993. Springer–Verlag.
97. V. Weber. Unification in Prolog by connectionist models. In P. Leong and M. Jabri, editors, *Proceedings of the Fourth Australian Conference on Neural Networks (ACNN'93)*, pages 5–8, Sydney, NSW, Australia, 1993. University of Sydney Electrical Engineering.
98. S. Wermter and R. Sun. *Hybrid Neural Systems*. Springer–Verlag, Heidelberg, 2000.
99. J. R. Williamson. Gaussian ARTMAP: A neural network for fast incremental learning of noisy multidimensional maps. *Neural Networks*, 9:881–897, 1996.
100. J. R. Williamson. A constructive, incremental–learning network for mixture modeling and classification. *Neural Computation*, 9:1517–1543, 1997.

101. J. R. Williamson. A neural model for self–organizing feature detectors and classifiers in a network hierarchy. Technical Report CAS/CNS-TR-98–033, Boston University, Boston, MA, 1998.
102. D. R. Wilson and T. R. Martinez. Improved heterogeneous distance functions. *Journal of Artificial Intelligence Research*, 6:1–34, 1997.
103. R. Wilson and T. R. Martinez. Heterogeneous radial basis function networks. In *Proceedings of the International Conference on Neural Networks (ICNN'96)*, volume 2, pages 1263–1267, Washington DC, 1996.
104. L. A. Zadeh. Fuzzy sets. *Information and Control*, 8:338–353, 1965.

Chapter 4

An Algorithmic Approach to the Main Concepts of Rough Set Theory

Joaquim Quinteiro Uchôa and Maria do Carmo Nicoletti

Summary. The Rough Set Theory (RST) is a mathematical formalism for representing uncertainty, which can be considered an extension of the classical set theory. It has been used in many different research areas, including those related to inductive machine learning and reduction of knowledge in knowledge based systems. This chapter introduces the main concepts of the RST and presents a family of algorithms for implementing them.

Keywords: rough set theory, indiscernibility, uncertainty, knowledge-based systems, knowledge representation systems.

1 Introduction

The Rough Set Theory (RST) was proposed by Pawlak in 1982 [1], as an extension of the classical set theory, to be used for representing incomplete knowledge. Rough sets can be considered sets with fuzzy boundaries – sets that can not be precisely characterized using the available set of attributes.

During the last twenty years RST has been approached as a formal tool used in connection with many different areas of research. There has been investigation of the relation between RST and Dempster-Shafer Theory [2, 3] and the relation between rough sets and fuzzy sets [4, 5, 6]. RST has provided the necessary formalism and ideas for the development of few propositional machine learning systems, i.e., for classification systems [7, 8, 9, 10, 11]. It has been used for knowledge representation [12, 13], data mining [14, 15], discovering rules from examples [16], dealing with imperfect data [17, 18], reducing the knowledge representation [19, 20, 21], helping to solve control problems [22, 23, 24], knowledge acquisition [25], modeling database [26] and analysing attribute dependencies [27, 28], among many others. Several applications of RST are shown and discussed in [29, 30].

In spite of the vast number of papers and the availability of software on RST (see, for instance, the Rough Set Library [31]), most of the main concepts of this theory have not been described in algorithmic terms. This chapter presents and discusses the main concepts of RST (extracted from the various sources listed at the end) and a family of algorithms that implements them. It is organized as follows: Section 2 presents the basic concepts and notations related to a knowledge representation system and Section 3, the basic concepts and notations related to independence and reduction of attributes which will be used subsequently in Section 4, for presenting the family of algorithms. In Section 5 the guidelines for future work are stated and the conclusions are presented.

2 Knowledge Representation System

2.1 Basic Concepts

The basic concept of the Rough Set Theory is the notion of *approximation space*, which is an ordered pair $A = (U, R)$, where

- U: nonempty set of objects, called *universe*;
- R: equivalence relation on U, called *indiscernibility relation*. If $x, y \in U$ and xRy then x and y are *indistinguishable* in A.

Each equivalence class induced by R, i.e., each element of the quotient set $\tilde{R} = U/R$ is called an *elementary set* in A. An approximation space can be alternatively noted by $A = (U, \tilde{R})$. The notion of approximation space can be visualized in Figure 1. It is assumed that the empty set is also elementary for every approximation space A. A *definable set* in A is any finite union of elementary sets in A. For $x \in U$, let $[x]_R$ denote the equivalence class of R, containing x. For each $X \subseteq U$, X is characterized in A by a pair of sets — its *lower* and *upper approximation* in A, defined respectively as:

$$A_{low}(X) = \{x \in U \mid [x]_R \subseteq X\} , \tag{1}$$

$$A_{upp}(X) = \{x \in U \mid [x]_R \cap X \neq \varnothing\} . \tag{2}$$

The lower approximation of X in A is the greatest definable set in A contained in X and the upper approximation of X in A is the least definable set in A containing X. Figures 2 and 3 show the concepts of lower and upper approximations, respectively. A set is *definable* in A iff $A_{low}(X) = A_{upp}(X)$. For a set $X \subseteq U$, the

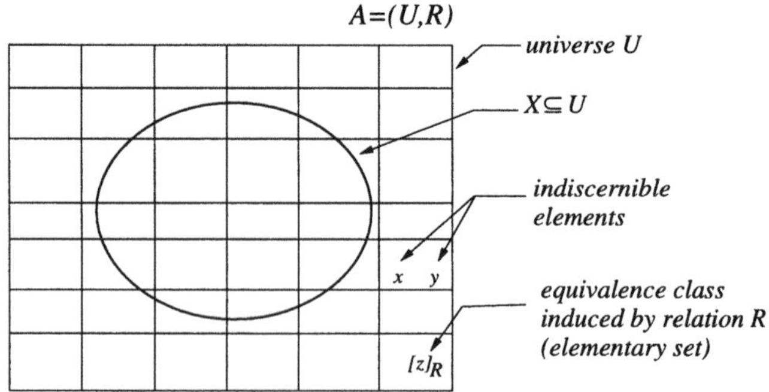

Figure 1. Approximation space $A = (U, R)$.

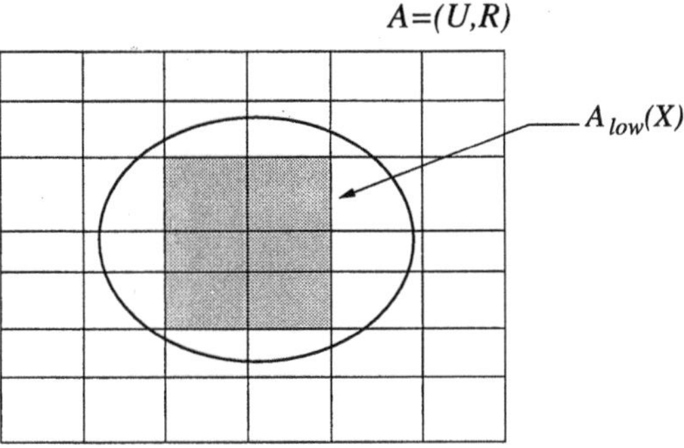

Figure 2. Lower approximation of X in $A = (U, R)$.

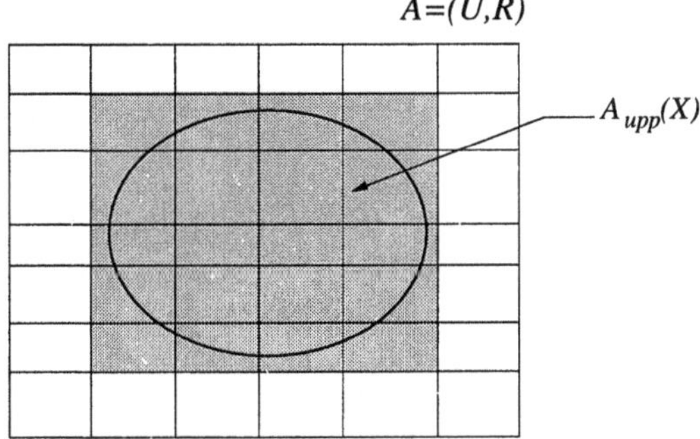

Figure 3. Upper approximation of X in $A = (U, R)$.

measure

$$\omega_A(X) = \frac{|A_{low}(X)|}{|A_{upp}(X)|} \tag{3}$$

defines the ***accuracy*** of X in A, and expresses the degree of definability of X in A. A ***rough set*** in A is the family of all subsets of U having the same lower and upper approximations. Another definition found in [32] states that: "a rough set is a representation of a given set X, by two subsets of the quotient set, which approach X as closely as possible from inside and outside respectively. That is, $\langle A_{low}(X), A_{upp}(X)\rangle$". Both definitions are shown to be equivalent in [33].

Given an approximation space $A = (U, R)$ and a set $X \subseteq U$, three different regions (illustrated in Figure 4) associated with X can be identified:

- ***positive region*** of X in A, $pos(X)$, defined by the union of all the elementary sets which are totally contained in X. The positive region of X in A coincides with the lower approximation of X in A, i.e., $pos(X) = A_{low}(X)$.
- ***negative region*** of X in A, $neg(X)$, defined by the union of all the elementary sets which have an empty intersection with X. The negative region of X in A is the union of those elementary sets which do not belong to the upper approximation of X in A, i.e., $neg(X) = U - A_{upp}(X)$.

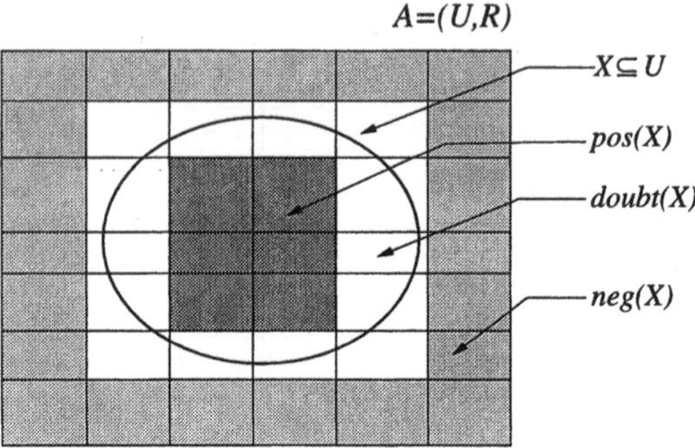

Figure 4. Regions of X in $A = (U, R)$.

- *doubtful region* of X in A, also known as *frontier* of X in A, defined as the set of elements of U which belong to the upper approximation of X in A but don't belong to the lower approximation of X in A, i.e., $doubt(X) = A_{upp}(X) - A_{low}(X)$.

In [4] is proposed an *approximate membership* function which 'measures' the degree of membership of any element of U to any of the regions defined by a set $X \subseteq U$. Given an approximation space $A = (U, R)$, $X \subseteq U$, and $x \in U$, the membership of x to X in A is defined by:

$$\mu_X(x) = \frac{|[x]_R \cap X|}{|[x]_R|} . \tag{4}$$

where $[x]_R$ is the equivalence class induced by the equivalence relation R, which contains x (and obviously all the elements equivalent to x, through R). It can be said that $\forall x \in U$, $[x]_R$ has at least one element (x itself). It is very important to observe that the membership function is dependent on the existing knowledge about the elements of the universe, i.e., the partition U/R. The above definition establishes an approximate membership of any element $x \in U$ to the set X, given the knowledge U/R. It helps 'measuring' to which extension the elements which belong to the equivalence class $[x]_R$ also belong to X, with respect to U/R, in an approximation space $A = (U, R)$.

Let $A = (U, R)$ be an approximation space and let $X \subseteq U$. The total number of elements of U which can be classified in two disjoint subsets, X and $U - X$ is the number of elements which do not belong to the doubtful region of X. This number is given by:

$$|U - doubt(X)| = |U - ((A_{upp}(X) - A_{low}(X))| = |U| - |A_{upp}(X) - A_{low}(X)| \tag{5}$$

The value:

$$\alpha_A(X) = \frac{|U| - |A_{upp}(X) - A_{low}(X)|}{|U|} \tag{6}$$

is defined as *discriminant index* of A with respect to X. This value gives a 'measure' of the degree of certainty to which an element of U belongs (or not) to X.

2.2 Knowledge Representation System

It is very common in Artificial Intelligence to express the knowledge about an object in a n-dimensional space as a n-dimensional vector of attribute-value pairs and a corresponding class. According to [34], such a *Knowledge Representation System* (KRS) can be formally expressed as $S = < U, Q, V, \rho >$, where:

- U: is a finite nonempty set of *objects* expressed through the assignment of some attributes and their values;
- $Q = C \cup \{\delta\}$ is a set of *attributes*, where C and δ are referred to as conditions and class (or decision), respectively;
- $V = \bigcup V_q$, $q \in Q$, is a set of *attribute values*, where V_q is the domain of attribute $q \in Q$;
- $\rho: U \times Q \rightarrow V$ is a *description function* such that $\rho(u, q) \in V_q$, for every $u \in U$ and $q \in Q$ (i.e., the function ρ assigns attribute values to each object u in U). A pair (q, v), $q \in Q$, $v \in V_q$, is called descriptor in S. The description function can also be noted by $\rho_u: Q \rightarrow V$, such that $\rho_u(q) = \rho(u, q) \in V_q$, $u \in U$, $q \in Q$.

From the above definition, it should be observed that:

- each attribute q has $|V_q|$ different values; that adds up to $\prod_{q \in Q} |V_q|$ possible descriptions in S;
- aiming at simplicity and assuming that objects are described by n attributes, the function ρ_u will be written as a sequence of attribute values such as $v_{1i_1}, v_{2i_2}, \ldots, v_{ni_n}, 1 \leq i_j \leq |V_{q_j}|, i \leq j \leq n$;
- objects $x, y \in U$ are *indiscernible* with relation to $q \in Q$ in S iff $\rho_x(q) = \rho_y(q)$, noted by $x\tilde{q}y$. It is obvious that $\tilde{q}$ is an equivalence relation. The relation $\tilde{q}$ is defined as the *indiscernibility relation* in S, with relation to attribute q;
- let $P \subseteq Q$ in S and let $x, y \in U$. The objects x and y are *indiscernible* with relation to P, noted by $x\tilde{P}y$, iff x and y are indiscernible with relation to each attribute $p \in P$. It could be observed that $\tilde{P}$ can be expressed by $\tilde{P} = \bigcap_{p \in P} \tilde{p}$. If $P = Q$, x and y are said to be indiscernible in S, which is noted by $x\tilde{S}y$ instead of $x\tilde{Q}y$. Obviously, $\tilde{P}$ is an equivalence relation in S for any subset of attributes $P \subseteq Q$. So, each knowledge representation system $S =< U, Q, V, \rho >$ defines uniquely an approximation space $A_S = (U, \tilde{S})$, where $\tilde{S}$ is the indiscernibility relation generated by S. The equivalence classes are called *elementary sets* or *atoms* of S;
- if $u \in U$ and ρ_u is the description of u in S, then it is assumed that ρ_u is also the description of the equivalence classes of the relation $\tilde{S}$ containing u;
- a subset $X \subseteq Q$ is *describable* in S iff X is definable in A_S. If X is undefinable in A_S, X is *nondescribable* in S;
- the description of a describable set in S consists of all descriptions of its elementary sets.

A KRS can be conveniently represented by a table, where each row represents the description of an object and each column, an attribute (generally, the last attribute is identified with the class of the object, i.e., the decision attribute).

3 Independence and Reduction of Attributes

The information in a knowledge representation system $S = < U, Q, V, \rho >$ is described by means of attributes and attributes values. A question is whether all available attributes are necessary to describe a subset of objects representing an expert concept. The RST can be used for investigating the dependence relation between attributes, aiming at the identification and elimination of redundant attributes.

The first step for identifying a minimal subset of essential attributes is established by the process of verifying the dependence relation among attributes. In order to do that, first few definitions and properties are necessary.

Let $S = < U, Q, V, \rho >$ a KRS and let $p, q \in Q$. The attribute p is said to be ***dependent*** on the attribute q in S if the value of p is determined by the value of q, i.e., if there is a dependency function $f: V_q \to V_p$, such that $\rho(x,p) = f(\rho(x,q))$. This definition can be easily expressed by means of the indiscernibility relation:

- the attribute p is dependent on attribute q in S, noted by $q \to p$ iff $\tilde{q} \subseteq \tilde{p}$ (notice that the symbol $\subseteq$ used between equivalence classes indicates a relation of "refinement" rather than the usual inclusion relation between sets);
- attributes p and q are called ***independent*** to each other in S iff neither $q \to p$ or $p \to q$ hold.

The concepts of dependence and independence between two attributes can easily be extended to a greater number of attributes $P \subseteq Q$:

- a subset of attributes $P \subset Q$ is ***dependent*** in S iff exists $M \subset P$ such that $\tilde{M} = \tilde{P}$ (notice that M is a proper subset of P). Otherwise, P is said ***independent*** in S;
- given $P \subset Q$ in S, a subset of attributes $M \subset P$ is said ***superfluous*** (or ***redundant***) in P iff $\tilde{P} = \tilde{N}$, where $N = P - M$.

Example 1. Let $S = < U, Q, V, \rho >$ the KRS given as example in [10, p. 474], where $U = \{x_1, x_2, x_3, x_4, x_5\}$, $Q = \{q_1, q_2, q_3, q_4\}$, $V_{q_1} = V_{q_2} = V_{q_3} = \{0, 1\}$, $V_{q_4} = \{0, 1, 2\}$ and ρ given by Table 1. The partitions of U, given by the indiscernibility relations are pictured in Figure 5.

Notice that $q_4 \to q_2$ and $q_4 \to q_1$ since $\tilde{q}_4 \subseteq \tilde{q}_2$ and $\tilde{q}_4 \subseteq \tilde{q}_1$. The attribute q_4 induces a partition on U that can be seen as a refinement of those induced by q_1 and q_2, as shows Figure 6.

Table 1. A KRS given by five elements described by four attributes.

U	q_1	q_2	q_3	q_4
x_1	0	0	0	0
x_2	0	1	0	2
x_3	1	1	0	1
x_4	1	1	0	1
x_5	0	1	1	2

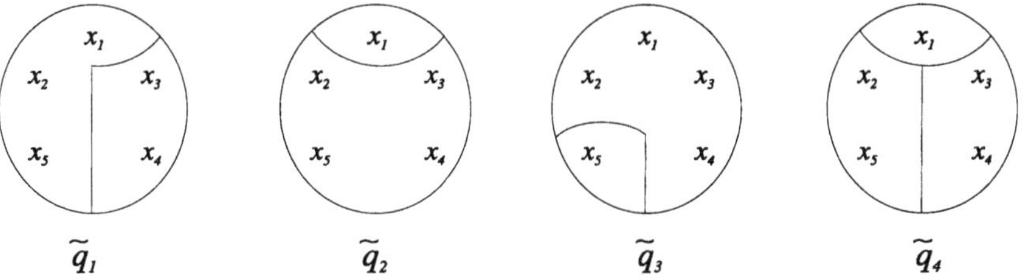

Figure 5. Four partitions of U induced by four indiscernibility relations: $\tilde{q_1}$, $\tilde{q_2}$, $\tilde{q_3}$, and $\tilde{q_4}$.

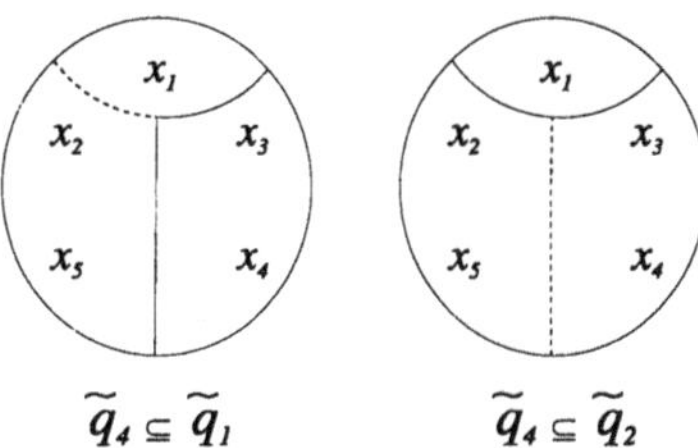

Figure 6. Relation $\tilde{q_4}$ is a refinement of relation $\tilde{q_1}$ and relation $\tilde{q_2}$.

Considering a KRS given by $S = < U, Q, V, \rho >$, and considering $P \subseteq Q$, where $P = \{p_1, p_2, p_3, \ldots, p_n\}$ and $P_i = P - \{p_i\}$, $1 \leq i \leq n$, the following properties hold:

1. P is dependent in S iff there exists $M \subset P$ such that M is superfluous in S;
2. if P is independent in S, then every subset of P is also independent in S;
3. if P is dependent in S, then every subset $M \subseteq Q$ such that $P \subseteq M$ is dependent in S;
4. P is independent in S iff $\tilde{P} \subset \tilde{P_i}$, $1 \leq i \leq n$;
5. P is independent in S iff $|U/\tilde{P_i}| < |U/\tilde{P}|$, $1 \leq i \leq n$;
6. if $P \subset Q$ is superfluous in Q and $\{p\}$ is superfluous in $Q - P$, then $P \cup \{p\}$ is superfluous in Q.

These properties can be used for checking dependence or independence of attributes. It is interesting, in many situations, to find a minimal subset of attributes which has the same power of the original set of attributes. This minimal subset is said to be a *reduct* of the initial set of attributes. Finding a reduct in a set of attributes allows the representation of a KRS in a reduced way, with a lesser number of attributes, and the same power of discernibility among the objects.

Formally, a subset of attributes $P \subseteq Q$ in a knowledge representation system $S = < U, Q, V, \rho >$ is said to be a reduct of Q in S iff $Q - P$ is superfluous in Q and P is independent in S. Of course, many reducts can exist in a given KRS. The intersection of all reducts is said to be the *core* of the original set of attributes.

The notion of a dependency relationship between two groups of attributes, as proposed in [12], is presented next. Let C and D be two sets of attributes in S, referred to as condition attributes and decision attributes, respectively. Let:

- $D' = \{D'_1, D'_2, \ldots, D'_n\}$, the family of elementary sets induced by $\tilde{D}$, i.e., $D' = U/\tilde{D}$
- $C' = \{C'_1, C'_2, \ldots, C'_m\}$, the family of elementary sets induced by $\tilde{C}$, i.e., $C' = U/\tilde{C}$

In the approximation space $A = (U, C)$, the ***positive region*** of D' is defined as:

$$pos(C, D) = \bigcup_{i=1}^{n} \{A_{C-low}(D'_i) | D'_i \in D'\}, \tag{7}$$

i.e., the union of the lower approximations (related to the space induced on U by C and, for this reason, noted as A_{C-low}) of all elementary sets induced by relation $\tilde{D}$.

In an approximation space $A = (U, C)$, the set of attributes D depends in ***degree*** κ $(0 \leq \kappa \leq 1)$ on the set of attributes C if:

$$\kappa(C, D) = \frac{|pos(C, D)|}{|U|} \tag{8}$$

As commented in [12, p. 201], "A dependency close to 1 gives reason to hypothesize that generally, there is a strong cause-effect relationship between attributes C and D, and a dependency close to 0 suggests weak, if any, cause-effect relationship between C and D".

It is possible to measure the relative contribution or significance of an individual attribute a which belongs to a reduct R, with respect to the dependency between D and R. In order to do that, in [12, p. 203] is defined the ***significance factor*** SGF as:

$$SGF(a, R, D) = \frac{\kappa(R, D) - \kappa(R - \{a\}, D)}{\kappa(R, D)} \tag{9}$$

assuming $\kappa(R, D) > 0$, where κ is the degree of dependency as defined previously. As stated in [12, p. 203]:

> Formally the significance factor reflects the relative degree of decrease of dependency level between R and D as a result of the removal of the attribute a from R. In practice, the stronger the influence of the attribute a is on the relationship between R and Q, the higher the value of the significance factor is.

4 The RST Family of Algorithms

The algorithms to be presented in this section have been designed taking into account the generic knowledge representation system $S = < E, Q, V, \rho >$ shown in Table 2 given by:

$$E = \{e_1, e_2, e_3, \ldots, e_m\}, |E| = m;$$
$$Q = \{q_1, q_2, q_3, \ldots, q_n\}, |Q| = n;$$
$$V_{q_j} = \{v_{j_1}, v_{j_2}, \ldots, v_{j_{i_j}}\}, |V_{q_j}| = i_j, 1 \leq j \leq n.$$

Table 2. A generic KRS.

E	q_1	q_2	q_3	...	q_n
e_1	v_{1e_1}	v_{2e_1}	v_{3e_1}	...	v_{ne_1}
e_2	v_{1e_2}	v_{2e_2}	v_{3e_2}	...	v_{ne_2}
e_3	v_{1e_3}	v_{2e_3}	v_{3e_3}	...	v_{ne_3}
...	...	...	...	...	...
e_m	v_{1e_m}	v_{2e_m}	v_{3e_m}	...	v_{ne_m}

The KRS represents m examples, each of them described by n attributes, where each attribute q_j can assume i_j different values, $j = 1, ..., n$. An example e_i is represented by a vector having n positions, each of them filled with a value of the attribute corresponding to that position, extracted from V_1 up to V_n respectively. An example e_i $(1 \leq i \leq m)$ is written as $[v_{1e_i}, v_{2e_i}, \ldots, v_{ne_i}]$, where each v_{je_i} corresponds to the value of attribute q_j in the example e_i.

The pseudocode of eleven algorithms will be presented next, taking into account the notation presented so far. These eleven algorithms will

1. generate the representation of a KRS, i.e., obtain its equivalence classes;
2. determine the lower approximation of a set;
3. determine the upper approximation of a set;
4. determine the precision of a set;
5. check for dependence/independence of attributes;
6. determine the reduct of a set of attributes;
7. determine the degree of membership of an element to a given set;
8. calculate the discriminant index of a set of elements with relation to a set of attributes;
9. determine the degree of dependency of an attribute with relation to a set of attributes;
10. calculate the significance factor of an attribute; and
11. determine the reducts of a given set of attributes with relation to the existing dependency between the given set and a given attribute.

Some auxiliary procedures used in these algorithms, such as number_of_elements and determines_first_not_used are left undefined; they can be easily implemented and what they do can be inferred from their names.

4.1 Algorithm 1 – Generating the Representation of a KRS

The generation of the representation of a KRS consists of the generation of the elementary sets of the KRS. The algorithm will generate the set of elementary sets, i.e., a set of those sets of examples which have the same value for all n attributes, $q_1, \ldots, q_n$.

Algorithm 1. Generating the representation of a KRS.

```
procedure rep(E,Q,R);
{
  Input: E = {e1, e2, ..., em}, set of m examples,
         described by n attributes , Q = {q1, ..., qn}
  Output: R = {E1, E2, ..., Es},
         representation of the KRS given by (E, Q)
}
begin
  {initialisation}
  i := 1;
  s := 0;
  R := {};
  for k := 1 to m do used[k] := false;
  {determination of elementary sets }
  while i <= m do
    begin
      s := s+1;
      Es := {ei};
      used[i] := true;
      j := i + 1;
      r := 1;
      while j <= m do
        begin
          while used[j] and j <= m do j := j + 1;
          if j <> m + 1 then
            while r <= n and v_rei = v_rej do r := r + 1;
          if r = n + 1 then
            begin
              Es := Es ∪ {ej};
              used[j] := true;
            end;
          r := 1;
          j := j + 1
        end;
      R := R ∪ Es;
      determines_first_not_used(i)
    end
end.
```

The input to Algorithm 1 is a KRS described by both, a set of examples $E = \{e_1, e_2, \ldots, e_m\}$ and a set of attributes $Q = \{q_1, \ldots, q_n\}$. The algorithm verifies, for each example e_i $(1 \leq i \leq m)$, if there are examples e_j $(1 \leq j \leq m)$ which are equal to e_i, with relation to all attribute values. The identical examples are gathered together in what constitutes an elementary set. After have gone through all examples, the algorithm returns in R, the set of all elementary sets of the given KRS.

4.2 Algorithm 2 – Constructing the Lower Approximation of a Given Set

The lower approximation of a given set X in a KRS consists of the set of all elementary sets of the KRS which are entirely contained in X.

Algorithm 2 expects as input: a) a KRS described by both, a set of examples $E = \{e_1, e_2, \ldots, e_m\}$ and a set of attributes $Q = \{q_1, \ldots, q_n\}$; b) a given set $X = \{x_1, x_2, \ldots, x_p\}$. Each $x_i \in X$, $1 \leq i \leq p$, is given as a vector of attribute

Algorithm 2. Constructing the lower approximation of a given set.

```
procedure low_approx(E,Q,X,L);
{
  Input: E = {e1, e2, ..., em} , set of m examples,
         described by n attributes , Q = {q1, ..., qn}
         X = {x1, x2, ..., xp} , given set
  Output: L = {El1, El2, ..., Elk} , 0 ≤ k ≤ |R| ,
          set of elementary sets that are entirely contained in X
}
begin
  {initialisation}
  rep(E,Q,R);
  num := number_of_elements(R);
  for i := 1 to num do used[i] := false;
  i := 1;
  L := {};
  {determination of lower approximation of X}
  while i <= p do
     begin
       {determination of the equivalence class containing xi}
       j := 1;
       while (j <= num) and (not( xi in Ej )) do j := j + 1;
       {verifying if Ej is entirely contained in X}
       if not used[j] then
          begin
            b := 1;
            contained := true;
            kj := number_of_elements(Ej);
            while (b <= kj ) and (contained) do
              if eb in X then b := b + 1
                 else contained := false;
            if contained then L := L ∪ Ej;
            used[j] := true;
            i := i + 1
          end
     end
end.
```

values $[v_{1x_i}, v_{2x_i}, \ldots, v_{nx_i}]$. Algorithm 2 gives as output the set L which contains the lower approximation of X. Algorithm 2 uses Algorithm 1, since for constructing the lower approximation of a given set in a KRS, the representation of the KRS is necessary. Algorithm 2 initially determines the representation of the KRS by calling rep(E,Q,R), whose output is the collection of the equivalence classes induced by Q on E. Next, for each one of the p elements $x_i \in X$, Algorithm 2 finds out to which equivalence class x_i belongs to. It verifies if every element of that class belongs to X. If they do, without exception, the corresponding equivalence class is added to the lower approximation of X. If there is a single element y such that $y \notin X$, then the elementary set to which y belongs to is not part of the lower approximation of X.

4.3 Algorithm 3 – Constructing the Upper Approximation of a Given Set

The upper approximation of a given set X in a KRS consists of the set of all elementary sets of the KRS which have a nonempty intersection with X.

Algorithm 3. Constructing the upper approximation of a given set.

```
procedure upp_approx(E,Q,X,U);
{
  Input: E = {e1, e2, ..., em}, set of m examples,
         described by n attributes , Q = {q1, ..., qn}
         X = {x1, x2, ..., xp}, given set
  Output: U = {El1, El2, ..., Elk}, 0 <= k <= |R|,
          set of elementary sets that have
          nonempty intersection with X
}
begin
  {initialisation}
  rep(E,Q,R);
  num := number_of_elements(R);
  for i := 1 to num do used[i] := false;
  i := 1;
  U := {};
  {determination of upper approximation of X}
  while i <= p do
    begin
      {determination of the equivalence class containing xi}
      j := 1;
      while (j <= num) and (not( xi in Ej )) do j := j + 1;
      {verifying if Ej is entirely contained in X}
      if not used[j] then
        begin
          U := U ∪ Ej;
          used[j] := true;
        end
      i := i + 1
    end
end.
```

Algorithm 3 expects as input: a) a KRS described by both, a set of examples $E = \{e_1, e_2, \ldots, e_m\}$ and a set of attributes $Q = \{q_1, \ldots, q_n\}$; b) a given set $X = \{x_1, x_2, \ldots, x_p\}$. Each $x_i \in X, 1 \leq i \leq p$, is given as a vector of attribute values $[v_{1x_i}, v_{2x_i}, \ldots, v_{nx_i}]$. Algorithm 3 gives as output the set U which contains the upper approximation of X. Algorithm 3 also uses Algorithm 1, since for constructing the upper approximation of a given set in a KRS, the representation of the KRS is necessary. Algorithm 3 initially determines the representation of the KRS by calling rep(E,Q,R), whose output $R = \{E_1, E_2, \ldots, E_s\}$ is the collection of the equivalence classes induced by Q. Next for each one of the p elements $x_i \in X$, the algorithm finds out to which equivalence class x_i belongs to and adds this equivalence class to the upper approximation U of X.

4.4 Algorithm 4 – Determining the Accuracy of a Given Approximation

The accuracy of a given set X in a KRS is obtainned by dividing the cardinality of its lower approximation by the cardinality of its upper approximation.

Algorithm 4 expects as input: a) a KRS described by both, a set of examples $E = \{e_1, e_2, \ldots, e_m\}$ and a set of attributes $Q = \{q_1, \ldots, q_n\}$; b) a given set $X = \{x_1, x_2, \ldots, x_p\}$. Each $x_i \in X, 1 \leq i \leq p$, is given as a vector of attribute values

Algorithm 4. Determining the accuracy of a given set.

```
procedure accuracy(E,Q,X,A);
{
  Input: E = {e1, e2, ..., em}, set of m examples,
         described by n attributes, Q = {q1, ..., qn}
         X = {x1, x2, ..., xp}, given set
  Output: A = accuracy of X
}
begin
        low_approx(E,Q,X,L);
        upp_approx(E,Q,X,U);
        A := number_of_elements(L) / number_of_elements(U)
end.
```

$[v_{1x_i}, v_{2x_i}, \ldots, v_{nx_i}]$. Algorithm 4 gives as output, the accuracy A of X. Algorithm 4 uses Algorithm 2 and Algorithm 3, since the determination of the accuracy of a given set in a KRS is based on its lower and upper approximations.

4.5 Algorithm 5 – Determining if a Set of Attributes is Dependent or Not

Algorithm 5 is based on the property 5, Section 3, which states that a set of attributes $P = \{p_1, \ldots, p_j\}$, $|P| = j$, is independent if all of its proper subsets $P_i = P - \{j_i\}$, $1 \leq i \leq j$, induces partitions which have a lesser number of elementary sets than the one induced by P.

Algorithm 5. Determining the dependence or independence of a set of attributes.

```
procedure dependence(E,Q,P,DEP);
{
  Input: E = {e1, e2, ..., em}, set of m examples,
         described by n attributes, Q = {q1, ..., qn}
          P = {p1, ..., pj}, subset of attributes (P ⊆ Q)
  Output: DEP = true : P is dependent in Q;
          DEP = false : P is independent in Q
}
begin
   {initialisation}
   DEP := false;
   i := 1;
   j := number_of_elements(P);
   {finding the representation of KRS using P}
   rep(E,P,RepP);
   cardRepP := number_of_elements(RepP);
   while not(DEP) and i <= j do
      begin
         Pi := P - {ji};
         {finding the representation of KRS using Pi}
         rep(E,Pi,RepPi);
         cardRepPi := number_of_elements(RepPi);
         if cardRepP = cardRepPi then DEP := true;
         i := i + 1
      end
end.
```

Algorithm 5 expects as input: a) a KRS described by both, a set of examples $E = \{e_1, e_2, \ldots, e_m\}$ and a set of attributes $Q = \{q_1, \ldots, q_n\}$; b) the subset of attributes, $P \subseteq Q$, given by $P = \{p_1, \ldots, p_j\}$, which will be checked for dependency. In order to verify if all proper subset P_i of P induces partitions with a lesser number of elements than the one induced by P, the algorithm first determines the representation of the KRS, using P. Then, for each i, $1 \leq i \leq j$, finds each $P_i = P - \{j_i\}$, as well as the representation of the KRS induced by P_i. If there is no P_i whose representation has the same number of elementary sets as the one induced by P, P is independent, otherwise, is dependent. The output of the Algorithm 5 is a boolean variable DEP, which will be true or false depending on P being dependent or not respectively.

4.6 Algorithm 6 – Finding the Reducts of a Set of Attributes

The algorithm for finding the reducts of a set of attributes is based on: the procedure for verifying the independence of a set of attributes; on the property 6, in Section 3, and on the possibility of eliminating superfluous attributes, step by step.

Algorithm 6. Determining the reducts of a set of attributes (*procedure reducts*).

```
procedure reducts(E,Q,RED)
{
  Input: E = {e1, e2, ..., em}, set of m examples,
         described by n attributes, Q = {q1, ..., qn}
  Output: RED, set containing every reduct of Q
}
begin
   {initialisation}
   rep(E,Q,R);
   cardR := number_of_elements(R);
   dependence(E,Q,Q,DEP);
   {if the set is independent then the reduct is the original set}
   if not(DEP) then RED := {Q}
      else
         begin
            reducts1(E,Q,CardR, RED);
            {excludes repeated reducts in RED}
            excludes_repeated(RED);
         end
end.
```

Algorithm 6 expects as input a KRS described by both, a set of examples $E = \{e_1, e_2, \ldots, e_m\}$ and a set of attributes $Q = \{q_1, \ldots, q_n\}$. The reducts are returned in RED. The algorithm is invoked by reducts(E,Q,RED), that checks first if Q is independent. In an affirmative case, the returned value is a set containing only Q. If Q is dependent, the procedure calls reducts1 (wich is recursive), responsible for searching reducts in subsets of Q. The goal of the excludes_repeated(RED) call is to exclude eventual repetitions of reducts in the set of reducts RED.

The procedure reducts1, by its turn, expects as input: a) a set of examples $E = \{e_1, e_2, \ldots, e_m\}$ described by $P = \{p_1, \ldots, p_j\}$ (a subset of Q that is known to have a reduct of Q) and b) CardR, cardinality of the representation of original system described by E and Q. The procedure returns the found reducts in RED1. In order to

Algorithm 6. Determining the reducts of a set of attributes (*procedure reducts1*).

```
procedure reducts1(E,P,CardR,RED1);
{
  Input: E = {e1, e2, ..., em}, set of m examples,
         described by j attributes , P = {p1, ..., pj} , P ⊆ Q
         CardR, cardinality of the representation
         of the original KRS, given by E and Q
  Output: RED1, set containing every reduct of P
}
begin
   {checking if P is dependent}
   dependence(E,P,P,DEP1);
   if DEP1 then
      begin
        {if dependent, searches for reducts in subsets of P
         with the same power of representation as the original system}
        RED1 := {};
        j := number_of_elements(P);
        i := 1;
        while i <= j do
           begin
              Pi := P - {pi};
              rep(E,Pi,R1);
              CardR1 := number_of_elements(R1);
              {checks if P has the same power of representation as
               the original set of attributes}
              if CardR1 = CardR then
                 begin
                    reducts1(E,Pi,CardR,RED1i);
                    RED1 = RED1 ∪ RED1i;
                 end;
              i := i + 1
           end
        end
        {If P is independent then P is a reduct }
        else
           RED1 := {P}
end.
```

do that, it first checks if P is dependent. In an affirmative case, it searches for reducts in subsets of P with the same representavity of the original set of attributes Q. In a negative case, P itself is a reduct and is added to RED1.

4.7 Algorithm 7 – Degree of Membership of an Element to a Given Set

The following algorithm implements the approximate membership function as defined in Section 2.1. It determines the value of the approximate membership of an element $x \in U$ to a given set $X \subseteq U$. The algorithm returns a value which gives the degree of 'certainty' to which $x \in X$. The following two properties are used: $\mu_X(x) = 0 \Leftrightarrow x \in neg(X)$ and $\mu_X(x) = 1 \Leftrightarrow x \in pos(X)$. When $x \in doubt(X)$, the algorithm first determines the representation of the SRC (through rep(E,Q,R)) and then finds the elementary set E_i, $1 \leq i \leq |R|$, to which x belongs to. The membership degree is returned in Memb.

Algorithm 7. Degree of membership of an element to a given set.

```
procedure membership(E,Q,X,x,Memb)
{
  Input: E = {e_1, e_2, ..., e_m}, set of m examples,
         described by n attributes, Q = {q_1, ..., q_n}
         X = {x_1, x_2, ..., x_p}, given set
         x ∈ E, given element
  Output: Memb, membership of x in X
}
begin
  upp_approx(E,Q,X,UppX);
  {if x ∉ UppX, then Memb is 0}
  if (not ( x in UppX ) ) then
     Memb := 0;
  else
     begin
       low_approx(E,Q,X,LowX);
       {if x ∈ LowX, then Memb is 1}
       if (x in LowX) then
          Memb := 1;
       else
         {calcules membership in the doubtful region}
         begin
           rep(E,Q,R);
           cardR := number_of_elements(R);
           {search for elementary set to which x belongs to}
           found := false;
           i := 0;
           while ((i < cardR) and (not(found)) do
               begin
                 i := i + 1;
                 found := (x in E_i);
               end
           Intersec := E_i ∩ X;
           card1 := number_of_elements(Intersec);
           card2 := number_of_elements(E_i);
           Memb := card1 / card2;
         end
     end
end
```

Algorithm 7 expects as input: a) a KRS described by both, a set of examples $E = \{e_1, e_2, \ldots, e_m\}$ and a set of attributes $Q = \{q_1, \ldots, q_n\}$; b) a given set $X = \{x_1, x_2, \ldots, x_p\}$, $X \subseteq E$ and c) an element $x \in E$, whose membership degree to X is to be determined.

4.8 Algorithm 8 – Discriminant Index of a Set of Elements with Relation to a Set of Attributes

In an approximation space $A = (U, R)$, the discriminant index gives a measure of the degree of certainty related to the membership of any element $x \in U$ to a given set X. Algorithm 8 generalizes the concept of discriminant index introduced in Section 2, by determining the discriminant index of A with relation to X, taking into account a subset of attributes $P \subseteq Q$.

Algorithm 8. Discriminant index of a set of elements.

```
procedure discriminant(E,Q,X,P,Disc)
{
  Input: E = {e1, e2, ..., em} , set of m examples,
         described by n attributes , Q = {q1, ..., qn}
         X = {x1, x2, ..., xp} , given set
         P = {p1, p2, ..., pj} ⊆ Q
  Output: Disc , discriminant index of X with relation to P
}
begin
  low_approx(E,P,X,lowX);
  upp_approx(E,P,X,uppX);
  doubtX := uppX - lowX;
  m := number_of_elements(E);
  count := number_of_elements(doubtX);
  Disc := (m - count) / m;
end
```

Algorithm 8 expects as input: a) a KRS described by both, a set of examples $E = \{e_1, e_2, \ldots, e_m\}$ and a set of attributes $Q = \{q_1, \ldots, q_n\}$; b) a given set $X = \{x_1, x_2, \ldots, x_p\}$, $X \subseteq E$, whose discriminant index in relation to a subset $P \subseteq Q$ is to be determined and c) the subset of attributes, $P = \{p_1, p_2, \ldots, p_j\}$. In order to calculate the discriminant index is necessary first to determine the lower and upper approximations of X in the approximate space induced by P.

4.9 Algorithm 9 – Degree of Dependency of an Attribute with Relation to a Set of Attributes

The degree of dependency of a (decision) attribute with relation to a set of attributes (condition attributes) measures how well the value of the decision attribute can be determined, knowing the values of the condition attributes. The degree of dependency of a set of attributes in relation to another set of attributes was defined in Section 3.

Algorithm 9 expects as input: a) a KRS described by both, a set of examples $E = \{e_1, e_2, \ldots, e_m\}$ and a set of attributes $Q = \{q_1, \ldots, q_n\}$; b) a decision attribute $d \in Q$, whose dependency degree in relation to a subset of attributes $P \subseteq Q$ is to be determined.

4.10 Algorithm 10 – Significance Factor of an Attribute

As presented in Section 3, the significance factor of an attribute indicates its importance in a set of conditions with relation to the existing dependency between the decision attribute and the condition attributes.

Algorithm 10 expects as input: a) a KRS described by both, a set of examples $E = \{e_1, e_2, \ldots, e_m\}$ and a set of attributes $Q = \{q_1, \ldots, q_n\}$; b) a decision attribute $d \in Q$; c) a set of condition attributes, $P \subseteq Q$ and d) an attribute $a \in P$, whose significance factor is to be determined. Algorithm 10 uses the procedure dep_grade defined by Algorithm 9.

Algorithm 9. Degree of dependency of an attribute.

```
procedure dep_grade(E,Q,P,d,DepG)
{
  Input: E = {e1,e2,...,em}, set of m examples,
         described by n attributes, Q = {q1,...,qn}
         P = {p1,p2,...,pj} ⊆ Q
         d ∈ Q, a given attribute
  Output: DepG, dependency degree of d with relation to P
}
begin
  rep(E,{d},R);
  count := 0;
  card := number_of_elements(R);
  i := 0;
  while (i ≤ card) do
    begin
      {Ri is the i-th element of R}
      low_aprox(E,P,Ri,Low)
      count := count + number_of_elements(Low);
      i := i + 1;
    end;
  DepG := count / card;
end
```

Algorithm 10. Significance factor of an attribute.

```
procedure significance(E,Q,P,d,a,Sig)
{
  Input: E = {e1,e2,...,em}, set of m examples,
         described by n attributes, Q = {q1,...,qn}
         P = {p1,p2,...,pj} ⊆ Q
         d ∈ Q, decision attribute
         a ∈ P, given attribute
  Output: Sig, significance factor of a
}
begin
  dep_grade(E,Q,P,a,DEP);
  P1 := P = {a};
  dep_grade(E,Q,P1,a,DEP1);
  Sig := (DEP - DEP1) / DEP;
end
```

4.11 Algorithm 11 – Reducts of a Given Set of Attributes with Relation to the Existing Dependency Between the Given Set and a Given Attribute

This algorithm finds the subsets of the condition set which have the same representational power as the original set; the search is directed by the existing dependency between a decision attribute and the condition attributes.

Algorithm 11 expects as input: a) a KRS described by both, a set of examples $E = \{e_1, e_2, \ldots, e_m\}$ and a set of attributes $Q = \{q_1, \ldots, q_n\}$; b) a decision attribute $d \in Q$ and c) a set of condition attributes, $P \subseteq Q$. The determination of the reducts is done recursively, using reducts_dep(E,Q,P1,d,ReD1), where $P1$ is a subset of P containing one less element than P. The algorithm is strongly based on

Algorithm 11. Determining the accuracy of a given set.

```
procedure reducts_dep(E,Q,P,d,RED)
{
  Input: E = {e1, e2, ..., em}, set of m examples,
         described by n attributes, Q = {q1, ..., qn}
         P = {p1, p2, ..., pj} ⊆ Q
         d ∈ Q, decision attribute
  Output: RED, reducts of P with relation to the dependency
          between d and P
}
begin
  dep_grade(E,Q,P,d,DEP);
  RED := {};
  has_reduct := false;
  {checking if P has a subset with the same representational power}
  j := number_of_elements(P);
  i := 0;
  if (j ≥ 1) then
    begin
      i := i + 1;
      while i ≤ j do
        begin
          P1 := P − {pi};
          dep_grade(E,Q,P1,d,DEP1);
          {if P1 has the same representation power as P then
            P1 has a reduct of P}
          if (DEP = DEP1) then
            begin
              reducts_dep(E,Q,P1,d,RED1);
              RED := RED ∪ RED1;
              has_reduct := true;
            end
        end
    end
  {if P does not have a subset with the same representational power,
    then P itself is the only reduct of P}
  if (not(has_reduct)) then
    RED := {P};
end
```

the determination of the dependence degree and, consequently, uses the procedure grau_dep described by Algorithm 9.

5 Conclusions

The basic concepts of Rough Set Theory (RST) have been used in many different ways for many different purposes. The main goal of this work is to introduce these concepts so to subsidize the presentation of a family of algorithms that implements them. We believe that by expressing these concepts in a procedural way, using a pseudocode very close to a high level programming language, we are contributing for a better understanding of RST and at the same time, easing the work of those who intend to investigate RST. The presented algorithms have been implemented by us in C++ and they run under Windows and Linux.

Acknowledgments

The authors wish to acknowledge FAPESP (Fundação de Amparo à Pesquisa do Estado de São Paulo - Brazil) and Leonie C. Pearson for her help and useful comments.

References

1. Pawlak, Z. (1982): Rough Sets. International Journal of Information and Computer Science 11(5), 341–356
2. Skowron, A.; Grzymala-Busse, J. (1994): From Rough Set Theory to Evidence Theory. In: Yager, R., Fedrizzi, M., Kacprzyk, J. (Eds.): Advances in the Dempster-Shafer Theory of Evidence. John Wiley & Sons, 193–236
3. Wong, S. K. M.; Lingras, P (1989): The Compatibility View of Shafer-Dempster Theory Using the Concept of Rough Set. Res, Z. W. (Ed.): Methodologies for Intelligent Systems 4, North Carolina, Charlotte, 33–42
4. Pawlak, Z. (1994): Hard and Soft Sets. In: Ziarko, W. P. (Ed.): Rough Sets, Fuzzy Sets and Knowledge Discovery. Springer-Verlag, 130–135
5. Wygralak, M. (1989): Rough Sets and Fuzzy Sets - Some Remarks on Interrelations. Fuzzy Sets and Systems 29, 1989, 241–243
6. Pawlak, Z; Skowron, R. (1994): Rough Membership Functions. In: Yager, R., Fedrizzi, M., Kacprzyk, J. (Eds.): Advances in the Dempster-Shafer Theory of Evidence. John Wiley & Sons, 251–271
7. Pawlak, Z. (1985): On Learning - a Rough Set Approach. Lecture Notes in Computer Science 208, Springer Verlag, 197–227
8. Wong, S. K. M.; Ziarko, W., Ye, R. L. (1986): Comparison of Rough Set and Statistical Methods in Inductive Learning. Int. Journal Man-Machine Studies 24, 53–72
9. Grzymala-Busse, J. W. (1992): LERS — A System for Learning from Examples based on Rough Sets, Chapter 1. In: Slowinski, R. (Ed.): Intelligent Decision Support, Kluwer Academic Publishers, 3–18
10. Pawlak, Z. (1984): Rough Classification. Int. Journal of Man-Machine Studies 469–483
11. Shan, N.; Ziarko, W. (1994): An Incremental Learning Algorithm for Constructing Decision Rules. In: Ziarko, W. P. (Ed.): Rough Sets, Fuzzy Sets and Knowledge Discovery. Springer-Verlag, 326–334
12. Ziarko, W. (1991): The Discovery, Analysis, and Representation of Data Dependencies in Databases. In: Piatestsky-Shapiro, G., Frawley, W. (Eds.): Knowledge Discovery in Databases. AAII Press/MIT Press, 195–209
13. Orlowska, E.; Pawlak, Z. (1984): Expressive Power of Knowledge Representation Systems. Int. Journal of Man-Machine Studies 20, 485–500
14. Aasheim, O. T.; Solheim, H. G. (1996): Rough Set as a Framework for Data Mining. Project Report of Knowledge Systems Group - Faculty of Computer Systems and Telematics. Trondheim, Norwegian University of Science and Technology

15. Deogun, J. S. et al. (1997): Data Mining: Trends in Research and Development. In: Lin, T. Y., Cercone, N. (Eds.): Rough Sets and Data Mining: Analysis for Imprecise Data. Boston, Kluwer Academic, 9–45
16. Mrózek, A. (1992): A New Method for Discovering Rules from Examples in Expert Systems. Int. Journal of Man-Machine Studies 36, 127–143
17. Szladow, A.; Ziarko, W. (1993): Rough Sets: Working with Imperfect Data. AI Expert, July, 36–41
18. Grzymala-Busse. J. (1988): Knowledge Acquisition under Uncertainty - a Rough Set Approach. Journal of Intelligent and Robotics Systems 1, 3–16
19. Pawlak, Z.; Wong, S. K. M.; Ziarko, W. (1988): Rough Sets: Probabilistic versus Deterministic Approach. International Journal of Man-Machine Studies, Vol. 29, 81–95
20. Grzymala-Busse, J. W. (1986): On the Reduction of Knowledge Representation Systems. Proceedings of the 6th International Workshop on Expert Systems and their Applications, Avignon, France, Vol. 1, 463–478
21. Jelonek, J.; Krawiec, K.; Slowinski, R. (1994): Rough Set Reduction of Attributes and Their Domains For Neural Networks. Computational Intelligence 2(5), 1–10
22. Slowinski, R. (1995): Rough Set Approach to Decision Analysis, AI Expert 1995, 19-25
23. Ohrn, A. (1993): Rough Logic Control: a New Approach to Automatic Control? Technical Report, Trondheim, Norway, April 1993
24. Pawlak, Z. (1997): Rough Real Functions and Rough Controllers. In: Lin, T. Y., Cercone, N. (Eds.): Rough Sets and Data Mining: Analysis for Imprecise Data. Boston, Kluwer Academic, 139–147
25. Hu, X.; Cercone, N.; Han, J. (1994): An Attribute-oriented Rough Set Approarch for Knowledge Discovery in Databases. In: Ziarko, W. P. (Ed.): Rough Sets, Fuzzy Sets and Knowledge Discovery. Springer-Verlag, 90–99
26. Beaubouef, T.; Petry, F. E. (1994): A Rough Set Model for Relational Databases. In: Ziarko, W. P. (Ed.): Rough Sets, Fuzzy Sets and Knowledge Discovery. Springer-Verlag, 100–107
27. Grzymala-Busse, J. W.; Mithal, S. (1991): On the Best Choice of the Best Test for Attribute Dependency in Programs for Learning From Examples. International Journal of Software Engineering and Knowledge Engineering 1(4), 413–438
28. Mrózek, A. (1989): Rough Sets and Dependency Analysis Among Attributes in Computer Implementations of Expert Inference Models. Int. Journal of Man-Machine Studies 30, 457–473
29. Slowinski, R. (Ed.) (1992): Intelligent Decision Support - Handbook of Advances and Applications of the Rough Set Theory. Kluwer Academic
30. Pal, S. K.; Skowron A. (Eds.) (1999): Rough Fuzzy Hybridization - A New Trend in Decision-Making. Springer-Verlag
31. Gawrys, M.; Sienkiewicz, J. (1993): Rough Set Library User's Manual (version 2.0), September 1993, Institute of Computer Science, Warsaw, University of Technology, Warsaw, Poland

32. Klir, G. J.; Yuan, B. (1995): Fuzzy Sets and Fuzzy Logic - Theory and Applications. Prentice Hall
33. Nicoletti, M. C.; Uchôa, J. Q. (1997): The Use of Membership Functions for Characterizing the Main Concepts of Rough Set Theory (in Portuguese). São Carlos/Brazil, Technical Report 005/97, DC-UFSCar
34. Pawlak, Z. (1981): Information systems: theoretical foundations. Information Systems 6(3), 205-218

Chapter 5

Automated Case Selection from Databases Using Similarity-Based Rough Approximation

Liqiang Geng and Howard J. Hamilton

Summary. Knowledge acquisition for a case-based reasoning system from domain experts is a bottleneck in the system development process. It would be useful to derive representative cases automatically from larger, available databases rather than acquiring them from domain experts. Case selection is a branch of data mining that aims at choosing representative cases from large data sets for future case-based reasoning tasks. This chapter presents two algorithms using similarity-based rough set theory to derive cases automatically from available databases. The first algorithm, SRS1, requires the user to choose the similarity thresholds for the objects in a database, while the second algorithm, SRS2, can automatically select proper similarity thresholds. These algorithms can handle noise and inconsistent data in the database and select a reasonable number of the representative cases from the database. The algorithms were implemented and the experimental results showed that their classification accuracy was similar to that of well-known machine learning systems, such as rule induction systems, and neural network systems, and other state of art instance selection algorithms RT3 and ICF.

Keywords: case selection, similarity, rough sets, classification.

1 Introduction

Case-based reasoning (CBR) is a computerized method that attempts to study solutions that were used to solve problems in the past so as to solve current problems by analogy or association [10]. The CBR approach has some advantages over the rule-based approach in unstructured and difficult-to-understand problem domains, such as diagnosis, planning and classification. It can also deal with complex types of data that rule-based system cannot handle, such as images, sets and text. See for example [4, 5, 9, 17]. The success of these systems suggests that CBR-based systems are well suited to unstructured domains. Moreover, case bases are maintainable because new cases can be added and existing ones modified or deleted easily. However, knowledge acquisition for a CBR system is still a labor-intensive task. The classification accuracy of the developed CBR system is also dependent on the quality of the cases obtained from the domain experts.

Techniques for automatically deriving high quality cases from an available database are being studied by many CBR researchers. In this area, many issues have been and are being tackled, such as how to deal with noise and inconsistent data, how to reduce the number of the selected cases while maintaining classification accuracy, how to make cases more compact by eliminating irrelevant features, and how to increase the classification accuracy of the derived casebase.

If the cases have a simple structure with values for all of a set of attributes, such as a table in a relational database, then the traditional machine learning methods can be applied to obtain classification models. These approaches include *decision tree* learning systems, such as C4.5 [19]; rule learning systems, such as LEM2 [6]; and neural network models [20]. We now briefly review these approaches and then summarize other related research.

C4.5 is a recursive algorithm that builds a decision tree from the data available. The training set corresponds to the root node of the tree. If all training examples belong to the same class, the decision tree is a single leaf node. Otherwise, *information gain* is used as a heuristic to select an attribute, which partitions the training set into subsets according to the possible values of the attribute. These subsets form the child nodes for the root node. This partitioning process continues recursively until no node can be expanded further. The result is the decision tree.

LEM2 is based on rough sets theory, where a rough set is defined based on a lower approximation containing all objects certainly belonging to set and an upper approximation containing all objects possibly belonging to the set. Before the algorithm is applied, either the lower approximation or the upper approximation of the data set is calculated, depending on the type of rules desired. This process transforms inconsistent data into consistent data. Then for each decision class, the algorithm finds a local covering for the lower/upper approximation. An attribute-value pair is selected as part of the local covering based on heuristics, such as its prespecified priority or its coverage. This selection process continues until the conjunction of the attribute-value pairs covers only positive examples. This conjunction forms the conditional part for a classification rule. After obtaining such a rule, the covered examples are removed from the training set. This process iterates until the training set is empty.

An *artificial neural network* (ANN) is an information-processing paradigm inspired by the way the interconnected, parallel structure of the mammalian brain processes information. The essence of the ANN paradigm is the novel structure of the information processing system. It is composed of a large number of highly interconnected processing elements that are analogous to neurons. These elements (*neurons*) are tied together with weighted connections that are analogous to synapses. In the learning process, the examples are fed to the network and output is obtained. Based on the error between the actual output and the supposed output, the algorithm adjusts the connection weights (*synapses*). These connection weights store the knowledge necessary to solve specific problems.

Although these methods and algorithms have many advantages and researchers have developed many variations and descendents of these methods, they still can only be applied to well-structured domains.

Case based reasoning can deal with complex data types, whenever a similarity measure between data items can be defined. Case selection algorithms aim at selecting representative cases for future classification tasks. Case selection algorithms can be classified as either incremental or batch. An *incremental approach* decides whether to keep each new case in the casebase whenever a new case is encountered, while a *batch approach* creates an initial casebase from a data set or updates the casebase when its size grows to a certain threshold.

Aha et al. proposed a series of incremental instance-based learning algorithms (IBL), called IB1, IB2, IB3 and IB4 [1]. IB1 is similar to the k-nearest neighbor (k-NN) clustering algorithm, because it stores all the instances in the casebase and classifies unseen instances to the same class as the most similar instance in the casebase. IB2 reduces the size of the casebase by storing only the cases that cannot be correctly classified by the existing casebase. IB3 keeps track of the frequency with which stored cases, when chosen as one of the current object's most similar stored cases, matched the current object's decision value. If the matching frequency is lower than a given threshold, the case is discarded. IB4 incorporates weight learning and thus can tolerate irrelevant features better than IB3. The IBL algorithms have $O(mn^2)$ worst-case time complexity for m attributes and n instances. Due to their incremental nature, the casebase generated by the IBL algorithms is strongly influenced by the order in which the objects are stored. No heuristics are used to select instances from a global point of view.

To overcome this limitation, competence measures can be used to order the instances before the selection process begins. Smyth et al. propose *relative coverage* to measure the competence of instances [22]. This measure is not only based on the coverage of an instance, but also based on how many other instances cover the covered instances. This ordering process transforms IBL from an incremental method into a batch method.

Similarly, Wilson et al. [24, 25] propose a series of batch case pruning algorithms named *RT*1, *RT*2, and *RT*3 (also called *Integrated Decremental Instance-Based Learning*). Instead of selecting good instances for inclusion, *RTx* selects bad ones for exclusion. *RT*1 first constructs lists of neighbors and lists of associates for each instance. Then it checks to see if the absence of each instance i will improve the classification accuracy of its *associates*, which are the instances that have i as one of their nearest neighbors. If it does, this instance is eliminated, and the lists of neighbors and associates are updated to delete this instance from their lists. *RT*2 does not update the associate lists; instead it keeps the deleted instances in the associate lists. Thus, the deleted instances continue to be considered during leave-one-out cross-validation. Furthermore, *RT2* ranks the instances according to their distances to the border in descending order. This sorting attempts to delete central points and keep border points. *RT3* attempts to eliminate noisy instances before the pruning process by deleting all instances that cannot be correctly classified by their nearest neighbors.

Brighton et al. [2] proposed a similar case pruning algorithm *ICF*. This algorithm uses the same method to identify and delete noisy instances as *RT*3. Then it calculates the reachable and coverage sets of each instance, which correspond to neighbors and associates in *RT*3. The difference between the reachable set and the neighbor set is that the neighbor set has fixed cardinality, while the reachable set has variable cardinality, which is determined by the *enemy* of the instance, i.e., the nearest neighbor of the instance that has a different decision value. The heuristic used here is that if the cardinality of the reachable set is greater than the cardinality of the coverage set, the instance is deleted. Although *RT3* makes only one pass, *ICF* performs multiple passes until no further instances can be removed.

In addition to the case selection or case deletion strategies mentioned above, other approaches have also been proposed for maintaining large casebases. For example, Yang et al. [26] propose that a clustering algorithm be used to divide a large casebase into several smaller ones. When a new case arrives, information gain is used to identify the most similar casebase cluster, and then the standard case retrieval method is used to obtain the most similar cases from this cluster. Essentially, this method tends to maintain a layered casebase. It can improve the speed of case retrieval, but it may miss relevant cases and it does not reduce storage costs.

Hybrid case selection methods are also being studied. Li et al. show that k-NN and its variants perform poorly on nominal attributes [12]. To improve accuracy on data sets containing a mixture of interval-based and nominal attributes, they combine k-NN with the DeEPs algorithm. DeEPs classifies a new instance based on the support of subsets of the attribute-value pairs of the test instance. If the total support for the subsets of the test instance which occur only in positive training instances is greater than the total support for those which occur only in negative instances, then the test instance is classified as positive, and vice versa. Using a similarity function, they define the neighborhood of a test instance. If the neighborhood contains training data, they apply k-NN to make a classification decision; otherwise, they use DeEPs [13].

Soft computing methods are also applied to case selection problems. Pal et al. use fuzzy set theory and rough set theory to induce representative cases from a database [16]. Their algorithm first translates the numeric data into fuzzy linguistic terms and then derives fuzzy rules from the database using a rough set approach. The fuzzy rules are then mapped to representative cases. Cao et al. use a similar method to derive representative cases [3]. They partition data into clusters, use a rough-fuzzy approach to obtain fuzzy adaptation rules for each cluster, and then select representative rules by applying these adaptation rules. As described, these algorithms can only deal with numeric data, which limits their applicability in case-based reasoning. Also these methods do not handle noisy or inconsistent data, and they require that many parameters and sub-algorithms be selected. The parameters and sub-algorithms include the number of the fuzzy terms for each attribute, thresholds for de-fuzzification, and fuzzy membership functions. Although these factors strongly influence the quality of the selected cases, they are difficult to choose. We believe that case generation should rely as much as possible on the available data rather than on information required from users.

In this chapter, two case selection algorithms are proposed. A similarity-based rough set method (SRS1) is suggested to select a reasonable number of representative cases from an available data set. It satisfies the following requirements: it selects the high quality center points, it has a small number of user-specified parameters and it explicitly handles noisy and inconsistent data. We also propose its variant (SRS2) that can determine the similarity threshold automatically. Section 2 discusses some basic concepts of similarity-based rough approximation, and Section 3 introduces the concept of similarity measure. Sections 4 and 5 describe the proposed algorithms SRS1 and SRS2, respectively. Section 6 presents an illustrative example. In Section 7, the results of testing the algorithms on well-known data sets are presented. Section 8 gives conclusions and directions for future work.

2 Similarity-Based Rough Sets

Rough sets [18] are a mathematical tool for dealing with vagueness and uncertainty in areas of artificial intelligence and cognitive sciences, such as data mining, decision making, and pattern recognition [15, 8, 27]. The concept of rough sets is based on the assumption that every object of the universe can be represented by all information available about it. Objects represented by the same information are considered to be *indiscernible*. All the indiscernible objects form an elementary set, i.e., granule knowledge about the universe. If a given set of objects is a union of some elementary sets, it is referred to as a *crisp set*; otherwise it is a *rough set*. A rough set can be represented by a pair of crisp sets, called the *lower* and the *upper approximation*. The lower approximation contains all objects that surely belong to the set, and the upper approximation contains all objects that possibly belong to the set, with respect to the given knowledge.

Rough sets based on the indiscernibility relation can only deal with nominal attributes in the decision table [18]. Continuous attributes must be discretized into intervals and then each of these intervals must be translated into qualifiers before rough set theory can be applied. The discretization methods adopted greatly influence the quality of the classification results. A typical CBR based system usually involves more complex data types. Similarity measures are usually used for retrieving appropriate cases from the casebase. Therefore, we adopted the similarity relation-based rough set approach for selecting representative cases. Similarity relation-based rough set is an extension of the standard rough set approach, which replaces the indiscernibility relation with a similarity relation in the approximation process [21].

It is widely accepted that *similarity relation* should have *reflexive* properties, i.e., any object should be similar to itself. However, it is controversial whether similarity relation should be *symmetric*. Recently, more and more researchers tend to exclude the similarity property [21]. We argue that whether symmetry should be a property of the similarity relation depends on the context of application. Theoretically, we adopt the definition that similarity relation is reflexive, but not necessarily symmetric. However, practically, we might incorporate symmetric property for similarity relation, since this property can enhance computational efficiency, and any properties that hold for asymmetrical similarity relations also hold for symmetric ones.

Given a finite non-empty set U of objects, called the *universe*, a binary relation R defined on $U \times U$ is a similarity relation if and only if aRa, where $a \in U$. From this definition, we can represent the relation R as a similarity graph and define a similarity class for each object $x \in U$. The *similarity class* of x, denoted by $R(x)$, is the set of objects that are similar to x.

$$R(x) = \{y \in U \mid yRx\} \tag{1}$$

The *rough approximation* of a set $X \subseteq U$ is a pair of sets called lower and upper approximations of X, denoted by $R_*(X)$ and $R^*(X)$ respectively, where

$$R_*(X) = \{x \in X \mid R(x) \subseteq X\} \tag{2}$$

$$R^*(X) = \cup_{x \in X} R(x) \tag{3}$$

The lower approximation $R_*(X)$ of a set X is the set of objects whose similarity class belongs to X and the set consists of elements that can certainly be classified as elements of X, while the *upper approximation* $R^*(X)$ of X is the union of the similarity classes of objects in X and the elements in this set can possibly be classified as elements of X.

To make the approximation more robust, we make two extensions to the definition of the lower approximation.

First, to enable it to deal with noise in a database, we propose the following *extended lower approximation* of a similarity-based rough set,

$$R_{*ext}(X) = \{x \in X \mid card(R'(x)) \,/\, card(R(x)) \geq ct\}, \tag{4}$$

where $R'(x) = \{y \mid y \in R(x) \text{ and } d(y) = d(x)\}$ refers to the set of the objects that are similar to object x and have the same decision value as x. The *consistency threshold* $ct \in [0, 1]$ defines the degree to which noisy data are tolerated. It is a parameter that must be specified by the user for the SRS1 and SRS2 algorithms. $R_{*ext}(X)$ is the set of objects which are classified, with certainty of at least ct, as elements of concept X according to the similarity relation R. If ct equals 1, $R_{*ext}(X)$ is identical to the standard similarity based lower approximation $R_*(X)$.

Secondly, we think that in some cases, the quality of an object depends not only on the cardinality of the sets $R(x)$ and $R'(x)$, but also on the similarity measure between the object and its similar objects in sets $R(x)$ and $R'(x)$. Therefore, we propose the following extended *similarity-sensitive lower approximation*,

$$R_{*ss} = \{x \in X \mid \sum_{x_1 \in R'(x)} s(x, x_1) / \sum_{x_2 \in R(x)} s(x, x_2) \geq ct\}, \tag{5}$$

where $s(x, x_1)$ and $s(x, x_2) \in [0, 1]$ denote similarity degree between object x and objects x_1 and x_2 respectively. In the rest of the chapter, we use the first extension $R_{*ext}(X)$ and its relevant parameters to present our algorithms and examples. This can be easily replaced with the second extension $R_{*ss}(X)$ and its relevant parameters.

To conduct casebase generation, a source of data is represented as a *decision table*, i.e., a two-dimensional table where each row represents an object, and each column represents an attribute. The similarity relation is defined on the objects in the decision table where the cases are derived. Let $T = (U, A \cup \{d\})$ be a decision table where U is the universe, A is a set of condition attributes and d is a decision attribute [18]. In our current study, the condition attributes are restricted to be continuous, ordinal, or nominal and the decision attribute is nominal. Let V_a be a set of values of attribute a, where $a \in A$, $r(d)$ be the number of decision values, d_i be the *ith* decision value, and $Y_i = \{x \in U \mid d(x) = d_i\}$ be the set of objects that have the *ith* decision value in the decision table. $POS(R,\{d\}) = \cup_{i=1}^{r(d)} R_{*ext}(Y_i)$ is called the *positive region* of the partition $\{Y_i \mid i = 1,\ldots,r(d)\}$.

The coefficient $r(R, \{d\}) = card(POS(R, \{d\})) / card(U)$ is called the quality of approximation of the classification. It expresses the ratio of objects that can be correctly classified to all the objects in the decision table. The objects that cannot be classified are considered as inconsistent objects.

3 Similarity Measure

To obtain the similarity class for every object, a similarity measure should be defined for each attribute. This definition depends on the type of attributes under study. We classify the attributes into the following categories: interval-scaled attributes, ordinal attributes, nominal attributes, fuzzy attributes, set attributes, text attributes, and non-text attributes, such as images, multimedia data, etc. Many similarity measures have been proposed for these attributes, see for example [7, 23]. The most widely used categories are discussed as follows.

1. *Interval-scaled attributes* are continuous measurements on a roughly linear scale, such as length and height. For this kind of attribute, the similarity measure between object x and y $S_a(x, y)$ can be defined as:

$$S_a(x, y) = 1 - |a(x) - a(y)| \,/\, |a_{max} - a_{min}| \tag{6}$$

   where x and y are the objects, a_{max}, a_{min} denote the minimum and maximum values of attribute a respectively, and $a(x)$, $a(y)$ denote the value of attribute a for object x and y respectively.
2. *Ordinal attributes* represent a set of states that are ordered in a meaningful sequence. Assuming that the values of attribute a are ordered in ascending order as $v_1, v_2,\ldots v_n$, and the objects x and y have the values v_i and v_j for attribute a, then S_a is defined as:

$$S_a(x, y) = 1 - |i - j| \,/\, (n - 1). \tag{7}$$

3. *Nominal attributes* have unordered discrete values. The following similarity measure is proposed:

$$S_a(x,y) = 1 - \sum_{k=1}^{r(d)} \frac{|P(d=k, a=a(x)) - P(d=k, a=a(y))|}{r(d)P(d=k)}, \tag{8}$$

where $P(d = k, a = a(x))$ is the probability that the value of conditional attribute a is equal to the value of attribute a of object x and the decision is k, and $P(d = k, a = a(y))$ is likewise for object y.

4. *Fuzzy attributes* are a kind of nominal attributes that use fuzzy membership functions to define the semantics of the nominal terms. For example, the attribute *age* can take values of fuzzy terms *young*, *middle-aged* and *old*, each of which corresponds to a fuzzy membership function. If the attribute domain is discrete, the similarity measure can be defined as:

$$S_a(x,y) = 1 - \sum_{i=1}^{n} \frac{|u_A(t_i) - u_B(t_i)|}{n} \tag{9}$$

where A and B are the fuzzy terms of attribute a for objects x and y respectively and u_A and u_B are the corresponding fuzzy membership functions.

If the domain involves a continuous interval $[t_0, t_1]$, we define the similarity measure as:

$$S_a(x,y) = 1 - \int_{t_0}^{t_1} \frac{|u_A(t) - u_B(t)|}{t_1 - t_0} dt \tag{10}$$

5. *Set attributes* take sets for their values. For example, since an employee can have more than one child, the attribute *children* in an *employee* table could be a set attribute. One similarity measure for a set attribute is defined as:

$$S_a(x,y) = \frac{|v_x \cap v_y|}{|v_x \cup v_y|} \tag{11}$$

where v_x and v_y are attribute a's value for object x and y respectively.

6. *Text attributes* have text or documents for their values. A text value can be processed by information retrieval techniques to produce a set of keywords. Then we can use a similarity measure suited to set attributes to compute the similarity between two text values.
7. *Image attributes* have pictures in some predefined format as their values. They can be processed with the specific purpose distance functions.

After obtaining the similarity measure for each attribute, we aggregate them to define the global similarity measure on the set of objects by taking their product:

$$S(x, y) = \prod_{a\in A} S_a(a(x), a(y)), \tag{12}$$

or by taking the minimum similarity measure of the individual attribute:

$$S(x, y) = \mathrm{Min}_{a\in A}\, S_a(a(x), a(y)), \tag{13}$$

or by taking the algebraic average:

$$S(x, y) = \sum_{a\in A} S_a(a(x), a(y))/n, \tag{14}$$

where n is the number of the attributes;
or by taking the geometric average:

$$S(x,y)=\sqrt{\frac{\sum_{a\in A}S_a^2(x,y)}{n}} \tag{15}$$

or by taking their weighted sum:

$$S(x,y)=\sum_{a\in A} w_a S_a(a(x),\ a(y)). \tag{16}$$

where w_a is the weight for the attribute a, with $\sum_{a\in A} W_a = 1$

We say that object x is *similar* to object y if and only if the similarity measure between these two objects is greater than or equal to a similarity threshold st, i.e., xRy if and only if $S(x,y) \geq st$. The similarity threshold st determines the granularity of classification; a higher st value indicates a more refined classification of the data into clusters of representative cases. In algorithm SRS1, this parameter must be specified by the user, while in SRS2, this parameter is automatically determined by the algorithm.

4 SRS1 Algorithm

Based on the concept of similarity measure and rough set theory described above, we propose the following algorithm to find representative cases from a database. First, in addition to the similarity thresholds st and consistency threshold ct mentioned in Sections 2 and 3, some variables used in the algorithm are defined as follows.

1. *SimilarNo*(i): the number of objects similar to the *ith* object O_i, including O_i itself.
2. *SimilarClassNo*(i): the number of objects similar to O_i that have the same decision value as O_i, including O_i itself.
3. *Consistency*(i): the fraction of the number of similar objects that have the same decision value as O_i. *Consistency*(i) = *SimilarClassNo*(i) / *SimilarNo*(i). If *Consistency*(i) is less than ct, O_i can be regarded as inconsistent with the database.
4. $r(R, \{d\})$: the quality of classification, i.e., the fraction of the all objects that are consistent.

The steps of the algorithm are shown in Figure 1.

First, according to the similarity measures and the similarity threshold st, we construct the similarity relation graph of the objects, where each node represents an object in the data set. We then delete the inconsistent nodes according to the value of *Consistency* and threshold ct. Next we select the most representative node in terms of *Consistency* and *SimilarClassNo* and delete all of its similar nodes. The process iterates until the node set is empty. Meanwhile we determine the ratio $r(R, \{d\})$ of the objects that can be classified correctly to the whole set of objects. If the ratio is too low, which means too many objects cannot be classified, the threshold for the similarity measure is increased to refine the granularity of the clusters. In Section 7, our experiments demonstrated that with a larger similarity threshold, more objects could be classified. But at the same time, the number of selected representative objects increases, which might give better classification accuracy.

Let m denote the number of the condition attributes and n denote the number of the objects in the decision table. The time complexity for creating a similarity relation graph is $O(mn^2)$, for computing *SimilarClassNo, SimilarNo* and *Consistency*, and selecting nodes is $O(n^2)$, and for deleting inconsistent nodes is $O(n)$. Overall, the complexity of the SRS1 is $O(mn^2)$. If $m << n$, the complexity approximates $O(n^2)$.

Algorithm SRS1(*st, ct*)
1. Based on *st*, compute the similarity measures between all pairs of objects and create a similarity matrix, which defines a similarity relation graph, with a node for each object and an arc between each pair of objects with nonzero similarity.
2. For each node *i* in the relation graph {
   Compute *SimilarNo*(*i*), *SimilarClassNo*(*i*), *and Consistency*(*i*).
   }
3. Compute $r(R, \{d\})$.
4. Delete nodes that are considered inconsistent, i.e., those which satisfy *Consistency*(*i*) < *ct*
5. Select isolated nodes, i.e., nodes whose *SimilarNo* is 1, and place them in the casebase.
6. While the node set is not empty {
   Select a node with the maximum value for *Consistency*. If there is a tie, select one with the maximum value for *SimilarClassNo*. If there is a tie again, randomly select one node. Insert it into the casebase.
   Delete the selected node and all nodes adjacent to it.
   }

Figure 1. Algorithm SRS1.

5 SRS2: Automatic Determination of Similarity Threshold

It is often difficult for users to determine an appropriate value for the similarity threshold *st*. Therefore this threshold is usually obtained through trial and error. We propose that a more precise and robust definition of similarity can be inferred directly from the data sets. Krawiec et al. [11] proposed a method to derive the discretization intervals for each conditional attribute in the decision rules. Their method is strongly local in that it computes the intervals for each conditional attribute independently from the rest of the conditional attributes. In our method, the entire set of conditional attributes is simultaneously considered. We take into account the similarity between objects. The algorithm is presented in Figure 2.

```
Algorithm SRS2(ct)
1. Compute the similarity measure between all pairs of objects and create a
   similarity matrix, which defines a similarity graph.
2. For each object O_i, do {
       Create an ordered list by sorting all objects in descending order accord-
       ing to their similarity recorded in the similarity matrix to object O_i
       SimilarNo(i) = 0;
       SimilarClassNo(i) = 0;
       SimilarSet(i) = {};
       For each object O_j in the ordered list, do the following {
           SimilarNo(i) = SimilarNo(i) + 1;
           SimilarSet(i) = SimilarSet(i) ∪ O_j;
           If the decision value of object O_j equals the decision value of O_i
               SimilarClassNo(i) = SimilarClassNo(i) + 1;
           Consistency(i) = SimilarClassNo / SimilarNo;
           If Consistency(i) < ct
               Break;
       }
       For each object O_j in SimilarSet(i), in ascending order in terms of the
       similarity measure, do the following {
           If the decision value of object O_j is not equal to the decision value of
           O_i {
               SimilarSet(i) = SimilarSet(i) - O_j;
               SimilarNo(i) = SimilarNo(i) - 1;
               Consistency(i) = SimilarClassNo(i) / SimilarNo(i);
           }
           Else if Consistency(i) < ct {
               SimilarSet(i) = SimilarSet(i) - O_j;
               SimilarNo(i) = SimilarNo(i) - 1;
               SimilarClassNo(i) = SimilarClassNo(i) - 1;
               Consistency(i) = SimilarClassNo(i) / SimilarNo(i);
           }
           Else
               Break;
       }
   }
3. While the set of the objects is not empty {
       Select isolated nodes to the casebase, i.e., nodes whose SimilarNo is 1.
       Select a node with the maximum value for Consistency. If there is a tie,
       select one with the maximum value for SimilarClassNo. If there is a tie
       again, randomly select one node. Insert it into the casebase.
       Delete the selected node and all nodes adjacent to it.
   }
```

Figure 2. Algorithm SRS2.

This algorithm assumes that the similarity threshold for each object is unique and that these thresholds can be determined on the basis of the consistency threshold *ct* provided by the users. In this algorithm, we first compute the similarity matrix for the given set of objects. Then for each object we sort the other objects that are similar to it in descending order according to the similarity measure. The similarity threshold for O_i should satisfy the consistency condition *SimilarClassNo(i) / SimilarNo(i)* $\geq ct$. Therefore, we select the similar objects in the ordered list one by one as long as the ratio is greater than or equal to the given threshold *ct*. Then, for the selected objects, we sort them in ascending order according to the similarity to O_i. In this order, we delete the objects one by one until the current object in the list has the same decision value as O_i, and at the same time, the consistency condition holds. In this way, we obtain the set of similar objects to O_i. Finally we construct the similarity relation graph, which in this case is a directed graph. Then we obtain the representative cases with the same method as in SRS1 described in Section 4.

Let *m* denotes the number of the condition attributes and *n* denotes the number of the objects in the decision table. The time complexity for creating a similarity matrix is $O(mn^2)$, that for sorting nodes according to similarity measure is $O(n^2 logn)$, that for computing *SimilarClassNo, SimilarNo* and *Consistency*, and for selecting nodes is $O(n^2)$, and that for deleting the inconsistent nodes is $O(n)$. Overall, the complexity of the SRS1 algorithm is $O(mn^2 logn)$. If $m << n$, the complexity approximates $O(n^2 logn)$.

6 An Illustrative Example

In this section, we present an example to illustrate the two algorithms. Table 1 has been adapted from [21] and consists of information about eleven companies. The condition attributes are *asset* (*a*), *profit* (*p*), and *type of product* (*t*), and the decision attribute is *credit* (*c*). *Asset* and *profit* are interval scaled values, *product* is a nominal attribute, and *credit* is a nominal attribute.

Table 1. Decision table.

Company	Asset	Profit	Type of Product	Credit
1	105	67	Computer software	Bad
2	54	75	Automobile	Good
3	80	93	Automobile	Bad
4	64	80	Automobile	Good
5	92	92	Computer hardware	Good
6	96	102	Computer hardware	Good
7	111	65	Computer software	Bad
8	58	70	Automobile	Good
9	74	77	Automobile	Bad
10	105	105	Computer hardware	Good
11	85	82	Automobile	Bad

Here we define the similarity measures for the condition attributes:

$$S_a(V_{ai}, V_{aj}) = 1 - |V_{ai} - V_{aj}| / |V_{amax} - V_{amin}| ;$$

$$S_p(V_{pi}, V_{pj}) = 1 - |V_{pi} - V_{pj}| / |V_{pmax} - V_{pmin}| ;$$

$$S_t(V_{ti}, V_{tj}) = 1, \text{ if } V_{ti} = V_{tj} ; S_t(V_{ti}, V_{tj}) = 0, \text{ otherwise;}$$

$$S(O_i, O_j) = S_a(V_{ai}, V_{aj}) * S_p(V_{pi}, V_{pj}) * S_t(V_{ti}, V_{tj}).$$

where O_i and O_j denote objects i and j respectively, V_{ai} and V_{aj} denote the values for O_i and O_j for attribute a respectively, V_{amax} and V_{amin} denote the maximum and minimum values for attribute a in the decision table. Similar notation is defined for attributes p and t. S_a, S_p, and S_t denote the similarity measure between objects in terms of the attributes a, p, and t, respectively. S denotes the overall similarity measure between objects.

By calculating the similarity measure for all pairs of objects, we obtain the similarity matrix shown in Table 2.

6.1 Example of Algorithm SRS1

With the similarity matrix shown in Table 2, we can apply algorithm SRS1 to the example in Table 1. If the similarity threshold *st* is set to 0.65, then we obtain the relation graph shown in Figure 3. The objects in the left area are the companies with *bad credit*, while those in the right area are the companies with *good credit*.

Table 2. Similarity matrix.

Object	1	2	3	4	5	6	7	8	9	10	11
1	1	0	0	0	0	0	0.85	0	0	0	0
2		1	0.3	0.72	0	0	0	0.81	0.62	0	0.38
3			1	0.38	0	0	0	0.26	0.54	0	0.66
4				1	0	0	0	0.67	0.76	0	0.6
5					1	0.7	0	0	0	0.52	0
6						1	0	0	0	0.78	0
7							1	0	0	0	0
8								1	0.6	0	0.37
9									1	0	0.71
10										1	0
11											1

If we set the consistency threshold *ct* to 0.7, the *SimilarNo, SimilarClassNo* and *Consistency* values for each object can be computed as shown in Table 3. Take object 9 for example. The similar objects for object 9 are objects 4, 9, and 11, in other words, similarity measures between object 9 and these three objects are greater than *st*, therefore *SimilarNo*(9) is calculated as 3. Among these similar objects, objects 9 and 11 have the same decision value, *bad credit*, as object 9. Object 4 has a different decision value. Therefore, *SimilarClassNo* is calculated as 2. Then *Consistency*(9) is calculated as 2/3 (0.67). Similarly, we obtain *Consistency*(4) = 0.75, and for the other nodes, *Consistency* = 1. For node 9, we can say that if a node is similar to it, it can be classified into the *bad credit* group only with a probability of 0.67, which is less than the threshold *ct*, 0.7. Hence we conclude that node 9 is an inconsistent object and should be eliminated from the graph. We then examine the nodes with the maximum value for *Consistency*. In this case nine nodes tie, and therefore we select a node with the maximum value for *SimilarClassNo*, i. e., one node from among the nodes 2, 6, 8, and 11. Since there are more than one candidate nodes, we randomly select one of them, say node 2, into the casebase. We then delete the nodes connected to node 2, i.e., node 8 and node 4. Next we select node 6 and delete node 5 and node 10. In the third iteration, we select node 11 and delete nodes 3 and 9. Finally we select node 1 and delete node 7. In all, nodes 1, 2, 6, and 11 are selected and placed in the casebase.

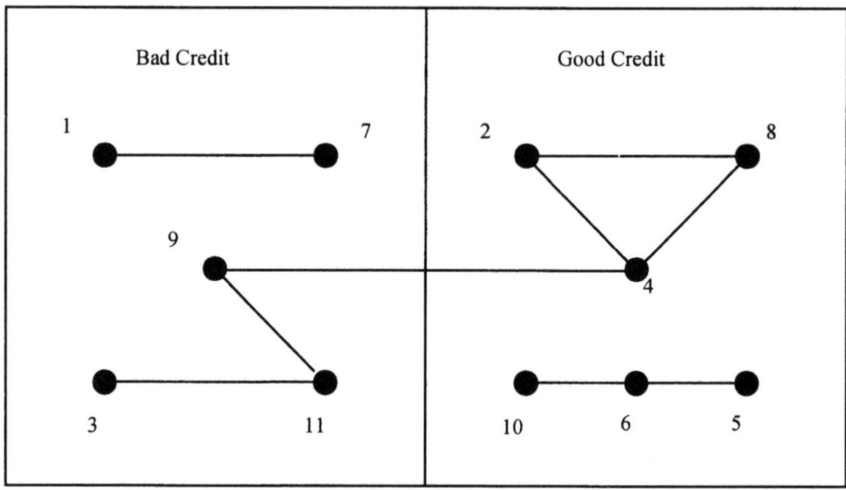

Figure 3. Similarity relation graph.

6.2 Example of Algorithm SRS2

The consistency threshold is defined as *ct* = 0.7. For each object, we sort all the objects in descending order according to the similarity measure, and identify the set of similar objects according to *ct*. Table 4 presents the set of similar object and other parameters for each object. Using the same selection strategy described for step 4 of the SRS1 algorithm, objects 2, 3, 5, and 1 are selected as the representative cases.

Table 3. Sets of similar objects with associated *SimilarNo*, *SimilarClassNo*, and *Consistency* values for example of SRS1.

Object	*SimilarNo*	*SimilarClassNo*	*Consistency*	Above *ct*
1	2	2	1	Yes
2	3	3	1	Yes
3	2	2	1	Yes
4	4	3	0.75	Yes
5	2	2	1	Yes
6	3	3	1	Yes
7	2	2	1	Yes
8	3	3	1	Yes
9	3	2	0.67	No
10	2	2	1	Yes
11	3	3	1	Yes

Table 4. Sets of similar objects with associated *SimilarNo*, *SimilarClassNo* and *Consistency* values for example of SRS2.

Object	Set of Similar Objects	*SimilarClassNo*	*SimilarNo*	*Consistency*
1	1, 7	2	2	1
2	2, 8, 4	3	3	1
3	3, 11, 9	3	3	1
4	4	1	1	1
5	5, 6, 10	3	3	1
6	6, 10, 5	3	3	1
7	7, 1	2	2	1
8	8, 2, 4	3	3	1
9	9	1	1	1
10	10, 6, 5	3	3	1
11	11, 9, 3	3	3	1

7 Experimental Results

SRS1 and SRS2 algorithms were implemented in Java and experiments were conducted on a PC with 128Mb memory and 733MHz CPU speed. Each run of the two algorithms on all testing data sets takes less than one minute.

In the experiments, we first applied the algorithms to three well-known data sets: Iris (Fisher's Iris Plant Database), Glass (Glass Identification Database) and Pima (Pima Indians Diabetes Database) to study the interdependency between the classification accuracy, storage and the parameters *ct* and *st*. These data sets are available from the University of California Repository of Machine Learning Databases [14] and are widely used by the machine learning community for empirical analysis of algorithms. The Iris data set has four condition attributes, three decision values and 150 objects; the Glass data set has nine condition attributes, six decision values and 214 objects; and the Pima data set has eight condition attributes, two decision values and 768 objects.

We adopted 10-fold cross validation method to evaluate the accuracy of the learned classification systems. The classification accuracy of the derived casebase is obtained in the following way. We divided the data set into 10 equal subsets. Then selected one subset as the testing set, and the union of the other nine sets as training set. We repeated this process a total of ten times. After generating the casebase from the training set, the system then classifies the testing set according to the similarity-based casebased reasoning algorithm.

The experimental results of SRS1 on the Iris data set are shown in Table 5. The first number in each grid of the table represents the number of the selected cases, the second one represents the average classification accuracy on the test data set, and the third one represents the standard deviation of the classification accuracy. Experimental runs were conducted using SRS1 with different similarity and consistency thresholds. The results for the run(s) with the highest classification accuracy are shown in boldface. From the experiments, it was observed that when the similarity threshold *st* increases, the granularity of the classification becomes finer, the number of the selected cases increases, and the classification accuracy usually, but not always, increases. When the consistency threshold *ct* increases, the system tolerates less noisy data and identifies more objects as noise. Therefore, the number of the selected cases decreases. For the Iris data set, we did not observe any interdependency between *ct* and the classification accuracy.

Table 5. Experimental results of running SRS1 on the Iris data set.

ct\st	**0.5**	**0.6**	**0.7**	**0.8**	**0.9**
0.5	8, 92.7, 7.6	13, 94.7, 5.0	20, 98.0, 4.3	40, 95.3, 4.3	92, 93.3, 4.2
0.6	8, 92.0, 7.8	13, 96.0, 4.4	**20, 98.0, 4.3**	40, 96.0, 3.3	92, 93.3, 4.2
0.7	7, 88.7, 7.9	13, 96.0, 4.4	**20, 98.0, 4.3**	40, 96.0, 3.3	92, 93.3, 4.2
0.8	6, 89.3, 6.8	12, 96.7, 3.3	**20, 98.0, 4.3**	40, 96.0, 3.3	92, 93.3, 4.2
0.9	6, 90.0, 5.4	10, 95.3, 6.0	18, 96.7, 4.5	38, 96.0, 3.3	92, 93.3, 4.2
1.0	6, 90.7, 5.3	10, 94.0, 6.3	18, 96.7, 4.5	38, 96.0, 3.3	92, 93.3, 4.2

Table 6 presents the experimental results of SRS2 on the three data sets. As in Table 5, the first number in each grid represents the number of the selected cases, and the second and third numbers represent the classification accuracy and standard deviation of the derived casebase, respectively. However, for SRS2, only one parameter, the consistency threshold *ct,* requires definition. When *ct* increases, it is observed that the number of the selected cases increases. This result is in contrast to the experiments with SRS1, where the number of the selected cases decreases when *ct* increases. This is because in SRS1, the similarity threshold and consistency threshold are given by the user and they have no inter-dependency, while in SRS2, the similarity threshold is determined directly by the consistency threshold. When the consistency threshold increases, the tolerance for noisy data decreases. Therefore, in the process of selection, the granularity of the clusters of cases becomes finer, and more objects are selected. In this experiment, we also observe that the optimal *ct* value depends on the data set itself.

Table 6. Experimental results of SRS2 on three data sets.

ct\data set	Iris	Glass	Pima
0.5	5, 78.7, 9.3	40, 47.6, 8.5	3, 34.2, 6.0
0.6	5, 79.3, 9.2	52, 47.7, 11.1	4, 39.7, 9.6
0.7	4, 75.3, 9.5	56, 53.7, 5.1	20, 59.0, 10.5
0.8	6, 73.3, 8.4	61, 62.1, 7.2	44, 65.5, 9.1
0.9	5, 88.0, 11.1	**76, 63.5, 7.5**	70, 68.5, 7.3
1.0	**10, 96.7, 3.3**	75, 63.2, 6.3	**127, 73.0, 6.6**

The performance of SRS1 and SRS2 was compared with other classification systems. Table 7 compares the classification accuracy obtained using the SRS1 and SRS2 algorithms with those of four well known data mining systems: the tree induction algorithm C4.5 [19], layered Artificial Neural Network (ANN) [20], Instance Based Learning 3 (IB3) [1] and rule induction algorithm LEM2 [6]. The data in the first four rows were obtained from [11]. The data of the last two rows are the best average accuracy for SRS1 and SRS2 obtained using the testing methods described above in this section.

Table 7. Comparison of the accuracy with other classification algorithms.

Algorithm	Iris	Glass	Pima
C4.5	95.5	67.9	70.8
ANN	95.3	65.0	**76.4**
IB3	96.7	65.4	68.2
LEM2	94.0	66.8	62.0
SRS1	**98.0**	**69.7**	74.2
SRS2	96.7	63.5	73.0

From Table 7, it can be seen that the system classification accuracy for both SRS1 and SRS2 algorithms approximate that of other well-known classification systems, provided the parameters are properly selected.

Finally we compare the performance of SRS1 and SRS2 with other state of art case selection algorithms, RT3 and ICF. Table 8 summarizes the classification accuracy (Acc) obtained and storage required using RT3, ICF, SRS1, and SRS2, on nine data sets. The results for RT3 and ICF were obtained from [2].

Table 8. Comparison of accuracy and storage with *RT3* and *ICF*.

Dataset	**RT3**		**ICF**		**SRS1**		**SRS2**	
	Acc	Storage	Acc	Storage	Acc	Storage	Acc	Storage
Balance-scale	83.4	18.2	81.5	14.7	**90.0**	**5.2**	79.1	**5.2**
Breast cancer-w	95.3	3.1	95.1	4.3	**96.4**	3.3	95.1	**2.2**
Ecoli	**82.8**	15.8	81.3	14.0	76.5	**6.6**	73.6	31.5
Glass	69.1	**23.3**	69.6	31.4	**69.7**	47.7	63.5	39.4
Iris	93.6	16.0	92.6	42.0	**98.0**	14.8	96.7	**7.4**
Pima	71.0	22.4	69.2	**17.2**	**74.2**	18.7	73.0	18.4
Voting	93.8	7.4	91.2	8.9	**94.0**	**1.3**	92.2	7.9
Wine	86.4	15.4	83.8	12.0	**95.5**	8.8	94.3	**8.1**
Zoo	87.1	26.1	92.4	52.8	**96.1**	26.4	89.3	**11.0**

8 Conclusion

In this chapter, we proposed two similarity-based rough set algorithms called SRS1 and SRS2 for selecting representative cases from data sets. These algorithms are elegant in that they need only one or two parameters to be specified before the selection process. SRS1 requires the user to input the consistency threshold and the similarity threshold. SRS2 requires only the consistency threshold from the user, because it can automatically choose an appropriate similarity threshold for each object. Therefore, it requires fewer runs and saves the user's time. Furthermore, SRS2 generates fewer cases than SRS1. Both SRS1 and SRS2 can deal with noise and inconsistent data in the database. The selected cases are representative in that they can cover all of the data in the data set in terms of the similarity relation, and at the same time, these cases are not similar to each other. The experimental results indicate that the classification accuracy for both systems is comparable to classical machine learning systems.

Another advantage of these algorithms is that they are highly scalable. We can divide a large data set into smaller ones, and select the cases from those subsets by applying these algorithms. Then, collecting all the selected cases together, we apply the algorithms again to select cases for the second round. Take SRS1 for example. Suppose there are m attributes and n objects in the data set. We divide it evenly into p subsets. Each subset has $q = [n/p]$ objects. So, applying SRS1 to each subset, we have complexity $O(mq^2)$. For all the subsets, complexity is $O(mnq)$. Then suppose we selected n' cases in total, then working on it, the complexity for the second round is $O(mn'^2)$. In total, the complexity is $O(mnq+mn'^2)$. When q and

n' are far less than n, the complexity is improved for $O(mn^2)$.

In the experiments reported here, we only used accuracy first case selection strategy. In the future we will adopt other heuristics, such as coverage first strategy or dynamic selection strategy, and try more similarity measures. We will apply these algorithms to data sets containing complex data types. We will also incorporate attribute reduction and weight assignment algorithms to make the generated cases more compact.

References

1. Aha, D.W., Kibler, D., and Albert, M.K., Instance based learning algorithms. *Machine Learning*, 6, 37-66, 1991.
2. Brighton, H. and Mellish, C., Advances in instance selection for instance-based learning algorithms. *Data Mining and Knowledge Discovery*, 6, 153-172, 2002.
3. Cao, G., Shiu, S., and Wang, X., A fuzzy-rough approach for case base maintenance. *Proceedings of the Fourth International Conference on Case-Based Reasoning*, 118-130, 2001.
4. Chan, C., Chen, L., and Geng, L., Knowledge engineering for an intelligent case-based system for help desk operations. *Expert Systems with Applications*, 18, 125-132, 2000.
5. Ganesan, K., Khoshgoftaar, T.M., and Allen, E.B., Case-based software quality prediction. *International Journal of Software Engineering and Knowledge Engineering*, 10 (2), 139-152, 2000.
6. Grzymala-Busse, J.W., LERS - a system for learning from examples based on rough sets. Slowinski, R. (ed.), *Intelligent Decision Support*, Kluwer Academic, 3-18, 1992.
7. Han, J. and Kamber, K. *Data Mining: Concepts and Techniques*. Morgan Kaufmann, 2000.
8. Hu, X. and Cercone, N., Rough sets similarity-based learning from databases. *Proceedings of the First International Conference on Knowledge Discovery and Data Mining*. 162-167, 1995.
9. Jurisica, I. and Glasgow, J., Extending case-based reasoning by discovering and using image features in IVF. *Proceedings of the 16th ACM SAC2001 Symposium on Applied Computing*, 52-59, 2000.
10. Ketler, K. Case based reasoning: an introduction. *Expert Systems with Applications*, 6, 3-8, 1993.
11. Krawiec, K., Slowinski, R., and Vanderpooten, D., Learning decision rules from similarity based rough approximations. Polkowski, L, Skowron, A. (eds.), *Rough Sets in Knowledge Discovery*, vol. 2, Heidelberg; New York: Physica-Verlag, 37-54, 1998.
12. Li, J., Dong, G., and Ramamohanarao, K., Instance-based classification by emerging patterns. *Proceedings of the Fourth European Conference on Principles and Practice of Knowledge Discovery in Databases*. 191-200, 2000.

13. Li, J., Ramamohanarao, K., and Dong, G., Combining the strength of pattern frequency and distance for classification. *Proceedings of the Fifth Pacific-Asia Conference on Knowledge Discovery and Data Mining*, 455-466, 2001.
14. Merz, C.J. and Murphy, P.M., UCI Repository of machine learning databases. University of California, Irvine, Department of Information and Computer Science, 1996.
15. Mrozek, A. and Skabek, K. Rough sets in economic applications. Polkowski, L., Skowron, A. (eds.), *Rough Sets in Knowledge Discovery*, vol. 2, Heidelberg; New York: Physica-Verlag 238-271, 1998.
16. Pal, S. K. and Mitra, P., Case generation: a rough-fuzzy approach. *Proceedings of Workshop Program at the 4th International Conference on Case-Based Reasoning*, 236-242, 2001.
17. Park, H., Oh, J., Jeong, D, and Park, K., Automated sleep stage scoring using hybrid rule- and case-based reasoning. *Computers and Biomedical Research*, 33(5), 330-349, 2000.
18. Pawalk, Z., *Rough Sets: Theoretical Aspects of Reasoning about Data*, Kluwer Academic, Dordrecht, 1991.
19. Quinlan, J.R., *C4.5: Programs for Machine Learning*. Morgan Kaufmann, San Mateo CA, 1988.
20. Rumelhart, D.E., Hinton, G.E., and Williams, R.J., Learning internal representations by error propagation. Rumelhart, D.E, McClelland, J.L. and the PDP Research Group (eds.), *Parallel Distributed Processing: Explorations in the Microstructure of Cognition*, MIT Press, Cambridge, MA, 318-362, 1986.
21. Slowinski, R. and Vanderpoonten, D., Similarity relation as a basis for rough approximations. *Advances in Machine Intelligence and Soft Computing*, 4, 17-33, 1997.
22. Smyth, B. and McKenna, E., Building compact competent case-bases. *Proceedings of the Third International Conference on Case-based Reasoning*, 329-342, 1999.
23. Wang, W., New similarity measures on fuzzy sets and on elements. *Fuzzy Sets and Systems*, 85, 305-309, 1997.
24. Wilson, D.R. and Martines, T.R., An integrated instance-based learning algorithm. *Computational Intelligence*, 16(1), 1-28, 2000.
25. Wilson, D.R. and Martines, T.R., Instance pruning techniques. *Proceedings of the Fourteenth International Conference on Machine Learning* 404-411, 1997.
26. Yang, Q. and Wu J., Keep it simple: a case-base maintenance policy based on clustering and information theory. *Proceedings of the Thirteenth Canadian AI Conference*, 102-114, 2000.
27. Ziarko, W., Probabilistic decision tables in the variable precision rough set model. *Computational Intelligence*, 17(3), 593-603, 2001.

Chapter 6

An Induction Algorithm with Selection Significance Based on a Fuzzy Derivative

Musa A. Mamedov and John Yearwood

Summary. In this chapter we introduce a new version of the machine learning algorithm, FDM, based on a new notion of the fuzzy derivative. The main idea is to describe the influence of the change of one parameter on another. In this algorithm we generate sets of classification rules. We define a coefficient of significance for every single rule. This coefficient describes for a test example a degree of membership of the class predicted by the rule. A new example is classified into a class for which its total degree of membership is maximal. In this way the effect of a single non-informative rule having occurred by chance is decreased due the coefficient of significance. The fuzzy derivative method is mainly used to study systems with qualitative features, but it also can be used for systems with quantitative features. The algorithm is applied to classification problems and comparisons made with other techniques.

Keywords: fuzzy derivative, fuzzy sets, data classification, rule induction, adverse drug reaction.

1 Introduction

In this chapter we develop a machine learning algorithm, FDM, introduced in [11], [12]. This method is based on the notion of a fuzzy derivative introduced in [10]. This definition is quite different from the notions used in the literature, where fuzzy derivatives are defined by using an ordinary derivative of some functions (see [4], [7], [8]). In [10] the notion of fuzzy derivative is used as a mathematical tool for describing a relationship between two parameters. The main idea is to describe the influence of the change of one parameter on another and from our point of view this is a more natural development of the fuzzy set idea introduced by L.Zadeh [17]. In this chapter we study the relationship between more than two parameters and the fuzzy derivative describes the influence of the changes of some combination of parameters on a chosen single parameter.

Every classification method has its own criterion for classification. For example, for methods based on the idea of clustering this criterion may be defined by the distance of a new observation to the centers of clusters [1]. In this work we use another criterion. Our aim is not only to classify a new observation but also to explain why a test point belongs to a particular class. We assume that the classification can be explained by the high or low levels of some features for the test point. To realize this idea we need to solve the following problems:

- to find a discriminating (informative) set of features;
- to determine whether the level of the feature (under consideration) is high or low?

These problems are solved by using the fuzzy derivative. We introduce a method based on incremental separation of sets. At every stage a set of features whose covariability distinguishes the classes is calculated (some of these features may coincide). Therefore the discriminating set of features emerges as a sequence of subsets of the set of all features.

We now consider the second problem. It is meaningful to assume that the given level of a feature cannot be taken as a reason for the classification of a point without knowledge of its previous state. High and low levels would be defined by the difference between the present and previous levels. If the level of a particular feature for some observation has increased (starting from the initial state) then we say that this feature has a high level (relative to the previous level). Similarly we can define a low level: if the level has decreased from the previous level we say that this is a low level. Therefore, in our approach the high and low levels are presented as increases and decreases in levels of a feature. Unfortunately the databases, which are considered below, do not contain information about the previous states of the features. This is why we take the initial level of the chosen feature as the same number (the centroid) for all observations which are presented at each stage. Indeed, the result of classification will be better if we can define the initial state more correctly. Taking the initial states as the centroids we define the high and low levels of the features.

It is important to note that the definition of high and low levels is changed stage by stage, according to the initial point (the centroid) by which the notions of increase and decrease are defined. Therefore in this way we define the high and low levels in the context of the situation (i.e., relative to the set of observations). The same level may be high for one case and low for another.

2 The Fuzzy Derivative and Some Calculation Methods

Let us consider n objects and assume that the states of these objects can be described by scalar variables x_i, $i = 1, ..., n$ The increase and decrease of these variables indicates changes in the objects.

Let x_i^0 be the initial states of the objects.

Definition 1. If $x_i > x_i^0$ ($x_i < x_i^0$, respectively) we say that for the i-th object the parameter x_i increases (decreases, respectively).

The events x_i increases and decreases will be denoted by $x_i \uparrow$ and $x_i \downarrow$, respectively.

Consider an event E related to the parameters $x_1, ..., x_i$. For example, this event can be $E = (x_1 \uparrow, x_2 \uparrow, ..., x_i \uparrow)$, which means that all parameters are increasing. We will study the influence of this event on another parameter x_j.

As a result of the event E the parameter x_j may either increase or decrease. To determine the influence we have to define the degree of these events; that is we need to define both

1. the degree of the increase of x_j, and
2. the degree of the decrease of x_j.

Of course the degree of these events depends on the initial state $(x_1^0, ..., x_i^0, x_j^0)$. We define the degree as a number in the interval [0,1]. This number will be considered as a value of a membership function. Taking different initial points we can define fuzzy sets on the $(i+1)$ dimensional space.

Therefore, we define the degree of the increase of x_j and the degree of the decrease of x_j as fuzzy sets, which are denoted by $d(E, x_j \uparrow)$ and $d(E, x_j \downarrow)$, respectively.

We assume that if, for example, $d(E, x \uparrow) = 1$ then the influence is highest and if $d(E, x \uparrow) = 0$ then the influence is lowest.

Clearly at every initial point $(x_1, ..., x_i, x_j)$ we have

$$0 \leq d(E, x_j \uparrow) + d(E, x_j \downarrow) \leq 2.$$

If $d(E, x_j \uparrow) + d(E, x_j \downarrow) = 1$, this is a case of probability. But in the applications below we see that as a rule the sum of $d(E, x_j \uparrow) + d(E, x_j \downarrow)$ does not equal 1 and this is the main factor which creates the "forces" producing changes in the system (see Proposition 3 [11]).

Therefore the fuzzy sets $d(E, x_j \uparrow)$ and $d(E, x_j \downarrow)$ completely describe the influence of the event E on x_j. Since this influence characterizes the changes of the parameter x_j we call it a *fuzzy derivative* and denote it by $\partial x_j / \partial E$;

$$\partial x_j / \partial E = (d(E, x_j \uparrow), d(E, x_j \downarrow)).$$

Now we give some examples to demonstrate the calculation of the fuzzy derivative. Note that there are no definite rules for calculation. We can suggest quite different methods that may be appropriate to the problem under consideration. Here we focus our attention on classification problems in datasets.

2.1 Consider data $(x_n, y_n, z_n), n = 1, ..., N$. Let (x^0, y^0, z^0) be an initial point.

For the sake of definiteness, we consider the event $E = (x \uparrow, y \downarrow)$ and study the influence of this event on the parameter z.

$$d(E, z \uparrow) = M_1/(M+1), \quad d(E, z \downarrow) = M_2/(M+1), \tag{1}$$

where

M is the number of points (x_n, y_n, z_n), satisfying $x_n > x^0$, and $y_n < y^0$;
M_1 is the number of points (x_n, y_n, z_n), satisfying $x_n > x^0$, $y_n < y^0$ and $z_n > z^0$;
M_2 is the number of points (x_n, y_n, z_n), satisfying $x_n > x^0$, $y_n < y^0$ and $z_n < z^0$.

Example 1. Let data (x_n, y_n, z_n) be as in Table 1.

Table 1. Data used in Example 1.

x	7	5	10	6	2	9	11	2	6	11	9	3	7	8	10	6	4	7	3
y	5	3	8	7	4	4	3	7	3	6	7	5	7	4	4	5	6	3	8
z	2	9	7	5	8	6	7	3	5	8	1	0	2	8	5	9	3	7	2

Consider an initial point $(x^0, y^0, z^0) = (7, 5, 2)$ and events $E_1 = (x \downarrow)$, $E_2 = (x \downarrow, y \uparrow)$. We calculate the values of the fuzzy derivatives $\partial y/\partial E_1$ and $\partial z/\partial E_2$ at the initial point $(7, 5, 2)$:

$$\partial y/\partial E_1 = (4/9,\ 3/9) = (0.44,\ 0.33),$$
$$\partial z/\partial E_2 = (3/4,\ 0/4) = (0.75,\ 0.00).$$

2.2. Consider a dataset $A = \{(a^n) = (a_1^n, a_2^n, ..., a_F^n),\ n = 1, ..., N\}$, where N is the number of all observations and F is the number of features (attributes). Assume that the dataset A consists of m classes: $A^i = \{(a^{i,n}) = (a_1^{i,n}, a_2^{i,n}, ..., a_F^{i,n}),\ n = 1, ..., N^i\}$, $\sum_{i=1,m} N^i = N$.

We now present methods for calculating the fuzzy derivatives at a certain point (the centroid). The fuzzy derivatives in these applications describe the influence of a combination of features on a particular single feature.

First we calculate the centroid (x_j), $j = 1, ..., F$ of the set A. We use the following formulas:

$$x_j = (a_j^1 + a_j^2 + ... + a_j^N)/N; \tag{2}$$
$$x_j = (a_j^1 + a_j^2 + ... + a_j^N)/(N+1). \tag{3}$$

It is interesting that the formula (3) sometimes gives better results than (2).

Take any combination of features, say $(j_1, j_2, ..., j_l)$, and a feature $j \in \{1, 2, ..., K\}$. Let E be a given event related to the chosen combination of features.

For each class A^i we calculate the influence of the event E on j as the values of fuzzy derivatives

$$(\partial j/\partial E)^i = (d^i(E, j\uparrow), d^i(E, j\downarrow)),\quad i = 1, ..., m,$$

computed at the centroid $(x_{j_1}, x_{j_2}, ..., x_{j_l}, x_j)$, that is taken as an initial point. We present two different methods that we call separate and combined.

Separate calculation.

$$d^i(E, j\uparrow) = \frac{K_{i,1}}{K_i + 1},\quad d^i(E, j\downarrow) = \frac{K_{i,2}}{K_i + 1},\qquad i = 1, ..., m; \tag{4}$$

Combined calculation.

$$d^i(E, j\uparrow) = \frac{K_{i,1}}{K + 1},\quad d^i(E, j\downarrow) = \frac{K_{i,2}}{K + 1},\qquad i = 1, ..., m. \tag{5}$$

Here $K = \sum_{i=1,n} K_i$ and K_i is the number of observations $a \in A^i$ satisfying E; $K_{i,1}$ is the number of observations $a \in A^i$ satisfying E and an inequality $a_j > x_j$; $K_{i,2}$ is the number of observations $a \in A^i$ satisfying E and an inequality $a_j < x_j$.

Note that in the applications to data classification we can apply both the separate and combined calculations. For example, in [11] for the liver-disorder database, better results were obtained by the separate calculation, but for the heart disease database using the combined calculation gave higher accuracy in the classification. Therefore in applications we should use different methods related to the problems under consideration to calculate fuzzy derivatives. In this work we use the combined calculation.

3 Applications to Classification Problems

The main idea of the classification algorithm is to examine the different way in which features change together over the classes. That is, rather than discriminate between the classes on the basis of the values of features we discriminate on the basis of the way in which features influence each other over the classes. The approach does not attempt to consider the variation of all features simultaneously but rather select the subsets of features that display the maximum difference in their fuzzy derivatives between the training sets.

Machine learning algorithms in general have some common points. First of all we should note the kind of classification rules generated by these algorithms. Classification rules can be defined either as a decision tree or as a set of classification rules (decision lists). In the second case classification rules may be either ordered or unordered. For example, the ID3 algorithm produces decision trees and for classification a path from the root of the decision tree to a leaf node is traced. Some machine learning algorithms (such as AQ algorithms, CN2 and so on) induce a set of classification rules for the classification. In the second case we need to solve a very important problem about the reliability and sensibility of each induced new rule. In order to solve this problem statistical ideas are used. For example CN2 [6] uses the information-theoretic entropy measure to evaluate rule quality and also uses the likelihood ratio statistic [9] to test significance.

In the algorithm FDM2 we use the following. The Algorithm produces ordered classification rules in the form "if ... then predict class ...." In the generation of each rule first we define a dominant (most common) class and then step by step we add new conjunctive terms until the (confidence) number $\alpha = 100n_c/(n+1)$ is greater than a given number for minimal accuracy for the training set. Here n_c is the number of observations in the dominant class and n is the number of all observations which are covered in this stage. Therefore we use an AQ evaluation function as search heuristic. Having chosen a classification rule (complex) we remove all examples that it covers from the training set. We repeat this process until no examples remain unclassified in the training set. In this manner we obtain a set of ordered classification rules.

Using an AQ evaluation function in producing a single classification rule may cause the generation of some non-informative rules which have occurred by chance. Then for every rule we define a number $D = n_c/(n-n_c+1)$. We call it a coefficient of significance for the chosen rule, or a degree of membership in the predicted dominant class. The algorithm FDM2 does not use other statistical methods for choosing more significant rules (for instance, CN2 uses a second evaluation function). FDM2 also generates other sets of classification rules. A new example is classified by each set of classification rules, but is classified to the class, which has maximal sum of degrees in prediction. Therefore, for classification of a new example FDM2 combines ideas used in both ordered and unordered classification rules.

Consider an example. Assume that 3 sets of ordered classification rules are generated and x is a test point. We also assume that by the first set of ordered classification rules the point x would be classified in class i by a rule which covers 100 positive and 0 negative (n_c = 100, $n - n_c$ = 0) examples. The coefficient of significance in this prediction is D_1 = 100/(0+1) = 100. Suppose by the second and third sets the point x would be classified in class j by rules which cover 50 positive and 1 negative examples (that is, n_c = 50, $n - n_c$ = 1 and D_2 = 50/(1+1) = 25), and 70 positive and 0 negative examples (that is, n_c = 70, $n - n_c$ = 0 and D_3 = 70/(0+1) = 70), respectively. After summing the coefficient of significances we have: the degree of prediction to the class i is 100, and the degree of prediction to the class j is 95 (= 25 + 70). Therefore, the point x would be classified in class i.

An alternative approach is to define the coefficient of significance as $ln(D + 1)$ instead of D. In this case, for this example, after summing the coefficient of significances we have: the degree of prediction to the class i is 4.6 (= $ln(100 + 1)$), and the degree of prediction to the class j is 7.5 (= ln(25+1) + ln(70+1)). Therefore, the point x would be classified in class j. Calculations have shown that the former approach (used here) provides the better results than if we use the logarithm.

3.1 The FDM2 Classification Algorithm

Consider a data (training) set A which consists of m classes A^i $(i = 1, ..., m)$. Let $J = \{1, 2, ..., F\}$ be a set of features.

The algorithm requires two parameters. The number $\alpha > 0$ is used for the minimal (or threshold) accuracy in the training phase. The number β is used in the definition of the dominant class. The algorithm generates sets of classification rules by taking different numbers α and β. Now we describe the FDM2 algorithm for set numbers α and β.

Step 1. Let $k = 1$ and $A^i_k = A^i, \ i = 1, ..., m$.

Step 2. For every fixed feature $j \in J$ we calculate the centroids of the sets A^i_k, $i = 1, ..., m$. Denote by j_{min} and j_{max} the minimal and maximal numbers of these centroids. Consider two intervals $J_1(j) = (-\infty, j_{min}]$ and $J_2(j) = [j_{max}, +\infty)$.

Denote by $N(J_p(j), A^i_k)$ the number of all observations of the set A^i_k that belong to the interval $J_p(j), p = 1, 2$. Let

$$N^0_p = N(J_p(j), A^1_k) + N(J_p(j), A^2_k) + ... + N(J_p(j), A^m_k) + 1,$$

$$N_p(j, i) = N(J_p(j), A^i_k)/N^0_p, \text{ and}$$

$$N^* = \max_{j=1,F} \max_{p=1,2} \max_{i=1,m} N(J_p(j), A^i_k)/N^0_p.$$

We choose j_k^1, J_k^1, and i_k, such that the number N_p^0 is maximal among all features $j \in \{1, ..., F\}$, intervals $J_p(j), (p = 1, 2)$ and classes $i \in \{1, ..., m\}$, for which $N_p(j, i) \geq N^* - \beta$.

Therefore the class i_k is the dominant class at this stage. We define the event $E(j_k^1)$ related to the interval J_k^1.

Step 3. Set $s = 3$. Denote by $A_k^i(s-2)$ the set of all observations in the class A_k^i for which $E(j_k^1, ..., j_k^{s-2})$ is satisfied.

Step 4. For every feature $j \in J$ we calculate a centroid (say x_j) of the set $A_k^1(s-2) \cup A_k^2(s-2) \cup ... \cup A_k^m(s-2)$. Denote $E = E(j_k^1, ..., j_k^{s-2})$.

We take a pair of features j^1, j^2 $(j^1 \neq j^2)$ and study the influence of j^1 on j^2 over the classes. First we define the events $E_+ = E \cap \{j^1 \uparrow\}$ and $E_- = E \cap \{j^1 \downarrow\}$.

Then we calculate the values of the fuzzy derivatives $(\partial j^2/\partial E_+)^i$ and $(\partial j^2/\partial E_-)^i$ at the point (x_{j^1}, x_{j^2}).

Let

$$(\partial j^2/\partial E_+)^i = (d_1^i(j^1, j^2), d_2^i(j^1, j^2)) \text{ and}$$
$$(\partial j^2/\partial E_-)^i = (d_3^i(j^1, j^2), d_4^i(j^1, j^2)),$$

with $i = 1, ..., m$. We are looking for the maximum discrimination of the class i_k from other classes. We choose a pair of features (j_k^{s-1}, j_k^s) and a number $l_k^s \in \{1, 2, 3, 4\}$ solving the following problem

$$\max_{j^1, j^2 \in J} \max_{l=1,2,3,4} f(i_k | \, d_l^1(j^1, j^2), d_l^2(j^1, j^2), ..., d_l^m(j^1, j^2)) \rightarrow \max .$$

We define $E(j_k^1, ..., j_k^s)$ corresponding to the solution (j_k^{s-1}, j_k^s) and l_k^s. Denote by $A_k^i(s)$ the set of all observations in the class $A_k^i(s-2)$ for which $E(j_k^1, ..., j_k^s)$ is satisfied. Let $N^i(s)$ be a number of observations in $A_k^i(s)$ and

$$N(s) = 1 + \sum_{i=1,m} N^i(s), \quad N_-(s) = 1 + \sum_{i=1,m; i \neq i_k} N^i(s).$$

If $100 N^{i_k}(s)/N(s) \geq \alpha$ or $s \geq F$ then go to *Step 5*. Otherwise we set $s = s + 2$ and go to *Step 4*.

Step 5. In this step we have the following classification rule: a new example a, for which $E(j_k^1, ..., j_k^s)$ is satisfied, belongs to the class i_k. The degree of membership in this class is defined as

$$D_k = \frac{N^i(s)}{N_-(s)}.$$

We denote by A_{k-}^i all observations in the class A_k^i satisfying this classification rule. Set $A_{k+1}^i = A_k^i \setminus A_{k-}^i$. The algorithm terminates if $A_k^i = \emptyset$ for all $i = 1, ..., m$. Otherwise we set $k = k + 1$ and go to *Step 2*.

Note that the function $f(i_k | \, c_1, ..., c_m)$ can be defined by different formulas. We use the following function

$$f(i_k | \, c_1, ..., c_m) = \sum_{i \neq i_k} (\max\{0, c_{i_k} - c_i\})^q.$$

It is well known that better results can be achieved if $q \geq 2$ (see, for example, [6]). This means that, for example, in the case of 4 classes with the first class being dominant, we prefer the distribution of (7, 3, 0, 0) to (7, 1, 1, 1). In the calculations below we take $q = 4$ which provides better results.

3.2 Arrangement of the Features According to Their Informativeness

Assume that A is a training set, which consists of m classes A^i, $i = 1, ..., m$, and $J = \{1, 2, ..., F\}$ is a set of features. We arrange the features according to their informativeness in two different ways.

$\mathcal{I}1$. We scale the training data A to $\mathcal{A}$ such that $x_i = x_j$ for all $i, j \in J$, where x_j is a midpoint of the set of j-th components of all examples in $\mathcal{A}$. Then for all $j \in J$ we calculate midpoints x_j^i corresponding to the classes $\mathcal{A}^i$, $i = 1, ..., m$. Let $d(j) = \sum_{i,k=1,m} |x_j^i - x_j^k|$. We arrange features in the form $(j_1, j_2, ..., j_F)$, such that $d(j_1) \geq d(j_2) \geq ... \geq d(j_F)$.

$\mathcal{I}2$. We apply the FDM2 algorithm to the training set A, taking different combinations of features $(j^1, j^2, ..., j^K)$. Assume that C and N stand for the numbers of correctly classified and misclassified points, respectively. Denote $P(j^1, j^2, ..., j^K) = C - N$.

We take any feature $j \in J$ and calculate the number $P(j)$ applying the classification algorithm to the training set A. We choose a feature j_1 such that the number $P(j_1)$ is maximal.

Then we consider the same problem with 2 features (j_1, j) where the first feature is fixed and the second feature $j \in J \setminus \{j_1\}$. We choose j_2 such that both the number $P(j_1, j_2)$ is maximal and $P(j_1, j_2) \geq P(j_1)$.

Assume that k features $(j_1, j_2, ..., j_k)$ are chosen. To define the next feature j_{k+1} we consider the same problem with $(k+1)$ features $(j_1, j_2, ..., j_k, j)$, where the first k features are fixed and $j \in J \setminus \{j_1, j_2, ... j_k\}$. We choose j_{k+1} such that both the number $P(j_1, j_2, ..., j_{k+1})$ is maximal and $P(j_1, j_2, ..., j_{k+1}) \geq P(j_1, j_2, ..., j_k)$.

We repeat this procedure until either all features are used or $P(j_1, j_2, ..., j_{k+1}) < P(j_1, j_2, ..., j_k)$ for some number $k < F$.

In this way, step by step, we arrange the features according to their informativeness:

$$(j_1, j_2, j_3, ..., j_{F'}), \quad (F' \leq F).$$

Sometimes we need to consider only informative features in the application to classification problems. So we shall consider our algorithms with different numbers of features. Experiments show that there may exist a number $S \leq F$ such that the algorithm with the first S features provides better results than if we take $S' \neq S$ features. In this case using superfluous features contribute to noise.

4 Results of Numerical Experiments

In this section we give results of the numerical experiments. We use the following notation:

- F' is the index of the last feature out of the ordered set of features;
- p_{tr} is the accuracy for the training set;
- p_{ts} is the accuracy for the test set.

The code has been written in Fortran-90 and numerical experiments have been carried out on a Pentium III PC with 800 MHz main processor. The results obtained by the algorithm FDM2 are presented in Tables 2 - 7. In these experiments the values for α (α = 95 - 100) are chosen because the datasets used are not very noisy. The values for β (between 1 and 5) in this work are chosen so that there are a large number of examples for the dominant class. This number depends on the number of features and the number of examples in the dataset.

In Tables 3, 5 and 7 we present the performance of some selected single sets of classification rules and for comparison results from combinations of sets of rules.

For comparison of the results of numerical experiments we present results obtained by other methods. We use the results presented in [15] and in particular, we present the results obtained by machine learning algorithms CN2, Cal5, C4.5, AC^2, NewID (Table 8).

4.1. Landsat satellite image (Satim) [16]. This database contains 6 classes, 36 attributes, 4,435 and 2,000 observations for the training set and the test set, respectively. We arrange the features according to their informativeness by $\mathcal{I}1$. In the calculations we take the first 27 features (plots 1, 2, 4) and all 36 features (plots 1, 2, 3, 4). The results are presented in Tables 2 and 3.

Table 2. The results obtained for the Satim database by FDM2 with α = 95 - 100 (* means that the centroids are calculated by (3)).

		$F'=27$		$F'=36$	
	β	p_{tr}	p_{ts}	p_{tr}	p_{ts}
*	1 - 5	99.8	89.5	99.8	89.5
	2 - 5	99.7	89.3	99.9	88.9
	1 - 5	99.9	88.8	100	89.8
	2 - 5	99.8	88.5	99.9	**90.0**

4.2. Letter recognition (Letter) [16]. This database contains 26 classes, 16 attributes, 15,000 and 5,000 observations for the training set and the test set, respectively. We arrange the features according to their informativeness by $\mathcal{I}1$ and obtain

$$J = (13, 15, 8, 9, 11, 7, 5, 12, 6, 10, 3, 14, 1, 16, 4, 2).$$

In the calculations we take the first 14 and all 16 features. The results are presented in Tables 4 and 5.

Table 3. The results obtained for Satim database by FDM2 for different numbers β and $\alpha = 95$ (the centroids are calculated by the formula (3)).

	$F' = 27$		$F' = 36$	
β	p_{tr}	p_{ts}	p_{tr}	p_{ts}
1	98.0	86.2	98.3	85.5
2	98.7	84.7	98.5	85.2
3	98.6	84.4	98.4	83.8
4	98.4	84.5	98.6	84.5
5	98.0	86.3	98.3	86.2
1 - 2	98.5	86.8	98.7	86.8
1 - 3	98.7	87.6	98.9	87.5
1 - 4	98.7	88.0	98.9	88.7
1 - 5	98.7	88.8	99.1	88.9

Table 4. The results obtained for the Letter database by FDM2 with $\alpha = 95$ - 100 (* means that the centroids are calculated by the formula (3)).

		$F' = 14$		$F' = 16$	
	β	p_{tr}	p_{ts}	p_{tr}	p_{ts}
*	0 - 5	99.9	**93.1**	99.9	92.6
*	1 - 5	99.9	92.9	99.9	92.7
*	1 - 7	99.9	93.0	99.9	92.9
	1 - 5	99.9	92.8	99.9	92.7
	1 - 7	99.9	92.9	99.9	92.9

Table 5. The results obtained for the Letter database by FDM2 for different numbers β and $\alpha = 100$ (the centroids are calculated by the formula (3)).

	$F' = 14$		$F' = 16$	
β	p_{tr}	p_{ts}	p_{tr}	p_{ts}
0	99.1	87.4	99.2	86.2
1	99.2	87.1	99.5	86.6
2	99.6	86.8	99.7	86.0
3	99.2	86.3	99.1	86.4
4	99.4	86.7	99.3	85.2
5	99.4	86.3	99.3	85.7
0 - 1	99.7	88.9	99.9	88.1
0 - 2	100	90.2	100	89.3
0 - 3	99.9	91.1	100	90.1
0 - 4	100	91.5	100	90.6
0 - 5	100	91.8	100	90.9

4.3. DNA [16]. This database contains 3 classes, 180 attributes, 2,000 and 1,186 observations for the training set and the test set, respectively. First we arrange the features according to their informativeness by $\mathcal{I}1$ and in the calculations we take the first 29, 31, 33 and all 180 features. Then we arrange the features according to their informativeness by $\mathcal{I}2$ and in the calculations we take the first 10 features - 90, 105, 85, 93, 94, 84, 82, 96, 95, 97. The results are presented in Tables 6 and 7.

Table 6. The results obtained for the DNA database by FDM2 with $\alpha = 95$ - 100 ($\mathcal{I}1$).

	$F' = 29$		$F' = 31$		$F' = 33$		$F' = 180$	
β	p_{tr}	p_{ts}	p_{tr}	p_{ts}	p_{tr}	p_{ts}	p_{tr}	p_{ts}
1 - 10	99.8	94.1	99.8	94.3	99.8	**94.4**	100	91.9
1 - 15	99.8	**94.4**	99.8	94.1	99.9	93.9	100	91.3

Table 7. The results obtained for the DNA database by FDM2 for $F' = 10$ ($\mathcal{I}2$).

β	α	p_{tr}	p_{ts}
0	100	95.3	94.1
1	100	95.3	94.0
0	95	95.2	94.1
1	95	95.2	94.1
0	95 - 100	95.3	94.3
0 - 5	100	95.2	94.2
0 - 5	95	95.2	94.2
1 - 5	95 - 100	95.3	94.3

Table 8. Accuracy for the test set obtained by other methods.

Data	Best result from [15]	CN2	Cal5	C4.5	AC^2	NewID
Satim	90.6	85.0	84.9	85.0	84.3	85.0
Letter	93.6	88.5	74.7	86.8	75.5	87.2
DNA	95.9	90.5	86.9	92.4	90.0	90.0

Therefore, for each database considered above the accuracy obtained by the algorithm FDM2 is close to the best result. For example, a set of 30 single sets of classification rules obtained for the numbers $\beta = 1 - 5$ and $\alpha = 95 - 100$ gives the following accuracy for the test set: for the Satim database - 89.8 (the best result from [15] is 90.6), for the Letter database - 92.7 (best - 93.6) and for the DNA database using just 10 features - 94.3 (best - 95.9).

5 Applications to the Australian Adverse Drug Reaction Database

A detailed description of this database (ADRAC), which contains about 140,000 records (involving 7416 different drugs and 1392 reactions), is given in [13].

Our main concern with Adverse Drug Reactions is to study associations between drugs and reactions. This is a very complicated problem and many approaches have been tried for the analysis of adverse reactions on the basis of databases collected from different countries.

In [13] the following approach is used:

1. "Meta-reactions" are defined as a combination of similar reactions in order to reduce the number of reactions.

2. For each drug a vector of degrees of association with a meta-reaction class is defined. The drug-reaction relations are presented by these vectors of degrees, called Hidden (Potential) reactions, because they give insight into potential reactions.

This approach was carried out on one part of the ADRAC data (called Card1), which contains all records having just Cardiovascular reactions. These reactions were grouped into 5 meta-reaction classes (the corresponding data was called Card2). Then for each drug (1750 drugs were used in Card2) a vector of degrees is calculated. Note that in the calculation of the vector of degrees different methods can be used and the study of *optimal* methods for the calculation is a very important problem.

In this work we examine the drug-reaction relationships, presented by the vectors of degrees calculated in [13], in terms of the prediction of *good* and *bad* outcomes. Note that in Card2 "outcome" information is presented by the feature "REACTION-OUTCOME" (see Section 9, [13]), which takes 8 different values. We define good and bad outcomes as follows. The values 1 (A - recovered without sequel), 2 (B - recovered with sequel), 5 (E - recovered without treatment) were taken as a good outcome and the values 3 (C - death maybe drug), 4 (D - death as a reaction), 6 (F - not yet recovered) were taken as a bad outcome. The records with outcomes 7 (N - unrelated death) and 8 (U - unknown) are not included.

Therefore we have two classes: the first class (good outcomes) contains 2859 records, the second class (bad outcomes) - 1159 records. We use 8 features: the first 3 features - age, weight and sex were included as information about the patient, the next 5 features are components of a vector of reactions.

The vector of reactions $D = (D_i)$, $i = 1, 2, ..., 5$, is used in three versions.

Predicted Potential Reactions. In this case we write the vector of degrees

$$D^p = (D_1^p, D_2^p, \cdots, D_5^p)$$

defined for each patient. We explain the definition of this vector.

In [13] for each drug d a vector of degrees - $\partial(d)$ of association with the 5 meta-reaction classes is defined. The vector $\partial(d) = (\partial_1(d), ..., \partial_5(d))$ has the following meaning. It was assumed that every drug can be the cause of some reactions to different degrees. The number $\partial_i(d)$ ($\partial_i(d) \in [0, 1]$) indicates the degree of the occurrence of the reaction (in fact meta-reaction) i. Therefore, the drug-reaction relationships are presented by the vectors $\partial(d)$ defined for each drug d and our aim here is to examine these relationships.

Now consider a patient a and let A be a set of drugs taken by this patient. The vector $D^p = (D_1^p, D_2^p, \cdots, D_5^p)$ is defined as follows:

$$D_i^p = \sum_{d \in A} \partial_i(d), \quad i = 1, 2, ..., 5.$$

Real Reactions. In this case this vector consists of the numbers 0 and 1, having just one feature with 1, which indicates the reaction that occurred with this patient. For example, if a patient has the second reaction (in fact meta-reaction) the vector of reactions is (0, 1, 0, 0, 0).

Combined vector degrees. Predicted Potential Reactions and Real reactions are combined as follows. Let $D^p = (D_i^p)$ and $D^r = (D_i^r)$ be the predicted potential and real observed reactions, respectively. A combined vector of degrees is defined as $D^c = (D_i^c)$, where

$$D_i^c = \begin{cases} D_i^p & \text{if } D_i^r = 0; \\ \max\{D_{max}, D_i^r\} & \text{if } D_i^r = 1. \end{cases}$$

Note that $D_{max} = \max\{D_1^p, D_2^p, ..., D_5^p\}$. For example, if $D^p = (1.2, 0.0, 0.1, 0.7, 0.0)$ and $D^r = (0, 0, 1, 0, 0)$ then $D^c = (1.2, 0.0, 1.2, 0.7, 0.0)$. Therefore in this case for every patient the reaction that occurred is presented with the maximal number in the vector of degrees of association.

The datasets related to these three versions will be denoted by **P, R, C**, respectively. Note that the classes in these datasets have some common points. That is, there are different records in both classes in which all 8 features coincide. We did not remove these records because our main aim is to examine drug-reaction associations presented by the vector degrees calculated in [13].

We use the following notations:

- p_{tr} is the accuracy for the training set;
- p_{ts} is the accuracy for the test set;
- n_{tr} is the number of records in the training set;
- n_{ts} is the number of records in the test set.

We applied the algorithm FDM2 for classification and in the calculations we take the parameter α as 5 different numbers - 60, 70, 80, 90, 100; the parameter β is taken as different numbers 1, 2, 3, 4, 5. Therefore, 5 x 5 = 25 different sets of classification rules are generated. In all cases the centroids are calculated by the formula (3). Training and test sets are chosen as follows. As a test set we choose all records from a particular year and a training set in this case contains all records from previous years. For example, if the records from the year 2000 are taken as a test set then the records from years 1972 - 1999 form a training set. In the calculations below the records from years 1998, 1999, 2000, 2001 are taken as test sets.

5.1 Informativeness of the Features

We arrange the features according to their informativeness by $\mathcal{I}1$. For the dataset **P** we obtain $J = (7, 2, 4, 5, 8, 3, 1, 6)$; for **R** - $J = (7, 4, 8, 2, 5, 3, 1, 6)$; for **C** - $J = (7, 2, 4, 8, 5, 3, 1, 6)$. In all cases the least informative features, in terms of prediction of good and bad outcomes, are the features 6 (the third meta-reaction), 1 (age) and 3 (sex).

Note that the potential reactions used in the data **P** and **C** contain information about patient age. This can be explained as follows. The potential reactions are defined by drugs taken. One assumption that was made in [13] is that dosage rates would, in general, be standardized by age. Therefore, the vector of potential reactions, calculated for some patient, contains information about dosage rates of drugs taken by this patient and so information about patient age. But it is interesting that this feature is not informative for all datasets **P, R** and **C.**

The feature "sex" is uninformative in all cases. Thus two features out of the three features about patient information (i.e., age, weight and sex) are not informative in terms of the prediction of reaction outcomes. The results presented below confirm this fact. The "weight" (second feature) is just one feature about patient information which plays an important role in the prediction of reaction outcomes.

Feature 7 (i.e., the fourth meta-reaction) is the most informative feature. There are 380 records from the first class (out of 2859) and 364 records from the second class (out of 1159) for which the fourth meta-reaction was observed. This is the most *dangerous* reaction, that most frequently results in a bad outcome.

5.2 The Results of the Classification

According to the informativeness of the features we classified examples from the test sets by taking different subsets of features.

First classification was done using all 8 features. The results obtained are presented in Table 9. Then classification was done based on different subsets of features. In all cases at least two features (age and weight) were used as information about patients.

1. Age, weight, sex and the 1-st, 2-nd, 4-th and 5-th meta-reactions (7 features). The third meta-reaction which was found most uninformative was not considered. Clearly in this case all records with the third meta-reaction should be removed. The number of records in the training and test sets corresponding to this case and the results of classification are presented in Table 10. The comparison of the results presented in Tables 9 and 10 shows that the prediction accuracy increases if we remove the 6-th feature (third meta-reaction). In general, this combination of features provides better accuracy in classification (see also Tables 11 and Table 12). It is interesting that for all test sets in Table 10 the better accuracy was obtained for the **P** data. This means that in prediction of reaction outcomes the predicted potential reactions work better than the real observed reactions and even the combined vector of degrees (which combine the real and predicted reactions).

2. Age, weight and all components of the vector of reactions (7 features). The results obtained are presented in Table 11. We compare the results presented in Tables 9 and 11. Better accuracies for years 1998 and 1999 are obtained if we remove the feature "sex". This fact emphasizes that in the prediction of reaction outcomes either this feature does not play an important role or this feature can be useful in combination with some other features (that are not considered here) describing individual characters of patients.

3. Age, weight, sex and the first 4 components of the vector of reactions (7 features). The last - fifth component of vector reactions (that is, the fifth meta-reaction) has been defined in [13] as a mixed meta-reaction, which contains at least two of the first 4 meta-reactions. The number of records with these mixed reactions is small enough by comparison with other reactions. Moreover in prediction, the mixed reactions could produce some noise because they contain different combinations of the first 4 meta-reactions. Therefore it is interesting to consider the case of the first four meta-reactions. In this case we have removed from the data, all records with the fifth meta-reaction. The results obtained are presented in Table 12.

The results presented in Tables 9 - 12 allow us to make the following conclusions.

Drug-reaction associations. In Table 11 the accuracy for **R** (real reactions) is better than for **P** (predicted reactions), but in Table 10 the accuracy for **P** is better than for **R**. In all other cases the real and predicted reactions perform similarly. Therefore, in terms of prediction of reaction outcomes, the predicted potential reactions can be considered as a sound method to present drug-reaction associations.

The Effect of Potential Reactions. One of the main advantages of the potential reactions is that information about potentially existing (but not occurring) reactions could be useful in different areas as well as in the prediction of reaction outcomes. An event, where using potential reactions provides better results in the prediction, is called an Effect of Potential (Hidden) Reactions [13]. We can observe this effect in the form that accuracy either for **P** or for **C** is, in almost all cases, better than for **R.**

Information about the patient. The prediction accuracy is not high. This situation can be explained by poor information about patients. As mentioned before just one feature - "weight" could play an important role in our analysis. Therefore, in the study of adverse drug reactions it is very important to have more detailed information about patients.

6 Conclusions

In this work we developed a new version of the machine learning algorithm FDM, introduced in [11], [12]. This version (called FDM2) is mainly directed at handling large datasets. We apply this algorithm to study Adverse Drug Reaction problems on the basis of the Australian Adverse Drug Reaction Database (ADRAC).

FDM2 induces sets of ordered rules and in the classification of a new example the effect of some non-informative complexes that have occurred by chance, is decreased by the use of a coefficient of significance. The results show that this method performs well in all cases considered in this paper, although other machine learning algorithms perform well only in some special cases (see for example [15]). Therefore, using

Table 9. The results of the classification obtained by all 8 features.

Data	Test Year	n_{tr}	n_{ts}	Default	p_{tr}	p_{ts}
P	1998	2885	285	63.2	91.0	64.9
R	1998	2885	285	63.2	79.7	64.9
C	1998	2885	285	63.2	89.0	67.0
P	1999	3170	307	66.4	90.8	69.1
R	1999	3170	307	66.4	79.1	69.7
C	1999	3170	307	66.4	89.9	70.0
P	2000	3477	347	57.3	90.8	69.1
R	2000	3477	347	57.3	79.1	69.7
C	2000	3477	347	57.3	89.9	70.0
P	2001	3824	194	57.2	90.7	65.5
R	2001	3824	194	57.2	78.5	66.0
C	2001	3824	194	57.2	88.9	64.9

Table 10. The results of the classification obtained by 7 features: age, weight, sex and 1-st, 2-nd, 4-th, 5-th meta-reactions.

Data	Test Year	n_{tr}	n_{ts}	Default	p_{tr}	p_{ts}
P	1998	2114	205	66.3	87.9	70.2
R	1998	2114	205	66.3	80.7	68.8
C	1998	2114	205	66.3	86.4	68.3
P	1999	2319	220	68.2	93.6	73.2
R	1999	2319	220	68.2	80.6	72.3
C	1999	2319	220	68.2	86.5	70.9
P	2000	2539	272	56.6	93.4	69.5
R	2000	2539	272	56.6	80.4	68.8
C	2000	2539	272	56.6	86.2	65.4
P	2001	2811	154	57.1	93.5	70.8
R	2001	2811	154	57.1	81.4	65.6
C	2001	2811	154	57.1	90.5	69.5

sets of classification rules rather than a single set of classification rules essentially increases performance of the algorithm.

In this work precise computation timing was not able to be carried out. Approximate learning times indicate that for one single set, times were 3 seconds (for $F' = 29$) and 120 seconds (for $F' = 180$) for the DNA dataset, 16 seconds for the Satim dataset and 55 seconds for the Letter dataset. Combined rule sets were multiples of the time for single sets.

We comment on some aspects of this classification algorithm. The algorithm con-

Table 11. The results of the classification obtained by 7 features: age, weight and vector reactions.

Data	Test Year	n_{tr}	n_{ts}	Default	p_{tr}	p_{ts}
P	1998	2885	285	63.2	82.7	63.9
R	1998	2885	285	63.2	78.0	67.0
C	1998	2885	285	63.2	82.1	67.4
P	1999	3170	307	66.4	81.3	71.3
R	1999	3170	307	66.4	76.9	69.1
C	1999	3170	307	66.4	81.8	69.7
P	2000	3477	347	57.3	81.0	63.1
R	2000	3477	347	57.3	76.2	65.7
C	2000	3477	347	57.3	80.6	66.6
P	2001	3824	194	57.2	79.8	62.4
R	2001	3824	194	57.2	75.3	62.4
C	2001	3824	194	57.2	80.1	62.4

Table 12. The results of the classification obtained by 7 features: age, weight, sex and the first 4 components of vector reactions.

Data	Test Year	n_{tr}	n_{ts}	Default	p_{tr}	p_{ts}
P	1998	2496	245	61.6	82.7	64.9
R	1998	2496	245	61.6	77.4	64.1
C	1998	2496	245	61.6	81.7	64.1
P	1999	2741	272	66.2	82.4	71.0
R	1999	2741	272	66.2	76.7	69.5
C	1999	2741	272	66.2	80.8	69.5
P	2000	3013	301	55.5	81.6	62.1
R	2000	3013	301	55.5	76.1	65.1
C	2000	3013	301	55.5	80.5	65.8
P	2001	3314	167	59.3	80.4	64.7
R	2001	3314	167	59.3	75.5	67.1
C	2001	3314	167	59.3	80.4	62.9

tains parameters (including α and β) which influence the accuracy of the classification and the choice of sensible or optimal values for these parameters will be a very important problem. Here we mention some of them.

1. One of the main parameters in this algorithm is the centroid used in the calculation of the fuzzy derivative. The calculations show that the accuracy essentially depends on the centroid. Changing it can increase or decrease the accuracy for each class of the database. Therefore to find an "optimal" centroid (or reference point)

will be a very important problem.

2. In this application we used only the value of the fuzzy derivative (membership functions) calculated by a global approach at the centroid. It will be interesting to calculate values of the membership functions at many points and to use these discrete membership functions in the data classification.

3. Finally we note a fact that is very important for classification. The main problem is to generate quite different sets of classification rules. In this work we generate different sets only changing the parameters α and β. But we do not know how the generated sets are different. In the future it will be very important to investigate the difference between these different generated sets.

References

1. Bagirov A. M., Rubinov A. M. and Yearwood J. (2002) A global optimization approach to classification. Optimization and Engineering (to appear).
2. Bennett K. P. and Mangasarian O. L. (1992) Robust linear programming discrimination of two linearly inseparable sets. Optimization Methods and Solware, **1,** 23-34.
3. Bradley P. S. and Mangasarian O. L. (1998) Feature selection via concave minimization and support vector machines. Machine Learning Proceedings of the Fifteenth International Conference (ICLML'98), J.Shavlik (ed.), Morgan Kaufmann, San Francisco, California, 82-90.
4. Buckley J. J., Yunaxia Qu. (1991) Fuzzy complex analysis I: Differentiation. Fuzzy Sets and Systems, **41**, 269-284.
5. Chen C. and Mangasarian O. L. (1995) Hybrid misclassification minimization. Mathematical Programming Technical Report, **95-05,** University of Wisconsin.
6. Clark P. and Niblet T. (1989) The CN2 Induction Algorithm. Machine learning, **3**, 261-283.
7. Dubois D. and Prade H. (1991) Towards Fuzzy differential calculus. Part 1: Integration of Fuzzy mappings. Fuzzy Sets and Systems, **8**, 1-17.
8. Ferraro M. and Foster D. N. (1987) Differentiation of Fuzzy Continuous Mappings on Fuzzy Topological Vector Spaces. Journal of Mathematical Analysis and Applications, **121**, 589-60.
9. Kalbfleish J. (1979) Probability and statistical inference, (Vol.2). New York: Springer-Verlag.
10. Mamedov M. A. (1994) Fuzzy derivative and dynamic systems. Proc. of the Intern. Conf. On Appl. of Fuzzy systems, ICAFS-94, Tabriz (Iran), Oct. 17-19, 122-126.
11. Mammadov M. A. (2001) Fuzzy derivative and its applications to data classification. University of Ballarat, Research Report, 01/13.
12. Mamedov M. A., Rubinov A. M. and Yearwood J. (2001) Sequential separation of sets and its applications to data classification. Proc. of the Post-gr. ADFA Conf. On Computer Science, Canberra (Aust.), July, 14, 75-80.
13. M.A.Mamedov and G. W. Saunders (2002) An Analysis of Adverse Drug Re-

actions from the ADRAC Database. Part 1: Cardiovascular group, *University of Ballarat School of Information Technology and Mathematical Sciences, Research Report 02/01,* February 2002, 1-48.
14. Mangasarian O. L. (1994) Misclassification minimization. Journal of Global Optimization. **5**, 309-323.
15. Michie D., Spiegelhalter D. J. and Taylor C. C. (eds.) (1994) Machine Learning, Neural and Statistical Classification. Ellis Horwood Series in Artificial Intelligence, London.
16. Murphy P. M. and Aha D. W. (1992) UCI repository of machine learning databases. www.ics.edu./ mlearn/MLRepository.html.
17. Zadeh L. A. (1965) Fuzzy sets. Inform. and Control, **8**, 338-353.

Chapter 7

Model and Fixpoint Semantics for Fuzzy Disjunctive Programs with Weak Similarity

Dušan Guller

Summary. In this chapter, we present declarative, model, and fixpoint semantics for fuzzy disjunctive programs with weak similarity - sets of graded strong literal disjunctions. We shall suppose that truth values constitute a complete Boolean lattice $\boldsymbol{L} = (L, \leq, \cup, \cap, \Rightarrow, \mathbf{0}, \mathbf{1})$. A graded strong literal disjunction is a pair (D, c) where D is a strong literal disjunction of the form $l_1 \dot{\vee} \cdots \dot{\vee} l_n$ and c is a truth value from the lattice $\boldsymbol{L}$. A graded disjunction can be understood as a means of the representation of incomplete and uncertain information; the incompleteness is formalised by its strong literal disjunction, while the uncertainty by its truth degree. Fuzzy disjunctive programs may contain the binary predicate symbol $\sim$ for weak similarity, which is the fuzzy counterpart of the classical equality. In the end, the mutual coincidence of the proposed semantics will be reached.

Keywords: disjunctive logic programming, multivalued logic programming, fuzzy logic, model theory, knowledge representation and reasoning.

1 Introduction

In knowledge representation and commonsense reasoning, we should be able to handle incomplete and uncertain information. Attempts to solve many real-world problems bring us to storing and retrieving incomplete and uncertain knowledge in deductive databases in order to represent an incomplete and uncertain model of the world, and to carry out a reasonable inference of new facts from this model. Expert systems, cognitive robots, mechatronic systems, sensor data, sound and image databases, temporal indeterminacy are only a few of the fields dealing with incomplete and uncertain information. In recent years, disjunctive and multivalued, annotated logic programming have been recognised as powerful tools for maintenance of such knowledge bases.

In disjunctive logic programming [71, 72, 73, 95], we allow the heads of clauses to be literal disjunctions in contrast to the case of Horn programs, where only single atoms (literals) may occur in the heads of clauses. At first, positive disjunctive programs were studied, among others in [68, 67, 70, 71]. A positive disjunctive program is a set of implications of the form

$$a_1 \vee \cdots \vee a_m \longleftarrow b_1 \wedge \cdots \wedge b_n$$

where a_i and b_j are atoms. If we infer from a given positive disjunctive program P a disjunction, e.g., $p \vee q$, this disjunction can be viewed as some kind of incomplete information. We know that $p \vee q$ is derivable from the program P, but we do not know whether p or q themselves are derivable. This is the main difference to the Horn case, where we are interested only in complete information represented by single atoms, in our case by p and q. So, disjunctions can be used for a suitable representation of incomplete information. For positive disjunctive programs, Minker, Rajasekar, and Zanon introduced minimal model, fixpoint, and procedural semantics (based on SLI-resolution) [70, 71]. Since we consider only disjunctions of atoms, no negated information can be inferred from positive disjunctive programs. Therefore, similarly to Horn programs, various rules for negation were proposed:

- *the Generalized Closed World Assumption* ($GCWA$) [68],
- *the Extended Generalized Closed World Assumption* ($EGCWA$) [114],
- *the Weak Generalized Closed World Assumption* ($WGCWA$) [97, 69],
- *the Negation as Finite Failure for non-Horn Programs Rule* ($NAFFNH$) [69], and
- *the Disjunctive Database Rule* (DDR) [97].

Positive disjunctive programs can be augmented by some kind of negation-by-default. Such programs are called normal disjunctive programs and they consist of clauses of the form

$$a_1 \vee \cdots \vee a_m \longleftarrow b_1 \wedge \cdots \wedge b_n \wedge not\, c_1 \wedge \cdots \wedge not\, c_k$$

where a_i, b_j, and c_o are atoms and not is a negation-by-default operator. Several semantics for normal disjunctive programs were introduced. Many of them are extensions of semantics for normal (Horn) programs:

- *the disjunctive well-founded semantics* ($DWFS$) [111],
- *the generalized disjunctive well-founded semantics* ($GWFS$) [7, 8],
- WF^3 [9],
- *the disjunctive perfect model semantics* ($DPMS$) [89, 25],
- *the stationary semantics* [91],
- *the disjunctive stable semantics* [28, 92, 24],
- *the disjunctive partial stable semantics* [90], and
- *the static semantics* [94].

Negation-by-default has a limited use, because *not a is true* does not mean the presence of some knowledges asserting the falsehood of a, but the absence of knowledges asserting the truth of a. Therefore, it is worth extending normal disjunctive programs by classical negation. Explicit use of classical negation allows to express *when an atom is false* in addition to *when an atom is true*, which enhances the expressive power of normal disjunctive programs. Such programs are called extended disjunctive programs and they consist of clauses of the form

$$l_1 \vee \cdots \vee l_m \longleftarrow l_{m+1} \wedge \cdots \wedge l_n \wedge not\, l_{n+1} \wedge \cdots \wedge not\, l_k$$

where l_i are literals. Hence, extended clauses contain the two kinds of negation: classical and default. Semantics of logic programs dealing with classical (strong) negation can be found in [28, 2, 90, 91], among others. In [72], a method how to obtain a semantics for extended disjunctive programs from a semantics for normal disjunctive programs, is described. This method is based on the prime-$\perp$ and neg transformations, which capture the intended meaning of original extended disjunctive programs after transformation to normal disjunctive programs. Another approach is described in [95]. Przymusinski introduced the Autoepistemic Logic of Knowledge and Beliefs ($AELB$), augmenting Moore's Autoepistemic Logic [78], as a unifying framework for various semantics of logic programs. Most semantics can be obtained by means of a suitable translation of logic programs into $AELB$. Multiple use of several kinds of negation is discussed in [5, 6, 74, 76, 75].

An interesting approach to uncertainty, vagueness, and impreciseness is that of multivalued and annotated logic programming. When we derive some formulae from a given program, they do not represent certain information in general, since some truth value or degree, e.g., from the unit interval $[0, 1]$, is assigned to each formula. So, derived formulae are only partly true in models of the given program. Roughly speaking, the main difference between multivalued and annotated logic programming is that implications in the multivalued approach have truth values (weights) associated with them as a whole in contrast to the annotated approach, where truth values are associated with each literal of an implication. This yields some advantages and disadvantages to both the approaches. Associating truth values with implications is intuitive and natural, yet the problem how truth values should be propagated across implications arises. On the other hand, annotated logics have a simpler formalism that is closer to classical logic. They use the classical definition of implication: $I \models F_1 \leftarrow F_2$ iff $I \models F_1$ or $I \not\models F_2$. Also many problems in nonmonotonic reasoning can be formulated more easily via non-monotonic annotated logics. This dichotomy may be partly removed by an appropriate generalisation of the concept of an annotation so that we are able to capture the propagation of truth values across implications. Annotated logics were introduced in [105], later developed in [10, 11, 45, 46, 47], and generalised in [48]. The multivalued approach can be roughly divided into the following subdomains:

- bilattice-based logics [29, 30, 26, 27, etc.]; and
- quantitative rule-based programs [79, 110, 96, 44, 102, 43, 103, etc.],
  - heuristic, non-numerical expressing uncertainty [14, 15, 23, etc.],
  - numerical expressing uncertainty and using, e.g., Dempster-Shafer calculus [43, 101, 20, etc.], using fuzzy logic [61, 60, 59, 12, 110, 102, etc.],
  - numerical expressing uncertainty and relying on probability theory [85, 4, 88, 80, 81, 82, 83, 84, 50, 51, 52, 53, 54, 55, 19, 22, 56, 57, etc.].

Recently, multi-adjoint logic programs have been introduced in [62, 63, 64, 65, 66, 113], where model, fixpoint, and procedural semantics for these programs have been proposed. Multi-adjoint logic programs allow simultaneous use of several implications in the rules and rather general connectives in the bodies. They generalise residuated and monotonic logic programs in [16, 17, 18].

Concerning the representation of uncertainty, we shall suppose that truth values constitute a complete Boolean lattice $\boldsymbol{L} = (L, \leq, \cup, \cap, \Rightarrow, \mathbf{0}, \mathbf{1})$. The truth functions of the connectives: $\wedge$ (conjunction), $\dot{\vee}$ (strong disjunction), $\vee$ (disjunction), $\rightarrow$ (implication), and $\neg$ (negation) are defined in Table 3.2 (p. 158) as usual.

In the present work, we shall aim to combine both the disjunctive and multivalued approaches and provide some formalisation of reasoning with incomplete and uncertain information represented by graded strong literal disjunctions. A graded strong literal disjunction is a pair (D, c) where D is a strong literal disjunction of the form $l_1 \dot{\vee} \cdots \dot{\vee} l_n$ and c is a truth value from the lattice $\boldsymbol{L}$. Also fuzzy disjunctive programs, from which we shall infer incomplete and uncertain information, can be viewed as sets of graded strong literal disjunctions. Similar approaches can be found in [59, 60, 61, 51, 56].

2 A Motivating Example

We now illustrate with a simple example how fuzzy disjunctive programs can work with incomplete and uncertain information. Suppose a robotic system which builds towers from cuboids with square ground-planes and from triangular prisms with triangle ground-planes, shown in Figure 1. Before or during the building of a tower, the robot has also to estimate whether the tower will be stable enough in order to avoid collapsing. Towers are always oriented towards four axes: x_1, x_2, x_3, and x_4, as outlined in Figure 3. Therefore, we can recognise one shape (square) of ground-planes of cuboids and four shapes (triangles) of ground-planes of prisms depending on their orientation towards the axes, see Figure 2.

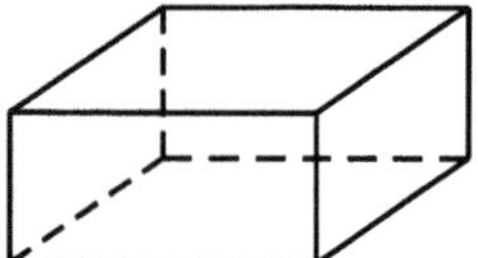

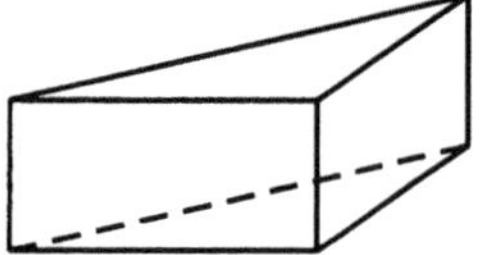

Figure 1. Cuboids and triangular prisms.

The stability of a tower can be computed out from the fixations of its components. The fixation, *fix*, of a figure is expressed by the set of the axes that belong to the shape of its ground-plane, Figure 2. This leads us to the idea of representing truth values by the powerset lattice $\boldsymbol{L} = (2^X, \subseteq, \cup, \cap, \emptyset, X)$ where X is the set of the axes $\{x_1, x_2, x_3, x_4\}$, and define a predicate $\mathit{fixed}(F)$ by the following graded rules:

$$\mathit{fixed}(F) \overset{\mathit{fix}_i}{\longleftarrow} \mathit{shape}(F, s_i) \text{ for } i = 1, \ldots, 5. \tag{1}$$

The predicate $\mathit{shape}(F, s_i)$ expresses that the figure F has the shape of its ground-plane s_i. In the following considerations, we shall write a graded disjunction or implication of the form (f, X) simply as f. By $\mathit{on}(F_1, F_2)$ we represent the fact (with the truth value X) that the figure F_1 lies on F_2. Then the stability of F_1 on F_2 can be computed as follows:

$$stab(F_1, F_2) \longleftarrow on(F_1, F_2) \wedge \mathit{fixed}(F_1) \wedge \mathit{fixed}(F_2). \tag{2}$$

The stability of a whole tower is defined by the transitive closure of the predicate $stab(F_1, F_2)$:

$$\begin{aligned} stab^*(F_1, F_2) &\longleftarrow stab(F_1, F_2), \\ stab^*(F_1, F_2) &\longleftarrow stab(F_1, F_3) \wedge stab^*(F_3, F_2). \end{aligned} \tag{3}$$

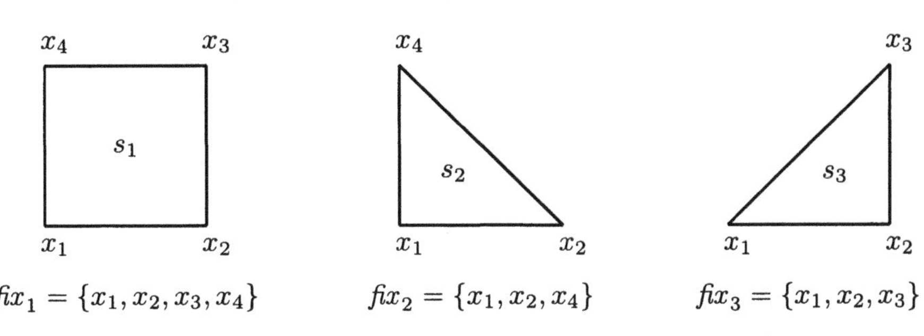

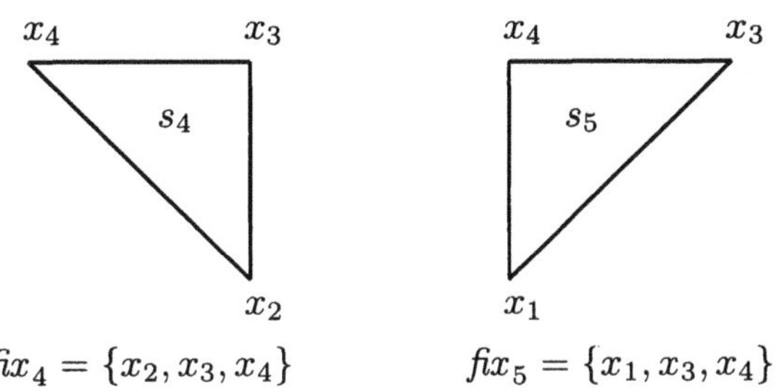

Figure 2. Shapes of ground-planes.

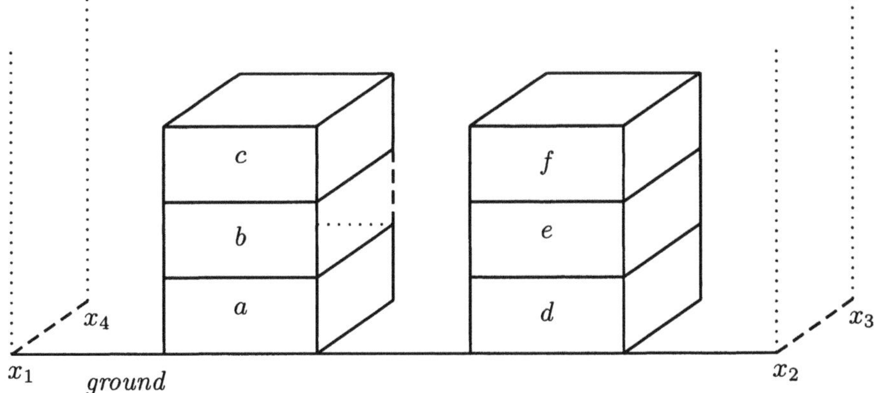

Figure 3. Built towers.

Suppose that the perception (vision) subsystem of our robot yields the following facts about the two towers in Figure 3:

$$\begin{array}{ll} on(a, ground), & on(d, ground), \\ on(b, a), & on(e, d), \\ on(c, b), & on(f, e), \end{array} \tag{4}$$

$$\begin{array}{ll} shape(ground, s_1), & \\ shape(a, s_1), & shape(d, s_1), \\ shape(b, s_2) \dot{\vee} shape(b, s_3), & shape(e, s_3), \\ shape(c, s_1), & shape(f, s_1). \end{array} \tag{5}$$

Note that the ambiguity of the shape of the ground-plane of b is outlined by the dotted and dashed lines in the figure b. The dotted line expresses the alternative $shape(b, s_2)$ and the dashed one the case $shape(b, s_3)$.

The inference mechanism can make the following derivations of the stabilities of the towers (by means of the fuzzy modus ponens rule):

$$\begin{array}{lr} fixed(ground), & (5), (1) \\ fixed(a), & (5), (1) \\ (fixed(b) \dot{\vee} shape(b, s_3), \{x_1, x_2, x_4\}), & (5), (1) \\ (fixed(b) \dot{\vee} fixed(b), \{x_1, x_2, x_4\} \cap \{x_1, x_2, x_3\}), & (1) \\ (fixed(b), \{x_1, x_2\}), & \\ fixed(c), & (5), (1) \\ stab(a, ground), & (4), (2) \\ (stab(b, a), \{x_1, x_2\}), & (4), (2) \\ (stab(c, b), \{x_1, x_2\}), & (4), (2) \\ (stab^*(c, ground), \{x_1, x_2\}); & (3) \end{array}$$

$$\begin{array}{lr} fixed(ground), & (5), (1) \\ fixed(d), & (5), (1) \\ (fixed(e), \{x_1, x_2, x_3\}), & (5), (1) \\ fixed(f), & (5), (1) \\ stab(d, ground), & (4), (2) \\ (stab(e, d), \{x_1, x_2, x_3\}), & (4), (2) \\ (stab(f, e), \{x_1, x_2, x_3\}), & (4), (2) \\ (stab^*(f, ground), \{x_1, x_2, x_3\}). & (3) \end{array}$$

We get that the stability of the first tower is less than the stability of the second one. So, the incomplete information about the shape of the ground-plane of b decreases the entire estimated stability for the first tower. By the example, we have outlined a way how incomplete and uncertain information can be together propagated via inference of new facts.

3 Basic Notions and Notation

3.1 Fixpoints

Consider a complete lattice $\boldsymbol{L} = (L, \leq, \cup, \cap, \mathbf{0}, \mathbf{1})$. Let $T : L \longrightarrow L$ be a mapping. We say T is monotonic iff for all $x, y \in L$, $x \leq y \Longrightarrow T(x) \leq T(y)$. T is said to be ω-continuous iff for all ω-chains $X \subseteq L$ ($X = \{x_i \mid i \in \omega\}$ is simply ordered by $\leq$), $T(\bigcup X) = \bigcup\{T(x) \mid x \in X\}$. We can define the α-th iteration power of T on $x \in L$, written as $T \uparrow \alpha(x)$, by transfinite recursion on α:

$$\begin{aligned} T \uparrow 0(x) &= x, \\ T \uparrow \alpha + 1(x) &= T(T \uparrow \alpha(x)), \\ T \uparrow \alpha(x) &= \textstyle\bigcup_{\beta<\alpha} T \uparrow \beta(x) \text{ for a limit ordinal } \alpha. \end{aligned}$$

$T \uparrow \alpha(\mathbf{0})$ will be abbreviated by $T \uparrow \alpha$. We say $x \in L$ is a fixpoint of T iff $x = T(x)$.

Proposition 1 (Fixpoint Theorem (Knaster, Tarski)**).**
Let $T : L \longrightarrow L$ be a mapping over a complete lattice $\boldsymbol{L} = (L, \leq, \cup, \cap, \mathbf{0}, \mathbf{1})$.

i) *If T is monotonic, then there exists the least fixpoint of T, denoted as $lfp(T)$.*
ii) *If T is ω-continuous, then $lfp(T) = T \uparrow \omega$.*

Proof: See [108]. □

3.2 Predicate Fuzzy Logic

Throughout the chapter, we shall use common notions of predicate fuzzy logic. Let $\mathcal{L}$ denote a predicate language. By $Var_{\mathcal{L}}$, $Pred_{\mathcal{L}}$, $Func_{\mathcal{L}}$, $Term_{\mathcal{L}}$, and $GTerm_{\mathcal{L}}$, we denote the sets of variables, predicate symbols, function symbols, terms and ground terms (without variables) of $\mathcal{L}$, respectively. We suppose that an arity $ar \geq 0$ is assigned to every predicate and function symbol. If the sets of predicate and function symbols are finite, i.e. $Pred_{\mathcal{L}} = \{p_1, \ldots, p_n\}$ and $Func_{\mathcal{L}} = \{f_1, \ldots, f_m\}$, we may display the language $\mathcal{L}$ as $\mathcal{L} = \{p_1, \ldots, p_n, f_1, \ldots, f_m\}$.

Let $e_1, \ldots, e_n$ be arbitrary expressions of $\mathcal{L}$ (terms, tuples of terms, or formulae). By $vars(e_1, \ldots, e_n)$ we denote the set of all the variables of $Var_{\mathcal{L}}$ occurring in these expressions.

We shall suppose that truth values (or degrees) of our fuzzy logic constitute a Brouwer lattice $\boldsymbol{L} = (L, \leq, \cup, \cap, \Rightarrow, \mathbf{0}, \mathbf{1})$ with a residuum $\Rightarrow$ where

- $\boldsymbol{L} = (L, \leq, \cup, \cap, \mathbf{0}, \mathbf{1})$ is a complete lattice;
- the greatest lower bound operator $\cup$ and the least upper bound operator $\cap$ are infinitely distributive, i.e. for all $K \subseteq L$, $a \in L$,

$$a \cup \bigcap K = \bigcap_{k \in K} (a \cup k),$$

$$a \cap \bigcup K = \bigcup_{k \in K} (a \cap k);$$

- the binary operation $\Rightarrow$ over L is non-increasing in the first argument and non-decreasing in the second one;
- the following condition of adjunction holds: for all $\alpha, \beta, \gamma \in L$,

$$\alpha \cap \beta \leq \gamma \text{ iff } \alpha \leq \beta \Rightarrow \gamma.$$

Our language $\mathcal{L}$ contains the following connectives: $\wedge$ (conjunction), $\dot{\vee}$ (strong disjunction), $\vee$ (disjunction), $\longrightarrow$ (implication), and $\neg$ (negation). The truth functions of the connectives are defined in Table 3.2 in the usual way. ($\bar{x}$, $\bar{y}$ denote the values of x and y, respectively.)

Table 1. Truth functions.

connective	truth function
$\wedge$	$\cap$
$\vee$	$\cup$
$\longrightarrow$	$\Rightarrow$
$\neg x$	$\bar{x} \Rightarrow \mathbf{0}$
$x \dot{\vee} y$	$((\bar{x} \Rightarrow \mathbf{0}) \cap (\bar{y} \Rightarrow \mathbf{0})) \Rightarrow \mathbf{0}$

Note here, if $\boldsymbol{L}$ is a Boolean lattice, then it is easy to see that $\bar{x} \Rightarrow \mathbf{0}$ is the complement operator, $((\bar{x} \Rightarrow \mathbf{0}) \cap (\bar{y} \Rightarrow \mathbf{0})) \Rightarrow \mathbf{0} = \bar{x} \cup \bar{y}$, and $\bar{x} \Rightarrow \bar{y} = (\bar{x} \Rightarrow \mathbf{0}) \cup \bar{y}$.

An $\boldsymbol{L}$-model, say $\mathfrak{A}$, for $\mathcal{L}$ is a structure $(\mathcal{U}_{\mathfrak{A}}, \{f^{\mathfrak{A}} \mid f \in Func_{\mathcal{L}}\}, \{p^{\mathfrak{A}} \mid p \in Pred_{\mathcal{L}}\})$ defined as follows:

- $\mathcal{U}_{\mathfrak{A}} \neq \emptyset$ is the universum (domain) of the model $\mathfrak{A}$;
- a function symbol $f \in Func_{\mathcal{L}}$ is interpreted as a function $f^{\mathfrak{A}} : \mathcal{U}_{\mathfrak{A}}^{ar(f)} \longrightarrow \mathcal{U}_{\mathfrak{A}}$;
- a predicate symbol $p \in Pred_{\mathcal{L}}$ is interpreted as a function, an $\boldsymbol{L}$-fuzzy relation, $p^{\mathfrak{A}} : \mathcal{U}_{\mathfrak{A}}^{ar(p)} \longrightarrow \boldsymbol{L}$.

A variable assignment in $\mathfrak{A}$ is a mapping $Var_{\mathcal{L}} \longrightarrow \mathcal{U}_{\mathfrak{A}}$ assigning an element of the universum $\mathcal{U}_{\mathfrak{A}}$ to each variable.

Let e be a variable assignment in $\mathfrak{A}$, t a term, and ϕ a formula of $\mathcal{L}$. We can assign an element of $\mathcal{U}_{\mathfrak{A}}$ to t and a truth value of $\boldsymbol{L}$ to ϕ, denoted as $||t||_e^{\mathfrak{A}}$ and $||\phi||_e^{\mathfrak{A}}$, respectively, in the following manner [118]:

- $t \equiv v \in Var_{\mathcal{L}}$, $||v||_e^{\mathfrak{A}} = e(v)$;
- $t \equiv f(t_1, \ldots, t_{ar(f)})$ where $f \in Func_{\mathcal{L}}$, t_i are terms of $\mathcal{L}$,
$$||f(t_1, \ldots, t_{ar(f)})||_e^{\mathfrak{A}} = f^{\mathfrak{A}}(||t_1||_e^{\mathfrak{A}}, \ldots, ||t_{ar(f)}||_e^{\mathfrak{A}});$$
- $\phi \equiv p(t_1, \ldots, t_{ar(p)})$ where $p \in Pred_{\mathcal{L}}$, t_i are terms of $\mathcal{L}$,
$$||p(t_1, \ldots, t_{ar(p)})||_e^{\mathfrak{A}} = p^{\mathfrak{A}}(||t_1||_e^{\mathfrak{A}}, \ldots, ||t_{ar(p)}||_e^{\mathfrak{A}});$$
- $\phi \equiv \neg\psi$ where ψ is a formula of $\mathcal{L}$, $||\neg\psi||_e^{\mathfrak{A}} = ||\psi||_e^{\mathfrak{A}} \Rightarrow \mathbf{0}$;
- $\phi \equiv \psi_1 \wedge \psi_2$ where ψ_1 and ψ_2 are formulae of $\mathcal{L}$,
$$||\psi_1 \wedge \psi_2||_e^{\mathfrak{A}} = ||\psi_1||_e^{\mathfrak{A}} \cap ||\psi_2||_e^{\mathfrak{A}};$$

- $\phi \equiv \psi_1 \dot{\vee} \psi_2$ where ψ_1 and ψ_2 are formulae of $\mathcal{L}$,
$$||\psi_1 \dot{\vee} \psi_2||_e^{\mathfrak{A}} = ((||\psi_1||_e^{\mathfrak{A}} \Rightarrow \mathbf{0}) \cap (||\psi_2||_e^{\mathfrak{A}} \Rightarrow \mathbf{0})) \Rightarrow \mathbf{0};$$
- $\phi \equiv \psi_1 \vee \psi_2$ where ψ_1 and ψ_2 are formulae of $\mathcal{L}$,
$$||\psi_1 \vee \psi_2||_e^{\mathfrak{A}} = ||\psi_1||_e^{\mathfrak{A}} \cup ||\psi_2||_e^{\mathfrak{A}};$$
- $\phi \equiv \psi_1 \rightarrow \psi_2$ where ψ_1 and ψ_2 are formulae of $\mathcal{L}$,
$$||\psi_1 \rightarrow \psi_2||_e^{\mathfrak{A}} = ||\psi_1||_e^{\mathfrak{A}} \Rightarrow ||\psi_2||_e^{\mathfrak{A}};$$
- $\phi \equiv \forall x\, \psi$ where ψ is a formula of $\mathcal{L}$, $||\forall x\, \psi||_e^{\mathfrak{A}} = \bigcap\{||\psi||_{e(x/u)}^{\mathfrak{A}} \mid u \in \mathcal{U}_{\mathfrak{A}}\}$;
- $\phi \equiv \exists x\, \psi$ where ψ is a formula of $\mathcal{L}$, $||\exists x\, \psi||_e^{\mathfrak{A}} = \bigcup\{||\psi||_{e(x/u)}^{\mathfrak{A}} \mid u \in \mathcal{U}_{\mathfrak{A}}\}$.

By a graded formula of $\mathcal{L}$ we mean a pair (ϕ, c) consisting of a formula ϕ of $\mathcal{L}$ and a truth value c of $\boldsymbol{L}$.

We say that a graded formula (ϕ, c) of $\mathcal{L}$ is true in the model $\mathfrak{A}$ wrt. a variable assignment e in $\mathfrak{A}$, written as $\mathfrak{A} \models (\phi, c)[e]$, iff $||\phi||_e^{\mathfrak{A}} \geq c$.

A graded formula (ϕ, c) of $\mathcal{L}$ is true in the model $\mathfrak{A}$, denoted as $\mathfrak{A} \models (\phi, c)$, iff for every variable assignment e in $\mathfrak{A}$, $||\phi||_e^{\mathfrak{A}} \geq c$, or equivalently $\bigcap\{||\phi||_e^{\mathfrak{A}} \mid e \text{ is a variable assignment in } \mathfrak{A}\} \geq c$. So, the truth value c determines the least truth degree in which the formula ϕ must be true in $\mathfrak{A}$.

A graded theory of $\mathcal{L}$ is a set of graded formulae of $\mathcal{L}$. $\mathfrak{A}$ is a model of a graded theory T, in symbols $\mathfrak{A} \models T$, iff $\mathfrak{A} \models (\phi, c)$ for every $(\phi, c) \in T$.

We say that a graded formula (ϕ, c) of $\mathcal{L}$ is a fuzzy logical consequence of a graded theory T of $\mathcal{L}$, written as $T \models (\phi, c)$, iff for every $\boldsymbol{L}$-model $\mathfrak{A}$ for $\mathcal{L}$, $\mathfrak{A} \models T$ implies $\mathfrak{A} \models (\phi, c)$.

A literal of $\mathcal{L}$ is either an atom of $\mathcal{L}$ or the negation of an atom of $\mathcal{L}$.

We say that a pair of literals l_1 and l_2 is contradictory iff l_1 is of the form a and l_2 of the form $\neg a$ for some atom a, or vice versa.

A strong literal disjunction D of $\mathcal{L}$ is a formula of the form

$$\neg(d_1 \wedge (d_2 \wedge \cdots \wedge d_n) \cdots)),$$

$n \geq 0$, such that d_i are literals of $\mathcal{L}$. We shall omit the parenthesis and write D in the form

$$d'_1 \dot{\vee} \cdots \dot{\vee} d'_n$$

where d'_i is an atom a if d_i is of the form $\neg a$; or conversely, a negation of an atom a when d_i is an atom a. If $n = 0$, the empty strong literal disjunction is denoted by $\square$. We put $||\square||_e^{\mathfrak{A}} = \mathbf{0}$ for any $\boldsymbol{L}$-model $\mathfrak{A}$ and variable assignment e.

Let D be a strong literal disjunction. By $\{\!|D|\!\}$ we denote the set of all the literals occurring in D. For example, $\{\!|a \dot{\vee} \neg b \dot{\vee} a \dot{\vee} \neg c|\!\} = \{a, \neg b, \neg c\}$.

A strong literal disjunction is said to be a tautology iff it contains a pair of contradictory literals. For example, $\neg a \dot{\vee} \neg b \dot{\vee} a$.

Let A be a set of literals. Then $\neg A = \{\neg a \mid a \textit{ is an atom and } a \in A\} \cup \{a \mid a \textit{ is an atom and } \neg a \in A\}$.

A strong literal disjunction D is called a subdisjunction of a strong literal disjunction D' (in other words, D subsumes D') iff $\{\!|D|\!\} \subseteq \{\!|D'|\!\}$, in symbols $D \sqsubseteq D'$.

A literal conjunction of $\mathcal{L}$ is a conjunction of the form

$$(l_1 \wedge (l_2 \wedge \cdots \wedge l_n) \cdots)),$$

$n \geq 0$, where $l_1, \ldots, l_n$ are literals of $\mathcal{L}$. We shall omit the parenthesis and write the conjunction in the form

$$l_1 \wedge \cdots \wedge l_n.$$

Let C be a literal conjunction. By $\{\!|C|\!\}$ we denote the set of all the literals occurring in C. For example, $\{\!|a \wedge \neg b \wedge a|\!\} = \{a, \neg b\}$.

A strong literal implication of $\mathcal{L}$ is an implication of the form

$$l_1 \dot{\vee} \cdots \dot{\vee} l_n \longleftarrow k_1 \wedge \cdots \wedge k_m,$$

$m, n \geq 0$, where l_i and k_j are literals. In case of $m = 0$, the implication $l_1 \dot{\vee} \cdots \dot{\vee} l_n \longleftarrow$ means (denotes) the disjunction $l_1 \dot{\vee} \cdots \dot{\vee} l_n$.

Let $D \longleftarrow C$ be a strong literal implication. By $\{\!|D \longleftarrow C|\!\}$ we denote the set $\{\!|D|\!\} \cup \neg \{\!|C|\!\}$. For example, $\{\!|a \dot{\vee} \neg b \longleftarrow d \wedge \neg a|\!\} = \{a, \neg b\} \cup \neg\{\neg a, d\} = \{a, \neg b, \neg d\}$.

We say that a strong literal implication $D \longleftarrow C$ is an implication form of another strong literal implication $D' \longleftarrow C'$ iff $\{\!|D \longleftarrow C|\!\} = \{\!|D' \longleftarrow C'|\!\}$. Note that such implications are fuzzy logically equivalent.

A strong literal disjunction D is a disjunctive form of a strong literal implication $D' \longleftarrow C'$ iff $\{\!|D|\!\} = \{\!|D' \longleftarrow C'|\!\}$. Moreover, D is a disjunctive factor of $D' \longleftarrow C'$ iff $\{\!|D|\!\} = \{\!|D' \longleftarrow C'|\!\}$ and D has exactly one occurrence of every literal in $\{\!|D|\!\}$. Note that such disjunction and implication are fuzzy logically equivalent.

For example, $\neg a \dot{\vee} b \longleftarrow c \wedge d$, $\neg a \dot{\vee} \neg c \longleftarrow \neg b \wedge d$, $\neg a \dot{\vee} b \dot{\vee} \neg c \dot{\vee} \neg d$ are implication forms of each other because $\{\!|\neg a \dot{\vee} b \longleftarrow c \wedge d|\!\} = \{\!|\neg a \dot{\vee} \neg c \longleftarrow \neg b \wedge d|\!\} = \{\!|\neg a \dot{\vee} b \dot{\vee} \neg c \dot{\vee} \neg d|\!\} = \{\neg a, b, \neg c, \neg d\}$. Moreover $\neg a \dot{\vee} b \dot{\vee} \neg c \dot{\vee} \neg d$ is a disjunctive form and also a disjunctive factor of the implications since it has exactly one occurrence of every literal in $\{\neg a, b, \neg c, \neg d\}$.

A graded strong literal disjunction of $\mathcal{L}$ is a pair (D, c) where D is a strong literal disjunction of $\mathcal{L}$ and c is a truth value from the lattice $\boldsymbol{L}$.

A graded strong literal implication of $\mathcal{L}$ is a pair (Im, c) where Im is a strong literal implication of $\mathcal{L}$ and c is a truth value from the lattice $\boldsymbol{L}$.

We say that a graded strong literal implication (Im, c) is an implication form of another graded strong literal implication (Im', c) iff Im is an implication form of Im'.

A graded strong literal disjunction (D, c) is a disjunctive form (factor) of a graded strong literal implication (Im', c) iff D is a disjunctive form (factor) of Im'.

3.3 Substitutions

We shall use the concepts of substitutions due to [104], with a much easier formalism than the standard notions in, e.g., [1, 49].

A substitution ϑ of $\mathcal{L}$ on a finite set of variables $X \subseteq Var_{\mathcal{L}}$ is a mapping $\vartheta : X \longrightarrow Term_{\mathcal{L}}$. X is called the domain of ϑ, in symbols $dom(\vartheta)$, and $range(\vartheta)$ is the set of all variables of $Var_{\mathcal{L}}$ occurring in terms $\vartheta(x)$ where $x \in X$. The set of all substitutions of $\mathcal{L}$ is denoted as $Subst_{\mathcal{L}}$.

By id we mean the identity mapping $Var_{\mathcal{L}} \longrightarrow Var_{\mathcal{L}} : id(x) = x$.

The application of a substitution ϑ of $\mathcal{L}$ to a term $t \in Term_{\mathcal{L}}$ where $dom(\vartheta) \supseteq vars(t)$, denoted as $t\vartheta$, is defined inductively:

i) If $t = x \in Var_{\mathcal{L}}$, then $t\vartheta = \vartheta(x)$.
ii) If $t = f(t_1, \ldots, t_{ar(f)})$, $f \in Func_{\mathcal{L}}$, then $t\vartheta = f(t_1\vartheta, \ldots, t_{ar(f)}\vartheta)$.

Note that $t\vartheta \in Term_{\mathcal{L}}$ and it is defined unambiguously.

We can extend the definition to atoms, tuples, and sets. For an atom $p(t_1, \ldots, t_{ar(p)})$, $p(t_1, \ldots, t_{ar(p)})\vartheta$ denotes the atom $p(t_1\vartheta, \ldots, t_{ar(p)}\vartheta)$. Let $e_1, \ldots, e_n$ be terms or atoms. We denote the tuple $(e_1\vartheta, \ldots, e_n\vartheta)$ and the set $\{e_1\vartheta, \ldots, e_n\vartheta\}$ by the expressions $(e_1, \ldots, e_n)\vartheta$ and $\{e_1, \ldots, e_n\}\vartheta$, respectively.

Let ϑ, η of $\mathcal{L}$ be substitutions where $range(\vartheta) \subseteq dom(\eta)$. The composition $\vartheta \circ \eta$ is defined as the mapping $\vartheta \circ \eta : dom(\vartheta) \longrightarrow Term_{\mathcal{L}}$ where for all $x \in dom(\vartheta)$, $\vartheta \circ \eta(x) = \vartheta(x)\eta$; and $range(\vartheta \circ \eta) = range(\eta|_{range(\vartheta)})$.

A substitution ϑ of $\mathcal{L}$ is said to be an instance of a substitution η of $\mathcal{L}$ iff there exists a substitution γ of $\mathcal{L}$ such that $\vartheta = \eta \circ \gamma$.

A substitution ϑ of $\mathcal{L}$ is a variable renaming iff

i) $\vartheta(x)$ is a variable of $Var_{\mathcal{L}}$ for all $x \in dom(\vartheta)$, and
ii) $\vartheta(x) \neq \vartheta(y)$ for all $x \neq y$ where $x, y \in dom(\vartheta)$.

Let ϑ and ϑ' be substitutions of $\mathcal{L}$. ϑ' is a regular extension of ϑ iff

i) $dom(\vartheta) \subseteq dom(\vartheta')$,
ii) $\vartheta'|_{dom(\vartheta)} = \vartheta$, and
iii) $\vartheta'|_{dom(\vartheta')-dom(\vartheta)}$ is a renaming such that
$$range(\vartheta) \cap range(\vartheta'|_{dom(\vartheta')-dom(\vartheta)}) = \emptyset.$$

If the renaming is an identity substitution, we say ϑ' is an identity regular extension of ϑ.

Let ϕ be an open formula (quantifier free). An open formula ϕ' is a variant of ϕ iff there exists a variable renaming ρ such that $\phi\rho = \phi'$.
Let (ϕ, c) be a graded open formula, ϕ is an open formula. A graded open formula (ϕ', c) is a variant of (ϕ, c) iff ϕ' is a variant of ϕ.

Let S be a finite set of expressions of $\mathcal{L}$. A substitution θ of $\mathcal{L}$ is called a unifier for S iff $S\theta$ is a singleton. The unifier θ for S is said to be a most general unifier, *mgu*, for S iff for every unifier ϑ of $\mathcal{L}$ for S, there exists a substitution γ of $\mathcal{L}$ such that $\vartheta|_{vars(S)} = \theta|_{vars(S)} \circ \gamma$. If S consists of elements $e_1, \ldots, e_n$, we often write $mgu(e_1, \ldots, e_n)$ to denote some most general unifier for S.

Proposition 2. *Let S be a finite non-empty set of terms, atoms of $\mathcal{L}$, or tuples of them. If there exists an unifier ϑ of $\mathcal{L}$ for S, then there exists an mgu θ of $\mathcal{L}$ for S with $range(\theta) \subseteq dom(\theta) = vars(S)$.*

Proof: A slight modification of the proof in [1] (pp. 5-7). □

4 Similarity

The language of the classical predicate calculus is often extended by the binary predicate symbol $=$ for equality. This symbol has a special treatment. We wish $=$ to be interpreted by the equality relation $=$. Therefore, we add to the axioms of the predicate calculus the following schemes of axioms to get the predicate calculus with equality:

$$x = x, \tag{6}$$

$$x_1 = y_1 \wedge \cdots \wedge x_n = y_n \longrightarrow f(x_1, \ldots, x_n) = f(y_1, \ldots, y_n), \tag{7}$$

$$x_1 = y_1 \wedge \cdots \wedge x_n = y_n \wedge p(x_1, \ldots, x_n) \longrightarrow p(y_1, \ldots, y_n) \tag{8}$$

where p is an n-ary predicate symbol (including also $=$), f is an n-ary function symbol, and x_i, y_j are some pairwise different variables.

From these axioms, we immediately obtain the reflexivity, symmetry, and transitivity of the predicate symbol $=$:

$$x = x, \tag{reflexivity}$$

$$x = y \longrightarrow y = x, \tag{symmetry}$$

$$x = y \wedge y = z \longrightarrow x = z. \tag{transitivity}$$

So, a binary relation interpreting $=$ must be an equivalence relation. Moreover, by the scheme (7), we conclude that such a binary relation is also congruent with every function interpreting a function symbol in the language. Let T be a theory in the predicate calculus with equality. The symbol $=$ may be interpreted in first-order models of T by congruence relations. Nevertheless if T is consistent, there exists a model of T such that the symbol $=$ is indeed interpreted by the equality relation $=$. By Gödel's completeness theorem, there exists a model, say $\mathfrak{A}$, for T. Let $=_\mathfrak{A}$ be the equivalence relation interpreting the symbol $=$ in $\mathfrak{A}$. Then it is a simple matter to check that the quotient model $\mathfrak{A}/_{=_\mathfrak{A}}$ is a model of T. However, the quotient equivalence relation $=_\mathfrak{A} /_{=_\mathfrak{A}}$ is the equality relation $=$ on $\mathcal{U}_\mathfrak{A}/_{=_\mathfrak{A}}$ where $\mathcal{U}_\mathfrak{A}$ is the universum of $\mathfrak{A}$.

Our aim is now to extend these considerations to fuzzy logic and build up the fuzzy counterpart to equality. We introduce a notion of weak similarity. Denote by $\sim$ the binary predicate symbol for weak similarity. We wish that the following schemes of axioms hold in every $\boldsymbol{L}$-model for a language with weak similarity:

$$(x \sim x, \mathbf{1}), \tag{9}$$

$$(\neg(x_1 \sim y_1 \wedge \cdots \wedge x_n \sim y_n \wedge \neg(f(x_1, \ldots, x_n) \sim f(y_1, \ldots, y_n))), \mathbf{1}), \tag{10}$$

$$(\neg(x_1 \sim y_1 \wedge \cdots \wedge x_n \sim y_n \wedge p(x_1, \ldots, x_n) \wedge \neg p(y_1, \ldots, y_n)), \mathbf{1}) \tag{11}$$

where p is an n-ary predicate symbol (including also $\sim$), f is an n-ary function symbol, and x_i, y_j are some pairwise different variables.

The first axiom is obvious. An element x is similar to itself in truth degree $\mathbf{1}$. The second and third schemes can be viewed as some conditions of consistency. We do not want to have the strong conjunction $x_1 \sim y_1 \wedge \cdots \wedge x_n \sim y_n \wedge \neg(f(x_1, \ldots, x_n) \sim f(y_1, \ldots, y_n))$ true in some truth degree different from $\mathbf{0}$ (the second scheme); and also we do not wish to have $x_1 \sim y_1 \wedge \cdots \wedge x_n \sim$

$y_n \wedge p(x_1, \ldots, x_n) \wedge \neg p(y_1, \ldots, y_n))$ true in some truth degree different from $\mathbf{0}$ (the third scheme). The schemes can be reformulated as graded strong literal disjunctions. Denote the following set of axioms as

$$\begin{aligned} WSim_{\mathcal{L}} = &\{(x \sim x, \mathbf{1})\} \cup \\ &\{(\neg(x_1 \sim y_1) \dot{\vee} \cdots \dot{\vee} \neg(x_n \sim y_n) \dot{\vee} f(x_1, \ldots, x_n) \sim f(y_1, \ldots, y_n), \mathbf{1}) \\ &\quad | f \in Func_{\mathcal{L}},\ ar(f) = n > 0\} \cup \\ &\{(\neg(x_1 \sim y_1) \dot{\vee} \cdots \dot{\vee} \neg(x_n \sim y_n) \dot{\vee} \neg p(x_1, \ldots, x_n) \dot{\vee} p(y_1, \ldots, y_n), \mathbf{1}) \\ &\quad | p \in Pred_{\mathcal{L}},\ ar(p) = n\} \end{aligned}$$

where x_i and y_j are pairwise different variables; and we suppose that the weak similarity symbol $\sim$ belongs to $Pred_{\mathcal{L}}$.

From the axioms, we immediately get the reflexivity, weak symmetry, and weak transitivity of the predicate symbol $\sim$:

$$(x \sim x, \mathbf{1}), \qquad \text{(reflexivity)}$$

$$(\neg(x \sim y) \longleftarrow \neg(y \sim x), \mathbf{1}), \qquad \text{(weak symmetry)}$$

$$(\neg(x \sim y \wedge y \sim z) \longleftarrow \neg(x \sim z), \mathbf{1}). \qquad \text{(weak transitivity)}$$

A binary $\boldsymbol{L}$-fuzzy relation $\sim$ on U (a mapping $U \times U \longrightarrow \boldsymbol{L}$) is a weak similarity relation iff the following properties hold:

$$(u \sim u) = \mathbf{1}, \qquad \text{(reflexivity)}$$

$$((u_1 \sim u_2) \Rightarrow \mathbf{0}) = ((u_2 \sim u_1) \Rightarrow \mathbf{0}), \qquad \text{(weak symmetry)}$$

$$((u_1 \sim u_2 \sqcap u_2 \sim u_3) \Rightarrow \mathbf{0}) \geq ((u_1 \sim u_3) \Rightarrow \mathbf{0}), \qquad \text{(weak } \sqcap\text{-transitivity)}$$

$$((u_1 \sim v_1 \sqcap \cdots \sqcap u_n \sim v_n) \Rightarrow \mathbf{0}) \geq ((f(u_1, \ldots, u_n) \sim f(v_1, \ldots, v_n)) \Rightarrow \mathbf{0}). \qquad \text{(weak } \sqcap\text{-congruence)}$$

It is a simple matter to show that in every $\boldsymbol{L}$-model of $WSim_{\mathcal{L}}$, for a language with weak similarity, the predicate symbol $\sim$ must be interpreted by a weak similarity relation.

Since $(a \longrightarrow \neg\neg a, \mathbf{1})$ holds, the notion of weak similarity generalises the well-known notion of similarity [42, 86, 98, 109, 115, 66, 118], denoted as $\approx$ and axiomatised by the following schemes of axioms:

$$(x \approx x, \mathbf{1}), \tag{12}$$

$$(x_1 \approx y_1 \wedge \cdots \wedge x_n \approx y_n \longrightarrow f(x_1, \ldots, x_n) \approx f(y_1, \ldots, y_n), \mathbf{1}), \tag{13}$$

$$(x_1 \approx y_1 \wedge \cdots \wedge x_n \approx y_n \wedge p(x_1, \ldots, x_n) \longrightarrow p(y_1, \ldots, y_n), \mathbf{1}) \tag{14}$$

where p is an n-ary predicate symbol (including also $\approx$), f is an n-ary function symbol, and x_i, y_j are some pairwise different variables.

From the axioms, we immediately get the reflexivity, symmetry, and transitivity of the predicate symbol $\approx$:

$$(x \approx x, \mathbf{1}), \qquad \text{(reflexivity)}$$

$$(x \approx y \longrightarrow y \approx x, \mathbf{1}), \qquad \text{(symmetry)}$$

$$(x \approx y \wedge y \approx z \longrightarrow x \approx z, \mathbf{1}). \qquad \text{(transitivity)}$$

Hence, it is easy to see that in every $\boldsymbol{L}$-model for a language with similarity, the predicate symbol $\approx$ must be interpreted by a similarity relation. A binary $\boldsymbol{L}$-fuzzy relation $\approx$ on U (a mapping $U \times U \longrightarrow \boldsymbol{L}$) is a similarity relation iff the following properties hold:

$$(u \approx u) = \mathbf{1}, \qquad \text{(reflexivity)}$$

$$(u_1 \approx u_2) = (u_2 \approx u_1), \qquad \text{(symmetry)}$$

$$(u_1 \approx u_2 \cap u_2 \approx u_3) \leq (u_1 \approx u_3), \qquad \text{($\cap$-transitivity)}$$

$$(u_1 \approx v_1 \cap \cdots \cap u_n \approx v_n) \leq (f(u_1, \ldots, u_n) \approx f(v_1, \ldots, v_n)). \qquad \text{($\cap$-congruence)}$$

A (weak) similarity relation is called a (weak) fuzzy equality iff the following condition strengthens the reflexivity:

$$(u \sim u') = \mathbf{1} \text{ iff } u = u', \qquad \text{(strengthened reflexivity)}$$

$$(u \approx u') = \mathbf{1} \text{ iff } u = u'. \qquad \text{(strengthened reflexivity)}$$

Note that if $(\neg\neg a \longrightarrow a, \mathbf{1})$ holds, then both the notions of similarity coincide. Analogously to the 2-valued case, it can be proved the following statement [115]:

Theorem 1. *Consider a fuzzy logic where $(\neg\neg a \longrightarrow a, \mathbf{1})$ holds. Let T be a consistent theory with (weak) similarity, then there exists an $\mathbf{L}$-model of T such that the predicate symbol $(\sim) \approx$ is interpreted as a (weak) fuzzy equality relation.*

Proof: Let $\mathfrak{A}$ be an $\boldsymbol{L}$-model of T. Let $\sim_{\mathfrak{A}}$ be the weak similarity relation and also the similarity relation interpreting the symbols $\sim$ and $\approx$ in $\mathfrak{A}$. As we have claimed above; if $(\neg\neg a \longrightarrow a, \mathbf{1})$ holds, both the notions of similarity coincide. Consider the following equivalence $u \equiv u'$ iff $u \sim_{\mathfrak{A}} u' = \mathbf{1}$. Then the quotient model $\mathfrak{A}/_{\equiv}$ is a model for T. However, the quotient (weak) similarity relation $\sim_{\mathfrak{A}} /_{\equiv}$ is a (weak) fuzzy equality relation on $\mathcal{U}_{\mathfrak{A}}/_{\equiv}$ where $\mathcal{U}_{\mathfrak{A}}$ is the universum of $\mathfrak{A}$. □

Definition 1. A fuzzy disjunctive program of $\mathcal{L}$ with weak similarity is a set of graded strong literal implications of $\mathcal{L}$ containing the set of weak similarity axioms $WSim_{\mathcal{L}}$.

5 Fuzzy Disjunctive L-Models

5.1 Motivation

Let P be a positive fuzzy disjunctive program, viewed as a set of graded strong disjunctions of atoms. Our next aim is to propose a suitable model semantics for P.

We can follow the traditional approach and develop a minimal model semantics for P. Suppose that the language $\mathcal{L}$ of P contains at least one constant symbol. We can define a Herbrand $\boldsymbol{L}$-model $\mathcal{H}$ for $\mathcal{L}$ as a set of graded ground atoms of $\mathcal{L}$ of the form (a, c) such that $c \neq \mathbf{0}$, and in the model, there does not exist a graded ground atom of the form (a, c') so that $c \neq c'$. It is a simple matter to see that such a set $\mathcal{H}$ of graded ground atoms uniquely determines an $\boldsymbol{L}$-model $\mathcal{H}^*$ with the universum consisting of all ground terms of $\mathcal{L}$ (which is not empty because of some constant belonging to $\mathcal{L}$). The function symbols of $\mathcal{L}$ are interpreted in $\mathcal{H}^*$ freely, i.e. $f^{\mathcal{H}^*}(t_1, \dots, t_{ar(f)}) = f(t_1, \dots, t_{ar(f)})$, $f \in \mathit{Func}_{\mathcal{L}}$, $t_i \in \mathit{GTerm}_{\mathcal{L}}$; and for each predicate symbol $p \in \mathit{Pred}_{\mathcal{L}}$, we have an $\boldsymbol{L}$-fuzzy relation $p^{\mathcal{H}^*}$ defined as:

$$p^{\mathcal{H}^*}(t_1, \dots, t_{ar(p)}) = \begin{cases} c, \ (p(t_1, \dots, t_{ar(p)}), c) \in \mathcal{H}; \\ \mathbf{0}, \ \mathit{else}. \end{cases}$$

In the following text, we shall identify $\mathcal{H}$ with $\mathcal{H}^*$.

We say that a Herbrand $\boldsymbol{L}$-model $\mathcal{H}$ of P for $\mathcal{L}$ is minimal iff there do not exist a graded ground atom $(a, c) \in \mathcal{H}$ and $c' \in \boldsymbol{L}$ such that $c' < c$ and $(\mathcal{H} - (a, c)) \cup (a, c')$ is a model of P. By virtue of minimal Herbrand $\boldsymbol{L}$-models of P, we can propose the minimal model semantics for P: $\mathcal{MS}(P) = \{d \mid d$ *is a graded ground strong disjunction of atoms; and for every minimal Herbrand model* $\mathcal{H}$ *of* P *for* $\mathcal{L}$, $\mathcal{H} \models d\}$. An important fact holds: the graded ground strong disjunctions of atoms satisfied in all the minimal Herbrand models of the program P contain those which are fuzzy logical consequences of P. This fact ensures the inclusion of the declarative semantics $\mathcal{DS}(P) = \{d \mid d$ *is a graded ground strong disjunction of atoms and* $P \models d\}$ in the minimal model semantics of P.

Unfortunately, the minimal model semantics has a serious flaw: that is, *the set of minimal Herbrand* $\boldsymbol{L}$*-models for* P *can be uncountable.* We can represent this set by an infinite tree. In contrast to the 2-valued case, if the carrier set L is uncountable, also the sets of nodes and edges labeled by graded ground atoms will be uncountable (even though we suppose a countable language). The uncountable set of branches forms the minimal Herbrand $\boldsymbol{L}$-models, as shown in Figure 4.

However, the definition of program semantics as an uncountable (set-theoretic) object is not altogether appropriate in computer science. For instance, if $\boldsymbol{L}$ is the unit interval $[0, 1]$ and the language of a program is countable, then the cardinality of the set of nodes and edges of the tree of minimal Herbrand $\boldsymbol{L}$-models is $2^{\aleph_0}$ (the power of the continuum), and so are the cardinalities of the set of branches and of the set of minimal Herbrand $\boldsymbol{L}$-models.

In order to confirm our considerations, let us investigate the minimal model semantics of the following simple positive fuzzy disjunctive program $P = \{(p_0(f(x)) \dot{\vee} p_1(f(x)), \mathbf{1})\}$. The set of minimal Herbrand models for P contains

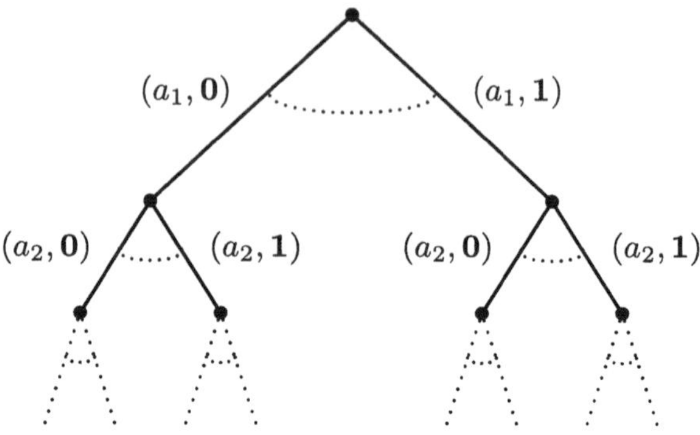

Figure 4. Tree of minimal Herbrand $\boldsymbol{L}$-models.

models of the form:

$$\bigcup_{n\geq 1} (\{(p_0(f^n(a)), s(n)) \mid s(n) \neq \mathbf{0}\} \cup \{(p_1(f^n(a)), s(n) \Rightarrow \mathbf{0}) \mid s(n) \Rightarrow \mathbf{0} \neq \mathbf{0}\})$$

where $s : \mathbb{N}^+ \longrightarrow \boldsymbol{L}$ (a is a new constant added to the language of P). Such a model, say $\mathfrak{A}$, is indeed a minimal Herbrand $\boldsymbol{L}$-model of P:

i) For an arbitrary variable assignment e, $e(x) = f^n(a)$, $n \geq 0$,

$$||p_0(f(x)) \dot{\vee} p_1(f(x))||_e^{\mathfrak{A}} = $$
$$((||p_0(f(x))||_e^{\mathfrak{A}} \Rightarrow \mathbf{0}) \cap (||p_1(f(x))||_e^{\mathfrak{A}} \Rightarrow \mathbf{0})) \Rightarrow \mathbf{0} =$$
$$((s(n+1) \Rightarrow \mathbf{0}) \cap ((s(n+1) \Rightarrow \mathbf{0}) \Rightarrow \mathbf{0})) \Rightarrow \mathbf{0} = \mathbf{0} \Rightarrow \mathbf{0} = \mathbf{1}.$$

So, P is satisfied in $\mathfrak{A}$.

ii) For any $n \geq 1$, neither $(p_0(f^n(a)), s(n))$ nor $(p_1(f^n(a)), s(n) \Rightarrow \mathbf{0})$ can be replaced in the model by $(p_0(f^n(a)), c)$ or $(p_1(f^n(a)), d)$, respectively, where $c < s(n)$, $d < s(n) \Rightarrow \mathbf{0}$. Otherwise from

$$||p_0(f(x)) \dot{\vee} p_1(f(x))||_{e(x/f^{n-1}(a))}^{\mathfrak{A}} = \mathbf{1},$$

there would exist either the pair c, $s(n) \Rightarrow \mathbf{0}$ or the pair $s(n)$, d of complementary elements such that $c \neq s(n)$ or $d \neq s(n) \Rightarrow \mathbf{0}$, which is a contradiction because an element in a Boolean lattice has exactly one complement. Hence, this model is also minimal.

Every minimal model of the above form is uniquely determined by a function $s : \mathbb{N}^+ \rightarrow \boldsymbol{L}$. In other words, there is a one-to-one mapping on the set of these minimal Herbrand $\boldsymbol{L}$-models onto the set of functions on $\mathbb{N}^+$ into $\boldsymbol{L}$, denoted as ${}^{\mathbb{N}^+}\boldsymbol{L}$. Since the lattice $\boldsymbol{L}$ contains at least two elements, e.g., $\mathbf{0}$, $\mathbf{1}$; ${}^{\mathbb{N}^+}\boldsymbol{L}$ is uncountable. So, in general, the set of minimal Herbrand $\boldsymbol{L}$-models for P is indeed uncountable.

We shall try to overcome this difficulty and propose a model semantics for fuzzy disjunctive programs on the basis of one model, instead of the class of minimal Herbrand $\boldsymbol{L}$-models. At the same time, we would like to guarantee the equivalence of this model semantics to the above mentioned declarative semantics, based on fuzzy logical entailment. The demand of one-model semantics implies an extension of the notion of a Herbrand $\boldsymbol{L}$-model, which is in a manner 'complete', to some kind of 'incomplete' model. Take the simple program $P = \{(p \dot{\vee} q, c)\}$, $c \neq \mathbf{0}$. We see that neither $P \models (p, c_1)$ for any $c_1 \neq \mathbf{0}$, nor $P \models (q, c_2)$ for any $c_2 \neq \mathbf{0}$, but $P \models (p \dot{\vee} q, c)$ for $c \neq \mathbf{0}$. In accordance with the intended equivalence to the declarative semantics, we expect that our proposed model, say $\mathfrak{M}$, will satisfy: $\mathfrak{M} \not\models (p, c_1)$ for any $c_1 \neq \mathbf{0}$, $\mathfrak{M} \not\models (q, c_2)$ for any $c_2 \neq \mathbf{0}$, but $\mathfrak{M} \models (p \dot{\vee} q, c)$ for $c \neq \mathbf{0}$ - that is, it will be 'incomplete' in contrast to any 'complete' Herbrand $\boldsymbol{L}$-model of P, where both p and q must be true in some truth degrees c_1 and c_2, respectively, such that $((c_1 \Rightarrow \mathbf{0}) \cap (c_2 \Rightarrow \mathbf{0})) \Rightarrow \mathbf{0} \geq c \neq \mathbf{0}$; hence, either $c_1 \neq \mathbf{0}$ or $c_2 \neq \mathbf{0}$.

5.2 Formal Treatment

At first, we informally explain one way how to extend classical 2-valued Herbrand models and develop a model semantics for disjunctive programs viewed as sets of literal implications of the form: $l_1 \vee \cdots \vee l_n \longleftarrow k_1 \wedge \cdots \wedge k_m$ where l_i and k_j are literals. Since we do not yet consider any kind of negation-by-default, literal implications of a disjunctive program may be rewritten on equivalent literal disjunctions without changing the logical meaning of the program. We introduce a class of disjunctive models for literal implications and disjunctions, proposed in [33] and formerly also called clausal models.

Roughly speaking, a disjunctive model for $\mathcal{L}$, say $\mathfrak{M}$, is a set of ground literal disjunctions of $\mathcal{L}$ augmented by a set of new pairwise different constants representing elements of the non-empty universum of $\mathfrak{M}$. A similar notion called the minimal model state is introduced in [68]. The model has to satisfy several technical yet natural conditions:

i) it does not contain tautologies, the empty disjunction, and superdisjunctions;
ii) it is closed under binary resolution.

The domain $\mathcal{U}_{\mathfrak{M}}$ of $\mathfrak{M}$ is a non-empty set. A function symbol $f \in Func_{\mathcal{L}}$ is interpreted as a function $f^{\mathfrak{M}} : \mathcal{U}_{\mathfrak{M}}^{ar(f)} \longrightarrow \mathcal{U}_{\mathfrak{M}}$.

A literal implication $l_1 \vee \cdots \vee l_n \longleftarrow k_1 \wedge \cdots \wedge k_m$ is satisfied with respect to a variable assignment e in $\mathfrak{M}$ iff there exists $b \in \mathfrak{M}$ such that b subsumes the ground literal disjunction $D = \neg k_1[e] \vee \cdots \vee \neg k_m[e] \vee l_1[e] \vee \cdots \vee l_n[e]$ in $\mathcal{L}$ augmented by a set of new pairwise different constants representing elements of the universum of $\mathfrak{M}$; or D is a tautology. An implication Im is true in $\mathfrak{M}$, in symbols $\mathfrak{M} \models^{\vee} Im$, iff Im is satisfied under all variable assignments in $\mathfrak{M}$. As we consider first-order logic with equality, we lay the following condition on the model: all the equality axioms of $\mathcal{L}$ must be true in $\mathfrak{M}$, in symbols $\mathfrak{M} \models^{\vee} Eq_{\mathcal{L}}$.

We now explain the above mentioned conditions for a disjunctive model more in detail. From the definition of a literal implication being true in a disjunctive model, it is evident that superdisjunctions and tautologies are unnecessary; therefore, we do not include them in a model. That is if there existed some b' in a model subsuming a given instantiated disjunction D, and moreover, there was some b in the model subsuming b', then b should subsume also D, and the unnecessary b' might be omitted. Since we explicitly test in the definition of disjunctive model whether a given instantiated disjunction is a tautology, tautologies may be also removed from a model. If a model contained the empty disjunction, then it could be straightforward to deduce that all literal implications would be true; in other words, the model would be 'inconsistent'. In the following treatment, we shall not consider trivial 'inconsistent' disjunctive models; therefore, we shall suppose that disjunctive models do not include the empty disjunction.

The condition that a disjunctive model is closed under binary resolution is all-important. Take the set $S = \{p \vee q, \neg q \vee r\}$. We see that $S \models p \vee r$, but no disjunction in S subsumes $p \vee r$. In order to make S a disjunctive model and achieve the equivalence to the declarative semantics, we must add $p \vee r$ to S and in this manner close S under binary resolution. In general, by closing a set of ground literal disjunctions under binary resolution, we augment this set with 'essential' tautological consequences in the form of ground literal disjunctions. They are really essential, because all tautological consequences in the form of ground literal disjunctions are either superdisjunctions of those in the augmented set or tautologies. Therefore this condition is necessary to the correct definition of disjunctive model.

We are in position to gather our considerations to the formal definition:

Definition 2. A disjunctive model $\mathfrak{M}$ for $\mathcal{L}$ is a pair $(\mathcal{U}_{\mathfrak{M}}, \mathcal{B}_{\mathfrak{M}})$ defined as follows:

- $\mathcal{U}_{\mathfrak{M}} \neq \emptyset$ is the universum (domain) of the model $\mathfrak{M}$.
- A function symbol $f \in Func_{\mathcal{L}}$ is interpreted as a function $f^{\mathfrak{M}} : \mathcal{U}_{\mathfrak{M}}^{ar(f)} \longrightarrow \mathcal{U}_{\mathfrak{M}}$.
- The base $\mathcal{B}_{\mathfrak{M}}$ of the model $\mathfrak{M}$ is a set of finite sets of triples of the form $(s, p, (u_1, \ldots, u_{ar(p)}))$ where $s \in \{0, 1\}$, $p \in Pred_{\mathcal{L}}$, and $u_i \in \mathcal{U}_{\mathfrak{M}}$. Such a triple will be denoted as $\overline{p(u_1, \ldots, u_{ar(p)})}$ when $s = 1$, or $\overline{\neg p(u_1, \ldots, u_{ar(p)})}$ when $s = 0$. This triple is understood as the 'semantic counterpart' of the ground literal $p(u_1, \ldots, u_{ar(p)})$ of the language $\mathcal{L} \cup \mathcal{U}_{\mathfrak{M}}$ (new added constants). A finite set of triples will be denoted as $\overline{l_1} \,\overline{\vee} \cdots \overline{\vee}\, \overline{l_n}$ where $\overline{l_i}$ are triples - the 'counterparts' of the ground literals l_i. This set of triples is understood as the 'semantic counterpart' of the ground literal disjunction $l_1 \vee \cdots \vee l_n$. The base $\mathcal{B}_{\mathfrak{M}}$ satisfies the following conditions:

  i) $\nexists\, b \in \mathcal{B}_{\mathfrak{M}}$ containing a pair of contradictory triples, i.e. triples of the forms $\overline{l}$ and $\overline{\neg l}$.

  ii) $\emptyset \notin \mathcal{B}_{\mathfrak{M}}$.

  iii) $\neg(\exists\, b, b' \in \mathcal{B}_{\mathfrak{M}}\; b \subset b')$.

  iv) If $\overline{l} \in b$, $\overline{\neg l} \in b'$, and $b, b' \in \mathcal{B}_{\mathfrak{M}}$, then either $\exists\, b'' \in \mathcal{B}_{\mathfrak{M}}\; b'' \subseteq (b - \{\overline{l}\}) \cup (b' - \{\overline{\neg l}\})$, or $(b - \{\overline{l}\}) \cup (b' - \{\overline{\neg l}\})$ contains a pair of contradictory triples.

- A literal implication of $\mathcal{L}$ $l_1 \vee \cdots \vee l_n \longleftarrow k_1 \wedge \cdots \wedge k_m$ is satisfied with respect to a variable assignment e in $\mathfrak{M}$ iff there exists $b \in \mathcal{B}_{\mathfrak{M}}$ such that $b \subseteq \overline{D[e]} = \overline{\neg k_1[e]} \mathbin{\overline{\vee}} \cdots \mathbin{\overline{\vee}} \overline{\neg k_m[e]} \mathbin{\overline{\vee}} \overline{l_1[e]} \mathbin{\overline{\vee}} \cdots \mathbin{\overline{\vee}} \overline{l_n[e]}$; or $\overline{D[e]}$ contains a pair of contradictory triples.
- An literal implication Im of $\mathcal{L}$ is true in $\mathfrak{M}$, in symbols $\mathfrak{M} \models^{\vee} Im$ iff Im is satisfied under all variable assignments in $\mathfrak{M}$.
- A set T of literal implications of $\mathcal{L}$ is true in $\mathfrak{M}$, in symbols $\mathfrak{M} \models^{\vee} T$, iff $\mathfrak{M} \models^{\vee} Im$ for every $Im \in T$.
- We write $T \models^{\vee} Im$ iff for every disjunctive model $\mathfrak{M}$ of $\mathcal{L}$, $\mathfrak{M} \models^{\vee} T$ implies $\mathfrak{M} \models^{\vee} Im$.
- $\mathfrak{M} \models^{\vee} Eq_{\mathcal{L}}$.

When applying disjunctive models, we obtain the following advantages [33]:

i) For any disjunctive program P, we have the unique model $\mathfrak{M}$ in place of the set (perhaps uncountable) of minimal models.
ii) The model $\mathfrak{M}$ is 'characteristic' in a manner of speaking: any literal disjunction is true in $\mathfrak{M}$ if and only if this disjunction is a logical consequence of the program P, which captures the coincidence between the model semantics $\mathcal{MS}(P) = \{d \mid d \textit{ is a literal disjunction, } \mathfrak{M} \models^{\vee} d\}$ and the declarative semantics $\mathcal{DS}(P) = \{d \mid d \textit{ is a literal disjunction, } P \models d\}$.
iii) The model $\mathfrak{M}$ is constructible by an operator $\mathcal{C}_P$ associated with the program P, proposed in [33].

Consider other simple examples of disjunctive models:

- Let $P = \{p(f(a)) \vee \neg q(g(b)),\ q(g(b)) \vee r(c),\ r(c)\}$. As we see, the third disjunction subsumes the second one; therefore, the corresponding model $\mathfrak{M}$ for P is without the second disjunction: $\mathfrak{M} = (GTerm_{\{f,g,a,b,\}}, \{\overline{p(f(a))} \mathbin{\overline{\vee}} \overline{\neg q(g(b))}, \overline{r(c)}\})$.
- A second example: Let $P = \{p(x) \vee q(x),\ \neg q(f(z,v)) \vee r(z)\}$. In order to construct the model $\mathfrak{M}$ for this program, we have to infer the consequence $p(f(t_1,t_2)) \vee r(t_1)$. Then the model consists of these ground disjunctions: $\mathfrak{M} = (GTerm_{\{f,a\}}, \{\overline{p(t)} \mathbin{\overline{\vee}} \overline{q(t)} \mid t \in GTerm_{\{f,a\}}\} \cup \{\overline{\neg q(f(t_1,t_2))} \mathbin{\overline{\vee}} \overline{r(t_1)} \mid t_1, t_2 \in GTerm_{\{f,a\}}\} \cup \{\overline{p(f(t_1,t_2))} \mathbin{\overline{\vee}} \overline{r(t_1)} \mid t_1, t_2 \in GTerm_{\{f,a\}}\})$.

In conclusion, we state the connection theorem [33], which relates disjunctive models to first-order models. The theorem shows that the relation of logical entailment $\models$ and the relation $\models^{\vee}$ from Definition 2 have the same meaning over literal implications.

Theorem 2 (Connection Theorem). *Let T be a set of literal implications of $\mathcal{L}$ and C be a literal implication of $\mathcal{L}$.*

$$T \models^{\vee} C \textit{ if and only if } T \models C.$$

Proof: See [33]. □

We now explain a way how to extend Herbrand $\boldsymbol{L}$-models and develop a model semantics for fuzzy disjunctive programs with weak similarity. We introduce a concept of a fuzzy disjunctive $\boldsymbol{L}$-model, which generalises the notion of a disjunctive model to the case of graded strong literal implications.

Definition 3. A fuzzy disjunctive $\boldsymbol{L}$-model $\mathfrak{M}$ for $\mathcal{L}$ is a pair $(\mathcal{U}_{\mathfrak{M}}, \mathcal{B}_{\mathfrak{M}})$ defined as follows:

- $\mathcal{U}_{\mathfrak{M}} \neq \emptyset$ is the universum (domain) of the model $\mathfrak{M}$.
- A function symbol $f \in Func_{\mathcal{L}}$ is interpreted as a function $f^{\mathfrak{M}} : \mathcal{U}_{\mathfrak{M}}^{ar(f)} \longrightarrow \mathcal{U}_{\mathfrak{M}}$.
- The base $\mathcal{B}_{\mathfrak{M}}$ is a set of pairs of the form $(\overline{D}, c)$.
  $\overline{D}$ is a finite set of triples of the form $(s, p, (u_1, \ldots, u_{ar(p)}))$ where $s \in \{0, 1\}$, $p \in Pred_{\mathcal{L}}$, $u_i \in \mathcal{U}_{\mathfrak{M}}$, and c is a truth value in $\boldsymbol{L}$. Such a triple will be denoted as $\overline{p(u_1, \ldots, u_{ar(p)})}$ if $s = 1$, or $\overline{\neg p(u_1, \ldots, u_{ar(p)})}$ for $s = 0$, and understood as the 'semantic counterpart' of the ground literal $(\neg)\, p(u_1, \ldots, u_{ar(p)})$ of the language $\mathcal{L} \cup \mathcal{U}_{\mathfrak{M}}$ (new added constants).
  A finite set of triples will be often denoted as $\overline{l_1} \,\bar{\dot\vee}\, \cdots \,\bar{\dot\vee}\, \overline{l_n}$ or $\overline{\dot\bigvee_{i=1\ldots n} l_i}$ where $\overline{l_i}$ are triples. Such a set can be understood as the 'semantic counterpart' of the ground strong literal disjunction $l_1 \dot\vee \cdots \dot\vee l_n$ of the language $\mathcal{L} \cup \mathcal{U}_{\mathfrak{M}}$.
  So, a pair $(\overline{D}, c)$ of a finite set of triples and of a truth value $c \in \boldsymbol{L}$ is understood as the 'semantic counterpart' of the graded ground strong literal disjunction $(l_1 \dot\vee \cdots \dot\vee l_n, c)$ of the language $\mathcal{L} \cup \mathcal{U}_{\mathfrak{M}}$.
  The base $\mathcal{B}_{\mathfrak{M}}$ has to satisfy the following conditions:
  i) $\forall\, (\overline{D}, c) \in \mathcal{B}_{\mathfrak{M}}\ (\overline{D} \neq \emptyset \wedge c \neq \mathbf{0})$;
  ii) $\neg(\exists\, (\overline{D}, c), (\overline{D}, c') \in \mathcal{B}_{\mathfrak{M}}\ c \neq c')$;
  iii) $\nexists\, (\overline{D}, c) \in \mathcal{B}_{\mathfrak{M}}$ such that $\overline{D}$ contains a pair of contradictory triples, i.e. triples of the forms $\overline{l}$ and $\overline{\neg l}$;
  iv) $\forall\, (\overline{D}, c), (\overline{D'}, c') \in \mathcal{B}_{\mathfrak{M}}\ (\overline{D} \subset \overline{D'} \longrightarrow c < c')$;
  v) if $\exists\, (\overline{D}, c), (\overline{D'}, c') \in \mathcal{B}_{\mathfrak{M}}$ and there exists a triple $\overline{l}$ such that $\overline{l} \in \overline{D}$ and $\overline{\neg l} \in \overline{D'}$, then either
  $$c \cap c' \leq \bigcup \{c'' \mid (\overline{D''}, c'') \in \mathcal{B}_{\mathfrak{M}},\ \overline{D''} \subseteq (\overline{D} - \{\overline{l}\}) \cup (\overline{D'} - \{\overline{\neg l}\})\},$$
  or $(\overline{D} - \{\overline{l}\}) \cup (\overline{D'} - \{\overline{\neg l}\})$ contains a pair of contradictory triples.
- A variable assignment in $\mathfrak{M}$ is a mapping $Var_{\mathcal{L}} \longrightarrow \mathcal{U}_{\mathfrak{M}}$ assigning an element of the universum $\mathcal{U}_{\mathfrak{M}}$ to each variable.
  Let e be a variable assignment in $\mathfrak{M}$, t a term, $p(t_1, \ldots, t_{ar(p)})$ an atom, and $l_1 \dot\vee \cdots \dot\vee l_n$, $n \geq 0$, a strong literal disjunction of $\mathcal{L}$. We can assign an element of $\mathcal{U}_{\mathfrak{M}}$ to t, denoted as $||t||_e^{\mathfrak{M}}$ in the following manner:
  – $t \equiv v \in Var_{\mathcal{L}}$, $||v||_e^{\mathfrak{M}} = e(v)$;
  – $t \equiv f(t_1, \ldots, t_{ar(f)})$ where $f \in Func_{\mathcal{L}}$, t_i are terms of $\mathcal{L}$,
  $$||f(t_1, \ldots, t_{ar(f)})||_e^{\mathfrak{M}} = f^{\mathfrak{M}}(||t_1||_e^{\mathfrak{M}}, \ldots, ||t_{ar(f)}||_e^{\mathfrak{M}});$$

By $\overline{(\neg)p(t_1,\ldots,t_{ar(p)})[e]}$ we denote the triple $\overline{(\neg)p(||t_1||_e^{\mathfrak{M}},\ldots,||t_{ar(p)}||_e^{\mathfrak{M}})}$; and by $\overline{(l_1 \dot\vee \cdots \dot\vee l_n \longleftarrow k_1 \wedge \cdots \wedge k_m)[e]}, m, n \geq 0$, we denote the set of triples $\overline{l_1[e]} \bar{\dot\vee} \cdots \bar{\dot\vee} \overline{l_n[e]} \bar{\dot\vee} \overline{\neg k_1[e]} \bar{\dot\vee} \cdots \bar{\dot\vee} \overline{\neg k_m[e]}$.

- Let Im be a strong literal implication of $\mathcal{L}$ and e a variable assignment in $\mathfrak{M}$. We can define the truth value of Im wrt. e as follows:
$$||Im||_e^{\mathfrak{M}} = \begin{cases} \mathbf{1}, & \overline{Im[e]} \textit{ contains a pair of contradictory triples;} \\ \bigcup\{c' \mid (\overline{D'}, c') \in \mathcal{B}_{\mathfrak{M}}, \overline{D'} \subseteq \overline{Im[e]}\}, & \textit{else.} \end{cases}$$
We define the truth value of Im in $\mathfrak{M}$ as the infimum
$$||Im||^{\mathfrak{M}} = \bigcap\{||Im||_e^{\mathfrak{M}} \mid e \textit{ is a variable assignment in } \mathfrak{M}\}.$$
- Let (Im, c) be a graded strong literal implication of $\mathcal{L}$. (Im, c) is true in $\mathfrak{M}$ wrt. a variable assignment e, denoted as $\mathfrak{M} \models^{\vee} (Im, c)[e]$, iff $||Im||_e^{\mathfrak{M}} \geq c$. Similarly, (Im, c) is true in $\mathfrak{M}$, in symbols $\mathfrak{M} \models^{\vee} (Im, c)$, iff $||Im||^{\mathfrak{M}} \geq c$.
We say that $\mathfrak{M}$ is a model of a set T of graded strong literal implications of $\mathcal{L}$, written as $\mathfrak{M} \models^{\vee} T$, iff $\mathfrak{M}$ satisfies every graded strong literal implication in T.
We write $T \models^{\vee} (Im, c)$ iff for every fuzzy disjunctive $\boldsymbol{L}$-model $\mathfrak{N}$ for $\mathcal{L}$, $\mathfrak{N} \models^{\vee} T$ implies $\mathfrak{N} \models^{\vee} (Im, c)$.
Let T' be a set of graded strong literal implications of $\mathcal{L}$. We write $T \models^{\vee} T'$ iff for every fuzzy disjunctive $\boldsymbol{L}$-model $\mathfrak{N}$ for $\mathcal{L}$, $\mathfrak{N} \models^{\vee} T$ implies $\mathfrak{N} \models^{\vee} T'$.
- $\mathfrak{M} \models^{\vee} WSim_{\mathcal{L}}$.

Let D be a strong literal disjunction and e a variable assignment. Similarly to disjunctive models, we can assign the truth value for D with respect to e as follows: if $\overline{D[e]}$ contains a pair of contradictory triples, say $\overline{l[e]}$, $\overline{\neg l'[e]}$ (the literals l, $\neg l'$ belong to D), i.e. $\overline{D[e]}$ is a 'tautology', then we explicitly set the truth value on $\mathbf{1}$; this is the only possible truth value of D wrt. e in every $\boldsymbol{L}$-model, since $\mathbf{1} = ||l \dot\vee \neg l'||_e \leq ||D||_e$. For this reason, we do not need include such 'tautologies' to the base $\mathcal{B}_{\mathfrak{M}}$, the condition iii).

When $\overline{D[e]}$ does not contain contradictory triples, we collect all pairs in $\mathcal{B}_{\mathfrak{M}}$ of the form $(\overline{D'}, c')$ such that $\overline{D'} \subseteq \overline{D[e]}$ and compute the supremum of their truth values c'. This is a slight generalisation of the case of disjunctive models, where it is sufficient to find out one subset of $\overline{D[e]}$, since all subsets in the base of a disjunctive model have the largest truth value $\mathbf{1}$ (we are in 2-valued first-order logic). For this reason, we do not include to $\mathcal{B}_{\mathfrak{M}}$ pairs with the truth value $\mathbf{0}$, the condition i). This condition also says that $\mathcal{B}_{\mathfrak{M}}$ does not contain a pair $(\emptyset, c)$ for any $c \neq \mathbf{0}$, for otherwise our model $\mathfrak{M}$ would be in a manner of speaking 'inconsistent': all strong literal disjunctions would be true in some non-zero truth degree.

In the definition, we allow for a given finite set of triples only one truth value, the condition ii), which is a technical constraint for simplifying the base $\mathcal{B}_{\mathfrak{M}}$. When $\mathcal{B}_{\mathfrak{M}}$ contains some pairs $(\overline{D}, c)$, $(\overline{D'}, c')$ such that $\overline{D} \subset \overline{D'}$, then c' must yield an 'non-zero' increase to the truth value of $\overline{D}$, else the pair $(\overline{D'}, c')$ would be unnecessary, the condition iv).

An interesting property is formulated by the condition v). This condition postulates a resolution inference rule and determines the least truth value of a resolvent as the greatest lower bound $\cap$ of the truth values of the input sets of triples. Via the condition v), the base $\mathcal{B}_{\mathfrak{M}}$ is closed under this resolution rule, which yields to $\mathcal{B}_{\mathfrak{M}}$ 'essential' consequences of already comprised pairs. Also, the following interesting fact holds: if $\mathcal{B}_{\mathfrak{M}}$ contains pairs $(\{\bar{l}\}, c)$ and $(\{\overline{\neg l}\}, c')$, where $\bar{l}$ and $\overline{\neg l}$ are contradictory triples, then $c \cap c' = \mathbf{0}$, for otherwise, by the condition v), $\mathcal{B}_{\mathfrak{M}}$ should contain a pair $(\emptyset, c'')$, $c'' \neq \mathbf{0}$. In other words, such pairs do not imply inconsistency. Indeed, consider the following graded theory $\{(p, c),\ (\neg p, c')\}$ where $c \cap c' = \mathbf{0}$. The theory has a $\boldsymbol{L}$-model $\mathfrak{A}$ such that $p^{\mathfrak{A}} = c$. Then $||\neg p||^{\mathfrak{A}} \geq c'$, since $c \cap c' = \mathbf{0}$ and $c' \leq c \Rightarrow \mathbf{0}$ by the condition of adjunction. So, both the graded contradictory literals are true in $\mathfrak{A}$. Hence, the theory is consistent.

We now illustrate the definition with simple examples. Suppose that $X = \{x_1, x_2, x_3\}$ and the language $\mathcal{L}$ consists of the unary predicate symbols p, q. Consider the Boolean lattice $\boldsymbol{L} = (2^X, \subseteq, \cup, \cap, \emptyset, X)$.

- Let $\mathfrak{M} = (\{u_1, u_2\},\ \{(\{\overline{p(u_1)}\}, \{x_1\}),\ (\{\overline{p(u_2)}\}, \{x_1, x_2\}),\ (\{\overline{q(u_1)}\}, \{x_3\}),\ (\{\overline{q(u_2)}\}, \{x_1, x_2\})\})$. Consider the atoms $(p(v_1), \{x_1\})$ and $(q(v_1), \{x_1\})$. We can see that $||p(v_1)||^{\mathfrak{M}} = \{x_1\} \cap \{x_1, x_2\} = \{x_1\}$ and $||q(v_1)||^{\mathfrak{M}} = \{x_3\} \cap \{x_1, x_2\} = \emptyset$. So, $\mathfrak{M} \models^{\vee} (p(v_1), \{x_1\})$, but $\mathfrak{M} \not\models^{\vee} (q(v_1), \{x_1\})$.
- Let $\mathfrak{M} = (\{u_1, u_2\},\ \{(\{\overline{p(u_1)}\}, \{x_1\}),\ (\{\overline{p(u_2)}\}, \{x_1, x_2\}),\ (\overline{p(u_1)}\ \bar{\dot{\vee}}\ \overline{q(u_1)}, \{x_1, x_2\})\})$. Take the graded strong disjunctions of atoms $(p(v_1), \{x_1\})$, $(p(v_1), \{x_1, x_2\})$, and $(p(v_1) \dot{\vee} q(v_1), \{x_1, x_2\})$. Then $||p(v_1)||^{\mathfrak{M}} = \{x_1\} \cap \{x_1, x_2\} = \{x_1\}$ and $||p(v_1) \dot{\vee} q(v_1)||^{\mathfrak{M}} = \{x_1, x_2\} \cap \{x_1, x_2\} = \{x_1, x_2\}$. Hence $\mathfrak{M} \models^{\vee} (p(v_1), \{x_1\})$, but $\mathfrak{M} \not\models^{\vee} (p(v_1), \{x_1, x_2\})$. However, $\mathfrak{M} \models^{\vee} (p(v_1) \dot{\vee} q(v_1), \{x_1, x_2\})$.
- Let $\mathfrak{M} = (\{u_1\},\ \{(\{\overline{p(u_1)}\}, \{x_1\}),\ (\{\overline{\neg p(u_1)}\}, \{x_2\}),\ (\overline{p(u_1)}\ \bar{\dot{\vee}}\ \overline{q(u_1)}, \{x_2, x_3\}), (\{\overline{q(u_1)}\}, \{x_1, x_2\})\})$. Consider the following graded strong literal disjunctions $(p(v_1), \{x_1\})$, $(\neg p(v_1), \{x_2\})$, $(p(v_1) \dot{\vee} \neg p(v_1), X)$, and $(p(v_1) \dot{\vee} \neg p(v_1), X)$. We can check that $||p(v_1)||^{\mathfrak{M}} = \{x_1\}$; $||\neg p(v_1)||^{\mathfrak{M}} = \{x_2\}$; $||p(v_1) \dot{\vee} \neg p(v_1)||^{\mathfrak{M}} = X$, since the set $\overline{p(u_1)}\ \bar{\dot{\vee}}\ \overline{\neg p(u_1)}$ contains a pair of contradictory literals; $||p(v_1) \dot{\vee} q(v_1)||^{\mathfrak{M}} = \{x_1\} \cup \{x_2, x_3\} \cup \{x_1, x_2\} = X$. So, all the considered graded strong literal disjunctions are true in $\mathfrak{M}$. Note that $||p(v_1)||^{\mathfrak{M}} \cap ||\neg p(v_1)||^{\mathfrak{M}} = \emptyset$; for otherwise $(\emptyset, c)$, $c \neq \emptyset$, should belong to $\mathcal{B}_{\mathfrak{M}}$, which would violate the condition i) of Definition 3.

In the end, we state the theorem which relates fuzzy disjunctive $\boldsymbol{L}$-models to $\boldsymbol{L}$-models.

Theorem 3 (Fuzzy Connection Theorem). *Suppose that the lattice of truth values $\boldsymbol{L}$ is a Boolean one with the complement operator $x' = (x \Rightarrow \mathbf{0})$. Let T be a set of graded strong literal implications of $\mathcal{L}$ and (C, c) be a graded strong literal implication of $\mathcal{L}$.*

$$T \models^{\vee} (C, c) \text{ if and only if } T \models (C, c).$$

Proof: The following identities hold in the lattice $\boldsymbol{L}$: Let $\alpha \in \boldsymbol{L}$ and $K \subseteq \boldsymbol{L}$, then

$$\bigcup K \Rightarrow \mathbf{0} = \bigcap_{k \in K} (k \Rightarrow \mathbf{0}),$$
$$(\alpha \Rightarrow \mathbf{0}) \Rightarrow \mathbf{0} = \alpha,$$
$$\bigcap K \Rightarrow \mathbf{0} = \bigcup_{k \in K} (k \Rightarrow \mathbf{0}).$$

From the antimonotonicity of $\Rightarrow \mathbf{0}$, we get $\bigcup K \Rightarrow \mathbf{0} \leq \bigcap_{k \in K}(k \Rightarrow \mathbf{0})$. By the condition of adjunction, we have $\bigcup K \Rightarrow \mathbf{0} \geq \bigcap_{k \in K}(k \Rightarrow \mathbf{0})$ is equivalent to $\bigcup K \cap \bigcap_{k \in K}(k \Rightarrow \mathbf{0}) \leq \mathbf{0}$. However, $\bigcup K \cap \bigcap_{k \in K}(k \Rightarrow \mathbf{0}) = \bigcup_{k \in K}(k \cap \bigcap_{k \in K}(k \Rightarrow \mathbf{0})) \leq \bigcup_{k \in K}(k \cap (k \Rightarrow \mathbf{0})) \leq \bigcup_{k \in K} \mathbf{0} = \mathbf{0}$. Hence, the first identity follows from the fact that $\boldsymbol{L}$ is a Brouwer lattice. The second identity is a axiom of the Boolean lattice $\boldsymbol{L}$. The third identity follows from the previous ones: $\bigcap K \Rightarrow \mathbf{0} = (\bigcap_{k \in K}((k \Rightarrow \mathbf{0}) \Rightarrow \mathbf{0})) \Rightarrow \mathbf{0} = (\bigcup_{k \in K}(k \Rightarrow \mathbf{0})) \Rightarrow \mathbf{0} \Rightarrow \mathbf{0} = \bigcup_{k \in K}(k \Rightarrow \mathbf{0})$.

($\Longrightarrow$) Suppose $\mathfrak{A}$ is an $\boldsymbol{L}$-model of $\mathcal{L}$. We can construct a fuzzy disjunctive $\boldsymbol{L}$-model $\mathfrak{M}$ of $\mathcal{L}$ such that for every graded strong literal implication (Im, c_{Im}) of $\mathcal{L}$,

$$\mathfrak{M} \models^{\vee} (Im, c_{Im}) \Longleftrightarrow \mathfrak{A} \models (Im, c_{Im}). \tag{15}$$

We let $\mathfrak{M} = (\mathcal{U}_{\mathfrak{A}}, \mathcal{B}_{\mathfrak{M}})$, the domain of $\mathfrak{M}$ is the same as that of $\mathfrak{A}$. A function symbol $f \in Func_{\mathcal{L}}$ is interpreted as the function $f^{\mathfrak{M}} = f^{\mathfrak{A}}$, and

$$\mathcal{B}_{\mathfrak{M}} = \{(\{\overline{p(u_1, \ldots, u_{ar(p)})}\}, c_p) \mid p^{\mathfrak{A}}(u_1, \ldots, u_{ar(p)}) = c_p \geq \mathbf{0}\} \cup$$
$$\{(\{\overline{\neg p(u_1, \ldots, u_{ar(p)})}\}, c_p \Rightarrow \mathbf{0}) \mid p^{\mathfrak{A}}(u_1, \ldots, u_{ar(p)}) = c_p, (c_p \Rightarrow \mathbf{0}) \geq \mathbf{0}\}.$$

It is easy to see that the conditions i) - v) in the definition 3 are satisfied. Since $\mathcal{B}_{\mathfrak{M}}$ consists only of singletons, it meets the conditions i) - iv) trivially. Also the condition v) is satisfied evidently; because, in $\mathcal{B}_{\mathfrak{M}}$ there exist only pairs of the form $(\{\overline{p(u_1, \ldots, u_{ar(p)})}\}, c_p)$ and $(\{\overline{\neg p(u_1, \ldots, u_{ar(p)})}\}, c_p \Rightarrow \mathbf{0})$. So, $\mathbf{0} = c_p \cap c_p \Rightarrow \mathbf{0} = \bigcup \emptyset = \bigcup \{c'' \mid (\overline{D''}, c'') \in \mathcal{B}_{\mathfrak{M}}, \overline{D''} \subseteq \emptyset\}$. $\mathfrak{M}$ is indeed a fuzzy disjunctive $\boldsymbol{L}$-model for $\mathcal{L}$.

Let $(l_1 \dot{\vee} \cdots \dot{\vee} l_n \longleftarrow k_1 \wedge \cdots \wedge k_m, c_l)$ be a strong literal implication of $\mathcal{L}$ and e a variable assignment in $\mathfrak{A}$ and also in $\mathfrak{M}$; since, the domains are the same. Then the following are equivalent:

- $\mathfrak{A} \models (l_1 \dot{\vee} \cdots \dot{\vee} l_n \longleftarrow k_1 \wedge \cdots \wedge k_m, c_l)[e]$ $\Longleftrightarrow$
- $\mathfrak{A} \models (\neg(\neg l_1 \wedge \cdots \wedge \neg l_n) \longleftarrow k_1 \wedge \cdots \wedge k_m), c_l)[e]$ $\Longleftrightarrow$
- $\mathfrak{A} \models (\neg(\neg l_1 \wedge \cdots \wedge \neg l_n \wedge k_1 \wedge \cdots \wedge k_m), c_l)[e]$ $\Longleftrightarrow$
- $\mathfrak{A} \models (||\neg l_1||_e^{\mathfrak{A}} \cap \cdots \cap ||\neg l_n||_e^{\mathfrak{A}} \cap ||k_1||_e^{\mathfrak{A}} \cap \cdots \cap ||k_m||_e^{\mathfrak{A}}) \Rightarrow \mathbf{0} \geq c_l$ $\Longleftrightarrow$
- $\mathfrak{A} \models ||\neg l_1||_e^{\mathfrak{A}} \Rightarrow \mathbf{0} \cup \cdots \cup ||\neg l_n||_e^{\mathfrak{A}} \Rightarrow \mathbf{0} \cup$ $||k_1||_e^{\mathfrak{A}} \Rightarrow \mathbf{0} \cup \cdots \cup ||k_m||_e^{\mathfrak{A}} \Rightarrow \mathbf{0} \geq c_l$ $\Longleftrightarrow$

- $\mathfrak{A} \models ||l_1||_e^{\mathfrak{A}} \cup \dots \cup ||l_n||_e^{\mathfrak{A}} \cup ||\neg k_1||_e^{\mathfrak{A}} \cup \dots \cup ||\neg k_m||_e^{\mathfrak{A}} \geq c_l \iff$
- $\bigcup\{c \mid (\{\overline{l[e]}\}, c) \in \{(\{\overline{l_1[e]}\}, c_1), \dots, (\{\overline{l_n[e]}\}, c_n), (\{\overline{\neg k_1[e]}\}, d_1), \dots, (\{\overline{\neg k_m[e]}\}, d_m)\}\} \geq c_l \iff$
- $\mathfrak{M} \models (l_1 \dot{\vee} \cdots \dot{\vee} l_n \dot{\vee} \neg k_1 \dot{\vee} \cdots \dot{\vee} \neg k_m, c_l)[e] \iff$
- $\mathfrak{M} \models (l_1 \dot{\vee} \cdots \dot{\vee} l_n \longleftarrow k_1 \wedge \cdots \wedge k_m, c_l)[e]$.

Hence,

$$\mathfrak{A} \models (l_1 \dot{\vee} \cdots \dot{\vee} l_n \longleftarrow k_1 \wedge \cdots \wedge k_m, c_l) \iff \mathfrak{M} \models^{\vee} (l_1 \dot{\vee} \cdots \dot{\vee} l_n \longleftarrow k_1 \wedge \cdots \wedge k_m, c_l).$$

We conclude: if an $\boldsymbol{L}$-model, say $\mathfrak{A}$, satisfies $\mathfrak{A} \models T$, then we can construct a fuzzy disjunctive $\boldsymbol{L}$-model $\mathfrak{M}$ of $\mathcal{L}$ such that the property 15 holds. Hence $\mathfrak{M} \models^{\vee} T$, which implies $\mathfrak{M} \models^{\vee} (Im, c_{Im})$ (the premise of this direction); therefore by the property 15, $\mathfrak{A} \models (Im, c_{Im})$. Thus $T \models C$. This direction is proved.

($\Longleftarrow$) Assume $\mathfrak{M}$ is a fuzzy disjunctive $\boldsymbol{L}$-model of $\mathcal{L}$ such that $\mathfrak{M} \models^{\vee} T$, but $\mathfrak{M} \not\models^{\vee} (C, c)$. This time we construct an $\boldsymbol{L}$-model $\mathfrak{A}$ of $\mathcal{L}$ for which the propositions

$$\mathfrak{A} \models T \text{ and } \mathfrak{A} \not\models (C, c) \text{ hold.} \tag{16}$$

Let $\mathfrak{M} = (\mathcal{U}_{\mathfrak{M}}, \mathcal{B}_{\mathfrak{M}})$. From the fact $\mathfrak{M} \not\models^{\vee} (C, c)$, there exists some variable assignment e^* in $\mathfrak{M}$ such that $\mathfrak{M} \not\models^{\vee} (C, c)[e^*]$.

Denote by $Atoms_{\mathcal{B}_{\mathfrak{M}}}$ the set of triples of the form $(1, p, (v_1, \dots, v_{ar(p)}))$, called atomic triples, for which triples of the form $(s, p, (v_1, \dots, v_{ar(p)}))$ occur in $\mathcal{B}_{\mathfrak{M}}$. That is,

$$Atoms_{\mathcal{B}_{\mathfrak{M}}} = \{(1, p, (v_1, \dots, v_{ar(p)})) \mid \exists\, s, st, c_s\, (s, p, (v_1, \dots, v_{ar(p)})) \in st, (st, c_s) \in \mathcal{B}_{\mathfrak{M}}\}.$$

Let S be a set of triples. We denote:

$$Atoms_S = \{(1, p, (v_1, \dots, v_{ar(p)})) \mid \exists s\, (s, p, (v_1, \dots, v_{ar(p)})) \in S\}.$$

Consider a set $M = \{(t, c_t) \mid t \in A, c_t \in \boldsymbol{L}\}$ where A is a set of atomic triples. Let $(\overline{D}, c)$ be a pair of a finite set of triples and a truth value from $\boldsymbol{L}$ such that $Atoms_{\overline{D}} \subseteq A$. We put

$$||\overline{D}||^M = (\bigcap\{c_t \mid \exists t\, (t, c_t) \in M, \neg t \in \overline{D}\} \cap \bigcap\{c_t \Rightarrow \mathbf{0} \mid \exists t\, (t, c_t) \in M, t \in \overline{D}\}) \Rightarrow \mathbf{0},$$

$$M \models (\overline{D}, c) \text{ iff } c \leq ||\overline{D}||^M.$$

Notice that

$$||\overline{D}||^M = \bigcup\{c_t \mid \exists t\, (t, c_t) \in M, t \in \overline{D}\} \cup \bigcup\{c_t \Rightarrow \mathbf{0} \mid \exists t\, (t, c_t) \in M, \neg t \in \overline{D}\}.$$

Denote $Atoms_M = A$.

We next construct a set $Mod_{\mathcal{B}_{\mathfrak{M}}} = \{(t, c_t) \mid t \in Atoms_{\mathcal{B}_{\mathfrak{M}}}, c_t \in \boldsymbol{L}\}$ with the property

$$\forall (\overline{D}, c) \in \mathcal{B}_{\mathfrak{M}} \; Mod_{\mathcal{B}_{\mathfrak{M}}} \models (\overline{D}, c). \tag{17}$$

The set $Mod_{\mathcal{B}_{\mathfrak{M}}}$ will be the 'core' of the intended model $\mathfrak{A}$.

We proceed as follows: Let $\overline{C[e^*]}$ be a finite set of triples of the form $l_1^* \,\bar{\vee} \cdots \bar{\vee}\, l_z^*$, $z \geq 0$, where l_i^* are pairwise different triples. We first define a set $Mod_{\overline{C[e^*]}}$ by recursion on $\alpha \leq z$:

1) $Mod_{\overline{C[e^*]}_0} = \emptyset$;
2) $Mod_{\overline{C[e^*]}_\alpha} = Mod_{\overline{C[e^*]}_{\alpha-1}} \cup \{(t_\alpha, c_\alpha)\}$, $\alpha > 0$, where
   - $Pos_\alpha = \{(\overline{D} \,\bar{\vee}\, l_\alpha^*, c_d) \mid (\overline{D} \,\bar{\vee}\, l_\alpha^*, c_d) \in \mathcal{B}_{\mathfrak{M}}, Atoms_{\overline{D}} \subseteq Atoms_{\{l_\beta^* \mid \beta < \alpha\}}\}$,
   - $c_\alpha^* = \bigcup\{(\|\overline{D}\|^{Mod_{\overline{C[e^*]}_{\alpha-1}}} \Rightarrow \mathbf{0}) \cap c_d \mid (\overline{D} \,\bar{\vee}\, l_\alpha^*, c_d) \in Pos_\alpha\}$,
   - $(t_\alpha, c_\alpha) = (l_\alpha^*, c_\alpha^*)$ if l_α^* is of the form $(1, p, (v_1, \ldots, v_{ar(p)}))$, otherwise $(t_\alpha, c_\alpha) = (\neg l_\alpha^*, c_\alpha^* \Rightarrow \mathbf{0})$.

$$Mod_{\overline{C[e^*]}} = Mod_{\overline{C[e^*]}_z}.$$

By the definition of fuzzy disjunctive $\boldsymbol{L}$-model, $\overline{C[e^*]}$ does not contain a pair of contradictory triples, for otherwise $\mathfrak{M} \models^{\vee} (C, c)[e^*]$ would hold; therefore, $Mod_{\overline{C[e^*]}}$ does not contain pairs of the form (t, c_t), (t, c_t') such that $c_t \neq c_t'$. Let

$$c^* = \bigcup\{c' \mid (\overline{C'}, c') \in \mathcal{B}_{\mathfrak{M}}, \overline{C'} \subseteq \overline{C[e^*]}\}.$$

Since $\mathfrak{M} \not\models^{\vee} (C, c)[e^*]$, $c \not\leq c^*$.

Using the Well Ordering Principle, we can arrange the set $Atoms_{\mathcal{B}_{\mathfrak{M}}} - Atoms_{\overline{C[e^*]}}$ by some ordinal γ in a sequence:

$$a_0, a_1, a_2, \ldots, a_\alpha, \ldots \qquad (\alpha < \gamma).$$

Next, by transfinite recursion, we define sets Mod_α for all $\alpha \leq \gamma$:

1) $Mod_0 = Mod_{\overline{C[e^*]}}$;
2) $Mod_{\alpha+1} = Mod_\alpha \cup \{(a_\alpha, c_\alpha)\}$ where
   - $Pos_\alpha = \{(\overline{D} \,\bar{\vee}\, a_\alpha, c_d) \mid (\overline{D} \,\bar{\vee}\, a_\alpha, c_d) \in \mathcal{B}_{\mathfrak{M}}, Atoms_{\overline{D}} \subseteq \{a_\beta \mid \beta < \alpha\} \cup Atoms_{\overline{C[e^*]}}\}$,
   - $c_\alpha = \bigcup\{(\|\overline{D}\|^{Mod_\alpha} \Rightarrow \mathbf{0}) \cap c_d \mid (\overline{D} \,\bar{\vee}\, a_\alpha, c_d) \in Pos_\alpha\}$;
3) $Mod_\alpha = \bigcup_{\beta < \alpha} Mod_\beta$ for a limit ordinal α.

It is straightforward to see that the sets Mod_α, $\alpha \leq \gamma$, form an increasing chain:

$$Mod_0 \subseteq Mod_1 \subseteq \cdots \subseteq Mod_\alpha \subseteq \cdots \qquad (\alpha \leq \gamma).$$

For all $\alpha \leq \gamma$, if $\beta \leq \alpha$, then $Mod_\beta \subseteq Mod_\alpha$.

The proof is by induction on $\alpha \leq \gamma$.

Let $\alpha = 0$: Then β must be also 0 and the statement holds trivially.

In case that α is a successor ordinal: If $\beta \leq \alpha$, we can distinguish two cases: $\beta = \alpha$ and $\beta \leq \alpha - 1$. In the first case, the statement holds trivially. Suppose the second one. Then, by hypothesis, $Mod_\beta \subseteq Mod_{\alpha-1}$, but by the recursive definition of Mod_α, $Mod_{\alpha-1} \subseteq Mod_\alpha$. So, the statement is true.

Suppose that α be a limit ordinal: If $\beta \leq \alpha$, we again have two cases: $\beta = \alpha$ and $\beta < \alpha$. The first one holds evidently. In the second case, by the definition of Mod_α, we obtain $Mod_\beta \subseteq \bigcup_{\beta<\alpha} Mod_\beta = Mod_\alpha$.

Next, by induction on $\alpha \leq \gamma$, we can prove the following statement:
For all $\alpha \leq \gamma$,

i) Mod_α does not contain pairs of the form (t, c_t), (t, c_t') such that $c_t \neq c_t'$;

ii) $Atoms_{Mod_\alpha} = \{a_\beta \mid \beta < \alpha\} \cup Atoms_{\overline{C[e^*]}}$;

iii) for all $(\overline{D}, c_d) \in \mathcal{B}_{\mathfrak{M}}$ such that $Atoms_{\overline{D}} \subseteq Atoms_{Mod_\alpha}$, $Mod_\alpha \models (\overline{D}, c_d)$, and $||\overline{C[e^*]}||^{Mod_\alpha} \leq c^*$.

Let $\alpha = 0$:

i) $Mod_0 = Mod_{\overline{C[e^*]}}$. So, as mentioned above, $Mod_{\overline{C[e^*]}}$ does not contain pairs of the form (t, c_t), (t, c_t') such that $c_t \neq c_t'$; hence, the first property holds.
ii) Also by the definition of $Mod_{\overline{C[e^*]}}$, $Atoms_{Mod_0} = Atoms_{Mod_{\overline{C[e^*]}}} = Atoms_{\overline{C[e^*]}}$.
iii) At first, we prove by induction on $n \leq z$ the statement:
   1) For all $(\overline{D}, c_d) \in \mathcal{B}_{\mathfrak{M}}$ such that $Atoms_{\overline{D}} \subseteq Atoms_{\{l_\beta^* \mid \beta \leq n\}}$,

$$Mod_{\overline{C[e^*]}_n} \models (\overline{D}, c_d).$$

   2) $||l_1^* \mathbin{\bar{\dot{\vee}}} \cdots \mathbin{\bar{\dot{\vee}}} l_n^*||^{Mod_{\overline{C[e^*]}_n}} \leq c^*$.

   The base case $n = 0$ is evident:
   1) Since $Atoms_{\{l_\beta^* \mid \beta \leq 0\}} = \emptyset$ and $(\emptyset, c_d) \notin \mathcal{B}_{\mathfrak{M}}$ for any $c_d \in \boldsymbol{L}$, the statement holds trivially.
   2) $||l_1^* \mathbin{\bar{\dot{\vee}}} \cdots \mathbin{\bar{\dot{\vee}}} l_0^*||^{Mod_{\overline{C[e^*]}_0}} = ||\emptyset||^{\emptyset} = \mathbf{0} \leq c^*$.

   Suppose $n > 0$ and the statement holds for $n - 1$.
   1) Let $(\overline{D}, c_d) \in \mathcal{B}_{\mathfrak{M}}$ satisfy the condition $Atoms_{\overline{D}} \subseteq Atoms_{\{l_\beta^* \mid \beta \leq n\}}$.
   If $Atoms_{\overline{D}} \subseteq Atoms_{\{l_\beta^* \mid \beta \leq n-1\}}$, by hypothesis, we get $Mod_{\overline{C[e^*]}_{n-1}} \models (\overline{D}, c_d)$. However, $Mod_{\overline{C[e^*]}_{n-1}} \subseteq Mod_{\overline{C[e^*]}_n}$; hence, $||\overline{D}||^{Mod_{\overline{C[e^*]}_n}} = ||\overline{D}||^{Mod_{\overline{C[e^*]}_{n-1}}} \geq c_d$. So, $Mod_{\overline{C[e^*]}_n} \models (\overline{D}, c_d)$.

   If not, we have to consider two cases:

   a) $\overline{D} = \overline{D'} \mathbin{\bar{\dot{\vee}}} l_n^*$,

   b) $\overline{D} = \overline{D'} \mathbin{\bar{\dot{\vee}}} \neg l_n^*$ where $Atoms_{\overline{D'}} \subseteq Atoms_{\{l_\beta^* \mid \beta \leq n-1\}}$.

By the definition of $Mod_{\overline{C[e^*]}_n}$, we get

a)

$$\begin{aligned}
&||\overline{D'}\ \bar{\dot{\vee}}\ l_n^*||^{Mod_{\overline{C[e^*]}_n}} &=\\
&||\overline{D'}||^{Mod_{\overline{C[e^*]}_n}} \cup c_n^* &=\\
&||\overline{D'}||^{Mod_{\overline{C[e^*]}_n}} \cup \bigcup\{(||\overline{D''}||^{Mod_{\overline{C[e^*]}_{n-1}}} \Rightarrow \mathbf{0}) \cap c''_d \\
&\qquad\qquad | (\overline{D''}\ \bar{\dot{\vee}}\ l_n^*, c''_d) \in Pos_n\} &\geq\\
&||\overline{D'}||^{Mod_{\overline{C[e^*]}_{n-1}}} \cup (||\overline{D'}||^{Mod_{\overline{C[e^*]}_{n-1}}} \Rightarrow \mathbf{0}) \cap c_d &=\\
&(||\overline{D'}||^{Mod_{\overline{C[e^*]}_{n-1}}} \cup (||\overline{D'}||^{Mod_{\overline{C[e^*]}_{n-1}}} \Rightarrow \mathbf{0})) \cap \\
&\qquad\qquad (||\overline{D'}||^{Mod_{\overline{C[e^*]}_{n-1}}} \cup c_d) &=\\
&\mathbf{1} \cap (||\overline{D'}||^{Mod_{\overline{C[e^*]}_{n-1}}} \cup c_d) &=\\
&||\overline{D'}||^{Mod_{\overline{C[e^*]}_{n-1}}} \cup c_d &\geq\\
&c_d.
\end{aligned} \tag{18}$$

b)

$$\begin{aligned}
&||\overline{D'}\ \bar{\dot{\vee}}\ \neg l_n^*||^{Mod_{\overline{C[e^*]}_n}} &=\\
&||\overline{D'}||^{Mod_{\overline{C[e^*]}_n}} \cup (c_n^* \Rightarrow \mathbf{0}) &=\\
&||\overline{D'}||^{Mod_{\overline{C[e^*]}_n}} \cup ((\bigcup\{(||\overline{D''}||^{Mod_{\overline{C[e^*]}_{n-1}}} \Rightarrow \mathbf{0}) \cap c''_d \\
&\qquad\qquad | (\overline{D''}\ \bar{\dot{\vee}}\ l_n^*, c''_d) \in Pos_n\}) \Rightarrow \mathbf{0}) &=\\
&||\overline{D'}||^{Mod_{\overline{C[e^*]}_n}} \cup \bigcap\{||\overline{D''}||^{Mod_{\overline{C[e^*]}_{n-1}}} \cup (c''_d \Rightarrow \mathbf{0}) \\
&\qquad\qquad | (\overline{D''}\ \bar{\dot{\vee}}\ l_n^*, c''_d) \in Pos_n\} &=\\
&\bigcap\{||\overline{D'}||^{Mod_{\overline{C[e^*]}_{n-1}}} \cup ||\overline{D''}||^{Mod_{\overline{C[e^*]}_{n-1}}} \cup (c''_d \Rightarrow \mathbf{0}) \\
&\qquad\qquad | (\overline{D''}\ \bar{\dot{\vee}}\ l_n^*, c''_d) \in Pos_n\} &\geq\\
&\bigcap\{c_d \cap c''_d \cup (c''_d \Rightarrow \mathbf{0}) \mid (\overline{D''}\ \bar{\dot{\vee}}\ l_n^*, c''_d) \in Pos_n\} &=\\
&\bigcap\{(c_d \cup (c''_d \Rightarrow \mathbf{0})) \cap (c''_d \cup (c''_d \Rightarrow \mathbf{0}) \mid (\overline{D''}\ \bar{\dot{\vee}}\ l_n^*, c''_d) \in Pos_n\} &=\\
&\bigcap\{(c_d \cup (c''_d \Rightarrow \mathbf{0})) \cap \mathbf{1} \mid (\overline{D''}\ \bar{\dot{\vee}}\ l_n^*, c''_d) \in Pos_n\} &=\\
&\bigcap\{(c_d \cup (c''_d \Rightarrow \mathbf{0})) \mid (\overline{D''}\ \bar{\dot{\vee}}\ l_n^*, c''_d) \in Pos_n\} &\geq\\
&c_d.
\end{aligned} \tag{19}$$

To show the inequality

$$\bigcap\{||\overline{D'}||^{Mod_{\overline{C[e^*]}_{n-1}}} \cup ||\overline{D''}||^{Mod_{\overline{C[e^*]}_{n-1}}} \cup (c''_d \Rightarrow \mathbf{0}) \mid (\overline{D''} \,\bar{\dot{\vee}}\, l^*_n, c''_d) \in Pos_n\} \geq$$

$$\bigcap\{c_d \cap c''_d \cup (c''_d \Rightarrow \mathbf{0}) \mid (\overline{D''} \,\bar{\dot{\vee}}\, l^*_n, c''_d) \in Pos_n\},$$

we have to distinguish two cases: $\overline{D'} \cup \overline{D''}$ contains a pair of contradictory triples, say l_1 and l_2, or not. In the first case, we get

$$||\overline{D'}||^{Mod_{\overline{C[e^*]}_{n-1}}} \cup ||\overline{D''}||^{Mod_{\overline{C[e^*]}_{n-1}}} \geq$$
$$||\{l_1\}||^{Mod_{\overline{C[e^*]}_{n-1}}} \cup ||\{l_2\}||^{Mod_{\overline{C[e^*]}_{n-1}}} = c_l \cup (c_l \Rightarrow \mathbf{0}) = \mathbf{1} \geq c_d \cap c''_d.$$

In the second case, by the definition of fuzzy disjunctive $\boldsymbol{L}$-model,

$$\bigcup\{c^\#_d \mid (\overline{D\#}, c^\#_d) \in \mathcal{B}_{\mathfrak{M}}, \overline{D\#} \subseteq \overline{D'} \cup \overline{D''}\} \geq c_d \cap c''_d.$$

Since for all $\overline{D\#}$, $Atoms_{\overline{D\#}} \subseteq Atoms_{\{l^*_\beta \mid \beta \leq n-1\}}$; by induction hypothesis, we have $||\overline{D\#}||^{Mod_{\overline{C[e^*]}_{n-1}}} \geq c^\#_d$. This yields

$$||\overline{D'}||^{Mod_{\overline{C[e^*]}_{n-1}}} \cup ||\overline{D''}||^{Mod_{\overline{C[e^*]}_{n-1}}} \geq$$
$$\bigcup\{||\overline{D\#}||^{Mod_{\overline{C[e^*]}_{n-1}}} \mid (\overline{D\#}, c^\#_d) \in \mathcal{B}_{\mathfrak{M}}, \overline{D\#} \subseteq \overline{D'} \cup \overline{D''}\} \geq$$
$$\{c^\#_d \mid (\overline{D\#}, c^\#_d) \in \mathcal{B}_{\mathfrak{M}}, \overline{D\#} \subseteq \overline{D'} \cup \overline{D''}\} \geq c_d \cap c''_d.$$

Hence, the inequality is proved.
So, in both the cases a) and b), $Mod_{\overline{C[e^*]}_n} \models (\overline{D}, c_d)$.

2) We next prove $||l^*_1 \,\bar{\dot{\vee}} \cdots \bar{\dot{\vee}}\, l^*_n||^{Mod_{\overline{C[e^*]}_n}} \leq c^*$.

$$\begin{aligned}
&||l^*_1 \,\bar{\dot{\vee}} \cdots \bar{\dot{\vee}}\, l^*_n||^{Mod_{\overline{C[e^*]}_n}} &= \\
&||l^*_1 \,\bar{\dot{\vee}} \cdots \bar{\dot{\vee}}\, l^*_{n-1}||^{Mod_{\overline{C[e^*]}_n}} \cup c^*_n &= \\
&||l^*_1 \,\bar{\dot{\vee}} \cdots \bar{\dot{\vee}}\, l^*_{n-1}||^{Mod_{\overline{C[e^*]}_{n-1}}} \cup \bigcup\{(||\overline{D''}||^{Mod_{\overline{C[e^*]}_{n-1}}} \Rightarrow \mathbf{0}) \cap c''_d \\
&\qquad \mid (\overline{D''} \,\bar{\dot{\vee}}\, l^*_n; c''_d) \in Pos_n\} &\leq \\
&c^*.
\end{aligned} \tag{20}$$

Since by induction hypothesis, $||l^*_1 \,\bar{\dot{\vee}} \cdots \bar{\dot{\vee}}\, l^*_{n-1}||^{Mod_{\overline{C[e^*]}_{n-1}}} \leq c^*$, it remains to show:

$$\bigcup\{(||\overline{D''}||^{Mod_{\overline{C[e^*]}_{n-1}}} \Rightarrow \mathbf{0}) \cap c''_d \mid (\overline{D''}\ \overline{\dot{\vee}}\ l^*_n, c''_d) \in Pos_n\} \quad =$$

$$\bigcup\{(||\overline{D''}||^{Mod_{\overline{C[e^*]}_{n-1}}} \Rightarrow \mathbf{0}) \cap c''_d \mid (\overline{D''}\ \overline{\dot{\vee}}\ l^*_n, c''_d) \in Pos_n, \exists l^*_i\ 1 \le i \le n-1, \neg l^*_i \in \overline{D''}\} \cup$$

$$\bigcup\{(||\overline{D''}||^{Mod_{\overline{C[e^*]}_{n-1}}} \Rightarrow \mathbf{0}) \cap c''_d \mid (\overline{D''}\ \overline{\dot{\vee}}\ l^*_n, c''_d) \in Pos_n, \overline{D''} \subseteq l^*_1\ \overline{\dot{\vee}} \cdots \overline{\dot{\vee}}\ l^*_{n-1}\} \le$$

$$\bigcup\{(||\{\neg l^*_i\}||^{Mod_{\overline{C[e^*]}_{n-1}}} \Rightarrow \mathbf{0}) \cap c''_d \mid (\overline{D''}\ \overline{\dot{\vee}}\ l^*_n, c''_d) \in Pos_n, \exists l^*_i\ 1 \le i \le n-1, \neg l^*_i \in \overline{D''}\} \cup$$

$$\bigcup\{c''_d \mid (\overline{D''}\ \overline{\dot{\vee}}\ l^*_n, c''_d) \in Pos_n, \overline{D''} \subseteq l^*_1\ \overline{\dot{\vee}} \cdots \overline{\dot{\vee}}\ l^*_{n-1}\} \quad \le$$

$$\bigcup\{||\{l^*_i\}||^{Mod_{\overline{C[e^*]}_{n-1}}} \mid (\overline{D''}\ \overline{\dot{\vee}}\ l^*_n, c''_d) \in Pos_n, \exists l^*_i\ 1 \le i \le n-1, \neg l^*_i \in \overline{D''}\} \cup$$

$$\bigcup\{c''_d \mid (\overline{D''}\ \overline{\dot{\vee}}\ l^*_n, c''_d) \in Pos_n, \overline{D''} \subseteq l^*_1\ \overline{\dot{\vee}} \cdots \overline{\dot{\vee}}\ l^*_{n-1}\} \quad \le$$

$$\bigcup\{c^* \mid (\overline{D''}\ \overline{\dot{\vee}}\ l^*_n, c''_d) \in Pos_n, \exists l^*_i\ 1 \le i \le n-1, \neg l^*_i \in \overline{D''}\} \cup \bigcup\{c^* \mid (\overline{D''}\ \overline{\dot{\vee}}\ l^*_n, c''_d) \in Pos_n, \overline{D''} \subseteq l^*_1\ \overline{\dot{\vee}} \cdots \overline{\dot{\vee}}\ l^*_{n-1}\} \le$$

$$c^*. \tag{21}$$

Notice that if $(\overline{D''}\ \overline{\dot{\vee}}\ l^*_n, c''_d) \in Pos_n$ and $\overline{D''} \subseteq l^*_1\ \overline{\dot{\vee}} \cdots \overline{\dot{\vee}}\ l^*_{n-1}$, then $\overline{D''}\ \overline{\dot{\vee}}\ l^*_n \subseteq \overline{C[e^*]}$; and hence, $c''_d \le c^*$.
Thus the point 2) is proved.

We are now in position to conclude the point iii): Using the above statement, for all $(\overline{D}, c_d) \in \mathcal{B}_{\mathfrak{M}}$ such that $Atoms_{\overline{D}} \subseteq Atoms_{Mod_0} = Atoms_{\overline{C[e^*]}}$, $Mod_0 = Mod_{\overline{C[e^*]}} = Mod_{\overline{C[e^*]}_z} \models (\overline{D}, c_d)$, and moreover $||\overline{C[e^*]}||^{Mod_0} = ||l^*_1\ \overline{\dot{\vee}} \cdots \overline{\dot{\vee}}\ l^*_z||^{Mod_{\overline{C[e^*]}_z}} \le c^*$.

In case that α is a successor ordinal:

i) Mod_α does not contain pairs of the form (t, c_t), (t, c'_t), $c_t \neq c'_t$; since by hypothesis, $Mod_{\alpha-1}$ does not contain such pairs, $Atoms_{Mod_{\alpha-1}} = \{a_\beta \mid \beta < \alpha - 1\} \cup Atoms_{\overline{C[e^*]}}$, and we add to $Mod_{\alpha-1}$ only one pair $(a_{\alpha-1}, c_{\alpha-1})$ so that $a_{\alpha-1}$ does not belong to $Atoms_{Mod_{\alpha-1}}$.

ii) $Atoms_{Mod_\alpha} = Atoms_{Mod_{\alpha-1}} \cup \{a_{\alpha-1}\} =$ (by hypothesis) $= \{a_\beta \mid \beta < \alpha\} \cup Atoms_{\overline{C[e^*]}}$.

iii) Let $(\overline{D}, c_d) \in \mathcal{B}_{\mathfrak{M}}$ meet the condition $Atoms_{\overline{D}} \subseteq Atoms_{Mod_\alpha} = \{a_\beta \mid \beta < \alpha\} \cup Atoms_{\overline{C[e^*]}}$.

If $Atoms_{\overline{D}} \subseteq \{a_\beta \mid \beta < \alpha - 1\} \cup Atoms_{\overline{C[e^*]}}$, by induction hypothesis, $Mod_{\alpha-1} \models (\overline{D}, c_d)$. However, $Mod_{\alpha-1} \subseteq Mod_\alpha$ and Mod_α meets the point i); hence, $||\overline{D}||^{Mod_\alpha} = ||\overline{D}||^{Mod_{\alpha-1}} \geq c_d$. So, $Mod_\alpha \models (\overline{D}, c_d)$.

If not, we have to consider two cases:

a) $\overline{D} = \overline{D'} \,\dot{\overline{\vee}}\, a_{\alpha-1}$,

b) $\overline{D} = \overline{D'} \,\dot{\overline{\vee}}\, \neg a_{\alpha-1}$ where $Atoms_{\overline{D'}} \subseteq \{a_\beta \mid \beta < \alpha - 1\} \cup Atoms_{\overline{C[e^*]}}$.

By the definition of Mod_α, we get

a)

$$\begin{aligned}
&||\overline{D'} \,\dot{\overline{\vee}}\, a_{\alpha-1}||^{Mod_\alpha} && = \\
&||\overline{D'}||^{Mod_\alpha} \cup c_{\alpha-1} && = \\
&||\overline{D'}||^{Mod_\alpha} \cup \bigcup\{(||\overline{D''}||^{Mod_{\alpha-1}} \Rightarrow \mathbf{0}) \cap c''_d \mid (\overline{D''} \,\dot{\overline{\vee}}\, a_{\alpha-1}, c''_d) \in Pos_{\alpha-1}\} && \geq \\
&||\overline{D'}||^{Mod_{\alpha-1}} \cup (||\overline{D'}||^{Mod_{\alpha-1}} \Rightarrow \mathbf{0}) \cap c_d && = \\
&(||\overline{D'}||^{Mod_{\alpha-1}} \cup (||\overline{D'}||^{Mod_{\alpha-1}} \Rightarrow \mathbf{0})) \cap (||\overline{D'}||^{Mod_{\alpha-1}} \cup c_d) && = \\
&\mathbf{1} \cap (||\overline{D'}||^{Mod_{\alpha-1}} \cup c_d) && = \\
&||\overline{D'}||^{Mod_{\alpha-1}} \cup c_d && \geq \\
&c_d.
\end{aligned} \tag{22}$$

b)

$$\begin{aligned}
&||\overline{D'} \,\dot{\overline{\vee}}\, \neg a_{\alpha-1}||^{Mod_\alpha} && = \\
&||\overline{D'}||^{Mod_\alpha} \cup (c^*_{\alpha-1} \Rightarrow \mathbf{0}) && = \\
&||\overline{D'}||^{Mod_\alpha} \cup ((\bigcup\{(||\overline{D''}||^{Mod_{\alpha-1}} \Rightarrow \mathbf{0}) \cap c''_d \\
&\qquad \mid (\overline{D''} \,\dot{\overline{\vee}}\, a_{\alpha-1}, c''_d) \in Pos_{\alpha-1}\}) \Rightarrow \mathbf{0}) && = \\
&||\overline{D'}||^{Mod_\alpha} \cup \bigcap\{||\overline{D''}||^{Mod_{\alpha-1}} \cup (c''_d \Rightarrow \mathbf{0}) \mid (\overline{D''} \,\dot{\overline{\vee}}\, a_{\alpha-1}, c''_d) \in Pos_{\alpha-1}\} && =
\end{aligned}$$

$$\bigcap\{||\overline{D'}||^{Mod_{\alpha-1}} \cup ||\overline{D''}||^{Mod_{\alpha-1}} \cup (c''_d \Rightarrow \mathbf{0}) \mid (\overline{D''} \,\overline{\dot{\vee}}\, a_{\alpha-1}, c''_d) \in Pos_{\alpha-1}\} \geq$$

$$\bigcap\{c_d \cap c''_d \cup (c''_d \Rightarrow \mathbf{0}) \mid (\overline{D''} \,\overline{\dot{\vee}}\, a_{\alpha-1}, c''_d) \in Pos_{\alpha-1}\} =$$

$$\bigcap\{(c_d \cup (c''_d \Rightarrow \mathbf{0})) \cap (c''_d \cup (c''_d \Rightarrow \mathbf{0}) \mid (\overline{D''} \,\overline{\dot{\vee}}\, a_{\alpha-1}, c''_d) \in Pos_{\alpha-1}\} =$$

$$\bigcap\{(c_d \cup (c''_d \Rightarrow \mathbf{0})) \cap \mathbf{1} \mid (\overline{D''} \,\overline{\dot{\vee}}\, a_{\alpha-1}, c''_d) \in Pos_{\alpha-1}\} =$$

$$\bigcap\{(c_d \cup (c''_d \Rightarrow \mathbf{0})) \mid (\overline{D''} \,\overline{\dot{\vee}}\, a_{\alpha-1}, c''_d) \in Pos_{\alpha-1}\} \geq$$

$$c_d. \tag{23}$$

To show the inequality

$$\bigcap\{||\overline{D'}||^{Mod_{\alpha-1}} \cup ||\overline{D''}||^{Mod_{\alpha-1}} \cup (c''_d \Rightarrow \mathbf{0}) \mid (\overline{D''} \,\overline{\dot{\vee}}\, a_{\alpha-1}, c''_d) \in Pos_{\alpha-1}\} \geq$$

$$\bigcap\{c_d \cap c''_d \cup (c''_d \Rightarrow \mathbf{0}) \mid (\overline{D''} \,\overline{\dot{\vee}}\, a_{\alpha-1}, c''_d) \in Pos_{\alpha-1}\},$$

we have to distinguish two cases: $D' \cup D''$ contains a pair of contradictory triples, say l_1 and l_2, or not. In the first case, we get

$$||\overline{D'}||^{Mod_{\alpha-1}} \cup ||\overline{D''}||^{Mod_{\alpha-1}} \geq ||\{l_1\}||^{Mod_{\alpha-1}} \cup ||\{l_2\}||^{Mod_{\alpha-1}} = c_l \cup (c_l \Rightarrow \mathbf{0}) = \mathbf{1} \geq c_d \cap c''_d.$$

In the second case, by the definition of fuzzy disjunctive $\boldsymbol{L}$-model,

$$\bigcup\{c^{\#}_d \mid (\overline{D\#}, c^{\#}_d) \in \mathcal{B}_{\mathfrak{M}}, \overline{D\#} \subseteq \overline{D'} \cup \overline{D''}\} \geq c_d \cap c''_d.$$

Since for all $\overline{D\#}$, $Atoms_{\overline{D\#}} \subseteq \{a_\beta \mid \beta < \alpha - 1\} \cup Atoms_{\overline{C[e^*]}}$; by induction hypothesis, we have $||\overline{D\#}||^{Mod_{\alpha-1}} \geq c^{\#}_d$. This yields

$$||\overline{D'}||^{Mod_{\alpha-1}} \cup ||\overline{D''}||^{Mod_{\alpha-1}} \geq \bigcup\{||\overline{D\#}||^{Mod_{\alpha-1}} \mid (\overline{D\#}, c^{\#}_d) \in \mathcal{B}_{\mathfrak{M}}, \overline{D\#} \subseteq \overline{D'} \cup \overline{D''}\} \geq \{c^{\#}_d \mid (\overline{D\#}, c^{\#}_d) \in \mathcal{B}_{\mathfrak{M}}, \overline{D\#} \subseteq \overline{D'} \cup \overline{D''}\} \geq c_d \cap c''_d.$$

Hence, the inequality is proved.
So, in both the cases a) and b), $Mod_\alpha \models (\overline{D}, c_d)$.
We know that $||\overline{C[e^*]}||^{Mod_0} \leq c^*$. However, $Mod_0 \subseteq Mod_\alpha$ and Mod_α meets the point i); hence,

$$||\overline{C[e^*]}||^{Mod_\alpha} = ||\overline{C[e^*]}||^{Mod_0} \leq c^*.$$

Let α be a limit ordinal:

i) Suppose that $Mod_\alpha = \bigcup_{\beta<\alpha} Mod_\beta$ contains pairs of the form (t, c_t), (t, c'_t) such that $c_t \neq c'_t$. Hence, the set of ordinals $\beta < \alpha$ such that $(t, c_t) \in Mod_\beta$, is not empty. Let $\beta' < \alpha$ be an ordinal such that $(t, c_t) \in Mod_{\beta'}$. Also, the set of ordinals $\beta < \alpha$ such that $(t, c'_t) \in Mod_\beta$, is not empty. Let $\beta'' < \alpha$ be an ordinal such that $(t, c'_t) \in Mod_{\beta''}$. Denote by δ the maximum of β' and β''. Since the sets Mod_β constitute an increasing chain, $Mod_{\beta'} \subseteq Mod_\delta$ and $Mod_{\beta''} \subseteq Mod_\delta$. So, Mod_δ, $\delta < \alpha$, contains the pairs (t, c_t), (t, c'_t), $c_t \neq c'_t$, which is a contradiction with the induction hypothesis.

ii) $Atoms_{Mod_\alpha} = Atoms_{\bigcup_{\beta<\alpha} Mod_\beta} = \bigcup_{\beta<\alpha} Atoms_{Mod_\beta} =$ (by hypothesis) $= \bigcup_{\beta<\alpha}(\{a_\delta \mid \delta < \beta\} \cup Atoms_{\overline{C[e^*]}}) = \{a_\delta \mid \delta < \alpha\} \cup Atoms_{\overline{C[e^*]}}$.

iii) Let $(\overline{D}, c_d) \in \mathcal{B}_\mathfrak{M}$ satisfy the condition $Atoms_{\overline{D}} \subseteq Atoms_{Mod_\alpha}$. Since the set $Atoms_{\overline{D}}$ is finite, then there exists the greatest ordinal δ such that $a_\delta \in Atoms_{\overline{D}}$ By the previous point, $Atoms_{Mod_\alpha} = \{a_\beta \mid \beta < \alpha\} \cup Atoms_{\overline{C[e^*]}}$. So, $\delta < \alpha$ and $\delta + 1 < \alpha$ (α is a limit ordinal); hence, $Atoms_{\overline{D}} \subseteq Atoms_{Mod_{\delta+1}}$. By hypothesis, $Mod_{\delta+1} \models (\overline{D}, c_d)$. However, $Mod_{\delta+1} \subseteq Mod_\alpha$ and Mod_α meets the point i); hence, $||\overline{D}||^{Mod_\alpha} = ||\overline{D}||^{Mod_{\delta+1}} \geq c_d$. So, $Mod_\alpha \models (\overline{D}, c_d)$.

We know that $||\overline{C[e^*]}||^{Mod_0} \leq c^*$. However, $Mod_0 \subseteq Mod_\alpha$ (and Mod_α meets the point i); hence, $||\overline{C[e^*]}||^{Mod_\alpha} = ||\overline{C[e^*]}||^{Mod_0} \leq c^*$.

The entire statement is proved.
We put

$$Mod_{\mathcal{B}_\mathfrak{M}} = Mod_\gamma.$$

Since $Atoms_{\mathcal{B}_\mathfrak{M}} \subseteq \{a_\beta \mid \beta < \gamma\} \cup Atoms_{\overline{C[e^*]}}$ and, by the previous statement, $Atoms_{Mod_\gamma} = \{a_\beta \mid \beta < \gamma\} \cup Atoms_{\overline{C[e^*]}}$; by the point iii) of the statement, the property 17 holds:

$$\forall\, (\overline{D}, c) \in \mathcal{B}_\mathfrak{M}\ Mod_{\mathcal{B}_\mathfrak{M}} \models (\overline{D}, c).$$

Moreover,

$$||\overline{C[e^*]}||^{Mod_{\mathcal{B}_\mathfrak{M}}} \leq c^*. \tag{24}$$

We are in position to construct the model $\mathfrak{A}$. We let $\mathcal{U}_\mathfrak{A} = \mathcal{U}_\mathfrak{M}$. A function symbol $f \in Func_\mathcal{L}$ is interpreted as the function $f^\mathfrak{A} = f^\mathfrak{M}$. A predicate symbol $p \in Pred_\mathcal{L}$ is interpreted as an $\boldsymbol{L}$-fuzzy relation $p^\mathfrak{A} : \mathcal{U}_\mathfrak{A}^{ar(p)} \longrightarrow \boldsymbol{L}$ where

$$\forall\, u_1, \dots, u_{ar(p)} \in \mathcal{U}_\mathfrak{A}$$

$$p^\mathfrak{A}(u_1, \dots, u_{ar(p)}) = \begin{cases} c_p, & (\overline{p(u_1, \dots, u_{ar(p)})}, c_p) \in Mod_{\mathcal{B}_\mathfrak{M}}; \\ \mathbf{0}, & else. \end{cases}$$

Note that the definition is correct, because of the condition i) for $Mod_{\mathcal{B}_\mathfrak{M}}$. Let $p(t_1, \dots, t_{ar(p)})$ be an atom, $l_1 \dot\vee \cdots \dot\vee l_n$ a strong literal disjunction, and e a variable renaming such that $\{\overline{p(t_1, \dots, t_{ar(p)})[e]}\} \cup Atoms_{\overline{(l_1 \dot\vee \cdots \dot\vee l_n)[e]}} \subseteq \{a_\beta \mid \beta <$

$\gamma\} \cup Atoms_{\overline{C[e^*]}}$, then from the definition of $\mathfrak{A}$, we immediately get:

$$||\{\overline{p(t_1,\dots,t_{ar(p)})[e]}\}||^{Mod_{\mathcal{B}_{\mathfrak{M}}}} = \bigcup\{c_p \mid (\overline{p(t_1[e],\dots,t_{ar(p)}[e])}, c_p) \in Mod_{\mathcal{B}_{\mathfrak{M}}}\} = c_p = p^{\mathfrak{A}}(t_1[e],\dots,t_{ar(p)}[e]) = ||p(t_1,\dots,t_{ar(p)})||_e^{\mathfrak{A}};$$

$$||\overline{(l_1 \dot\vee \cdots \dot\vee l_n)[e]}||^{Mod_{\mathcal{B}_{\mathfrak{M}}}} = ||\overline{l_1[e]} \mathbin{\bar{\dot\vee}} \cdots \mathbin{\bar{\dot\vee}} \overline{l_n[e]}||^{Mod_{\mathcal{B}_{\mathfrak{M}}}} = ||\{\overline{l_1[e]}\}||^{Mod_{\mathcal{B}_{\mathfrak{M}}}} \cup \cdots \cup ||\{\overline{l_n[e]}\}||^{Mod_{\mathcal{B}_{\mathfrak{M}}}} = ||l_1||_e^{\mathfrak{A}} \cup \cdots \cup ||l_n||_e^{\mathfrak{A}} = ||l_1 \dot\vee \cdots \dot\vee l_n||_e^{\mathfrak{A}}. \quad (25)$$

In the next part, we prove the property 16. Let $(Im, c_i) \in T$ be a graded strong literal implication and e a variable assignment in $\mathfrak{A}$, and also in $\mathfrak{M}$; since the domains are the same. $\mathfrak{M} \models^{\vee} T$ therefore $\mathfrak{M} \models^{\vee} (Im, c_i)[e]$. Then we obtain two cases:

$$\bigcup\{c' \mid (\overline{D'}, c') \in \mathcal{B}_{\mathfrak{M}}, \overline{D'} \subseteq \overline{Im[e]}\} \geq c_i,$$

or $\overline{Im[e]}$ contains a pair of contradictory triples.

Suppose the first case. We have:

$$\begin{aligned}
&c_i && \leq\\
&\bigcup\{c' \mid (\overline{D'}, c') \in \mathcal{B}_{\mathfrak{M}}, \overline{D'} \subseteq \overline{Im[e]}\} && \leq\\
&\bigcup\{||\overline{D'}||^{Mod_{\mathcal{B}_{\mathfrak{M}}}} \mid (\overline{D'}, c') \in \mathcal{B}_{\mathfrak{M}}, \overline{D'} \subseteq \overline{Im[e]}\} && (25) =\\
&\bigcup\{||Im^*||_e^{\mathfrak{A}} \mid \exists Im^*\ \{\!\{Im^*\}\!\} \subseteq \{\!\{Im\}\!\}, \overline{Im^*[e]} = \overline{D'}, (\overline{D'}, c') \in \mathcal{B}_{\mathfrak{M}}, \overline{D'} \subseteq \overline{Im[e]}\} && \leq\\
&\bigcup\{||Im||_e^{\mathfrak{A}} \mid (\overline{D'}, c') \in \mathcal{B}_{\mathfrak{M}}, \overline{D'} \subseteq \overline{Im[e]}\} && \leq\\
&||Im||_e^{\mathfrak{A}}.
\end{aligned} \quad (26)$$

Let $\overline{Im[e]}$ contain a pair of contradictory triples. Then the triples are of the form $\overline{p(t_1,\dots,t_{ar(p)})[e]}$ and $\overline{\neg p(r_1,\dots,r_{ar(p)})[e]}$ where $||t_i||_e^{\mathfrak{M}} = ||r_i||_e^{\mathfrak{M}}$ and $p(t_1,\dots,t_{ar(p)}), \neg p(r_1,\dots,r_{ar(p)}) \in \{\!\{Im\}\!\}$. From the definition of $\mathfrak{A}$, we get $||t_i||_e^{\mathfrak{A}} = ||r_i||_e^{\mathfrak{A}}$. It is a simple matter to see that

$$||Im||_e^{\mathfrak{A}} \geq ||p(t_1,\dots,t_{ar(p)}) \dot\vee \neg p(r_1,\dots,r_{ar(p)})||_e^{\mathfrak{A}} = ||p(t_1,\dots,t_{ar(p)})||_e^{\mathfrak{A}} \cup ||\neg p(r_1,\dots,r_{ar(p)})||_e^{\mathfrak{A}} = p^{\mathfrak{A}}(||t_1||_e^{\mathfrak{A}},\dots,||t_{ar(p)}||_e^{\mathfrak{A}}) \cup (p^{\mathfrak{A}}(||r_1||_e^{\mathfrak{A}},\dots,||r_{ar(p)}||_e^{\mathfrak{A}}) \Rightarrow \mathbf{0}) = \mathbf{1} \geq c_i.$$

So, in both the cases $\mathfrak{A} \models (Im, c_i)[e]$. Whence $\mathfrak{A} \models (Im, c_i)$ and also $\mathfrak{A} \models T$.

Consider the graded strong literal implication (C, c) and the variable assignment e^* in $\mathfrak{M}$, and also in $\mathfrak{A}$. By 24, we get

$$||C||_{e^*}^{\mathfrak{A}} \overset{(25)}{=\!=} ||\overline{C[e^*]}||^{Mod_{\mathcal{B}_{\mathfrak{M}}}} \leq c^* \not\geq c.$$

Hence, $||C||^{\mathfrak{A}}_{e^*} \not\geq c$ and $\mathfrak{A} \not\models (C, c)[e^*]$. Thereby $\mathfrak{A} \not\models (C, c)$.

The property 16 holds. We conclude: if $T \not\models^{\vee} (C, c)$, then $T \not\models (C, c)$.

The proof is completed. □

From this theorem, it follows the equivalence between the declarative and model semantics, which will be shown in the next section.

Assumption 1 *Henceforth, we shall suppose that the lattice* $\mathbf{L}$ *is a Boolean one.*

6 Model and Fixpoint Semantics

In this section, we introduce the model and fixpoint semantics for fuzzy disjunctive programs with weak similarity. We proceed from the above mentioned declarative semantics:

Definition 4 (Declarative semantics). Let P be a fuzzy disjunctive program of $\mathcal{L}$ with weak similarity. $\mathcal{DS}(P) = \{d \mid d$ *is a graded strong literal disjunction of* $\mathcal{L}$ *and* $P \models d\}$.

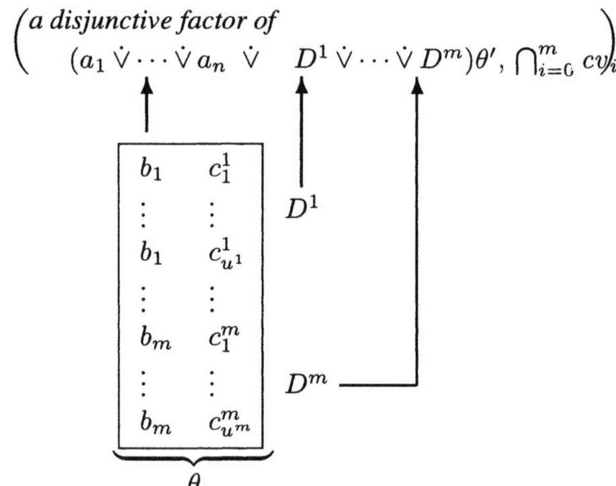

Figure 5. $\mathcal{C}_P$-operator.

6.1 Hyperresolution Operator

The fixpoint semantics is based on a generalised hyperresolution $\mathcal{C}_P$-operator, proposed in [33], which computes over the set of all graded strong literal disjunctions of $\mathcal{L}$, denoted as $Dis_{\mathcal{L}}$. To sketch how the operator works, let us consider Figure 5. An input comprises:

- an implication form $(a_1 \dot{\vee} \ldots \dot{\vee} a_n \longleftarrow b_1 \wedge \cdots \wedge b_m, cv_0)$ of some variant of a graded strong literal implication in the program P, and
- disjunctive forms $(c_1^i \dot{\vee} \cdots \dot{\vee} c_{u^i}^i \dot{\vee} D^i, cv_i)$, $i = 1 \ldots m$, of some variants of graded strong literal disjunctions in I being a subset of $Dis_{\mathcal{L}}$.

Each disjunctive form is divided into two parts. The first parts enter into the unification with the body of the implication form, and the rest together with the head of the implication create the output disjunction, which is instantiated by a regular extension θ' of the most general unifier θ of the columns. The output is formed of a disjunctive factor of the output disjunction and of the resulting truth value $\bigcap_{i=0}^{m} cv_i$.

The $\mathcal{C}_P$-operator is monotonic and ω-continuous; therefore, we can reach the least fixpoint in at most the ω-th step of the iteration of $\mathcal{C}_P$ from $\emptyset$. We next give a more formal treatment.

Definition 5. Let P be a fuzzy disjunctive program of $\mathcal{L}$. The closure operator $\mathcal{C}_P$ is defined as follows: $\mathcal{C}_P : 2^{Dis_{\mathcal{L}}} \longrightarrow 2^{Dis_{\mathcal{L}}}$,

$$\mathcal{C}_P(I) = \{(D, \bigcap_{i=0}^{m} cv_i) \mid$$

- *where D is a disjunctive factor of $(a_1 \dot{\vee} \cdots \dot{\vee} a_n \dot{\vee} \dot{\bigvee}_{i=1}^{m} D^i)\theta'$;*
- *$(c_1^i \dot{\vee} \cdots \dot{\vee} c_{u^i}^i \dot{\vee} D^i, cv_i)$ are disjunctive forms of some variants of graded strong literal disjunctions in I;*
- *$(a_1 \dot{\vee} \cdots \dot{\vee} a_n \longleftarrow b_1 \wedge \cdots \wedge b_m, cv_0)$ is an implication form of a variant of a graded strong literal implication in P;*
- *$vars(a_1 \dot{\vee} \cdots \dot{\vee} a_n \longleftarrow b_1 \wedge \cdots \wedge b_m) \cap vars(c_1^i \dot{\vee} \cdots \dot{\vee} c_{u^i}^i \dot{\vee} D^i) = \emptyset$, $vars(c_1^i \dot{\vee} \cdots \dot{\vee} c_{u^i}^i \dot{\vee} D^i) \cap vars(c_1^j \dot{\vee} \cdots \dot{\vee} c_{u^j}^j \dot{\vee} D^j) = \emptyset$ for $i \neq j$;*
- *$\theta = mgu((\underbrace{b_1, \ldots, b_1}_{u^1}, \ldots, \underbrace{b_m, \ldots, b_m}_{u^m}), (c_1^1, \ldots, c_{u^1}^1, \ldots, c_1^m, \ldots, c_{u^m}^m))$, $dom(\theta) = vars(b_1, \ldots, b_m, c_1^1, \ldots, c_{u^1}^1, \ldots, c_1^m, \ldots, c_{u^m}^m)$; and*
- *θ' is its regular extension to $vars(a_1 \dot{\vee} \cdots \dot{\vee} a_n \longleftarrow b_1 \wedge \cdots \wedge b_m) \cup \bigcup_{i=1}^{m} vars(c_1^i \dot{\vee} \cdots \dot{\vee} c_{u^i}^i \dot{\vee} D^i)\}$.*

The operator has obvious properties, as mentioned above.

Lemma 1. *For a given fuzzy disjunctive program P of $\mathcal{L}$, $\mathcal{C}_P$ is monotonic and ω-continuous.*

Proof: Immediately from Definition 5. □

6.2 Characterisation Theorem

We now recall the definition of characteristic disjunctive model from [33]. In the Motivation of the previous section, we have claimed that one of our aims is to achieve the equivalence between the model and declarative semantics. Since the model semantics should be based on one model, this model has to satisfy the following important condition - namely, it has to be characteristic:

Definition 6. Let T be a set of graded strong literal implications of $\mathcal{L}$. A fuzzy disjunctive $\boldsymbol{L}$-model $\mathfrak{M}$ for $\mathcal{L}$ is a characteristic model of T if and only if for any graded strong literal implication Im of $\mathcal{L}$,

$$\mathfrak{M} \models^{\vee} Im \Longleftrightarrow T \models^{\vee} Im.$$

Note that this condition implies $\mathfrak{M} \models^{\vee} T$.

The following characterisation theorem will be used to establish the equivalence between the model and fixpoint semantics.

Theorem 4 (Characterisation Theorem). *Let P be a fuzzy disjunctive program of $\mathcal{L}$ with weak similarity.*

i) *If P has a fuzzy disjunctive $\mathbf{L}$-model for $\mathcal{L}$, then P has a characteristic model for $\mathcal{L}$.*
ii) *Let (D, c_d) be a graded strong literal disjunction of $\mathcal{L}$, $P \models^{\vee} (D, c_d)$ if and only if either $c_d \leq \bigcup\{c'_d \mid$ there exist $(D', c'_d) \in \mathcal{C}_P \uparrow \omega$ and a substitution ϑ of $\mathcal{L}$ such that $D'\vartheta \sqsubseteq D$ (D' subsumes D)$\}$; or D is a tautology.*

Proof: Consider the closure operator $\mathcal{C}_P$. Since $\mathcal{C}_P$ is ω-continuous, we reach the least fixpoint in at most the ω-th step of the iteration of the operator from $\emptyset$, denoted as $\mathcal{C}_P \uparrow \omega$.

In the proof, we shall also use a closure operator $\mathcal{CB}$ based on binary resolution. The operator $\mathcal{CB}$ is defined as follows: $\mathcal{CB} : 2^{Dis_{\mathcal{L}}} \longrightarrow 2^{Dis_{\mathcal{L}}}$,

$\mathcal{CB}(I) = I \cup \{(D, c_a \cap c_b) \mid$

- *where D is a disjunctive factor of $(A_1 \dot\vee B_1)\theta'$;*
- *$(A_1 \dot\vee A_2, c_a)$ and $(B_1 \dot\vee B_2, c_b)$ are disjunctive forms of some variants of graded strong literal disjunctions in I not sharing common variables, where A_i and B_j are some strong literal disjunctions;*
- *θ is an mgu for the set of literals $\{\!\{A_2\}\!\} \cup \neg\{\!\{B_2\}\!\}$ $dom(\theta) = vars(A_2, B_2)$;*
- *θ' is its regular extension to $vars(A_1 \dot\vee A_2,\ B_1 \dot\vee B_2)\}$.*

Immediately from the definition, we can see that $\mathcal{CB}$ is monotonic and ω-continuous. Denote by $P^{\vee}$ the set $\{D \mid$ *D is a disjunctive factor of a variant of a graded strong literal implication in P*$\}$. Since $\mathcal{CB}$ is ω-continuous, we reach the least fixpoint in at most the ω-th step of the iteration of the operator from $P^{\vee}$, denoted as $\mathcal{CB} \uparrow \omega(P^{\vee})$. It can be proved that

$$(D, c_d) \in \mathcal{CB} \uparrow \omega(P^{\vee}) \longrightarrow \exists\, \delta \in Subst_{\mathcal{L}}\, \exists\, (D', c_d) \in \mathcal{C}_P \uparrow \omega \; \{\!\!\{D\}\!\!\} = \{\!\!\{D'\delta\}\!\!\}, \tag{27}$$

$$\mathcal{C}_P \uparrow \omega \subseteq \mathcal{CB} \uparrow \omega(P^{\vee}). \tag{28}$$

Next, we show that

$$P \models^{\vee} \mathcal{C}_P \uparrow \omega. \tag{29}$$

Using the inclusion 28, we need to show that for any fuzzy disjunctive $\boldsymbol{L}$-model $\mathfrak{M}$ of P and $\alpha \in \omega$

$$\mathfrak{M} \models^{\vee} \mathcal{CB} \uparrow \alpha(P^{\vee}). \tag{30}$$

We proceed by induction on α.

The base case $\alpha = 0$ is trivial. $\mathcal{CB} \uparrow 0(P^{\vee}) = P^{\vee}$. We know that $\mathfrak{M} \models^{\vee} P$. By the definitions of fuzzy disjunctive $\boldsymbol{L}$-model and disjunctive factor, it is evident that for any graded strong literal implication (Im, c_i), any variant (Im^*, c_i) of (Im, c_i), and its disjunctive factor $(Im^{*\vee}, c_i)$, $\mathfrak{M} \models^{\vee} (Im, c_i)$ implies $\mathfrak{M} \models^{\vee} (Im^*, c_i)$ and $\mathfrak{M} \models^{\vee} (Im^{*\vee}, c_i)$. So, $\mathfrak{M} \models^{\vee} P^{\vee} = \mathcal{CB} \uparrow 0(P^{\vee})$.

Suppose $\alpha > 0$ and the statement holds for $\alpha - 1$. Let (D, c_d) be a graded strong literal disjunction such that $(D, c_d) \in \mathcal{CB} \uparrow \alpha(P^{\vee})$. By the definition of $\mathcal{CB}$, we have two cases: either $(D, c_d) \in \mathcal{CB} \uparrow \alpha - 1(P^{\vee})$, then by hypothesis, $\mathfrak{M} \models^{\vee} (D, c_d)$; or there exist:

1) $(A_2 \dot{\vee} A_1, c_a)$ and $(C_2 \dot{\vee} C_1, c_c)$, disjunctive forms of variants of graded strong literal disjunctions in $\mathcal{CB} \uparrow \alpha - 1(P^{\vee})$ not sharing common variables, where A_i, C_j are strong literal disjunctions;
2) θ, an mgu for the set of literals $\{\!\!\{A_1\}\!\!\} \cup \neg\{\!\!\{C_1\}\!\!\}$, $dom(\theta) = vars(A_1, C_1)$;
3) θ', its regular extension to $vars(A_2 \dot{\vee} A_1,\, C_2 \dot{\vee} C_1)$;

such that D is a disjunctive factor of $(A_2 \dot{\vee} C_2)\theta'$ and $c_d = c_a \cap c_c$.

By hypothesis and the definition of fuzzy disjunctive $\boldsymbol{L}$-model, it is straightforward to see that

1) $\mathfrak{M} \models^{\vee} (A_2 \dot{\vee} A_1, c_a)$, $\mathfrak{M} \models^{\vee} (C_2 \dot{\vee} C_1, c_c)$ and also
2) $\mathfrak{M} \models^{\vee} ((A_2 \dot{\vee} A_1)\theta', c_a)$, $\mathfrak{M} \models^{\vee} ((C_2 \dot{\vee} C_1)\theta', c_c)$ - their instances.

Let e be a variable assignment in $\mathfrak{M}$. We show that $\mathfrak{M} \models^{\vee} (D, c_d)[e]$. We have $\mathfrak{M} \models^{\vee} ((A_2 \dot{\vee} A_1)\theta', c_a)[e]$ and $\mathfrak{M} \models^{\vee} ((C_2 \dot{\vee} C_1)\theta', c_c)[e]$. By the definition of fuzzy disjunctive $\boldsymbol{L}$-model, we get the following finite sets of triples

$$\overline{A_2\theta'[e]}\ \bar{\dot{\vee}}\ \{\overline{a_1\theta'[e]}\},$$
$$\overline{C_2\theta'[e]}\ \bar{\dot{\vee}}\ \{\overline{c_1\theta'[e]}\}$$

where $\overline{a_1\theta'[e]}$ denotes the unique triple obtained from the disjunction $A_1\theta'$ of the identical unified literals; and similarly, $\overline{c_1\theta'[e]}$ denotes the unique triple obtained from the disjunction $C_1\theta'$ of the identical unified literals. Note that the triples $\overline{a_1\theta'[e]}$ and $\overline{c_1\theta'[e]}$ are contradictory. Consider the finite set of triples

$$\overline{D[e]} = \overline{A_2\theta'[e]} \,\bar{\dot{\vee}}\, \overline{C_2\theta'[e]}.$$

We can recognise a few cases:

$\overline{D[e]}$ contains a pair of contradictory triples. Then $||D||_e^{\mathfrak{M}} = 1$ and $\mathfrak{M} \models^{\vee} (D, c_d)[e]$.

$\overline{D[e]}$ does not contain a pair of contradictory triples.

1) If $\overline{A_2\theta'[e]} \,\bar{\dot{\vee}}\, \{\overline{a_1\theta'[e]}\}$ contains a pair of contradictory triples, then $\overline{A_2\theta'[e]}$ does not, for otherwise $\overline{D[e]} \supseteq \overline{A_2\theta'[e]}$ would also include this pair, which is a contradiction with the assumption. So, $\overline{a_1\theta'[e]}$ is one triple of this contradictory pair and $\overline{c_1\theta'[e]}$ the other triple. Hence, $\overline{c_1\theta'[e]} \in \overline{A_2\theta'[e]}$. This yields $\overline{D[e]} = \overline{A_2\theta'[e]} \,\bar{\dot{\vee}}\, \overline{C_2\theta'[e]} \supseteq \overline{C_2\theta'[e]} \,\bar{\dot{\vee}}\, \{\overline{c_1\theta'[e]}\}$. Therefore we get that $\overline{C_2\theta'[e]} \,\bar{\dot{\vee}}\, \{\overline{c_1\theta'[e]}\}$ also does not contain a pair of contradictory triples. By the definition of model,

$$c_d = c_a \cap c_c \leq c_c \leq ||C_2\theta' \dot{\vee} C_1\theta'||_e^{\mathfrak{M}} = \bigcup\{c'_d \mid (\overline{D'}, c'_d) \in \mathcal{B}_{\mathfrak{M}}, \overline{D'} \subseteq \overline{C_2\theta'[e]} \,\bar{\dot{\vee}}\, \{\overline{c_1\theta'[e]}\}\} \leq \bigcup\{c'_d \mid (\overline{D'}, c'_d) \in \mathcal{B}_{\mathfrak{M}}, \overline{D'} \subseteq \overline{D[e]}\} = ||D||_e^{\mathfrak{M}}.$$

Whence, $\mathfrak{M} \models^{\vee} (D, c_d)[e]$.

2) If $\overline{C_2\theta'[e]} \,\bar{\dot{\vee}}\, \{\overline{c_1\theta'[e]}\}$ contains a pair of contradictory triples, then $\overline{C_2\theta'[e]}$ does not, for otherwise $\overline{D[e]} \supseteq \overline{C_2\theta'[e]}$ would also include this pair, which is a contradiction with the assumption. So, $\overline{c_1\theta'[e]}$ is one triple of this contradictory pair and $\overline{a_1\theta'[e]}$ the other triple. Hence, $\overline{a_1\theta'[e]} \in \overline{C_2\theta'[e]}$. This yields $\overline{D[e]} = \overline{A_2\theta'[e]} \,\bar{\dot{\vee}}\, \overline{C_2\theta'[e]} \supseteq \overline{A_2\theta'[e]} \,\bar{\dot{\vee}}\, \{\overline{a_1\theta'[e]}\}$. Therefore we get that $\overline{A_2\theta'[e]} \,\bar{\dot{\vee}}\, \{\overline{a_1\theta'[e]}\}$ also does not contain a pair of contradictory triples. By the definition of model,

$$c_d = c_a \cap c_c \leq c_a \leq ||A_2\theta' \dot{\vee} A_1\theta'||_e^{\mathfrak{M}} = \bigcup\{c'_d \mid (\overline{D'}, c'_d) \in \mathcal{B}_{\mathfrak{M}}, \overline{D'} \subseteq \overline{A_2\theta'[e]} \,\bar{\dot{\vee}}\, \{\overline{a_1\theta'[e]}\}\} \leq \bigcup\{c'_d \mid (\overline{D'}, c'_d) \in \mathcal{B}_{\mathfrak{M}}, \overline{D'} \subseteq \overline{D[e]}\} = ||D||_e^{\mathfrak{M}}.$$

Whence, $\mathfrak{M} \models^{\vee} (D, c_d)[e]$.

3) Suppose that neither $\overline{A_2\theta'[e]} \,\bar{\dot{\vee}}\, \{\overline{a_1\theta'[e]}\}$ nor $\overline{C_2\theta'[e]} \,\bar{\dot{\vee}}\, \{\overline{c_1\theta'[e]}\}$ contain a pair of contradictory triples. We get

$$c_d = c_a \cap c_c \leq$$

$$||A_2\theta' \dot{\vee} A_1\theta'||_e^{\mathfrak{M}} \cap ||C_2\theta' \dot{\vee} C_1\theta'||_e^{\mathfrak{M}} =$$

$$\bigcup\{c'_d \mid (\overline{D'}, c'_d) \in \mathcal{B}_{\mathfrak{N}}, \overline{D'} \subseteq \overline{A_2\theta'[e]} \mathrel{\dot{\overline{\vee}}} \{\overline{a_1\theta'[e]}\}\} \cap \bigcup\{c'_d \mid (\overline{D'}, c'_d) \in \mathcal{B}_{\mathfrak{N}}, \overline{D'} \subseteq \overline{C_2\theta'[e]} \mathrel{\dot{\overline{\vee}}} \{\overline{c_1\theta'[e]}\}\} =$$

$$(\bigcup\{c'_d \mid (\overline{D'}, c'_d) \in \mathcal{B}_{\mathfrak{N}}, \overline{D'} \subseteq \overline{A_2\theta'[e]}\} \cup \bigcup\{c'_d \mid (\overline{D'}, c'_d) \in \mathcal{B}_{\mathfrak{N}}, \overline{D'} \subseteq \overline{A_2\theta'[e]} \mathrel{\dot{\overline{\vee}}} \{\overline{a_1\theta'[e]}\}, \overline{a_1\theta'[e]} \in \overline{D'}\}) \cap$$

$$(\bigcup\{c'_d \mid (\overline{D'}, c'_d) \in \mathcal{B}_{\mathfrak{N}}, \overline{D'} \subseteq \overline{C_2\theta'[e]}\} \cup \bigcup\{c'_d \mid (\overline{D'}, c'_d) \in \mathcal{B}_{\mathfrak{N}}, \overline{D'} \subseteq \overline{C_2\theta'[e]} \mathrel{\dot{\overline{\vee}}} \{\overline{c_1\theta'[e]}\}, \overline{c_1\theta'[e]} \in \overline{D'}\}) \leq$$

$$(\bigcup\{c'_d \mid (\overline{D'}, c'_d) \in \mathcal{B}_{\mathfrak{N}}, \overline{D'} \subseteq \overline{D[e]}\} \cup \bigcup\{c'_d \mid (\overline{D'}, c'_d) \in \mathcal{B}_{\mathfrak{N}}, \overline{D'} \subseteq \overline{A_2\theta'[e]} \mathrel{\dot{\overline{\vee}}} \{\overline{a_1\theta'[e]}\}, \overline{a_1\theta'[e]} \in \overline{D'}\}) \cap$$

$$(\bigcup\{c'_d \mid (\overline{D'}, c'_d) \in \mathcal{B}_{\mathfrak{N}}, \overline{D'} \subseteq \overline{D[e]}\} \cup \bigcup\{c'_d \mid (\overline{D'}, c'_d) \in \mathcal{B}_{\mathfrak{N}}, \overline{D'} \subseteq \overline{C_2\theta'[e]} \mathrel{\dot{\overline{\vee}}} \{\overline{c_1\theta'[e]}\}, \overline{c_1\theta'[e]} \in \overline{D'}\}) =$$

$$\bigcup\{c'_d \mid (\overline{D'}, c'_d) \in \mathcal{B}_{\mathfrak{N}}, \overline{D'} \subseteq \overline{D[e]}\} \cup$$

$$(\bigcup\{c'_d \mid (\overline{D'}, c'_d) \in \mathcal{B}_{\mathfrak{N}}, \overline{D'} \subseteq \overline{A_2\theta'[e]} \mathrel{\dot{\overline{\vee}}} \{\overline{a_1\theta'[e]}\}, \overline{a_1\theta'[e]} \in \overline{D'}\} \cap \bigcup\{c'_d \mid (\overline{D'}, c'_d) \in \mathcal{B}_{\mathfrak{N}}, \overline{D'} \subseteq \overline{C_2\theta'[e]} \mathrel{\dot{\overline{\vee}}} \{\overline{c_1\theta'[e]}\}, \overline{c_1\theta'[e]} \in \overline{D'}\}) =$$

$$\bigcup\{c'_d \mid (\overline{D'}, c'_d) \in \mathcal{B}_{\mathfrak{N}}, \overline{D'} \subseteq \overline{D[e]}\} \cup$$

$$\bigcup\{c'_d \cap c''_d \mid (\overline{D'}, c'_d), (\overline{D''}, c''_d) \in \mathcal{B}_{\mathfrak{N}}, \overline{D'} \subseteq \overline{A_2\theta'[e]} \mathrel{\dot{\overline{\vee}}} \{\overline{a_1\theta'[e]}\}, \overline{a_1\theta'[e]} \in \overline{D'}, \overline{D''} \subseteq \overline{C_2\theta'[e]} \mathrel{\dot{\overline{\vee}}} \{\overline{c_1\theta'[e]}\}, \overline{c_1\theta'[e]} \in \overline{D''}\} \leq$$

$$\bigcup\{c'_d \mid (\overline{D'}, c'_d) \in \mathcal{B}_{\mathfrak{N}}, \overline{D'} \subseteq \overline{D[e]}\} \cup$$

$$\bigcup\{\bigcup\{c'''_d \mid (\overline{D'''}, c'''_d) \in \mathcal{B}_{\mathfrak{N}}, \overline{D'''} \subseteq (\overline{D'} - \{\overline{a_1\theta'[e]}\}) \cup (\overline{D''} - \{\overline{c_1\theta'[e]}\}) \subseteq (\overline{A_2\theta'[e]} - \{\overline{a_1\theta'[e]}\}) \cup (\overline{C_2\theta'[e]} - \{\overline{c_1\theta'[e]}\}) \subseteq \overline{D[e]}\}$$
$$\mid (\overline{D'}, c'_d), (\overline{D''}, c''_d) \in \mathcal{B}_{\mathfrak{N}}, \overline{D'} \subseteq \overline{A_2\theta'[e]} \mathrel{\dot{\overline{\vee}}} \{\overline{a_1\theta'[e]}\}, \overline{a_1\theta'[e]} \in \overline{D'}, \overline{D''} \subseteq \overline{C_2\theta'[e]} \mathrel{\dot{\overline{\vee}}} \{\overline{c_1\theta'[e]}\}, \overline{c_1\theta'[e]} \in \overline{D''}\} \leq$$

$$\bigcup\{c'_d \mid (\overline{D'}, c'_d) \in \mathcal{B}_{\mathfrak{N}}, \overline{D'} \subseteq \overline{D[e]}\} \cup \bigcup\{c'''_d \mid (\overline{D'''}, c'''_d) \in \mathcal{B}_{\mathfrak{N}}, \overline{D'''} \subseteq \overline{D[e]}\} =$$

$$||D||^{\mathfrak{N}}_e. \tag{31}$$

Notice that for any $(\overline{D'}, c'_d), (\overline{D''}, c''_d) \in \mathcal{B}_{\mathfrak{N}}$, $\overline{D'} \subseteq \overline{A_2\theta'[e]}\ \dot{\vee}\ \{\overline{a_1\theta'[e]}\}$, $\overline{a_1\theta'[e]} \in \overline{D'}$, and $\overline{D''} \subseteq \overline{C_2\theta'[e]}\ \dot{\vee}\ \{\overline{c_1\theta'[e]}\}$, $\overline{c_1\theta'[e]} \in \overline{D''}$, by the definition of model,

$$\begin{aligned} c'_d \cap c''_d \leq \\ \bigcup\{c'''_d \mid (\overline{D'''}, c'''_d) \in \mathcal{B}_{\mathfrak{N}}, \\ \overline{D'''} \subseteq (\overline{D'} - \{\overline{a_1\theta'[e]}\}) \cup (\overline{D''} - \{\overline{c_1\theta'[e]}\}) \subseteq \\ (\overline{A_2\theta'[e]} - \{\overline{a_1\theta'[e]}\}) \cup (\overline{C_2\theta'[e]} - \{\overline{c_1\theta'[e]}\}) \subseteq \overline{D[e]}\} \leq \\ \bigcup\{c'''_d \mid (\overline{D'''}, c'''_d) \in \mathcal{B}_{\mathfrak{N}}, \overline{D'''} \subseteq \overline{D[e]}\}. \end{aligned}$$

It is evident that $(\overline{D'} - \{\overline{a_1\theta'[e]}\}) \cup (\overline{D''} - \{\overline{c_1\theta'[e]}\})$ does not contain a pair of contradictory triples, since neither does the superset $\overline{D[e]}$.
Thus $\mathfrak{N} \models^{\vee} (D, c_d)[e]$.

From the above considerations, we can conclude that $\mathfrak{N} \models^{\vee} (D, c_d)$. So, $\mathfrak{N} \models^{\vee} \mathcal{CB} \uparrow \alpha(P^{\vee})$. Since $\mathcal{CB} \uparrow \omega(P^{\vee}) = \bigcup_{\alpha \in \omega} \mathcal{CB} \uparrow \alpha(P^{\vee})$, we get

$$\mathfrak{N} \models^{\vee} \mathcal{CB} \uparrow \omega(P^{\vee}) \supseteq \mathcal{C}_P \uparrow \omega \qquad \text{(by 28)}.$$

Hence, the statement 29 is proved.

Further, consider the case $(\Box, c) \in \mathcal{C}_P \uparrow \omega$ for some $c > \mathbf{0}$. Then P has no fuzzy disjunctive $\boldsymbol{L}$-model for $\mathcal{L}$, for otherwise such a model, say $\mathfrak{N}$, should satisfy P and, by the previous statement, also $\mathcal{C}_P \uparrow \omega$. However, from the definition of model, this model could not satisfy $(\Box, c)$ because of the condition i): $||\Box||^{\mathfrak{N}}_e = \bigcup\{c' \mid (\overline{D'}, c') \in \mathcal{B}_{\mathfrak{N}}, \overline{D'} \subseteq \overline{\Box[e]} = \emptyset\} = \mathbf{0} \not\geq c$. So, P has no fuzzy disjunctive $\boldsymbol{L}$-model for $\mathcal{L}$, which contradicts the assumption of the statement i).

Suppose that $(\Box, c) \notin \mathcal{C}_P \uparrow \omega$ for $c > \mathbf{0}$. We construct a characteristic model $\mathfrak{M} = (\mathcal{U}_{\mathfrak{M}}, \mathcal{B}_{\mathfrak{M}})$ of P for $\mathcal{L}$ satisfying the condition:

For any graded strong literal disjunction (D, c_d) of $\mathcal{L}$,

$$\text{if } \mathfrak{M} \models^{\vee} (D, c_d), \text{ then } P \models^{\vee} (D, c_d). \tag{32}$$

Let $\mathcal{A}_{\mathfrak{M}} = \{a_n \mid n \in \omega\}$ where a_n are new pairwise different constant symbols not occurring in $\mathcal{L}$. We let $\mathcal{U}_{\mathfrak{M}} = GTerm_{Func_{\mathcal{L}} \cup \mathcal{A}_{\mathfrak{M}}}$. Since $\mathcal{A}_{\mathfrak{M}}$ contains the constants a_n, $\mathcal{U}_{\mathfrak{M}} \neq \emptyset$. A function symbol $f \in Func_{\mathcal{L}}$ is interpreted as the function $f^{\mathfrak{M}}$: $\mathcal{U}_{\mathfrak{M}}^{ar(f)} \longrightarrow \mathcal{U}_{\mathfrak{M}}$, $f^{\mathfrak{M}}(u_1, \ldots, u_{ar(f)}) = f(u_1, \ldots, u_{ar(f)})$ where $u_1, \ldots, u_{ar(f)} \in \mathcal{U}_{\mathfrak{M}}$. Let D be a ground strong literal disjunction of $\mathcal{L} \cup \mathcal{A}_{\mathfrak{M}}$. Denote $\overline{D} = \overline{D[e]}$ for some variable assignment e in $\mathfrak{M}$. Notice that since D is ground, $\overline{D}$ does not depend on e. We put

$\mathcal{B}_{\mathfrak{M}} = \{(\overline{B}, c_b) \mid$ • $\exists\,(D, c_d) \in \mathcal{C}_P \uparrow \omega$, $\exists\,\vartheta \in Subst_{\mathcal{L}\cup\mathcal{A}_{\mathfrak{M}}}$ *such that* $B = D\vartheta$ *is a ground instance of* D *in the language* $\mathcal{L} \cup \mathcal{A}_{\mathfrak{M}}$*;*

- $\overline{B}$ *does not contain a pair of contradictory triples;*
- $c_b = \bigcup\{c'_d \mid \exists\,(D', c'_d) \in \mathcal{C}_P \uparrow \omega,\ \exists\,\vartheta' \in Subst_{\mathcal{L}\cup\mathcal{A}_{\mathfrak{M}}},\ D'\vartheta'$ *is ground,* $\overline{D'\vartheta'} \subseteq \overline{B}\} > \mathbf{0}$*;*
- $\forall\,(D^*, c^*_d) \in \mathcal{C}_P \uparrow \omega$ *such that* $\exists\,\vartheta^* \in Subst_{\mathcal{L}\cup\mathcal{A}_{\mathfrak{M}}}$ *where* $B^* = D^*\vartheta^*$ *is a ground instance of* D^* *in the language* $\mathcal{L} \cup \mathcal{A}_{\mathfrak{M}}$ *and* $\overline{B^*} \subset \overline{B}$, $c_b > \bigcup\{c'_d \mid \exists\,(D', c'_d) \in \mathcal{C}_P \uparrow \omega,\ \exists\,\vartheta' \in Subst_{\mathcal{L}\cup\mathcal{A}_{\mathfrak{M}}},$ $D'\vartheta'$ *is ground,* $\overline{D'\vartheta'} \subseteq \overline{B^*}\}\}$.

It is straightforward to see that $\mathcal{B}_{\mathfrak{M}}$ satisfies the conditions i) - v) from the definition of fuzzy disjunctive $\boldsymbol{L}$-model.

The following auxiliary fact holds: Let $\overline{B}$ be a finite set of triples in $\mathfrak{M}$ not containing a pair of contradictory triples. Then

$$\bigcup\{c'_b \mid (\overline{B'}, c'_b) \in \mathcal{B}_{\mathfrak{M}}, \overline{B'} \subseteq \overline{B}\} =$$
$$\bigcup\{c'_d \mid \exists\,(D', c'_d) \in \mathcal{C}_P \uparrow \omega, \exists\,\vartheta' \in Subst_{\mathcal{L}\cup\mathcal{A}_{\mathfrak{M}}}, D'\vartheta' \text{ is ground}, \overline{D'\vartheta'} \subseteq \overline{B}\}. \tag{33}$$

Let (Im, c_i) be a graded strong literal implication in P and $(Im^\vee, c_i)$ some disjunctive factor of (Im, c_i). Then, by the definition of $\mathcal{C}_P$, $(Im^\vee, c_i) \in \mathcal{C}_P \uparrow 1 \subseteq \mathcal{C}_P \uparrow \omega$. Let e be a variable assignment in $\mathfrak{M}$. We show that $\mathfrak{M} \models^\vee (Im^\vee, c_i)[e]$. Consider the ground instance $Im^\vee e|_{vars(Im^\vee)}$ of $Im^\vee$. By the definition of $\mathfrak{M}$, it is a simple matter to show that for any term t of $\mathcal{L}$, $t[e] = t(e|_{vars(t)})$. Therefore $\overline{Im^\vee[e]} = \overline{Im^\vee e|_{vars(Im^\vee)}}$. However, by the definition of $\mathcal{B}_{\mathfrak{M}}$, either $\overline{Im^\vee e|_{vars(Im^\vee)}}$ contains a pair of contradictory triples; or if not,

$$\bigcup\{c_b \mid (\overline{B}, c_b) \in \mathcal{B}_{\mathfrak{M}}, \overline{B} \subseteq \overline{Im^\vee e|_{vars(Im^\vee)}}\} \overset{(33)}{=\!=}$$
$$\bigcup\{c'_d \mid \exists\,(D', c'_d) \in \mathcal{C}_P \uparrow \omega, \exists\,\vartheta' \in Subst_{\mathcal{L}\cup\mathcal{A}_{\mathfrak{M}}},$$
$$D'\vartheta' \text{ is ground}, \overline{D'\vartheta'} \subseteq \overline{Im^\vee e|_{vars(Im^\vee)}}\} \geq$$
$$\bigcup\{c'_d \mid \exists c'_d\,(Im^\vee, c'_d) \in \mathcal{C}_P \uparrow 1, e|_{vars(Im^\vee)} \in Subst_{\mathcal{L}\cup\mathcal{A}_{\mathfrak{M}}},$$
$$Im^\vee e|_{vars(Im^\vee)} \text{ is ground}, \overline{Im^\vee e|_{vars(Im^\vee)}} = \overline{Im^\vee e|_{vars(Im^\vee)}}\} \geq$$
$$c_i. \tag{34}$$

So, $\mathfrak{M} \models^\vee (Im^\vee, c_i)[e]$. Since $\overline{Im^\vee[e]} = \overline{Im[e]}$, by the definition of model, $\mathfrak{M} \models^\vee (Im, c_i)[e]$. This yields $\mathfrak{M} \models^\vee (Im, c_i)$. So, $\mathfrak{M} \models^\vee P$ and $\mathfrak{M} \models^\vee WSim_{\mathcal{L}}$. Hence, $\mathfrak{M}$ is indeed a fuzzy disjunctive $\boldsymbol{L}$-model of P for $\mathcal{L}$.

It remains to prove the main property 32 of $\mathfrak{M}$ and the point ii) of the theorem. Let (D, c_d) be a graded strong literal disjunction of $\mathcal{L}$. Suppose $\mathfrak{M} \models^\vee (D, c_d)$. Consider a variable assignment e^* such that for all $v \neq v' \in vars(D)$, $e^*(v) \neq e^*(v') \in \mathcal{A}_{\mathfrak{M}}$. So, we assign new pairwise different constants not occurring in $\mathcal{L}$ to the variables in

D. We can see that $\overline{D[e^*]} = \overline{De^*|_{vars(D)}}$. Since $\mathfrak{M} \models^{\vee} D[e^*]$, either $\overline{De^*|_{vars(D)}}$ contains a pair of contradictory triples; or

$$c_d \leq \bigcup \{c'_b \mid (\overline{B'}, c'_b) \in \mathcal{B}_{\mathfrak{M}}, \overline{B'} \subseteq \overline{De^*|_{vars(D)}}\}.$$

Assume the first case. Let $\overline{l_1^*}$ and $\overline{l_2^*}$ be contradictory triples belonging to $\overline{De^*|_{vars(D)}}$. Then $De^*|_{vars(D)}$ contains the contradictory literals l_1^* and l_2^*, i.e. $l_1^* = \neg l_2^*$. Since $v \neq v' \in vars(D)$, $e(v) \neq e(v') \in \mathcal{A}_{\mathfrak{M}}$; for all terms $t \neq t'$ and literals $k \neq k'$ of $\mathcal{L}$ such that $vars(t), vars(t'), vars(k), vars(k') \subseteq vars(D)$, we get $t(e^*|_{vars(D)}) \neq t'(e^*|_{vars(D)})$ and $k(e^*|_{vars(D)}) \neq k'(e^*|_{vars(D)})$. Denote by l_1 and l_2 literals in D such that $l_1^* = l_1(e^*|_{vars(D)})$ and $l_2^* = l_2(e^*|_{vars(D)})$, respectively. Then $l_1 = \neg l_2$. Hence, D also contains the contradictory literals l_1 and l_2, i.e. D is a tautology.

Suppose the second case

$$c_d \leq \bigcup \{c'_b \mid (\overline{B'}, c'_b) \in \mathcal{B}_{\mathfrak{M}}, \overline{B'} \subseteq \overline{De^*|_{vars(D)}}\}$$

and $\overline{De^*|_{vars(D)}}$ does not contain a pair of contradictory triples. By the fact 33, we get

$$c_d \leq \bigcup \{c'_d \mid \exists\, (D', c'_d) \in \mathcal{C}_P \uparrow \omega, \exists\, \vartheta' \in Subst_{\mathcal{L} \cup \mathcal{A}_{\mathfrak{M}}},$$
$$dom(\vartheta') = vars(D'), D'\vartheta' \text{ is ground}, \overline{D'\vartheta'} \subseteq \overline{De^*|_{vars(D)}}\}.$$

It is evident that $\{\!|D'\vartheta'|\!\} \subseteq \{\!|De^*|_{vars(D)}|\!\}$. Assume that ρ, $dom(\rho) = vars(D')$, is a variable renaming such that $range(\rho) \cap vars(D) = \emptyset$. Denote by D^* the variant $D'\rho$ of D'. Let $\vartheta^* = \rho^{-1} \circ \vartheta'$ be a substitution of $\mathcal{L} \cup \mathcal{A}_{\mathfrak{M}}$. Then $dom(\vartheta^*) = vars(D^*)$ and $D^*\vartheta^* = D'(\rho \circ \rho^{-1} \circ \vartheta') = D'\vartheta'$.

Let $\{\!|D'\vartheta'|\!\} = \{b_1, \ldots, b_n\}$. Denote by d_i, $i = 1, \ldots, n$, a literal in D such that $d_i(e^*|_{vars(D)}) = b_i$. Note that, by the above facts about $e^*|_{vars(D)}$, there exists the only one literal d_i in D satisfying this condition (perhaps with multiple occurrence). Denote by $d_1^{*i}, \ldots, d_{k^i}^{*\,i}$, $i = 1, \ldots, n$, all the literals in D^* such that $d_j^{*i}\vartheta^* = b_i$ for $j = 1, \ldots, k_i$. Since $D^*\vartheta^* = D'\vartheta'$, every literal in D^* is denoted as d_j^{*i} for some i, j. Hence $\{\!|D^*|\!\} = \{\!|\dot{\bigvee}_{i=1}^{n} \dot{\bigvee}_{j=1}^{k^i} d_j^{*i}|\!\}$. Consider the tuples

$$\begin{array}{l} (d_1^{*1}, \ldots, d_{k^1}^{*\,1}, \ldots, d_1^{*n}, \ldots, d_{k^n}^{*\,n}) \\ (\,d_1\,, \ldots,\; d_1\;, \ldots,\, d_n\,, \ldots,\; d_n\;) \end{array} \tag{35}$$

of $\mathcal{L}$. Then the substitution $\vartheta^* \cup e^*|_{vars(D)}$ of $\mathcal{L} \cup \mathcal{A}_{\mathfrak{M}}$, $dom(\vartheta^* \cup e^*|_{vars(D)}) = vars(D^*) \cup vars(D)$ ($dom(\vartheta^*) \cap vars(D) = \emptyset$), is a unifier for the tuples. Notice that $\{\!|D^*(\vartheta^* \cup e^*|_{vars(D)})|\!\} = \{d_1, \ldots, d_n\}(\vartheta^* \cup e^*|_{vars(D)}) = \{b_1, \ldots, b_n\}$. Hence, there exist an mgu θ of $\mathcal{L}$, $dom(\theta) = vars(35)$, and a substitution γ, $dom(\gamma) = range(\theta)$, such that $(\vartheta^* \cup e^*|_{vars(D)})|_{vars(35)} = \theta \circ \gamma$. Note that γ is a substitution of $\mathcal{L} \cup \mathcal{A}_{\mathfrak{M}}$. Consider some regular extension θ' of θ to $vars(D, D^*)$ (θ' is a substitution of $\mathcal{L}$). This yields

$$\vartheta^* \cup e^*|_{vars(D)} = \theta' \circ \gamma'$$

where $\gamma' = \gamma \cup (\theta'|_U{}^{-1} \circ (\vartheta^* \cup e^*|_{vars(D)})|_U)$, $dom(\gamma') = range(\theta')$, and $U = vars(D, D^*) - vars(35)$.

We now show that $\theta'|_{vars(D)}$ is a renaming. Let $v \neq v'$ be variables in $vars(D)$. Then $v(\theta' \circ \gamma') = v(e^*|_{vars(D)}) = a_v \in \mathcal{A}_{\mathfrak{M}}$, $v'(\theta' \circ \gamma') = v'(e^*|_{vars(D)}) = a_{v'} \in \mathcal{A}_{\mathfrak{M}}$, and $a_v \neq a_{v'}$. However, θ' is a substitution of $\mathcal{L}$; hence, $v\theta'$ must be a variable, for otherwise $v\theta'$ should be of the form $f(t_1, \dots, t_{ar(f)})$ where $f \in Func_{\mathcal{L}}$, and we could get $f(t_1, \dots, t_{ar(f)})\gamma' = f(t_1\gamma', \dots, t_{ar(f)}\gamma') \neq a_v \in \mathcal{A}_{\mathfrak{M}}$. For this reason, also $v'\theta'$ must be a variable. Moreover $v\theta' \neq v'\theta'$, for otherwise $a_v = v\theta'\gamma' = v'\theta'\gamma' = a_{v'}$, which is a contradiction. So, $\theta'|_{vars(D)}$ is indeed a variable renaming.

Since θ is an mgu for 35, $range(\theta'|_{vars(D^*)}) = range(\theta'|_{vars(d_1,\dots,d_n)}) \subseteq range(\theta'|_{vars(D)})$. Therefore $\theta'|_{vars(D^*)} \circ (\theta'|_{vars(D)})^{-1}$ is a substitution of $\mathcal{L}$ such that $\{\!\!\{D^*(\theta'|_{vars(D^*)} \circ (\theta'|_{vars(D)})^{-1})\}\!\!\} = \{\!\!\{\{d_1, \dots, d_n\}\theta(\theta'|_{vars(D)})^{-1})\}\!\!\} = \{d_1, \dots, d_n\}$. Moreover, let $\vartheta = \rho \circ \theta'|_{vars(D^*)} \circ (\theta'|_{vars(D)})^{-1}$, $dom(\vartheta) = vars(D')$, then $\vartheta \in Subst_{\mathcal{L}}$ and $\{\!\!\{D'\vartheta\}\!\!\} = \{\!\!\{D^*(\theta'|_{vars(D^*)} \circ (\theta'|_{vars(D)})^{-1})\}\!\!\} = \{d_1, \dots, d_n\} \subseteq \{\!\!\{D\}\!\!\}$. This yields

$$c_d \leq \bigcup \{c'_d \mid \exists\, (D', c'_d) \in \mathcal{C}_P \uparrow \omega, \exists\, \vartheta \in Subst_{\mathcal{L}}, D'\vartheta \sqsubseteq D\}. \tag{36}$$

Hence, we have shown: if (D, c_d) is a graded strong literal disjunction of $\mathcal{L}$ and $P \models^{\vee} (D, c_d)$, then $\mathfrak{M} \models^{\vee} (D, c_d)$ ($\mathfrak{M}$ is a fuzzy disjunctive $\boldsymbol{L}$-model of P), and consequently either the property 36 holds; or D is a tautology. So, the left-right direction of the point ii) of the theorem holds.

The property 32 and the right-left direction of the point ii) are proved as follows: Let $\mathfrak{M} \models^{\vee} (D, c_d)$. We show that $P \models^{\vee} (D, c_d)$. If D is a tautology, then evidently $P \models^{\vee} (D, c_d)$ by the definition of fuzzy disjunctive $\boldsymbol{L}$-model. If not, denote

$$C'_d = \{c'_d \mid \exists\, (D', c'_d) \in \mathcal{C}_P \uparrow \omega, \exists\, \vartheta \in Subst_{\mathcal{L}}, D'\vartheta \sqsubseteq D\},$$
$$c^*_d = \bigcup C'_d.$$

Since $\mathfrak{M} \models^{\vee} (D, c_d)$, by the above considerations, $c_d \leq c^*_d$. Note that this is also the assumption of the right-left direction of the point ii). We have proved that for every fuzzy disjunctive $\boldsymbol{L}$-model $\mathfrak{N}$, if $\mathfrak{N} \models^{\vee} P$, then $\mathfrak{N} \models^{\vee} (D', c'_d)$, $(D', c'_d) \in \mathcal{C}_P \uparrow \omega$, and also $\mathfrak{N} \models^{\vee} (D'\vartheta, c'_d)$ where $\vartheta \in Subst_{\mathcal{L}}$ and $D'\vartheta \sqsubseteq D$. This yields $\mathfrak{N} \models^{\vee} (D, c'_d)$ for every $c'_d \in C'_d$ because of $\overline{D'\vartheta[e]} \subseteq \overline{D[e]}$ for any variable assignment e in $\mathfrak{N}$. We next show that $\mathfrak{N} \models^{\vee} (D, c^*_d)$. Let e be a variable assignment in $\mathfrak{N}$. If $\overline{D[e]}$ contains a pair of contradictory triples, then clearly $\mathfrak{N} \models^{\vee} (D, c^*_d)[e]$. If not, denote

$$c^{\#}_d = \{c''_d \mid (\overline{D''}, c''_d) \in \mathcal{B}_{\mathfrak{N}}, \overline{D''} \subseteq \overline{D[e]}\}.$$

Since $\mathfrak{N} \models^{\vee} (D, c'_d)[e]$, $c'_d \in C'_d$, $c'_d \leq c^{\#}_d$. Hence, $c_d \leq c^*_d = \bigcup C'_d \leq c^{\#}_d$. So, $\mathfrak{N} \models^{\vee} (D, c^*_d)[e]$. We obtain that $\mathfrak{N} \models^{\vee} (D, c^*_d)$ and $\mathfrak{N} \models^{\vee} (D, c_d)$. Thus $P \models^{\vee} (D, c_d)$. The property 32 and the right-left direction of the point ii) of the theorem hold.

Let (Im, c_i) be a graded strong literal implication of $\mathcal{L}$ and $(Im^{\vee}, c_i)$ a disjunctive form of (Im, c_i). Then the following are equivalent:

$$\begin{array}{llll} \mathfrak{M} \models^{\vee} (Im, c_i) & \Longleftrightarrow & \mathfrak{M} \models^{\vee} (Im^{\vee}, c_i) & \Longleftarrow\!(32 \text{ and } \mathfrak{M} \models^{\vee} P)\!\Longrightarrow \\ P \models^{\vee} (Im^{\vee}, c_i) & \Longleftrightarrow & P \models^{\vee} (Im, c_i). & \end{array}$$

Hence, $\mathfrak{M}$ is a characteristic model of P for $\mathcal{L}$ and the point i) is proved.

Thus the proof is completed. □

6.3 Semantics

We are now in position to define the model and fixpoint semantics.

Definition 7 (Model semantics). Let P be a fuzzy disjunctive program of $\mathcal{L}$ with weak similarity.

- If $(\Box, c) \notin \mathcal{C}_P \uparrow \omega$ for $c \neq \mathbf{0}$, we form the following characteristic fuzzy disjunctive $\boldsymbol{L}$-model $\mathfrak{M} = (\mathcal{U}_{\mathfrak{M}}, \mathcal{B}_{\mathfrak{M}})$ for P (defined in the proof of Theorem 4):
  - Let $\mathcal{A}_{\mathfrak{M}} = \{a_n \mid n \in \omega\}$ where a_n are new pairwise different constant symbols not occurring in $\mathcal{L}$. We let $\mathcal{U}_{\mathfrak{M}} = GTerm_{Func_{\mathcal{L}} \cup \mathcal{A}_{\mathfrak{M}}}$. Since $\mathcal{A}_{\mathfrak{M}}$ contains constants a_n, $\mathcal{U}_{\mathfrak{M}} \neq \emptyset$. A function symbol $f \in Func_{\mathcal{L}}$ is interpreted as the function $f^{\mathfrak{M}} : \mathcal{U}_{\mathfrak{M}}^{ar(f)} \to \mathcal{U}_{\mathfrak{M}}$, $f^{\mathfrak{M}}(u_1, \ldots, u_{ar(f)}) = f(u_1, \ldots, u_{ar(f)})$ where $u_1, \ldots, u_{ar(f)} \in \mathcal{U}_{\mathfrak{M}}$.
  - We put

$$\mathcal{B}_{\mathfrak{M}} = \{(\overline{B}, c_b) \mid$$

- $\exists\,(D, c_d) \in \mathcal{C}_P \uparrow \omega$, $\exists\,\vartheta \in Subst_{\mathcal{L} \cup \mathcal{A}_{\mathfrak{M}}}$ *such that* $B = D\vartheta$ *is ground in* $\mathcal{L} \cup \mathcal{A}_{\mathfrak{M}}$*;*
- $\overline{B}$ *does not contain a pair of contradictory triples;*
- $c_b = \bigcup\{c'_d \mid \exists\,(D', c'_d) \in \mathcal{C}_P \uparrow \omega,\ \exists\,\vartheta' \in Subst_{\mathcal{L} \cup \mathcal{A}_{\mathfrak{M}}},$ $D'\vartheta'$ *is ground,* $\overline{D'\vartheta'} \subseteq \overline{B}\} > \mathbf{0}$*;*
- $\forall\,(D^*, c^*_d) \in \mathcal{C}_P \uparrow \omega$ *such that* $\exists\,\vartheta^* \in Subst_{\mathcal{L} \cup \mathcal{A}_{\mathfrak{M}}}$ *where* $B^* = D^*\vartheta^*$ *is ground in* $\mathcal{L} \cup \mathcal{A}_{\mathfrak{M}}$ *and* $\overline{B^*} \subset \overline{B}$, $c_b > \bigcup\{c'_d \mid \exists\,(D', c'_d) \in \mathcal{C}_P \uparrow \omega,\ \exists\,\vartheta' \in Subst_{\mathcal{L} \cup \mathcal{A}_{\mathfrak{M}}},\ D'\vartheta'$ *is ground,* $\overline{D'\vartheta'} \subseteq \overline{B^*}\}\}$.

  - $\mathcal{MS}(P) = \{d \mid d$ *is a graded strong literal disjunction of* $\mathcal{L}$, $\mathfrak{M} \models^{\vee} d\}$.
- Suppose that $(\Box, c) \in \mathcal{C}_P \uparrow \omega$ for some $c \neq \mathbf{0}$. In this case the program is inconsistent and the model does not exist. We put: $\mathcal{MS}(P) = \{d \mid d$ *is a graded strong literal disjunction of* $\mathcal{L}\}$.

Definition 8 (Fixpoint semantics). Let P be a fuzzy disjunctive program of $\mathcal{L}$ with weak similarity.

- If $(\Box, c) \notin \mathcal{C}_P \uparrow \omega$ for $c \neq \mathbf{0}$. $\mathcal{FS}(P) = \{(D, c_d) \mid (D, c_d)$ *is a graded strong literal disjunction of* $\mathcal{L}$ *and either* $c_d \leq \{c'_d \mid \exists\,(D', c'_d) \in \mathcal{C}_P \uparrow \omega, \exists\,\vartheta \in Subst_{\mathcal{L}}, D'\vartheta \sqsubseteq D\}$*; or* D *is a tautology*$\}$.

- Suppose that $(\Box, c) \in \mathcal{C}_P \uparrow \omega$ for some $c \neq \mathbf{0}$. In this case we put: $\mathcal{FS}(P) = \{d \,|\, d \textit{ is a graded strong literal disjunction of } \mathcal{L}\}$.

It can be proved the coincidence of these semantics. We close the chapter with the statement:

Corollary 1. *Let P be a fuzzy disjunctive program of $\mathcal{L}$ with weak similarity.*

$$\mathcal{DS}(P) = \mathcal{MS}(P) = \mathcal{FS}(P).$$

Proof: An immediate consequence of Theorems 3 and 4. Suppose that d is a graded strong literal disjunction of $\mathcal{L}$. By Theorem 3, we get:

$$d \in \mathcal{DS}(P) \Longleftrightarrow P \models d \Longleftrightarrow P \models^{\vee} d.$$

Assume that $(\Box, c) \in \mathcal{C}_P \uparrow \omega$ for some $c \neq \mathbf{0}$. Then by 29,

$$P \models^{\vee} \mathcal{C}_P \uparrow \omega$$

and P has no fuzzy disjunctive $\boldsymbol{L}$-model for $\mathcal{L}$, as shown in the proof of Theorem 4. Hence by the above considerations, for all graded strong literal disjunction d of $\mathcal{L}$, $d \in \mathcal{DS}(P)$. So, $\mathcal{DS}(P) = \{d \,|\, d \textit{ is a graded strong literal disjunction of } \mathcal{L}\}$. However, in this case $\mathcal{MS}(P) = \{d \,|\, d \textit{ is a graded strong literal disjunction of } \mathcal{L}\}$ and also $\mathcal{FS}(P) = \{d \,|\, d \textit{ is a graded strong literal disjunction of } \mathcal{L}\}$ by the definitions.

So, $\mathcal{DS}(P) = \mathcal{MS}(P) = \mathcal{FS}(P)$.

If $(\Box, c) \notin \mathcal{C}_P \uparrow \omega$ for $c \neq \mathbf{0}$, there exists the characteristic fuzzy disjunctive $\boldsymbol{L}$-model $\mathfrak{M}$ of P for $\mathcal{L}$ defined in the proof of Theorem 4, and $\mathcal{MS}(P) = \{d \,|\, d \textit{ is a graded strong literal disjunction of } \mathcal{L}, \mathfrak{M} \models^{\vee} d\}$. Since $\mathfrak{M}$ is a characteristic model of P, $\mathfrak{M} \models^{\vee} d \Longleftrightarrow P \models^{\vee} d$. By the above considerations, we get:

$$\begin{aligned} & d \in \mathcal{DS}(P) && \Longleftrightarrow \\ & P \models^{\vee} d && \Longleftrightarrow \\ & \mathfrak{M} \models^{\vee} d && \Longleftrightarrow \\ & d \in \mathcal{MS}(P). && \end{aligned}$$

Let (D, c_d) be a graded strong literal disjunction of $\mathcal{L}$. By Theorem 4 ii) and by the definition of $\mathcal{FS}(P)$, the following are equivalent:

- $(D, c_d) \in \mathcal{DS}(P)$. $\Longleftrightarrow$
- $P \models^{\vee} (D, c_d)$. $\xleftrightarrow{(4\text{ ii})}$
- Either $c_d \leq \{c'_d \,|\, \exists\, (D', c'_d) \in \mathcal{C}_P \uparrow \omega, \exists\, \vartheta \in Subst_{\mathcal{L}}, D'\vartheta \sqsubseteq D\}$; or D is a tautology. $\Longleftrightarrow$
- $(D, c_d) \in \mathcal{FS}(P)$.

Hence, $\mathcal{DS}(P) = \mathcal{MS}(P) = \mathcal{FS}(P)$ also in this case. □

7 Conclusion

In this chapter, we have proposed declarative, model, and fixpoint semantics for fuzzy disjunctive programs with weak similarity. We have introduced a class of fuzzy disjunctive L-models, a 'conservative' extension of known L-models. In terms of fuzzy disjunctive L-models and their characterisation theorem, the model and fixpoint semantics have been established. In the end, we have reached the mutual coincidence of these semantics. The proposed semantics, in conjunction with the procedural semantics in [41], can be used for a design of an inference subsystem dealing with incomplete and uncertain knowledge in, e.g., cognitive robots, expert, and mechatronic systems.

References

1. Apt, K. R. Introduction to logic programming. *Report CS-R8826, Centre for Mathematics and Computer Science, Amsterdam, The Netherlands*, 1986.
2. Alferes, J. J. and Pereira, L. M. On logic program semantics with two kinds of negation. *International Joint Conference and Symposium on Logic Programming, MIT Press*, 574-588, 1992.
3. Baaz, M. and Fermüller, Ch. G. Resolution-Based Theorem Proving for Many-valued Logics. *Journal of Symbolic Computation*, 19(4):353-391, 1995.
4. Bacchus, F. Representing and reasoning with probabilistic knowledge. *Research Report CS-88-31, University of Waterloo*, 1988.
5. Baral, C. Issues in Knowledge Representation: Semantics and Knowledge Combination. *PhD thesis, Dept. of Computer Science, University of Maryland, College Park, USA*, 1991.
6. Baral, C. L_3: A logic programming language with multiple default negations. *Proceedings of the Fifth International Symposium on AI*, 1992.
7. Baral, C., Lobo, J. and Minker, J. Generalized disjunctive well-founded semantics: Declarative semantics. *Proceedings of the Fifth International Symposium on Methodologies for Intelligent Systems*, 465-473, 1990.
8. Baral, C., Lobo, J. and Minker, J. Generalized disjunctive well-founded semantics: Procedural semantics. *Proceedings of the Fifth International Symposium on Methodologies for Intelligent Systems*, 456-464, 1990.
9. Baral, C., Lobo, J. and Minker, J. WF^3 : A semantics for negation in normal disjunctive logic programs. *Proceedings of the Sixth International Symposium on Methodologies for Intelligent Systems*, 459-468, 1991.
10. Blair, H. A. and Subrahmanian, V. S. Paraconsistent foundations for logic programming. *Journal of Non-Classical Logic*, 5(2):45-73, 1988.
11. Blair, H. A. and Subrahmanian, V. S. Paraconsistent logic programming. *Theoretical Computer Science*, 68:135-154, 1989.
12. Bonissone, P. Summarizing and propagating uncertain information with triangular norms. *International Journal of Approximate Reasoning*, 1:71-101, 1987.

13. Brewka, G. and Dix, J. Knowledge Representation with Logic Programs. *Handbook of Philosophical Logic, 2nd ed., vol. 6, chap. 6, Oxford University Press*, 2001.
14. Cohen, P. R. and Grinberg, M. R. A framework for heuristic reasoning about uncertainty. *Proc. IJCAI*, 355-357, Karlsruhe, Germany, 1983.
15. Cohen, P. R. and Grinberg, M. R. A theory of heuristic reasoning about uncertainty. *AI Magazine*, 4(2):17-23, 1983.
16. Damásio, C. V. and Pereira, L. M. A theory of logic programming. *Tech. Report, Dept. Computer Science, Univ. Nova de Lisboa*, 2000.
17. Damásio, C. V. and Pereira, L. M. Hybrid probabilistic logic programs as residuated logic programs. *Proc. Logics in Artificial Intelligence, LNAI 1919, Springer-Verlag*, 57-73, 2000.
18. Damásio, C. V. and Pereira, L. M. Monotonic and residuated logic programs. *Proc. of ECSQARU'01, LNAI 2143, Springer-Verlag*, 748-759, 2001.
19. Dekhtyar, A. and Subrahmanian, V. S. Hybrid probabilistic programs. *Journal of Logic Programming*, 43(3):187-250, 2000.
20. Dempster, A. P. A generalization of Bayesian inference. *Journal of the Royal Statistical Society*, B(30):205-247, 1968.
21. Dix, J. Semantics of Logic Programs: Their Intuitions and Formal Properties, An Overview. In *Logic, Action, and Information - Essays on Logic in Philosophy and Artificial Intelligence*, Fuhrmann, A. and Rott, H., eds., *De Gruyter*, 241-327, 1995.
22. Dix, J., Nanni, M. and Subrahmanian, V. S. Probabilistic agent reasoning. *Transactions of Computational Logic*, 1(2):208-246, 2000.
23. Doyle, J. Methodological simplicity in expert system construction: the case of judgements and reasoned assumptions. *AI Magazine*, 4(2):39-43, 1983.
24. Fernández, J. A., Lobo, J., Minker, J. and Subrahmanian, V. S. Disjunctive LP + integrity constraints = stable model semantics. *Annals of Mathematics and Artificial Intelligence*, 8(3-4):449-474, 1993.
25. Fernández, J. A. and Minker, J. Bottom-up computation of perfect models for disjunctive theories. *Journal of Logic Programming*, 25(1):33-51, 1995.
26. Fitting, M. C. Logic programming on a topological bilattice. *Fundamenta Informaticae*, 11:209-218, 1988.
27. Fitting, M. C. Bilattices and the semantics of logic programming. *Journal of Logic Programming*, 11:91-116, 1991.
28. Gelfond, M. and Lifschitz, V. Logic programs with classical negation. *Proceedings of the Seventh International Logic Programming Conference, MIT Press*, 579-597, Jerusalem, Israel, 1990.
29. Ginsberg, M. Multivalued logics: a uniform approach to reasoning in artificial intelligence. *Computational Intelligence*, 4:265-316, 1988.
30. Ginsberg, M. Bilattices and modal operators. *Proc. Intl. Conf. on Theoretical Aspects of Reasoning about Knowledge, Morgan Kaufmann*, 273-287, 1990.
31. Gottwald, S. Fuzzy Sets and Fuzzy Logic. *Vieweg*, 1993.
32. Guller, D. Semantics for disjunctive programs. *Proceedings of the Fifth Workshop on Logic, Language, Information and Computation*, 79-91, Sao Paulo,

1998.

33. Guller, D. On clausal models. *Proceedings of the Sixth Workshop on Logic, Language, Information and Computation*, 115-126, Rio de Janeiro, 1999.
34. Guller, D. One generalisation of Herbrand's theorem. *Proceedings of the Second Panhellenic Logic Symposium*, 112-118, Delphi, 1999.
35. Guller, D. Declarative semantics for quantified disjunctive programs. *Proceedings of the Seventh Workshop on Logic, Language, Information and Computation*, 89-102, Natal, 2000.
36. Guller, D. Model, fixpoint, and Herbrand semantics for quantified disjunctive programs. *Proceedings of the Logic Colloquium 2000*, Paris, 2000.
37. Guller, D. One formalisation of reasoning with incomplete and imprecise information. *Proceedings of the Int. Conf.: Emerging Technologies and New Challenges in Information Society*, 136-145, Aizu-Wakamatsu, 2000.
38. Guller, D. Semantics for fuzzy disjunctive programs. *Proceedings of the WSES Conference: Automation and Information: Theory and Applications*, Skiathos, 2001.
39. Guller, D. Semantics for fuzzy disjunctive programs with weak similarity. In *Hybrid Information Systems*, Abraham, A. and Köppen, M., eds., *Advances in Soft Computing, Springer-Verlag*, 285-300, 2002.
40. Guller, D. Procedural semantics for fuzzy disjunctive programs with weak similarity. *Proceedings of the Seventh International Symposium on Artificial Intelligence and Mathematics*, AI&M 9-2002, http://rutcor.rutgers.edu/~amai/aimath02, Fort Lauderdale, 2002.
41. Guller, D. Procedural semantics for fuzzy disjunctive programs. *Proceedings of the Ninth International Conference Logic for Programming, Artificial Intelligence, and Reasoning*, Baaz, M. and Voronkov, A., eds., *LNAI vol. 2514, Springer-Verlag*, 247-261, Tbilisi, 2002.
42. Hájek, P. Metamathematics of fuzzy logic. *Trends in Logic, Studia Logica Library, Kluwer Academic Publishers*, 1998.
43. Ishizuka, M. Inference methods based on extended Dempster and Shafer's theory with uncertainty/fuzziness. *New Generation Computing*, 1(2):159-168, 1983.
44. Ishizuka, M. and Kanai, N. PROLOG-ELF incorporating fuzzy logic. *New Generation Computing*, 3:479-486, 1985.
45. Kifer, M. and Li, A. On the semantics of rule-based expert systems with uncertainty. *2nd Intl. Conf. on Database Theory, LNCS vol. 326, Springer-Verlag*, 102-117, Bruges, Belgium, 1988.
46. Kifer, M. and Lozinskii, E. L. RI: a logic for reasoning with inconsistency. *4th Symposium on Logic in Computer Science*, 253-262, Asilomar, CA, 1989.
47. Kifer, M. and Lozinskii, E. L. A logic for reasoning with inconsistency. *Journal of Automated Reasoning, Kluwer Ac. Publ.*, 9(2):179-215, 1992.
48. Kifer, M. and Subrahmanian, V. S. Theory of the generalized annotated logic programming and its applications. *Journal of Logic Programming*, 12, 335-367, 1992.
49. Lloyd, J. W. Foundations of logic programming. *Springer-Verlag*, 1987.

50. Lukasiewicz, T. Probabilistic logic programming. *Proc. 13th Biennial European Conference on Artificial Intelligence, J. Wiley & Sons*, 388-392, 1998.
51. Lukasiewicz, T. Many-valued disjunctive logic programs with probabilistic semantics. *Proc. 5th International Conference LPNMR, LNAI vol. 1730, Springer-Verlag*, 277-289, El Paso, USA, 1999.
52. Lukasiewicz, T. Many-valued first-order logics with probabilistic semantics. *Proc. CSL-98, LNCS vol. 1584, Springer-Verlag*, 415-429, 1999.
53. Lukasiewicz, T. Probabilistic and truth-functional many-valued logic programming. *Proc. ISMVL, IEEE Computer Society*, 236-241, 1999.
54. Lukasiewicz, T. Local probabilistic deduction from taxonomic and probabilistic knowledge-bases over conjunctive events. *Int. Journal of Approximate Reasoning*, 21(1):23-61, 1999.
55. Lukasiewicz, T. Probabilistic deduction with conditional constraints over basic events. *Journal of Artificial Intelligence Research*, 10:199-241, 1999.
56. Lukasiewicz, T. Fixpoint characterizations for many-valued disjunctive logic programs with probabilistic semantics. *Proc. Logic Programming and Non-Monotonic Reasoning, LNAI vol. 2173, Springer-Verlag*, 336-350, 2001.
57. Lukasiewicz, T. Probabilistic logic programming with conditional constraints. *ACM Trans. Computational Logic*, 2(3):289-337, 2001.
58. Marks II, R. J. Fuzzy Logic Technology and Applications. *IEEE Technical Activities Board*, 1994.
59. Mateis, C. A quantitative extension of disjunctive logic programming. *PhD Thesis, Technical University of Vienna*, 1998.
60. Mateis, C. Extending disjunctive logic programming by t-norms. *Proc. 5th International Conference LPNMR, LNAI vol. 1730, Springer-Verlag*, 290-304, El Paso, USA, 1999.
61. Mateis, C. Quantitative disjunctive logic programming: semantics and computation. *AI Communications*, 13(4):225-248, 2000.
62. Medina, J., Ojeda-Aciego, M. and Vojtáš, P. A completeness theorem for multi-adjoint logic programming. *Proc. 10th IEEE International Conference on Fuzzy Systems, IEEE Press*, 2001.
63. Medina, J., Ojeda-Aciego, M. and Vojtáš, P. A multi-adjoint logic approach to abductive reasoning. *Proc. 17th International Conference on Logic Programming, LNAI vol. 2273, Springer-Verlag*, 269-283, 2001.
64. Medina, J., Ojeda-Aciego, M. and Vojtáš, P. Multi-adjoint logic programming with continuous semantics. *Proc. Logic Programming and Non-Monotonic Reasoning, LNAI vol. 2173, Springer-Verlag*, 351-364, 2001.
65. Medina, J., Ojeda-Aciego, M. and Vojtáš, P. A procedural semantics for multi-adjoint logic programming. *Proc. EPIA'01, LNAI vol. 2258, Springer-Verlag*, 290-297, 2001.
66. Medina, J., Ojeda-Aciego, M. and Vojtáš, P. Similarity-based unification: a multi-adjoint approach. *Technical Report MA/01/01, University of Málaga*, 2001.
67. Minker, J. and Zanon, G. An extension to linear resolution with selection function. *Inform. Process. Lett.*, 14(3):191-194, 1982.

68. Minker, J. On indefinite databases and the closed world assumption. *Proc. 6th Conference on Automated Deduction, LNCS vol. 138, Springer Verlag*, 292-308, New York, 1982.
69. Minker, J., Rajasekar, A. and Lobo, J. Weak generalized closed world assumption. *Journal of Automated Reasoning*, 5:293-307, 1989.
70. Minker, J. and Rajasekar, A. A fixpoint semantics for non-Horn logic programs. *Journal of Logic Programming*, 9(1):45-74, 1990.
71. Minker, J., Rajasekar, A. and Lobo, J. Foundations of disjunctive logic programming. *MIT Press, Cambridge, MA*, 1992.
72. Minker, J. and Ruiz, C. Semantics for disjunctive logic programs with explicit and default negation. *Fundamenta Informaticae*, 20(3/4):145-192, 1994.
73. Minker, J. Overview of disjunctive logic programming. *Annals of Mathematics and Artificial Intelligence*, 12:1-24, 1994.
74. Minker, J. and Ruiz, C. Mixing a default rule with stable negation. *Proceedings of the Fourth International Symposium on Artificial Intelligence and Mathematics*, 122-125, 1996.
75. Minker, J. and Ruiz, C. Combining closed world assumptions with stable negation. *Fundamenta Informaticae*, XX(1997):1-19, 1997.
76. Minker, J. and Ruiz, C. Logic knowledge bases with two default rules. *Annals of Mathematics and Artificial Intelligence*, 22(3-4):333-361, 1998.
77. Minker, J. and Seipel, D. Disjunctive Logic Programming: A Survey and Assessment. In *Computational Logic: Logic Programming and Beyond, Essays in Honour of Robert A. Kowalski, Part I.*, Kakas, A. C. and Sadri, F., eds., *LNCS vol. 2407, Springer-Verlag*, 472-511, 2002.
78. Moore, R. C. Semantic considerations on non-monotonic logic. *Journal of Artificial Intelligence*, 25:75-94, 1985.
79. Morishita, S. A unified approach to semantics of multi-valued logic programs. *Tech. Report RT 5006, IBM Tokyo*, 1990.
80. Ng, R. T. and Subrahmanian, V. S. Non-monotonic negation in probabilistic deductive databases. *Proc. Seventh Conf. Uncertainty in AI*, 249-256, Los Angeles, 1991.
81. Ng, R. T. and Subrahmanian, V. S. Probabilistic logic programming. *Information and Computation*, 101:150-201, 1992.
82. Ng, R. T. and Subrahmanian, V. S. Empirical probabilities in monadic deductive databases. *Proc. Eighth Conf. Uncertainty in AI*, 215-222, Stanford, 1992.
83. Ng, R. T. and Subrahmanian, V. S. A semantical framework for supporting subjective and conditional probabilities in deductive databases. *Journal of Automated Reasoning*, 10(2):191-235, 1993.
84. Ng, R. T. and Subrahmanian, V. S. Stable semantics for probabilistic deductive databases. *Information and Computation*, 110:42-83, 1994.
85. Nilsson, N. J. Probabilistic logic. *Artificial Intelligence*, 28:71-87, 1986.
86. Ovchinnikov, S. Similarity relations, fuzzy partitions, and fuzzy orderings. *Fuzzy Sets and Systems*, 40:107-126, 1991.
87. Pavelka, J. On fuzzy logic I, II, III. *Zeitschr. f. Math. Logik und Grund. der*

Math., 25:45-52, 119-134, 447-464, 1979.
88. Pearl, J. Probabilistic reasoning in intelligent systems - networks of plausible inference. *Morgan Kaufmann*, 1988.
89. Przymusinski, T. C. On the declarative semantics of deductive databases and logic programs. *Foundations of Deductive Databases and Logic Programming, Morgan Kaufmann*, 193-216, 1988.
90. Przymusinski, T. C. The well-founded semantics coincides with the three-valued stable semantics. *Fundamenta Informaticae*, 13(4):445-464, 1990.
91. Przymusinski, T. C. Stationary semantics for disjunctive logic programs and deductive databases. *Proceedings of the North American Conference on Logic Programming, MIT Press*, 459-477, 1990.
92. Przymusinski, T. C. Stable semantics for disjunctive programs. *New Generation Computing*, 9:401-424, 1991.
93. Przymusinski, T. C. Autoepistemic logic of knowledge and beliefs. *Technical report, University of California at Riverside*, 1994.
94. Przymusinski, T. C. Static semantics for normal and disjunctive logic programs. *Annals of Mathematics and Artificial Intelligence, Special Issue on Disjunctive Programs*, 14(2-4):323-357, 1995.
95. Przymusinski, T. C. Semantics of normal and disjunctive logic programs, a unifying framework. *Technical report, University of California at Riverside*, 1995.
96. Rine, D. C. Some relationships between logic programming and multiple-valued logic. *Proc. IEEE Intl. Symp. on Multiple-Valued Logic*, 160-163, 1986.
97. Ross, K. A. and Topor, R. W. Inferring negative information from disjunctive databases. *Journal of Automated Reasoning*, 4(2):397-424, 1988.
98. Ruspini, E. H. A theory of fuzzy clustering. *Proc. IEEE Conf. on Decision and Control*, 1378-1383, New Orleans, 1977.
99. Shoenfield, J. R. Mathematical Logic. *Addison-Wesley*, 1967.
100. Schweizer, B. and Sklar, A. Associative functions and abstract semi-groups. *Publ. Math. Debrecen*, 10:69-81, 1963.
101. Shafer, G. A mathematical theory of evidence. *Princeton University Press*, 1976.
102. Shapiro, E. Logic programs with uncertainties: a tool for implementing expert systems. *Proc. IJCAI, William Kauffman*, 529-532, 1983.
103. Shortliffe, E. Computer-based medical consultation: MYCIN. *Elsevier Science*, 1976.
104. Steiner, J. Assignments as a basis for unification: a predicate algebra for logic programming. *Proc. LOP'92*, 207-216, 1992.
105. Subrahmanian, V. S. On the semantics of quantitative logic programs. *Proc. 4th IEEE Symposium on Logic Programming, Computer Society Press*, 173-182, Washington DC, 1987.
106. Takeuti, G. Proof Theory. *2nd ed., North-Holland*, 1987.
107. Takeuti, G. and Titani, S. Fuzzy logic and fuzzy set theory. *Annals of Mathematical Logic*, 32:1-32, 1992.
108. Tarski, A. A lattice-theoretical fixed point theorem and its applications. *Pacific*

Journal of Mathematics, 83:157-167, 1955.

109. Valverde, L. On the structure of F-indistinguishability operators. *Fuzzy Sets and Systems*, 17, 313-328, 1985.
110. Van Emden, M. H. Quantitative deduction and its fixpoint theory. *Journal of Logic Programming*, 4(1):37-53, 1986.
111. Van Gelder, A., Ross, K. A. and Schlipf, J. S. The well-founded semantics for general logic programs. *Journal of the ACM*, 38(3):620-650, 1991.
112. Vojtáš, P. and Paulík, L. Soundness and completeness of non-classical extended SLD-resolution. *Proc. Extensions of Logic Programming, LNCS vol. 1050, Springer-Verlag*, 289-301, 1996.
113. Vojtáš, P. Fuzzy logic programming. *Fuzzy sets and systems*, 124(3):361-370, 2001.
114. Yahya, A. and Henschen, L. J. Deduction in non-Horn databases. *Journal of Automated Reasoning*, 1(2):141-160, 1985.
115. Zadeh, L. Fuzzy Similarity relations and fuzzy orderings. *Information Sciences*, 3:177-200, 1969.
116. Zadeh, L. Fuzzy Logic. *IEEE Computer*, 21(4):83-93, 1988.
117. Zimmermann, H. J. Fuzzy Set Theory and its Applications. *2nd ed., Kluwer Academic Publishers*, 1991.
118. Zlatoš, P. Two-levelled logic and model theory. *Colloquia Mathematica Societatis János Bolyai, 28. Finite Algebra and Multiple-Valued Logic*, 825-871, Szeged, Hungary, 1979.
119. Zlatoš, P. Unitary congruence adjunctions. *Colloquia Mathematica Societatis János Bolyai, 43. Lectures in Universal Algebra*, 587-647, Szeged, Hungary, 1983.

Chapter 8

An Automated Report Generation Tool for the Data Understanding Phase

Juha Vesanto and Jaakko Hollmén

Summary. To successfully prepare and model data, the data miner needs to be aware of the properties of the data manifold. In this chapter, the outline of a tool for automatically generating data survey reports for this purpose is described. Such a report is used as a starting point for data understanding, acts as documentation of the data, and can easily be redone if necessary. The main focus is on describing the cluster structure and the contents of the clusters. The described system combines linguistic descriptions (rules) and statistical measures with visualizations. Whereas rules and mathematical measures give quantitative information, the visualizations give qualitative information on the data sets, and help the user to form a mental model of the data based on the suggested rules and other characterizations.

Keywords: clustering, cluster description, visualization, descriptive rules, data analysis.

1 Introduction

The purpose of data mining is to find knowledge from databases where the dimensionality, complexity, or amount of data is prohibitively large for manual analysis. This is an interactive process which requires that the intuition and background knowledge of application experts are coupled with the computational efficiency of modern computer technology.

The CRoss-Industry Standard Process for Data Mining (CRISP-DM) [4] divides the data mining process to several phases. One of the first phases is data understanding, which is concerned with understanding the origin, nature and reliability of the data, as well as becoming familiar with the contents of the data through data exploration. Understanding the data is essential in the whole knowledge discovery process. Proper data preparation, selection of modeling tools and evaluation processes is only possible if the miner has a good overall idea, or a mental model, of the data.

The data exploration is usually done by interactively applying a set of data exploration tools and algorithms to get an overview of the properties of the data manifold. However, understanding a single data set is often not enough. Because of the iterative nature of the knowledge discovery process, several different data sets and preprocessing strategies need to be considered and explored. The task of data understanding is engaged again and again. Therefore, the tools used for data understanding should be as automated as possible.

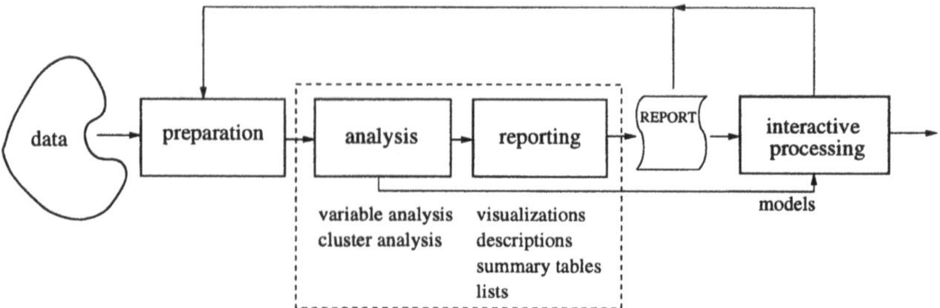

Figure 1. Creating understanding of the data in a survey cycle. The data is prepared and fed into the analysis system which generates the data survey report. Based on the findings in the report and possibly further insights based on interactive investigation, the data miner may either proceed with the next data mining phase, or prepare the data set in a better, alternative fashion and, with a push of a button, make a new report. The area within the dashed box corresponds to the implemented report generation system.

1.1 Automated Analysis of Table-Format Data

This chapter presents a selection of techniques and associated presentation templates to automate part of the data understanding process. The driving goal of the work has been to construct a framework where an overview and initial analysis of the data can be executed automatically, without user intervention. The motivation for the work has come from a number of data mining projects, in process industry for example [1], where we have repeatedly encountered the data understanding task when new data sets and/or preprocessing strategies have been considered.

While statistical and numerical program packages provide a multitude of techniques that are similar to those presented here, the novelty of our approach is to combine the techniques and associated visualizations into a coherent whole. In addition, using such program packages requires considerable time and expertise. An automated approach used in this chapter has a number of desirable properties:

- the analysis is easy to execute again (and again and ...),
- the required level of technical know-how of the data miner is reduced when compared to fully interactive data exploration, and
- the resulting report acts as documentation that can be referred to later.

Of course, an automatically performed analysis can never replace the flexibility and power inherent in an interactive approach. Instead, we consider the report generation system described here to provide an advantageous starting point for such interactive analysis (see Figure 1).

The nature of the report generation system imposes some requirements for the applied methods. The algorithms should be computationally light, robust, and require no user-defined parameters. They should also be generic enough so that their results are of interest in most cases. Naturally, what is interesting depends on the problem and the data domain. In the implemented system, the domain has been restricted to

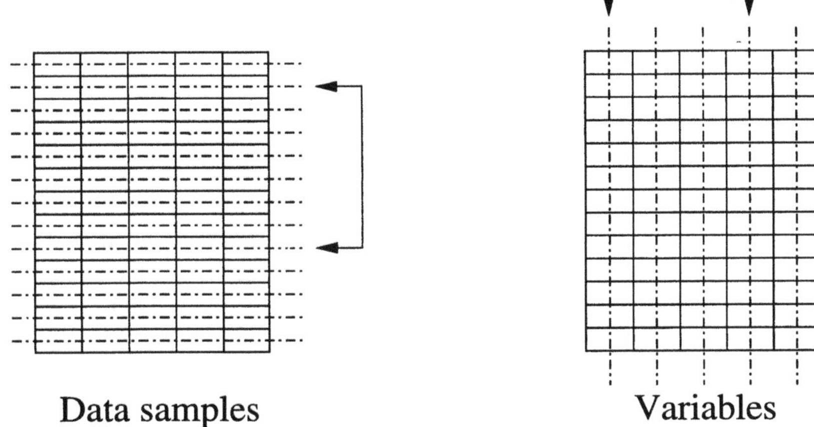

Figure 2. Table-format data can be investigated both in terms of samples and in terms of variables. Samples are analyzed to find groups of similar items. Variables are analyzed to find groups of related items. Sample analysis is covered in Section 2 and variable analysis in Section 3.

unsupervised analysis of unordered, numerical vector (table-format) data, see Figure 2. In contrast, supervised estimation of one or more output variables, analysis of purely categorical variables, and analysis of data sequences or time-series are not considered. In a more complete system also these, especially the issues in supervised estimation, should be addressed.

However, simply applying the analysis algorithms to the data is not enough: the knowledge must be transferred to the data miner. This is accomplished through a data survey report. The report consists of visualizations and numerical or linguistic descriptions organized according to a predefined template into summary tables and lists. Visualizations are a very important part of the report since they allow the display of large amounts of detailed information in a coherent manner, and give the data miner a chance to validate the quantitative descriptions with respect to the actual data.

1.2 Related Work

In [15], Pyle introduces the concept of data survey as a tool for getting a feel of the data manifold. The emphasis is on variable dependency analysis using entropy and related measures, and on identifying problems in the data. Unfortunately, only rather general guidelines are given, and cluster analysis is handled very briefly.

Cluster analysis, and especially characterization of the contents of the clusters is one of the key issues in this chapter. The clusters can be characterized by listing the variable values typical for each cluster using, for example, characterizing rules [18]. Another approach is to rank the variables in the order of significance [12, 16, 17]. In this chapter, both approaches are used.

In [14], a report generation system KEFIR is described. It automatically analyzes data in large relational databases, and produces a report on the key findings. The difference to our work is that KEFIR compares a data set and *a priori* given normative values, and tries to find and explain deviations between them, and thus requires considerable setting up. The system described in this chapter, on the contrary, is applied when the data miner starts with a single unfamiliar data set.

In this sense, the recent work by Bay and Pazzani [2] is much closer to certain parts of this work. They examine the problem of mining contrast sets, where the fundamental problem is to find out how two groups differ from each other, and propose an efficient search algorithm to find conjunctive sets of variables and values which are meaningfully different in the two groups. The difference to our work is that they are concerned with categorical variables, and want to find all relevant differences between two arbitrary groups. In this work, the data is numerical, and the two groups are always two clusters, or a cluster and the rest of the data.

In the implemented system, metric clustering techniques are used, so the input data needs to be numerical vector-data (which may have missing values). Another possibility would be to use conceptual clustering techniques [13], which inherently focus on descriptions of the clusters. However, conceptual clustering techniques are rather slow, while recent advances have made metric clustering techniques applicable to very large data sets [27, 7].

1.3 Contents

The chapter is organized as follows. In this section, the methodology and basic motivation for the implemented system has been described. In Sections 2 and 3, the analysis methods for samples and variables, respectively, are described. In Section 4, the overall structure of the data survey report is explained. The system data set collected by the authors is used throughout the chapter to illustrate the different aspects of the report and is more closely described in the Appendix at the end of this chapter. In Section 5, a publicly available insurance company data set [21] is analyzed in order to describe the use of the data survey report. The work is summarized in Section 6.

2 Sample Analysis

The relevant questions in terms of samples are: Are there natural groups, i.e., clusters, in the data? What kind of segments can be formed, and what are their properties?

2.1 Projection

A qualitative idea of the cluster structure in the data is acquired by visualizing the data using vector projection methods. Projection algorithms try to preserve distances or neighborhoods of the samples, and thus encode similarity information in the projection coordinates. There are many different kinds of projection algorithms, see for example [11], but in the proposed system, a classical linear projection based on principal component analysis (PCA) is used. In PCA, the directions are found which account for most of the variance in the data. This is done by calculating the eigenvectors $\mathbf{e}_1, ..., \mathbf{e}_d$ and corresponding eigenvalues $\lambda_1, ..., \lambda_d$ of the covariance matrix of the data, and ordering them by decreasing eigenvalues $\lambda_1 > \lambda_2 > ... > \lambda_d$. The first direction $\mathbf{e}_1$ accounts for most – specifically $\frac{\lambda_1}{\sum_i \lambda_i}100\%$ – of the variance in the data, the second for the second largest amount, and so on. By projecting data to the space spanned by the first few eigenvectors as much of the variance is preserved as possible. The sum of the corresponding eigenvalues gives the amount of variance preserved in the projection, and thus indicates the error made in the low-dimensional projection. For example, a projection to a 2-dimensional plane is defined as:

$$\mathbf{y} = \begin{bmatrix} \mathbf{e}_1^T \\ \mathbf{e}_2^T \end{bmatrix} \mathbf{x}. \tag{1}$$

The advantages of PCA projection include computational efficiency, and the ability to project new data points (e.g., cluster centers) easily. It also allows the use of a scree plot for easily understandable validation of the projection, see Figure 3a.

Like spatial coordinates, colors can also be used to encode similarity information [26, 23, 9]. In the implemented system, the colors are assigned from the hues on a color circle by a simple projection of cluster centroids onto the circle. A color coding can be constructed by defining a smooth coloring in a low-dimensional manifold, and projecting the data onto this manifold, for example as follows:

1. A 1-dimensional SOM (see Section 2.2 below) is trained using the data. The trained SOM forms a principal curve going through the data manifold.
2. A color from the color hue circle (from the HSV color model, see for example [26], with hue = ϕ, saturation = 1 and value = 1) is assigned to each map unit i of the 1-dimensional SOM. The colors can be assigned equidistant from each other $\phi_i = 2\pi i/M$ or the distances between neighboring prototypes can be taken into account: $\phi_i = 2\pi \sum_{j=1}^{i} \|\mathbf{m}_{i+1} - \mathbf{m}_i\| / \sum_{j=1}^{M-1} \|\mathbf{m}_{i+1} - \mathbf{m}_i\|$.
3. Each data sample picks the same color as its BMU.

These colors are used consistently in various visualizations throughout the report.

2.2 Clustering

Clustering algorithms, see for example [6], provide a more quantitative analysis of the natural groups that exist in the data. In real data, however, clusters are rarely compact, well-separated groups of objects – the conceptual idea that is often used to motivate clustering algorithms. Apart from noise and outliers, clustering may depend on the level of detail being observed. Therefore, instead of providing a single partitioning of the data, the implemented system constructs a cluster hierarchy, see Figure 3b. This may represent the inherent structure of the data set better than a direct partitioning. Equally important from data understanding point of view is that it also allows the data to be investigated at several levels of granularity.

2.2.1 Base Clusters

In the implemented system, the Self-Organizing Map (SOM) is used for clustering [10]. The SOM is a collection of prototype vectors $\mathbf{m}$, between which a neighborhood relation h is defined. The neighborhood relation defines a structured lattice, usually a two-dimensional, rectangular or hexagonal lattice of map units. After initializing the prototype vectors with, for example, random values, training takes place. Training a Self-Organizing Map from data is divided to two steps, which are applied alternately. First, a best-matching unit (BMU) or a winner unit b_i is searched, which minimizes the Euclidean distance between a data sample $\mathbf{x}_i$ and the map unit prototypes $\mathbf{m}_j$

$$b_i = \arg\min_j \|\mathbf{x}_i - \mathbf{m}_j\|. \tag{2}$$

Then, new prototypes are calculated as:

$$\mathbf{m}_j = \frac{\sum_{i=1}^{n} h_{b_i j}\mathbf{x}_i}{\sum_{i=1}^{n} h_{b_i j}}, \tag{3}$$

where $h_{b_i j}$ is the neighborhood strength between map units b_i and j, and n is the number of data samples. This is the batch training algorithm for the SOM.

After quantizing the data using a SOM with a few hundred map units, the map units are clustered. Thus, in the second phase only a few hundred objects need to be clustered instead of all the original data samples. This 2-phase strategy reduces the computational complexity of the clustering considerably [24]. In addition, the SOM is useful as a convenient projection of the data cloud, see Section 3.

To cluster the units of the SOM, a widely used technique is the U-matrix [20]. It is, in effect, a measure of the local probability density of the data in each map unit. Thus, the local minima of the U-matrix – map units for which the distance matrix value is lower than that of any of their neighbors – can be used to identify cluster centers. In [22], the rest of the map units were assigned to the cluster whose center was closest. This procedure is simple and fast, but it also makes the implicit assumption that the border between two clusters lies on the middle point between their cluster centers. We use an enhanced version based on region-growing. This procedure provides a partitioning of the map into a set of base clusters, the number of which is equal to the number of local minima on the distance matrix [25].

2.2.2 Cluster Hierarchy

Starting from the base clusters, some agglomerative clustering algorithm is used to construct the initial cluster hierarchy. Agglomerative clustering algorithms start from some initial set of c clusters and successively join the two clusters closest to each other (in terms of some distance measure), until there is only one cluster left. This produces a binary tree with $2c - 1$ clusters.

Since the binary structure does not necessarily reflect the properties of the data set, a number of the clusters in the initial hierarchy will be superfluous and need to

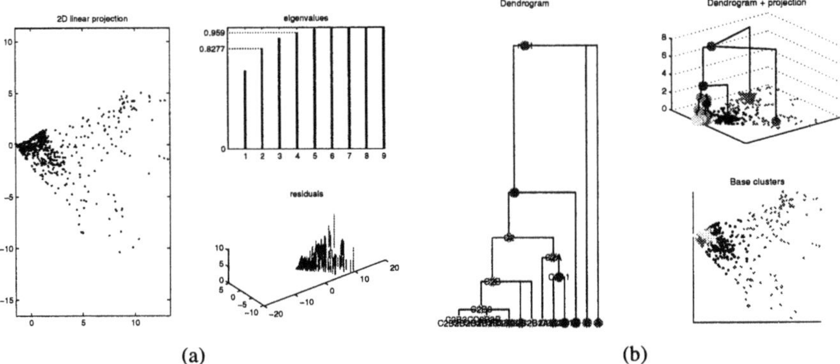

(a) (b)

Figure 3. Projection visualization (on left) and cluster hierarchy (on the right). The projection gives an idea of the shape of the data manifold. The figure is accompanied by the scree plot of eigenvalues, and a plot of projection residuals. These give an idea of the inherent dimensionality of the data set, and the reliability of the 2-dimensional projection. In this case, the projection covers 83% of the total variance. The 95% coverage limit would be reached in 4-dimensional projection (of the total of 9 dimensions). The cluster hierarchy is visualized using a dendrogram with colors and names of the clusters also indicated. The two smaller figures in the right panel (b) link the dendrogram to the projection visualization by showing the dendrogram starting from the 2-dimensional projection coordinates. Each point is colored with the color of the (base) cluster it belongs to.

be pruned out. This can be done by hand using some kind of interactive tool [3], or in an automated fashion using some cluster validity index to prune out the improper clusters. In the implemented system, the following procedure is applied:

1. Start from root (top level) cluster.
2. For the cluster c under investigation, generate different sub-cluster sets. A sub-cluster set may contain either sub-clusters of cluster c or sub-clusters of c's sub-clusters (sub-sub-clusters).
3. Each sub-cluster set defines a partitioning of the data in the investigated cluster. Investigate each partitioning using some clustering validity measure, for example Davies-Bouldin index [5] or other similar index [25].
4. Select the best sub-cluster set (for example the one with minimum I_{gap}), and prune the corresponding intermediate clusters.
5. Select an uninvestigated cluster (if any), and continue from step 2.

In Figure 3b, the clusters and cluster hierarchy are presented in three visualizations linked with each other and with the projection results.

2.3 Cluster Characterization

Descriptive statistics – for example means, standard deviations and histograms of individual variables – can be used to list the typical values for each cluster. Not all variables are equally interesting or important, though. Interestingness can be defined as deviation from the expected [14, 8]. It can be measured for each cluster as the difference between variable distributions in the cluster versus the whole data either using probability densities or some more robust measures, for example standard deviation [17]. Each cluster can then be characterized by using a list of the most important variables and their descriptive statistics.

Another frequently employed method is to form characterizing rules [19, 18] to describe the values in each cluster:

$$R_i : \mathbf{x} \in C_i \Leftrightarrow x_k \in [\alpha_k, \beta_k] \tag{4}$$

where x_k is the value of variable k in sample vector $\mathbf{x}$, C_i is the investigated cluster, and $[\alpha_k, \beta_k]$ is the range of values allowed for the variable according to the rule. These rules may be expressed in terms of single variables like R_i above, or be conjunctions of several variable-wise rules in which case the rule forms a hypercube in the input space.

The main advantage of such rules is that they are compact, simple and therefore easy to understand. The problem is of course that clusters often do not coincide well with the rules since the edges between clusters are not necessarily in parallel with the edges of the hypercube. In addition, the cluster may include some uncharacteristic points or outliers. Therefore the rules should be accompanied by validity information. The rules R_i can be divided to two different cases, characterizing rules (R_i^c) and differentiating rules (R_i^d):

$$\begin{aligned} R_i^c &: \quad \mathbf{x} \in C_i \Rightarrow x_k \in [\alpha_k, \beta_k] \\ R_i^d &: \quad x_k \in [\alpha_k, \beta_k] \Rightarrow \mathbf{x} \in C_i. \end{aligned}$$

The validity with respect to each case can be measured using confidence: $P_i^c = P(x_k \in [\alpha_k, \beta_k] \,|\, C_i)$ and $P_i^d = P(C_i \,|\, x_k \in [\alpha_k, \beta_k])$, respectively.

To form the characterizing rules – in effect to select the low and high limits of the range – one can use statistics of the values in the clusters [19]. Another approach is to optimize the rules with respect to their significance. The optimization can be interpreted as a two-class classification problem between the cluster and the rest of the data. The boundaries in the characterizing rule can be set by maximizing a function which gets its highest value when there are no misclassification's, for example:

$$s_1 = \frac{a+d}{a+b+c+d}, \tag{5}$$

$$s_2 = \frac{a}{a+b}\frac{a}{a+c}, \tag{6}$$

$$s_3 = \frac{a}{a+b+c}, \tag{7}$$

where a, b, c and d are from the truth table in Figure 4.

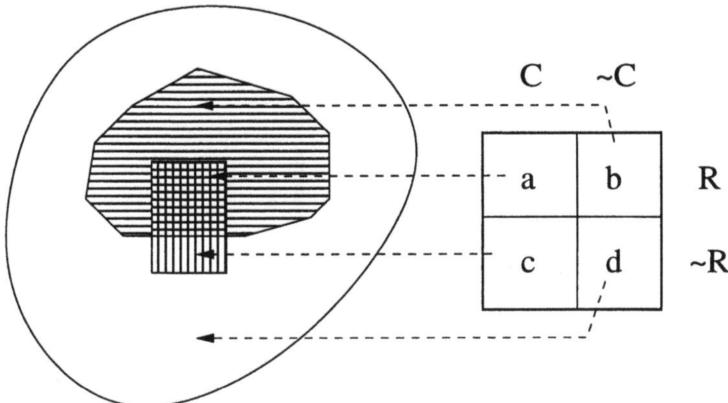

Figure 4. The four-way truth table of cluster C and rule R. The cluster is the horizontally shaded area, and the rule (or classification model) is the vertically shaded area. On the right is the corresponding confusion matrix: a is the number of samples which are in the cluster, and for which the rule is true. In contrast, d is number of samples which are out of the cluster, and for which the rule is false. Ideally, the off-diagonal elements in the matrix should be zero.

The first function s_1 is simply the classification accuracy. It has the disadvantage that if the number of samples in the cluster is much lower than in the whole data set (which is very often the case), s_1 is dominated by the need to classify most of the samples as false. Thus, the allowed range of values in the rule may vanish entirely. However, when characterizing the (positive) relationship between rule R and cluster C, the samples belonging to d are not really interesting. As pointed out in [2], traditional rule-based classification techniques are not well suited for the characterization task.

The two latter measures consider only cases a, b and c. The second measure s_2 is the product of the confidences $s_2 = P_i^c P_i^d$. The third measure is its approximation $s_3 \approx s_2$ when $a \gg b + c$. Compared to s_2, s_3 has the advantage of a clearer interpretation. It is the ratio of correctly classified samples when the case d is ignored, whereas s_2 is the product of two such ratios.

Apart from characterizing the internal properties of the clusters, it is important to understand how they differ from the other, especially neighboring clusters. For the neighboring clusters, the constructed rules may be quite similar, but it is still important to know what makes them different. To do this, rules can be generated using the same procedure as above, but taking only the two clusters into account. In this case, however, both clusters are interesting, and therefore s_1 should be used.

The report elements to describe the clusters are shown in Figure 5 and Table 1. The former shows the most significant rule visualized with projection and histograms, and the latter a summary table of variable values and associated descriptive rules of each variable, ordered by the significance s_2 of the variable.

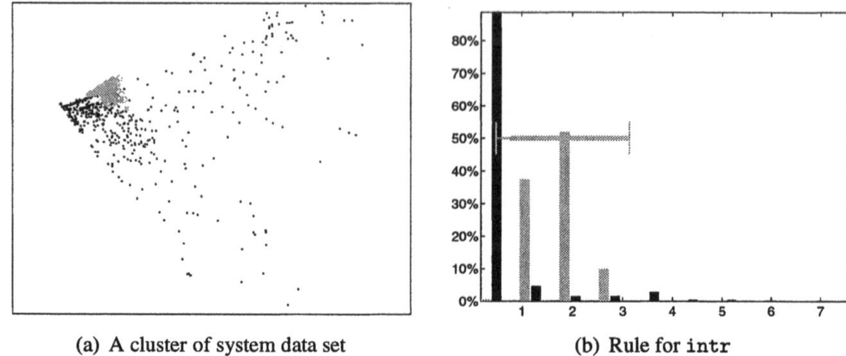

(a) A cluster of system data set (b) Rule for `intr`

Figure 5. A cluster of the system data set. In (a) the PCA projection of the data is shown, with the cluster indicated by the gray markers and the rest of the data with black markers. In (b), the histogram corresponding to the most significant variable is shown (see Table 1). The gray vertical bars are the histogram for the cluster, and the black the histogram for the rest of the data. The horizontal bar indicates the range allowed by the rule (thick part) and the real range of values in the cluster (thin part).

Table 1. Rule summary for the cluster in Figure 5 in the system data. Variables are listed in order of decreasing significance as measured by s_2. The columns in the middle indicate the properties for the variable-wise rules, and the columns on the right for a conjunctive rule formed of the indicated variables starting from the top. The "diff" columns are confidences in the differentiating property of the rule P^d and "char" column in the characterizing property P^c.

Variable	**Rule**	**diff**	**char**	s_2	**diff**	**char**	s_2
			single			**cumulative**	
`intr`	[0.78,3.1]	75%	99%	0.745	75%	99%	0.745
`idle`	[3.3,4.3]	65%	98%	0.639	87%	98%	0.856
`usr`	[1.2,2.3]	61%	98%	0.605	90%	98%	0.885
`sys`	[0.71,1.4]	69%	62%	0.425			
`blks/s`	< 0.82	22%	98%	0.217	100%	96%	0.96
`wio`	< 1.1	21%	100%	0.214			
`ipkts`	< 0.97	20%	100%	0.201			
`opkts`	< 0.97	20%	100%	0.2			
`wblks/s`	< 1.4	20%	100%	0.198			

3 Variable Analysis

The relevant questions with respect to variables are: What are the distribution characteristics of the variables? Are there pairs or groups of dependent variables? If so, how do they depend on each other?

The distributions of individual variables can be characterized by simple descriptive statistics, for example histograms. The histogram bins are formed either based on the unique values of the variable, if there are at most 10 unique values, or by dividing the range between minimum and maximum values of the variable to 10 equally sized bins.

Dependencies between variables are best detected from ordinary scatter plots, for example from a scatter plot matrix which consists of several subgraphs where each variable is plotted against each other variable. Of course, such visualization technique has the deficiency that the number of pairwise scatter plots increases quadratically with the number of variables. A more efficient, if less accurate, technique is to use component planes. A component plane is a colored scatter plot of the data, where the positions of the markers are defined using a projection such that similar data samples are close to each other. The color of the markers are defined by the values of a variable in the data samples. By using one component plane for each variable, the whole data set can be visualized, see Figure 6a. Relationships between variables can be seen as similar patterns in identical places in the component planes. The projection made by SOM works very well with this technique, since the projection focuses locally such that the behavior of the data can be seen irrespective of the local scale.

A more quantitative measure of the dependency between pairs of variables $\{i, j\}$ is the correlations coefficient c_{ij}:

$$c_{ij} = \frac{1}{\sigma_i \sigma_j} \sum_{k=1}^{n} (x_{ki} - \mu_i)(x_{kj} - \mu_j) \tag{8}$$

where μ_i and σ_i are the mean and standard deviations of variable i. It is robust, computationally light and can be efficiently visualized as a matrix, see Figure 6b. Selected correlations are also visualized on an association graph, see Figure 6c. In the implemented system, the correlation coefficients are also used as feature vectors for each variable: $\mathbf{v}_i = [\,|c_{i1}|, |c_{i2}|, ..., |c_{id}|\,]$. Using them and some clustering or projection method (see Section 2), the variables can be clustered (or ordered) to indicate groups of dependent variables [23]. In all three visualizations in Figure 6, the variables have been ordered such that related variables are near each other.

4 Data Survey Report

The system was implemented as a Matlab script. It was installed on a web server as a cgi-bin script, and thus it could be run remotely through any web browser. The user provided the data and optionally a few other parameters. The resulting report was provided both as a hypertext document in HTML and in printable form in PostScript/PDF.

The report starts with an overview part, where the top-level results of both variable and cluster analysis are given. The overview provides a quick look at the data. It is short, only 2-4 pages in length, and consists mainly of visualizations so that it can be understood at a glance.

Most of the elements of the overview have already been shown earlier in the chapter (Figures 3a and b, 6a and c). From Figure 3a, one can see that a two-dimensional

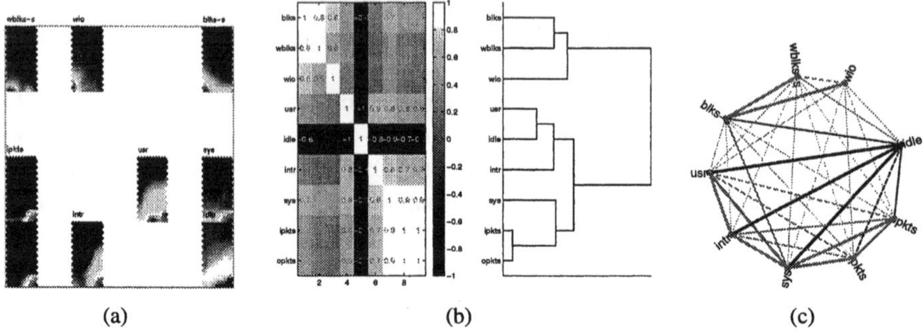

(a) (b) (c)

Figure 6. (a) Component planes. Relationships between variables can be seen as similar patterns on the component planes. (b) Correlation coefficient matrix, and the dendrogram resulting from clustering of the variables using an agglomerative clustering algorithm. (c) Association graph. On the association graph, high negative and positive correlations are shown with blue (in this case all links to the right-most node) and red (all other) lines, respectively.

projection preserves the structure of the system data set very well. Most of the data is tightly packed in a single region, and there is a fan of dispersed data. From Figure 3b, one can see that there are three main operational states in the system, one of which divides further to several sub-states. From Figure 6 one can see that there are two main groups of variables, those involved with disk operations, and the rest.

In addition, the overview part has a table of descriptive statistics for each variable, and a list of most significant rules for each cluster (not shown). From the latter one could see that the main operational states correspond to a normal operation state and two different high load states where either the amount of disk operations is high (`wio` > 2.1) or there is a lot of network traffic (`ipkts` > 1.9). Further investigation reveals that the normal operation state divides to several different types, for example totally idle state, state for mainly system operations, and state for mainly user operations.

The overview is followed by more detailed information of all variables, and clusters (individual cluster characterizations). These allow the reader to get immediately some further information on interesting details. For example, the cluster depicted in Table 1 and Figure 5 is mainly characterized by relatively high level of interruptions, and a moderate level of idle time.

The computational complexity of quantization using SOM is $\mathcal{O}(nmd)$, where n is the number of data points, m is the number of map units, and d is the vector dimension. The complexity of clustering the SOM is $\mathcal{O}((m-c)(c+1)d)$, where c is the number of base clusters. The computational complexity of the hierarchical clustering phase is $\mathcal{O}(c^2d)$. The search for the best rule is an optimization problem which can be solved, for example, using (fast) exhaustive search techniques like the ones introduced in [2]. In the implemented system, though, a much less complex greedy search is used which is linear in computational complexity with respect to the number of variables. Thus, its computational complexity is $\mathcal{O}(cd)$. The computational complexity of variable analysis is $\mathcal{O}(d^3)$ because of clustering the variables. Thus, the overall complexity is of the order of $\mathcal{O}((nm + c^2 + d^2)d)$.

The system runs in main memory, which of course limits the possible size of the data set. For a data set of size 4000 samples and 40 variables, the report generation takes 15 minutes on a Linux workstation with Pentium II 350 MHz processor and 256 MB of memory. The majority of the time is spent to writing the images to files, which is unfortunately slow in Matlab. The actual analysis only takes about 2 minutes. The heaviest part of the analysis is clustering, which scales linearly with the number of data samples. Thus, the system scales well to larger data sets, too.

5 Case Study: Caravan Insurance Policy Data Set

The second data set used in this chapter is a publicly available insurance company data set used in a recent CoIL challenge [21]. In this data mining competition, two separate tasks were given: predicting and explaining caravan insurance policy ownership. The samples represent customers and the variables different aspects of the customers or their behavior. We do not attempt to give an answer to either of the questions posed in the competition directly, but rather to demonstrate how our data survey report can be used in the initial phases of the data analysis project. In the following, we describe the generation of the report and give a rough idea how to make valuable observations about the data set with the help of the data survey report.

Creating a data survey report begins by inputting the data set to the data survey generator. As indicated in Section 4, the system has been implemented using a interface on a Web server, so the table-format data can be uploaded to the generator using a Web based interface. The data survey report itself is a hypertext document, which allows for navigating between overviews and detailed descriptions as well as between clusters and their sub-clusters with minimal effort. The report starts with an index document that lists all the names of the variables along with their summary statistics. Summary statistics used are the minimum, maximum, mean, standard deviation, and entropy of the recorded variables. In addition, number of missing values, number of unique values and whether all values have integer values are shown. This helps finding out possibly categorical data or discretized variables. This is the case in the insurance data set, which is easily seen from the report. The data set size of 9822 samples and 86 variables is visible on the report header.

The main document is followed by an overview. The overview contains a hierarchical table resembling a dendrogram, where clusters and the cluster hierarchy are represented. The entries in the table indicate how many samples are contained in each level of the cluster hierarchy. Also, the most descriptive rules are listed. For instance, two clusters near the top of the hierarchy are described by the descriptive rules. The descriptive rules for the cluster A are

$$\text{Number of third party insurance (agriculture)} >= 0.5$$
$$\text{Contribution family accidents insurance policies} < 2.5$$

and for the cluster B, the descriptive rules are

$$\text{Contribution third party insurance (agriculture)} < 1.5$$
$$\text{Contribution tractor policies} < 1.5$$
$$\text{Contribution trailer policies} < 1.5$$

The majority of the customers (97 %) belong to the cluster B, also indicated in the tables. Focusing on the B clusters in the hierarchy, it is valuable to know how each cluster differs from its peer clusters (in effect, the clusters which have the same parent in the cluster hierarchy). The report indicates a single most discriminative rule according to which each peer cluster differs from the present cluster. In the case of sub-clusters of cluster B called B1 and B2, the discriminative rule dictates the B1 to have a significant rule

$$\text{Contribution family accidents insurance policies} >= 1$$

in contrast to the present cluster B2. Projection of the cluster-specific data complements this view. Navigating back to the top level, projection visualization as in the Figure 3 is presented.

As the description above demonstrates, working with the data survey report is highly interactive, and the implementation as a hypertext document greatly facilitates working with the report. The user navigates through the document searching for suggestive information to be used in the later stages of the data analysis process. It is important to realize that the data survey report serves as a initial tool to gain understanding through suggestive information on the structure of the data manifold. The results should be used as an initial step, and care should be taken in inferring knowledge from the report.

6 Conclusion

In the initial phases of a data analysis project, the data miner should have some perception of the data, or a mental model of it, to be able to formulate models in the latter phases of the project successfully. Helping to reach this goal, an implemented system for automatically creating data survey reports on numerical table-format data sets has been described. The system applies a set of generic data analysis algorithms – variable and variable relation analysis, projection, clustering, and cluster description algorithms – to the data and writes a report which can be used as the starting point for data understanding and as a reference later on. The system integrates linguistic descriptions (rules) and statistical measures with visualizations. Visualizations provide qualitative information of the data sets, and give an overview of the data. The visualizations also help in assessing the validity of the proposed measures, clusters and descriptive rules. The report provides a coherent, organized document about the properties of the data, making it preferable to applying the same algorithms to the original data sets in an unorganized manner and in different formats.

In our experience, the implemented system succeeds in automating a lot of the initial effort done in the beginning of a typical data analysis project. In fact, the system has been built on top of our experience in many collaborative data analysis projects with industrial partners involving real-world data. In all, we feel that the current version should have general appeal to a wide variety of projects and should help in gaining an initial understanding for successful modeling in many domains.

Appendix

System Data

The first data set used throughout this chapter to illustrate the resulting data survey report is a simple 9-dimensional real-world data set. The *system data set* describes the operation of a single computer workstation in a networking environment. The workstation was used for daily activities of a research scientist ranging from computationally intensive (data analysis) tasks to editing of programs and publications. The data set has been collected by the authors at their research institution.

The number of variables recorded was 9. Four of the variables reflect the volumes of network traffic and five of them the CPU usage in relative measures. The variables for measuring the network traffic were `blks/s` (read blocks per second), `wblks/s` (written blocks per second), `ipkts` (the number of input packets) and `opkts` (the number of output packets). Correspondingly, the central processing unit activities were measured with variables `usr` (time spent in user processes), `sys` (time spent in system processes), `intr` (time spent handling interrupts), `wio` (CPU was idle while waiting for I/O), `idle` (CPU was idle and not waiting for anything). Whereas the network traffic is unconstrained (within reasonable bounds), the full capacity of the CPU is always divided between activities. Therefore, the five last measurements add up to the full, unit capacity of the CPU. In all, 1908 data vectors were collected.

References

1. Esa Alhoniemi, Jaakko Hollmén, Olli Simula, and Juha Vesanto, *Process Monitoring and Modeling Using the Self-Organizing Map*, Integrated Computer-Aided Engineering **6** (1999), no. 1, 3–14.
2. Stephen D. Bay and Michael J. Pazzani, *Detecting group differences: Mining contrast sets*, Data Mining and Knowledge Discovery **5** (2001), no. 3, 213–246.
3. Eric Boudaillier and Georges Hebrail, *Interactive Interpretation of Hierarchical Clustering*, Intelligent Data Analysis **2** (1998), no. 3.
4. Pete Chapman, Julian Clinton, Thomas Khabaza, Thomas Reinartz, and Rüdiger Wirth, *The CRISP-DM process model*, Tech. report, CRISM-DM consortium, March 1999, `http://www.crisp-dm.org`.
5. David L. Davies and Donald W. Bouldin, *A Cluster Separation Measure*, IEEE Trans. on Pattern Analysis and Machine Intelligence **PAMI-1** (1979), no. 2, 224–227.
6. Richard O. Duda, Peter E. Hart, and David G. Stork, *Pattern classification*, sec-

ond ed., John Wiley & Sons, 2001.
7. Sudipto Guha, Rajeev Rastogi, and Kyuseok Shim, *CURE: an efficient clustering algorithm for large databases*, Proceedings of SIGMOD International Conference on Management of Data (New York), ACM, 1998, pp. 73–84.
8. R. Hilderman and H. Hamilton, *Knowledge discovery and interestingness measures: A survey*, Tech. Report CS 99-04, Department of Computer Science, University of Regina, October 1999.
9. Johan Himberg, *Enhancing SOM-based data visualization by linking different data projections*, Intelligent Data Engineering and Learning (IDEAL'98) (L. Xu, L. W. Chan, and I. King, eds.), Springer, 1998, pp. 427–434.
10. Teuvo Kohonen, *Self-Organizing Maps*, 2nd ed., Springer Series in Information Sciences, vol. 30, Springer, Berlin, Heidelberg, 1995.
11. Andreas König, *A survey of methods for multivariate data projection, visualization and interactive analysis*, Proceedings of the 5th International Conference on Soft Computing and Information/Intelligent Systems (IIZUKA'98) (T. Yamakawa and G. Matsumoto, eds.), World Scientific, October 1998, pp. 55–59.
12. Krista Lagus and Samuel Kaski, *Keyword selection method for characterizing text document maps*, Proceedings of ICANN99, Ninth International Conference on Artificial Neural Networks, vol. 1, IEE, London, 1999, pp. 371–376.
13. R.S. Michalski and R. Stepp, *Automated construction of classifications: Conceptual clustering versus numerical taxonomy*, IEEE Transactions on Pattern Analysis and Machine Intelligence **5** (1983), 396–410.
14. G. Piatetsky-Shapiro and C. Matheus, *The interestingness of deviations*, Proceedings of KDD'94, July 1994, pp. 25–36.
15. Dorian Pyle, *Data Preparation for Data Mining*, Morgan Kaufmann Publishers, 1999.
16. Andreas Rauber and Dieter Merkl, *Automatic labeling of self-organizing maps: Making a treasure-map reveal its secrets*, Proceedings of the 3rd Pasific-Area Conference on Knowledge Discovery and Data Mining (PAKDD'99), 1999.
17. Markus Siponen, Juha Vesanto, Olli Simula, and Petri Vasara, *An approach to automated interpretation of SOM*, Proceedings of Workshop on Self-Organizing Map 2001 (Nigel Allinson, Hujun Yin, Lesley Allinson, and Jon Slack, eds.), Springer, June 2001, pp. 89–94.
18. A. Ultsch, *Self-organized feature maps for monitoring and knowledge acquisition of a chemical process*, Proceedings of International Conference on Artificial Neural Networks (ICANN) 1993, September 1993, pp. 864–867.
19. A. Ultsch, G. Guimaraes, D. Korus, and H. Li, *Knowledge extraction from artificial neural networks and applications*, Proceedings of Transputer-Anwender-Treffen / World-Transputer-Congress (TAT/WTC) 1993 (Aachen, Tagungsband), Springer Verlag, September 1993, pp. 194–203.
20. A. Ultsch and H.P. Siemon, *Kohonen's Self Organizing Feature Maps for Exploratory Data Analysis*, Proceedings of International Neural Network Conference (INNC'90) (Dordrecht, Netherlands), Kluwer, 1990, pp. 305–308.
21. P. van der Putten and M. van Someren (eds.), *Coil challenge 2000: The insurance company case*, Tech. Report 2000-09, Leiden Institute of Advanced Com-

puter Science, 2000.

22. A. Vellido, P.J.G Lisboa, and K. Meehan, *Segmentation of the on-line shopping market using neural networks*, Expert Systems with Applications **17** (1999), 303–314.
23. Juha Vesanto, *SOM-Based Data Visualization Methods*, Intelligent Data Analysis **3** (1999), no. 2, 111–126.
24. Juha Vesanto and Esa Alhoniemi, *Clustering of the Self-Organizing Map*, IEEE Transactions on Neural Networks **11** (2000), no. 2, 586–600.
25. Juha Vesanto and Mika Sulkava, *Distance matrix based clustering of the self-organizing map*, Proceedings of the Twelfth International Conference on Artificial Neural Networks (ICANN'02), Springer-Verlag, 2002, LNCS 2415, pp. 951–956.
26. Colin Ware, *Information visualization: Perception for design*, Morgan Kaufmann Publishers, 2000.
27. Tian Zhang, Raghu Ramakrishnan, and Miron Livny, *Birch: An efficient data clustering method for very large databases*, Proceedings of the 1996 ACM SIGMOD International Conference on Management of Data (Montreal, Canada), 1996, pp. 103–114.

Chapter 9

Finding Trigonometric Identities with Tree Adjunct Grammar Guided Genetic Programming

N.X. Hoai, R.I. McKay, and D. Essam

Summary. We introduce a new form of Grammar-Guided Genetic Programming using Tree Adjunct Grammars instead of Context Free Grammars. We apply it to a standard problem of finding trigonometric identities, and compare its performance with standard approaches. We analyze the fitness landscape of the problem, and gain some understanding of the relative performance of the standard and new representations.

Keywords: Genetic programming, tree adjunct grammars, context free grammars.

1 Introduction

Genetic programming (GP) can be viewed as a machine learning method, inducing a population of computer programs by evolutionary means (Banzhaf et al. 1998). Genetic programming has been used successfully in generating computer programs for solving a number of problems in a wide range of areas. In (Hoai and McKay 2001), we proposed a framework for a grammar-guided genetic programming system called Tree-Adjunct Grammar Guided Genetic Programming (TAG3P), which uses tree-adjunct grammars along with a context-free grammar to set language bias in genetic programming. The use of tree-adjunct grammars can be seen as a process of building context-free grammar guided programs in two dimensional space. In this chapter, we show some results of TAG3P on the trigonometric identity discovery problem. The organization of the remainder of the chapter is as follows. In Section 2, we give a brief overview of GP, grammar-guided GP (GGGP), tree-adjunct grammars (TAGs) and TAG3P. The problem of finding trigonometric identities will be given in Section 3. Section 4 contains the experiment and results of TAG3P on that problem. The nature of the search space is empirically analyzed, and bias by selective adjunction is introduced. The last section contains conclusions and future work.

2 Background

In this section, we first give a brief overview of GP and GGGP, then the definition of TAGs and the components of TAG3P.

2.1 Genetic Programming

GP is an evolutionary algorithm in which computer programs are the evolutionary targets. An early definition, model, techniques and problems can be found in (Koza 1992). For a good survey, (Banzhaf et al. 1998) is recommended. A basic GP system consists of five basic components (Koza 1992): program representation (called genome structure), procedure to initialize a population of programs, fitness evaluation, genetic operators, and parameters. In (Koza 1992) the structure of programs is the program structure tree; fitness of a program is evaluated by its performance; and the main operators are selection and crossover; parameters are population size, maximum number of generations and probabilities for genetic operators. In this chapter, we will refer to GP as in (Koza 1992) as canonical GP.

2.2 Grammar Guided Genetic Programming

One of the limitations of canonical GP is the closure requirement: that all variables, constants, arguments of functions, and return values from functions must be of the same data type, so that crossover can take place at any point of a program structure.

One of the early attempts to overcome this limitation was by Montana (1995). He proposed a strongly typed genetic programming system, in which all variables, constants, arguments of functions, return values of functions are given data types beforehand. The initialization and genetic operators are designed to respect the data type nodes so that only legal genomes are generated. The genome structure in (Montana 1995) is also a tree.

Another promising way to cope with closure uses syntactical constraints, as originally suggested by Koza (1992), though without an explicit formalism for defining syntactical constraints. The first practical implementation by Whigham (1995) used context-free grammars (CFGs) to define the target languages of programs.

Whigham (1996) also incorporated domain knowledge to change grammars to bias the search process; he demonstrated positive results on the 6-multiplexer problem and subsequent problems.

Gruau (1996) formally proved that using syntactical constrains reduces the search space size. He also used CFGs to describe target languages but did not limit the depth of derivation trees, allowing potentially extreme growth in tree size (Ratle and Sebag 2000).

Wong and Leung (1996) used logic grammars, combining logic programming and genetic programming. They incorporated domain knowledge into the grammars to guide the evolutionary process of logic programs. However, because they did not maintain explicit parse trees, their system suffers from ambiguity when it tries to generate programs from parse trees (Ryan et al. 1998).

Ryan et al. (1998) propose grammatical evolution (GE), which evolves programs in an arbitrary context-free language. Their system differs from Whigham's in that they do not evolve derivation trees directly. Instead, genomes in GE are binary strings representing eight-bit numbers; each number indicates the choice of production rule for the non-terminal symbol being processed. GE has been shown to outperform canonical GP in a number of problems (O'Neil and Ryan 1999a; O'Neil 1999).

One of the great advantages of GE is the linear (variable-length) genome structure, which allows GE to employ well-studied operators of genetic algorithms. The linear structure also reduces tree-based GP's well-known bias towards selection of leaves for operator sites (Banzhaf et al. 1998). Moreover, the implementation of GE above provides a type of many-to-one genotype-to-phenotype mapping, which can provide neutral evolution in the genome space.

A key weak point of GE is its difficulties in guaranteeing legal genomes, which creates some problems such as introns and unnatural multiple use of genes (Hoai and McKay 2001).

2.3 Tree-Adjunct Grammars

Tree-adjunct grammars (TAGs) are tree-rewriting systems, first introduced in (Joshi et al. 1975). TAGs have been used widely in natural language processing; and a comprehensive survey can be found in (Joshi and Schabes 1997). A TAG can be formally defined as follows (Joshi et al. 1975):

Definition: A tree-adjunct grammar comprises of 5-tuple (T, V, I, A, S), where

(i) T is a finite set of terminal symbols.

(ii) V is a finite set of non-terminal symbols; and $T \cap V = \varnothing$.

(iii) $S \in V$ is a distinguished symbol called the start symbol.

(iv) I is a finite set of finite trees called initial trees. An initial tree is defined as follows:
   - The root node is S.
   - All interior nodes are labeled by non-terminal symbols.
   - Each node on the frontier is labeled by a terminal symbol.

(v) A is a finite set of finite trees called auxiliary trees, which can be defined as follows:
   - Internal nodes are labeled by non-terminal symbols.
   - All but one node on the frontier are labeled by terminal symbols. The special non-terminal node on the frontier is called the foot node. It is required that the foot node is labeled by the same (non-terminal) symbol as the root node of the tree. We will also follow the convention in (Joshi and Schabes 1996) of marking a foot node with an asterisk (*).

The trees in $E = I \cup A$ are called elementary trees. In the literature, initial trees and auxiliary trees are usually denoted α and β respectively. A non-terminal (resp. terminal) node is one labeled by a non-terminal (resp. terminal) symbol. An elementary tree is called X-type if its root is labeled by the non-terminal symbol X. The key operation in a tree-adjunct grammar is the adjunction of trees.

Adjunction builds a new (derived) tree γ from an auxiliary tree β and a tree α (initial, auxiliary or derived). If tree α has a non-terminal node labeled A and β is an A-type tree then the adjunction of β and α to produce γ is as follows. First, the sub-tree α_1 rooted at A is temporarily disconnected from α. Next, β is attached to α to replace this sub-tree. Finally, α_1 is attached back to the foot node of β. γ is the final derived tree achieved from this process. Adjunction is illustrated in Figure 1:

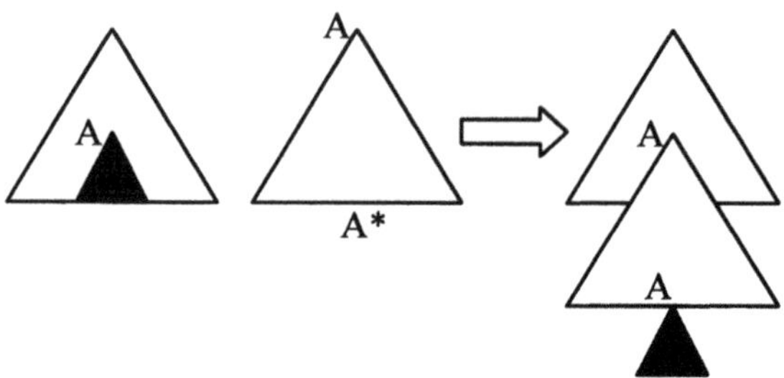

Figure 1. Adjunction.

The tree set of a TAG can be defined as follows (Joshi 1975):

$$T_G = \{ t \mid t \text{ is completed and } t \text{ is derived from some initial trees}\}.$$

Where a tree t completed if t is an initial tree and t has no non-terminal node on its frontier; and a tree t is said to be derived from a TAG G if and only if t results from a sequence adjunction (the derivation sequence) of the form: $\alpha\ \beta_1(a_1)\ \beta_2(a_2)\ldots\ \beta_n(a_n)$ (*), where n is an arbitrary integer, α , β_i (i = 1, 2, ..., n) are initial and auxiliary trees of G and a_i (i = 1, 2, ..., n) are node addresses where adjunctions take place. The language L_G generated by a TAG is then defined as the set of yields of all trees in T_G:

$$L_G = \{w \in T^* \mid w \text{ is the yield of some tree } t \in T_G\}.$$

The set of languages generated by TAGs (called TAL) is a superset of context-free languages; and is properly included in indexed languages (Joshi and Schabes 1997). One special class of tree-adjunct grammars (TAGs) is lexicalized tree-adjunct grammars (LTAG): each elementary tree of a LTAG must have at least one terminal node (called anchor). It has been proven that for any context-free grammar G, there exists a LTAG G_{lex} that generates the same language and tree set as G (G_{lex} is then said to strongly lexicalize G) (Joshi and Schabes 1992, 1997).

2.4 Tree Adjunct Grammar Guided Genetic Programming

In (Hoai and McKay, 2001), we proposed a grammar guided genetic programming system called TAG3P, which uses a CFG G and its corresponding LTAG G_{lex} to guide the evolutionary process. The main idea of TAG3P is to evolve the derivation sequence in G_{lex} (genotype) rather than evolve the derivation tree as in Whigham's system. This creates a many-to-one genotype-to-phenotype map. As in canonical GP, TAG3P comprises the following main components:

Program representation: linear derivation sequence of elementary trees of G_{lex} in the form (*). Currently, we add a restriction that the address a_i must be in β_{i-1}. Consequently, the genome structure in TAG3P is linear and length-variant. Although the language domain of this linear form is not yet fully determined, it is clearly very rich, and we have been able to find linear TAG representations for all the standard problems we have examined (Hoai, 2001).

Initialization procedure: A procedure for initializing a population is given in (Hoai and McKay 2001). For each individual, it selects a length at random; the genome is then generated from left to right, by choosing randomly first an α tree, then a sequence of β trees and adjoining addresses. We have shown that this procedure generates legal genomes of arbitrary finite lengths (Hoai and McKay 2001).

Fitness Evaluation: Fitness evaluation in TAG3P is the same as in canonical GP (Koza 1992).

Genetic operators: In (Hoai and McKay 2001), we proposed two types of crossover operators, namely one-point crossover and two-point crossover, and three types of mutation operators, which are replacement, insertion and deletion. The crossover operators in TAG3P are similar to those in genetic algorithms; however, the crossover point(s) are chosen so that only legal genomes are produced. In replacement, a gene is picked at random and replaced by a compatible gene to give a valid resultant genome. In insertion and deletion, a gene is inserted into or deleted from the genome respectively. With these operators, TAG3P produces only legal genomes. Selection in TAG3P is similar to canonical GP. Currently, reproduction is not employed by TAG3P.

Parameters: The parameters in TAG3P are minimum length of genomes - MIN_LENGTH, maximum length of genomes MAX_LENGTH, size of population - POP_SIZE, maximum number of generations – MAX_GEN and probabilities for genetic operators.

As discussed in (Hoai and McKay 2001), TAG3P combines some good aspects of other grammar guided GP systems. As in Whigham's model, it uses context-free grammars (along with the corresponding LTAGs) to guide the evolutionary process; the phenomes are also derivation trees of the context-free grammar that generates the target language of computer programs. Consequently, it can constrain and reduce the search space (as has been shown in (Gruau 1996)) and evolve programs in an arbitrary context-free language. It resembles GE in employing a genotype-to-phenotype map, where the genotypes are derivation sequences in the corresponding LTAG. The genome structure is also linear, which helps to reduce the known bias toward selecting a node near the leaves in tree-based structure (Banzhaf et al. 1998). However, unlike GE, TAG3P only produces legal genomes during its evolutionary process. Further intuitive advantages of TAG3P can be found in (Hoai and McKay 2001).

3 The Trigonometric Identities Problem

The problem of finding trigonometric identities can be stated as follows (Koza 1992). Given a trigonometric function in symbolic form, a system must try to discover alternative representations of the function. For example, the function cos(2x) has a representation (one of many well-known alternatives) as 1 - $2\sin^2(x)$. The problem has been tackled by symbolic methods in artificial intelligence by repeatedly applying logically sound transformations to the mathematical expression to get new expressions. Koza (1992) has shown that GP can solve the problem of representing cos(2x).

We use the same function set (consisting of only sin function and arithmetic operations) and terminal set (consisting of only variable x and constant 1.0) as (Koza 1992). The context-free grammar G and corresponding G_{lex} used to describe the problem are as follows:

The context-free grammar for the problem of finding trigonometric identities for function cos(2x): G = (N = {EXP, PRE, OP, VAR, NUM}, T = {X, sin, +, –, *, /, (,), 1.0}, P, {EXP}), where the rule set P is as follows:

EXP :- EXP OP EXP
EXP :- PRE (EXP)
EXP :- VAR | NUM
OP :- + | – | * | /
PRE :- sin
VAR :- X
NUM :- 1.0

The tree-adjunct grammar for the problem of finding trigonometric identities for function cos(2x): G_{lex} = (N = {EXP, PRE, OP, VAR, NUM}, T = {X, sin, +, –, *, /, (,), 1.0}, I, A, {EXP}), where I ∪ A is as in Figure 2.

4 Experiment and Results

We first give the comparative results of GP, GGGP and TAG3P on the problem of finding trigonometric identities. Then, the nature of the search space is also analyzed based on these results. Finally, we introduce a mechanism to implement search bias in TAG3P.

4.1 Experiment Setup

We applied GP, GGGP and TAG3P to the problem of finding trigonometric identities for function cos(2x) as in (Koza 1992). This function has three well-known alternative representations, namely, $1 - 2\sin^2(x)$, $\sin(\pi/2 - 2x)$, and sin(($\pi/2 + 2x$). The systems are given a set of input and output pairs, and they must evolve the alternative representations of the function. Table 1 below (which is adapted from (Koza 1992)) summarizes our experiment setup.

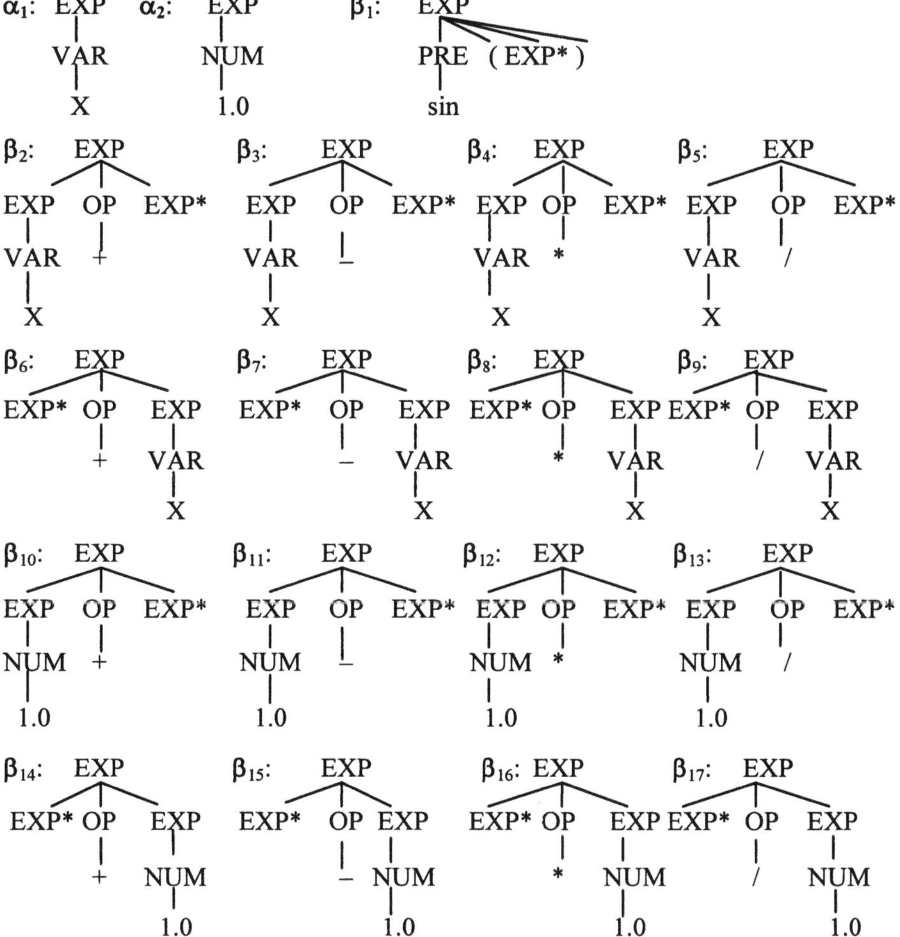

Figure 2. Elementary trees for G_{lex}.

4.2 Results

We conducted 50 runs for each system (150 in total). The results are summarized in table 2. GP failed to find a solution, while GGGP and TAG3P succeeded in finding the exact solution $1 - 2\sin^2(x)$ and the approximate solutions $\sin(\pi/2 - 2x)$ and $\sin((\pi/2 + 2x)$. Here we considered a solution S is an approximation for $\sin(\pi/2 - 2x)$ or $\sin(\pi/2 + 2x)$ if it has the form sin(A-2x) or sin(A+2x), where A is an approximation of $\pi/2$, S scores all 20 hits and has standardized fitness less than 0.02.

Approximate solutions arise from the system evolving constant expressions approximating $\pi/2$ from the only constant at hand (1.0). Here are some examples of such evolvable constants found by TAG3P:

$$1+\sin(\sin(\sin(\sin(\sin(\sin(1/\sin(\sin(\sin(\sin(\sin(\sin(\sin(\sin(1)))))))))))))) \approx 1.570727,$$
$$1+\sin(\sin(\sin(\sin(\sin(\sin(\sin(\sin(\sin(\sin(1+1))))+1)+1))))) \approx 1.570812, \text{ and}$$
$$1+1/(1+1/(1+1/(1+(1+1)))) = 11/7 \approx 1.571492.$$

Interestingly, the last is the standard school approximation of $\pi/2$.

Table 1. The experiment setup.

Objective	Find a function of one independent variable and one dependant variable that fits a given sample of 20 (x_i, y_i) data points, where the target function is cos(2X).
Terminal Operands	X (the independent variable), 1.0 (the constant).
Terminal Operators	The binary operators are +,-,*,/. The unary operator is the sin function.
Fitness Cases	20 sample points in the interval $[0..2\pi]$.
Raw fitness	The sum, taken over 20 fitness cases, of the errors.
Standardized Fitness	Same as raw fitness.
Hits	The number of fitness cases for which the error less than 0.01.
Genetic Operators	Tournament selection, one-point crossover and replacement for TAG3P. Tournament selection, normal crossovers and mutations for GP and GGGP.
Parameters	The crossover probability for GP, GGGP, and TAG3P is 0.9. The mutation probability for GP and GGGP is 0.1. Replacement probability for TAG3P is 0.05. Tournament size is 3. MAX_GEN is 200. POP_SIZE=500.
Success predicate	An individual scores 20 hits.

Table 2. The comparative results.

Number of solutions found	**GP**	**GGGP**	**TAG3P**
Exact solutions	0 (0%)	1 (2%)	2 (4%)
Approximate solutions	0 (0%)	9 (18%)	16 (32%)
Total	0 (0%)	10 (20%)	18 (36%)

Figures 3 and 4 depict, by generation, the standardized fitness of best–of–generation individuals and the average fitness of the population (which is vertically scaled by the Log10 function) of two successful runs of TAG3P. The first (Figure 3) stopped at generation 72 and found the exact solution $1-2\sin^2(x)$. The second (Figure 4) stopped at generation 54 and found function: $\sin(\sin(\sin(\sin(\sin(\sin(\sin(1/\sin(1)))))))+1-2x) \approx \sin(1.57383-2x) \approx \sin(\pi/2 - 2x)$. It is noted that reproduction and elitism are currently not used in TAG3P.

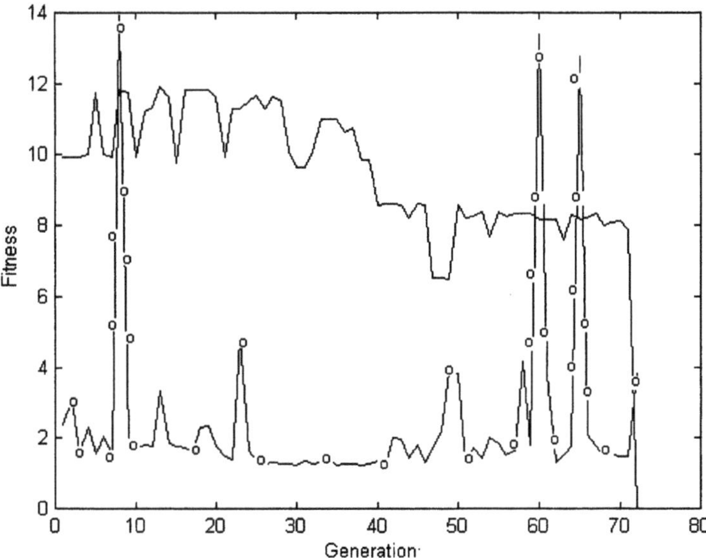

Figure 3. The best and average standardized fitness in one run that found the exact solution.

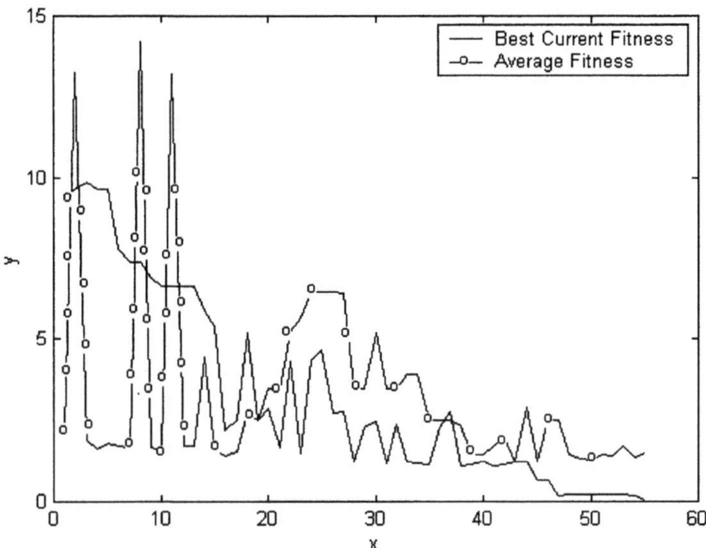

Figure 4. The best and average standardized fitness in one run that found an approximate solution.

Figure 5 shows the cumulative frequencies of GP, GGGP and TAG3P in our experiment.

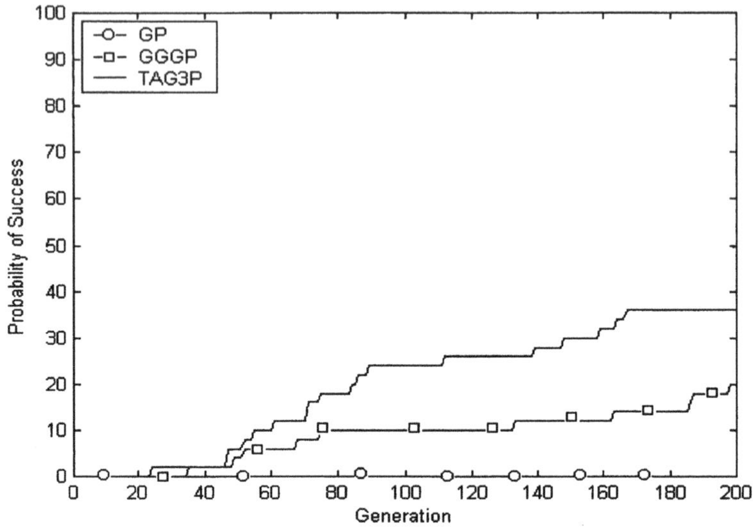

Figure 5. The cumulative frequencies of GP, GGGP and TAG3P.

4.3 Discussion

The results show that TAG3P outperforms GP and GGGP on the problem. Moreover, there is a tendency for GGGP and TAG3P to converge towards the approximate solutions (16/18 successful runs by TAG3P and 9/10 successful runs by GGGP).

As discussed in (Hoai 2001b) the reason for the bias towards the approximate solutions is that, in the representation space, the exact solution ($1-2\sin^2(x)$) is like a needle in a haystack surrounded by highly unfit expressions. Therefore the problem does not provide any useful building blocks. This conjecture can be partially proved based on the results of the runs that found the exact solutions. In Figure 3, the fitness of the best individual at each generation indicates that in the first 70 generations the population had been spread on a flat region of the search space before accidentally encountering the exact solution. Figure 6 also shows some expressions that are very 'near' to $1-2\sin^2(x)$ in the representation space but very poor in fitness.

In contrast, the approximate solutions have useful building blocks, as the constant $\pi/2$ needs to be evolved. One can see repeated patterns appearing in the evolved constants in the previous subsection. Consequently, the two approximate solutions act as attractors in the representation space.

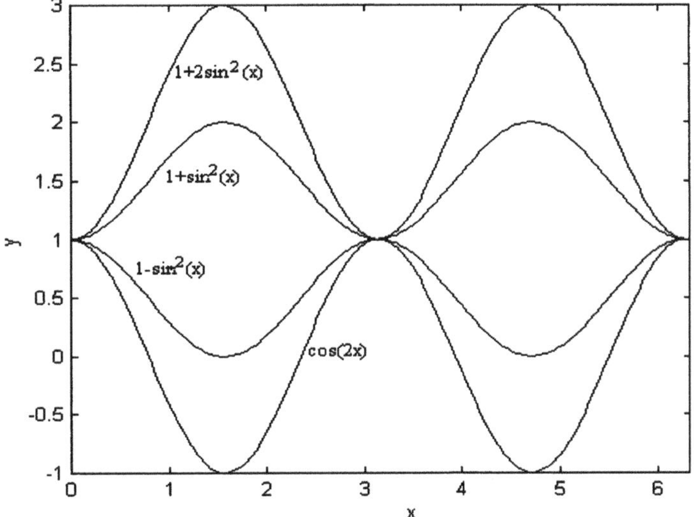

Figure 6. Some expressions similar to $1–2\sin^2(x)$ but poor in fitness.

Thus we believe the superior performance of TAG3P over GP and GGGP lies in its superior capability of preserving and combining building blocks (Hoai 2001a, 2001b, Hoai et al. 2001). We examined the runs and found that TAG3P preserved and replicated the following building blocks: sin(2x+1+t), sin(2x+sin(1/sin(1)+t), sin(2x+1+1/sin(1)+t), sin(1–2x+t), sin(sin(1/sin(1))–2x+t), and sin(1/sin(1)+1–2x+t), each the result of a sequence of beta trees adjoined together (where t is a parameter). They appeared in most generations of most of the 50 runs of TAG3P. After emerging in the population they tended to combine. Figure 7 shows some of the building block (with t = 0) approximations to cos(2x).

4.4 Bias Towards the Exact Solution with Selective Adjunctions

In order to bias the population towards to the exact solution, we use selective adjunctions with G_{lex}. TAGs with constraints (Joshi et al. 1975) are extended versions of TAGs, in which each node in an elementary tree is associated with a list of permitted beta trees, allowed to be adjoined into the node. We used selective adjunction (SA): only the pre-specified beta trees in the list attached to the node may be adjoined there.

TAG3P converges towards the approximate solutions by replicating and combining certain building blocks. If we want the population avoid the two attractors, these building blocks must not be avoided. We set SA constraints on the root nodes of the beta trees excluding β_1 from the list of the beta trees that can be adjoined. The effect is that the expression sin(x+t) can not be evolved.

We conducted 50 runs; 9 runs (18%) converged to the exact solution and no approximate solutions were found.

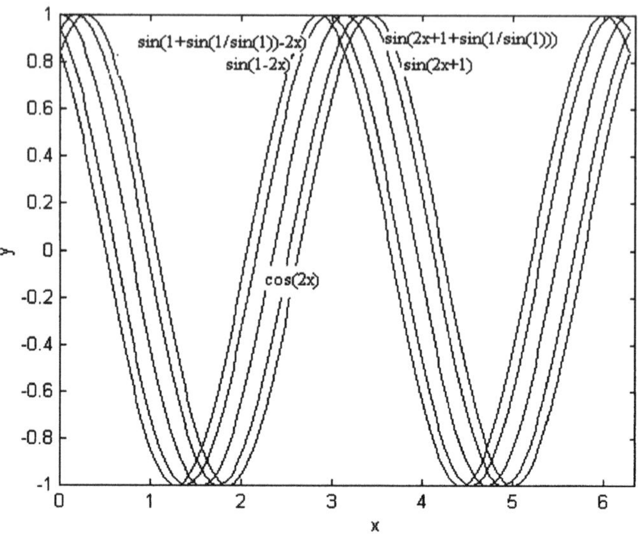

Figure 7. Some building blocks.

5 Conclusion and Future Work

In this chapter, we gave the results of TAG3P on a standard problem in GP, finding trigonometric identities. TAG3P outperforms GP and GGGP because of its capabilities in preserving and replicating building blocks. The nature of the search space of the problem was analyzed, and we introduced search bias in TAG3P through selective adjunctions.

In future, we will investigate more thoroughly the behavior of genetic operators in TAG3P, and theoretically explore the language domain of the linear form of derivation sequence. We are also developing an implementation of TAG3P using the most general form of derivation sequence.

References

Banzhaf W, Nordin P, Keller R E, and Francone FD (1998). *Genetic Programming: An Introduction.* Morgan Kaufmann Pub.

Gruau F (1996). On Using Syntactic Constraints with Genetic Programming. In P. J. Angeline and K. E. Kinnear Jr. (eds), *Advances in Genetic Programming*, pp 377-394, The MIT Press.

Hoai NX (2001a). Solving The Symbolic Regression Problem with Tree-Adjunct Grammar Guided Genetic Programming: The Preliminary Results. In the proceedings of The 5th Australasia-Japan Co-Joint Workshop on Evolutionary Computation, Dunedin, New Zealand, Nov. 2001, pp 52-61.

Hoai NX (2001b). Solving Trigonometric Identities with Tree Adjunct Grammar Guided Genetic Programming. In the proceedings of The First International Workshop on Hybrid Intelligent Systems, Adelaide, Australia, Dec, 2001, pp 339-352.

Hoai NX and McKay RI (2001). A Framework for Tree Adjunct Grammar Guided Genetic Programming. In H.A. Abbass and M. Barlow (eds), *Proceedings of the Post-graduate ADFA Conference on Computer Science* (PACCS'01), pp 93-99.

Hoai NX, McKay RI, and D. Essam. Solving Symbolic Regression with Tree Adjunct Grammar Guided Genetic Programming: The Comparative Results. To appear in the proceedings of IEEE Congress on Evolutionary Computation, Hawaii, USA, 2002.

Joshi AK and Schabes Y (1992). Tree-Adjoining Grammars and Lexicalized Grammars. In M. Nivat and A. Podelski (eds), *Tree Automata and Languages*, pp 409-431. Elsevier Science Publisher.

Joshi AK and Schabes Y (1997). Tree Adjoining Grammars. In Grzegorz Rozenberg and Arto Saloma (eds), *Handbook of Formal Laguages, Vol 3*, pp 69-123. Springer-Verlag, NY, USA.

Joshi AK, Levy LS, and Takahashi M (1975). Tree Adjunct Grammars. *Journal of Computer and System Sciences*, 10(1), pp136-163.

Koza J (1992). *Genetic Programming.* MIT Press.

Montana DJ (1995). Strongly Typed Genetic Programming. *Evolutionary Computation* 3(2), pp 199-230.

O'Neill M (1999). Automatic Programming with Grammatical Evolution. In *Proceedings of the Genetic and Evolutionary Computation Conference Workshop Program*, July 13-17, Orlando, Florida USA.

O'Neill M and Ryan C (1998). Grammatical Evolution: A Steady State Approach. In *Proceedings of the Second International Workshop on Frontiers in Evolutionary Algorithms 1998*, pp 419-423.

O'Neil M and Ryan C (1999a). Under the Hood of Grammatical Evolution. In W. Banzhaf, J.Daida, A.E. Eiben, M.H. Garzon, V. Hovana, M. Jakiela, and R.E. Smith (eds.). *GECCO-99*, Vol. 2, pp 1143-1148, Morgan Kaufmann Pub.

O'Neill M and Ryan C (1999b). Genetic Code Degeneracy: Implication for Grammatical Evolution and Beyond. In *Proceedings of the European Conference on Artificial Life*, pp 149-153, Springer-Verlag, Germany.

Ratle A and Sebag M (2000). Genetic Programming and Domain Knowledge: Beyond the Limitations of Grammar-Guided Machine Discovery. Ecole Polytechnique, France. (Available at http://www.researchindex.com. Accessed: 30, May, 2001).

Ryan C, Collin JJ , and O'Neill M (1998). Grammatical Evolution: Evolving Programs for an Arbitrary Language. *Lecture Note in Computer Science 1391, Proceedings of the First European Workshop on Genetic Programming*, pp 83-95, Springer-Verlag.

Whigham P (1995). Grammatically-based Genetic Programming. In *Proceedings of the Workshop on Genetic Programming: From Theory to Real-World Applications*, pp 33-41, Morgan Kaufmann Pub.

Whigham P (1996). Search Bias, Language Bias and Genetic Programming. In *Genetic Programming 1996*, pp 230-237, MIT Press.

Wong ML and Leung KS (1996). Evolving Recursive Functions for Even-Parity Problem Using Genetic Programming. In P.J. Angeline and K.E. Kinnear Jr. (eds.), *Advances in Genetic Programming*, pp 221-240, The MIT Press.

Part 2

Applications

Chapter 10

Modeling a Distributed Knowledge Management for Autonomous Cooperative Agents with Knowledge Migration

Noriko Etani

Summary. This chapter introduces modeling an autonomous agent and a cooperative system which consists of autonomous agents of guide activities in a laboratory. In multi-agent environment, each agent can work at common goals with globally cooperative behaviors. In order to construct a model integrating agent's behavior and cooperation among agents, we present two approaches for agent collaboration to resolve the above mentioned issues. For the first approach, we introduce social agency model for constructing a prototype system for guide activities in a laboratory. We,then, formalize the interaction between autonomous agents. For the second approach, we present an autonomous agent's architecture in social agency aimed at communicating with other agents in knowledge-level. The main contribution of this chapter has been to propose the agent's model to determine both agent's behavior and cooperation among agents allowing to express (1) cooperation, (2) adaptability, (3) mobility, and (4) transparency, and verify its model by developing the prototype system. Future research will indicate the scaling problem in different knowledge representation schemes between people and robots with sensors in a dynamic, unpredictable environment.

1 Introduction

This chapter will show modeling an autonomous agent and a cooperative system which consists of autonomous agents by the prototype system of guide activities in a laboratory. Information environments are composed of distributed, autonomous, heterogeneous components. Modules and processes of computer systems have a goal to realize autonomy, cooperation, adaptability, and transparency of knowledge to form cooperation among agents in multi-agent system. Then, we introduce social agency which is designed focusing on a role to achieve a common goal, and cooperative method which is knowledge migration with guiding authority between agents in its cooperative system based on ontology-based knowledge description following KQML [8]. This proposed model is realized and evaluated by a prototype system of guide activities in a laboratory using a mobile computer and an autonomous mobile robot. And an autonomous agent's architecture including a human interaction for an autonomous mobile robot is developed to compose by independent software components. As a result of executing guide activities in a laboratory, a common goal among agents have been achieved.

The rest of the chapter is organized as follows. In Section 2 we provide background and related work. In Section 3 we provide a social agency model for agent collaboration with knowledge migration. We show terminology and goals of this model. This is followed by descriptions of system overview, modeling, and evaluations. Section 4 provides autonomous agent's architecture with knowledge migration. We explain design of agent architecture, and evaluation. Section 5 provides conclusion about this work. Finally, future research in this chapter is presented in Section 6.

2 Background and Related Work

An agent is a module or a process which has autonomy, cooperation, adaptability, mobility and transparency of knowledge. Agents can work when they are constructed in the development of techniques for designing agents with the above-mentioned features. As a result, agents fulfill their rational, social, interactive, and adaptive functions. We describe the development based on the above-mentioned agency model. FRIEND21 is a national project name of "Future Personalized Information Environment Development", which aimed at an ideal figure of 21th century's human interface in Ministry of International Trade and Industry. It had started in 1988 and had finished in 1994. As a result of this project, human interface architecture, which was called agency model, was developed. This agency model is a executive environment model of meta-ware which is a method of interface design equipped with a real cognitive mechanism so that it may be a dynamic drive by a symbol suitable for a task. It is an integrated operating environment. A memory space of studio is mediated between autonomous modules which are called agents to communicate with each other [9]. In this chapter, data on shared memory is accessed by independent processes and control functions through socket-base communication in our autonomous agent's architecture of an autonomous mobile robot. If we follow a guideline of FRIEND21, behaviors of an autonomous mobile robot, a graphic character and a voice guidance are classified as meta-ware which presents an internal state of data.

3 Social Agency Model with Knowledge Migration

A prototype system for guide activities in a laboratory is constructed based on the multi-agent environment. Introducing mobile computing into the multi-agent environment, mobile agents between machines are able to enhance computing efficiency and services. To realize the robust model for required additional services, the following points are considered.

(1) to execute services on heterogeneous environments
(2) to develop independent software components

Especially, this section focuses on the basic loop determining both agent's behavior and cooperation among agents. These subsections give some backgrounds by providing terminology and goals, and describe its model, and its evaluation.

3.1 Terminology

The terminology of some background in social agency model is provided:

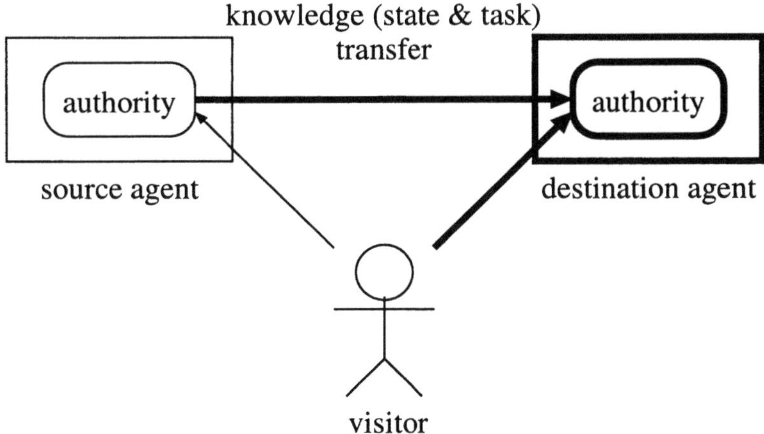

Figure 1. Transferring authority for guide activities by knowledge migration.

A **real-world agent** is software agent which has facilities to obtain information from the physical environment or to do something to the environment [13]. This agent can have an autonomous control as an autonomous agent which is defined that an autonomous agent is a system situated within and part of an environment that senses that environment and acts on it so as to effect what it senses in the future [1]. In this chapter, a real-world agent is a robotic software agent on an autonomous mobile robot.

Mobile agents move dynamically from one machine to another, transferring code, data, and especially authority to act on the owner's behalf within the network [12]. In this chapter, a mobile agent with authority can move between the mobile client and the mobile server.

Knowledge is state and task of a source agent in Figure 1. It is ontology-based knowledge description to control agents' behavior. And it contributes to the agent's duality.

Knowledge migration shown in Figure 1 is the act of transferring authority between agents. The transferred authority of the source agent includes the state of a guide and the task to guide a visitor in knowledge-level. After knowledge migration, the destination agent gets the authority and guides a visitor in a laboratory. Throughout this process, agents can share knowledge to achieve a goal of guide activities.

3.2 Goal

The goals of unifying agent's behavior and cooperation among agents include:

Cooperation is a goal of social agency when the multiple processes in the communication between a guide agent and a robotic software agent can work to achieve a common goal that is guiding a visitor to his destination.

Adaptability is a goal of social agency when its autonomous software agent can manage both its knowledge and other agents' migrated knowledge to execute its behavior in knowledge-level.

Mobility is a goal of social agency when mobile computer and autonomous mobile robot equipped with a network can guide a visitor in a laboratory by knowledge migration between mobile computer and autonomous mobile robot.

Transparency is a goal of knowledge migration because the communication and guide activities in a laboratory require to construct transparent knowledge boundaries between real space and virtual space which a computer generates in its display. Real space means the environment in which a visitor, a hand-held mobile computer, and an autonomous mobile robot exist. Virtual space means the graphical map to show that environment in the computer display.

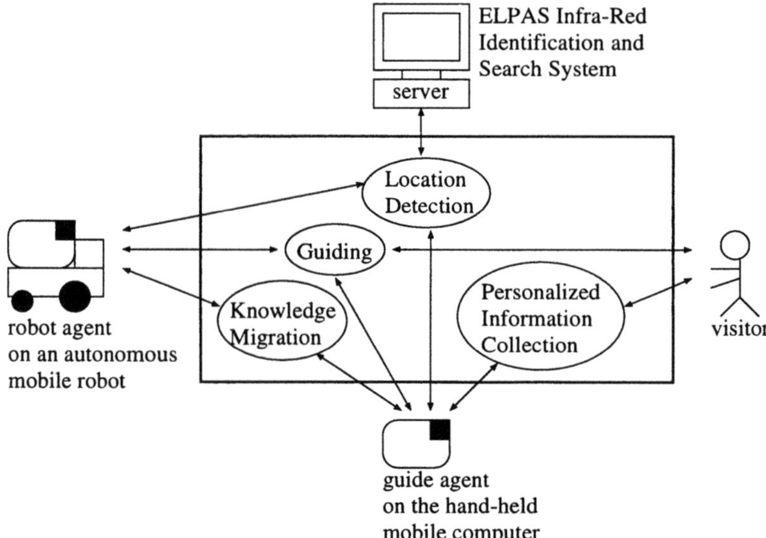

Figure 2. Overview of the agents' cooperation model for guide activities in multi-agent environment.

3.3 Overview

Figure 2 illustrates an overview of the agents' cooperation model for guide activities in multi-agent environment. It is composed of an infrared location system, its management server, a hand-held mobile computer and an autonomous mobile robot (Pioneer1 mobile robot) connected by wireless LAN.

3.4 Model

3.4.1 Component

This model's organization to manage cooperative knowledge is described as follows:

(1) Guide Agent(GA)
Figure 3 shows a guide agent on hand-held mobile computer. A guide agent displays a map in a laboratory and a graphic character to guide a visitor on the hand-held mobile computer. This mobile computer is "VAIO PCG-C1" made by Sony. The network connection between this mobile computer and other computers utilizes wireless "WaveLan," which operates at a 1.2 G Hz bandwidth. This system can transmit data at 1 megabit per second.

(2) Robotic Software Agent (RA)
Figure 4 shows a prototype autonomous mobile robot for guide activities. A robotic software agent assists visitors to utilizes an autonomous mobile robot made by ActivMedia. This autonomous mobile robot has seven sonars, an encoder, an electrical compass, and two motors. It is controlled by an operating system only used for this robot (PSOS), and this OS is installed in a control board on the robot. A client system terminal is connected to the OS. This terminal is a notebook type personal computer, "SOLO" made by Gateway, in which Red Hat Linux release 5.1 is installed. This client system receives a packet from PSOS including input from seven sonars, an encoder and a compass data. The data transmission rate is one packet per 100 m sec. This client system was developed by using Saphira Libraries to connect with PSOS.

(3) Location System (LS)
It is utilized to detect the location of the robotic software agent and the visitor's mobile computer. The location system can read the infrared emission from

Figure 3. Guide agent on hand-held mobile computer.

Figure 4. Prototype autonomous mobile robot for guide activities.

a badge put on a mobile robot and a mobile computer. The infrared location system's readers on the ceilings of the hallways detect the mobile computer's and mobile robot's location. This location information is updated on the location system's server.

(4) Visitor
A visitor is a person who visits a laboratory and has an interest in research and researchers.

3.4.2 Interaction

In the multi-agent environment, there are four kinds of interaction as follows:

(1) Personalized Information Collection
A visitor inputs his research interest on the hand-held mobile computer according to a guide agent's instruction.

(2) Knowledge Migration
Authority for guiding a visitor is transferred by knowledge migration between a guide agent and a robotic software agent.

(3) Guiding
A visitor is navigated by a guide agent and a robotic software agent to a visitor's destination in a laboratory.

(4) Location Detection
ELPAS infra-red identification and search system detects the physical locations of the mobile computer and the mobile robot.

3.4.3 Knowledge Migration

(1) Design

Figure 5 illustrates the object oriented modeling of a distributed knowledge for cooperative agents. A part of agent's knowledge is migrated to another agent's knowledge in order to decide another agent's behavior. This is a distributed knowledge sharing among autonomous cooperative agents. Figure 6 shows ontology-based knowledge sharing. Each agent has its own knowledge constrained by its role and its obligation. The location system detects a guide agent's and a robotic software agent's locations, and notifies each location to each agent. A guide agent collects visitor's personalized information which is context for guiding. This knowledge of a guide agent is transferred to a robotic software agent. After transferring, related knowledge to migrated one is combined and the robotic software agent's knowledge is newly formed. After that, the robotic software agent can decide its behavior in the environment and have authority to guide a visitor.

(2) Protocol

Using knowledge migration, authority of guiding is transferred from a guide agent to a robotic software agent. Figure 7 indicates a protocol and processes to execute authority. This protocol trace is described as follows.

1. to detect and avoid obstacles by using input from the seven sonars at one packet per 100 m sec.
2. to transmit a site number sent by the location system through the communication layer to the behavior layer.

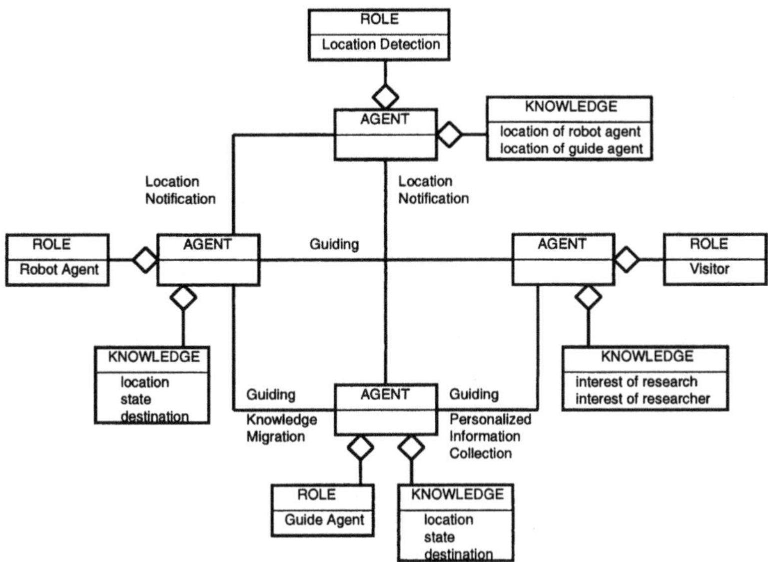

Figure 5. Object oriented modeling of distributed knowledge for cooperative agents.

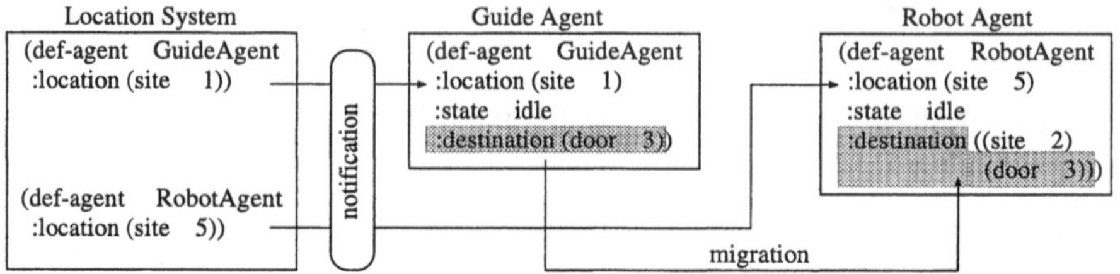

Figure 6. Ontology-based knowledge sharing and knowledge migration.

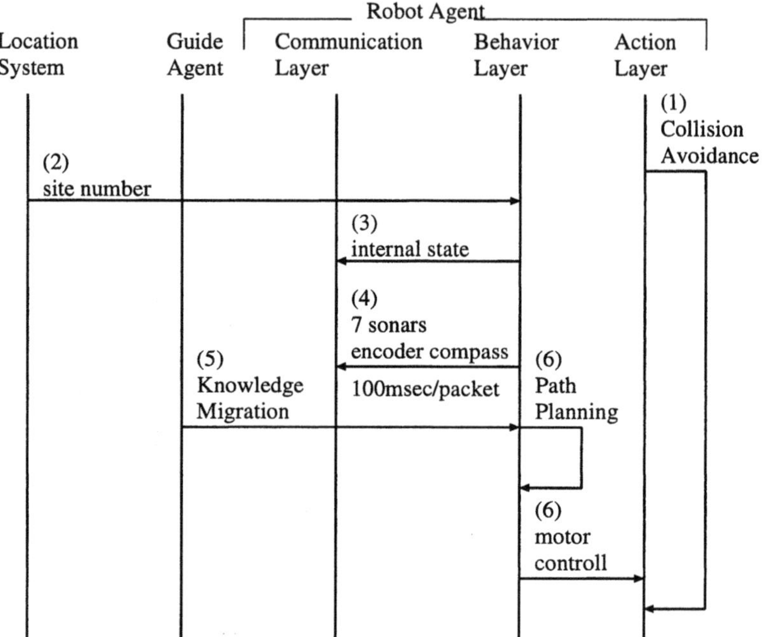

Figure 7. Protocol trace to execute authority which is migrated from guide agent.

3. to transmit an internal state from the behavior layer to the communication layer for updating this state.
4. to transmit input from seven sonars, a compass value and an encoder value from the action layer to the behavior layer at one packet per 100 m sec.
5. to migrate knowledge of a guide agent from the communication layer to the behavior layer.
6. to execute path planning in the behavior layer, and to transmit motor control from the behavior layer to the action layer.

3.4.4 Formalization

We focus on the relationship between agents to clarify knowledge retrieve for a distributed knowledge sharing. A formal representation of the agents' cooperation model [14] is introduced.

$$\textit{Agent System} = <\textit{Agents, Environment, Coupling}>,$$

where:

(1) $$\textit{Agent} = <\textit{State, Input, Output, Process}>$$

- State is the set of properties (values, true propositions) that completely describes the agent.
- Input and output are subsets of state whose variables are coupled to the environment.
- Process is an autonomously executing process that changes the agent's state.

(2) $$\textit{Environment} = <\textit{State, Process}>$$

- The environment has its own process that can change its state, independent of the actions of its embedded agents.

(3) *Coupling is a mapping of an agent's input and output from/to the environment's state.*

Based on this formalization [14], It shows precisely how agents' interaction and communication can be proven to guarantee the navigation for a visitor in knowledge-level. The formal model is defined by the following five elements.

Definition 1. The multi-agent model, M, is a structure:

$$M = (A, E, C),$$

where:

- $A = \{GA, RA\}$ is a set of agents;
- $E = \{LS\}$ is a set of environment which has its own process that can change its state, independent of the actions of A;
- C is coupling model that the agents have a mapping of input and output to or from the environment.

Definition 2. Coupling model, C, is a structure:

$$C = (I, O),$$

where:

- $I = \{INTEREST\}$ is a set of inputs from the environment;
- $O = \{DESTINATION\}$ is a set of outputs to the environment.

Definition 3. GA model, $A(GA)$, is a structure:

$$A(GA) = (S, I, O, P),$$

where:

- $S = \{$*attention, calculating, pushing, waiting*$\}$ is a set of GA's states [7];
- $I = \{$*INTEREST, SITE*$\}$ is a set of inputs to the environment;
- $O = \{$*KNOWLEDGE*$\}$ is a set of outputs to the environment;
- $P = \{$*instruction, guiding, migration*$\}$ is a set of processes that change the agent's states [7].

Definition 4. RA model, $A(RA)$, is a structure:

$$A(RA) = (S, I, O, P),$$

where:

- $S = \{$*idle, transmission, guiding, goal*$\}$ is a set of RA's states;
- $I = \{$*GA(KNOWLEDGE), SITE*$\}$ is a set of inputs to the environment;
- $O = \{$*DESTINATION*$\}$ is a set of outputs to the environment;
- $P = \{$*wandering, guiding*$\}$ is a set of processes that change the agent's states.

Definition 5. LS model, $E(LS)$, is a structure:

$$E(LS) = (S, P),$$

where:

- $S = \{$*GA(SITE), RA(SITE)*$\}$ is a set of LS's states;
- $P = \{$*detection, notification*$\}$ is a set of processes that change LS's states.

3.5 Evaluation

Social agency model is evaluated by guiding a visitor in a prototype system. Knowledge migration between a guide agent and a robotic software agent can guarantee the agent's duality. As for its duality including people, they conducted experiments, and proved a visitor's guiding context persistence in a laboratory [15].

4 Autonomous Agent's Architecture

This section will describe a design and its evaluation of real-time control architecture for an autonomous agent in guide activities. An autonomous agent is an autonomous mobile robot which is a robotic software entity. An autonomous agent decides its behavior from environment inputs, controls interface agent with warning voice and a life-like graphic character, and communicates with other agents in knowledge-level connected by TCP. This agent is introduced into a prototype system for guide activities in a laboratory to evaluate its real-time performance. To realize the robust model for required additional services, the following points are considered.

(1) to develop independent software components;
(2) to negotiate agent's behavior between software components using protocols.

4.1 Architecture

This section describes the implementation of determining both agent's behavior and cooperation among agents. This software architecture shown in Figure 8 has three layers of "communication layer," "behavior layer" and "action layer." The functions in each layer are described as follows.

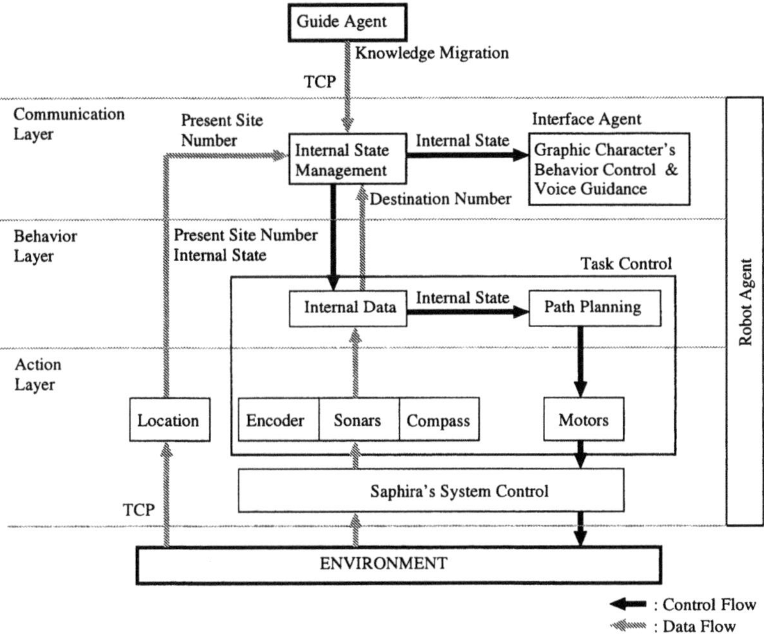

Figure 8. Agent's software architecture on autonomous mobile robot.

4.1.1 The Control Layers

(1) Communication Layer

This layer has two functions. One is to manage four internal states for a robotic software agent's behavior: "Idle," "Transmission," "Guiding" and "Goal" according to the interruption of its state change. Its interruption details describe in a next subsection. Another is to control interface agent which displays life-like graphical character and outputs a voice guidance according to these internal states.

(2) Behavior Layer

This layer has two functions. One is to manage several inputs which are a site number given by the location system, data from the seven sonars, a compass value, an encoder value, an internal state, and a door number as a destination. Another is to execute path planning to direct the real-world mobile agent's behavior.

(3) Action Layer

This is composed of Saphira control architecture, three input components from and one output component to Saphira control architecture, and the management system for input information from the location system by TCP.

Figure 8 illustrates the structure of a robotic software agent's software architecture within Pioneer1 mobile robot holding the Saphira's control architecture [11]. The robotic software agent's software architecture organizes multiple processes to the cooperative architecture. Running processes in each layer communicate with each other using UNIX sockets. These modules of encoder, sonars, compass inputs, and motors' outputs using Saphira Libraries are controlled in task switching manner using finite-state machine. Interface agent control of graphic character's behavior and voice guidance are implemented on Java applets in Web. Internal state data on shared memory is accessed by control functions (encoder, sonars, and compass), path planning, and internal state management. The robotic software agent receives its environment inputs through packet communication protocol, migrated knowledge, and a site number sent by the location system through TCP. It manages its present location and running direction, executes path planning, and carries out run operation to find a destination.

4.2 Protocol

Figure 9 illustrates an internal state transition model. This model is described as the following cycle.

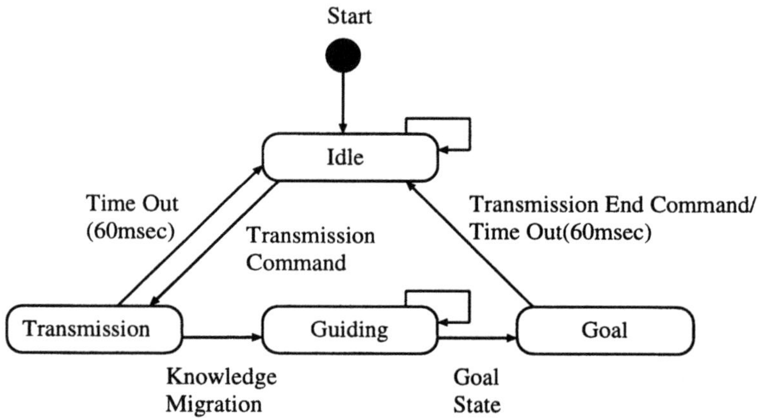

Figure 9. State transition to determine basic loop of agent's behavior.

1. In the idle state shown in Figure 10, the robotic software agent goes back and forth in a corridor, stops in front of each door and outputs a voice guidance.
2. In the transmission state shown in Figure 10, the robotic software agent meets visitors in the idle state. And the guide agent on the visitor's mobile computer migrates authority to the robotic software agent after sending a transmission com-

Figure 10. Idle, transmission, guiding.

mand from the guide agent to the robotic software agent. In the case of no visitor with a mobile computer or no need for migration, the robotic software agent continues in an idle state.
3. In the guiding state shown in Figure 10, the robotic software agent starts to guide the visitor to the destination.
4. In the goal state, the robotic software agent reaches its destination and sends a transmission end command to a guide agent. After that, the robotic software agent is in the idle state.

4.2.1 Transition of Robotic Software Agent's Behavior State

Following this internal state transition, path planning is executed in the behavior layer, and a forward, backward or rotation control command is sent to the action layer to control the two motors. Figure 11 illustrates the trigger protocol to transit an internal state in the communication layer as follows.

1. to send an internal state from the behavior layer to the communication layer for updating the robotic software agent's internal state.
2. to be in the idle state in the behavior layer.
3. to send a transmission command from a guide agent through the communication layer to the behavior layer.
4. to be in the transmission state in the behavior layer. If the agent cannot change its state to the idle state within 60 ms, its state will default to the idle state (back to above 2.).
5. to transfer knowledge for guiding from a guide agent through the communication layer to the behavior layer.
6. to be in the guiding state in the behavior layer.
7. to execute path planning, and to run forward a destination location.
8. to send a goal state through the communication layer from the behavior layer to a mobile computer.
9. to send a transmission end command from a guide agent through the communica-

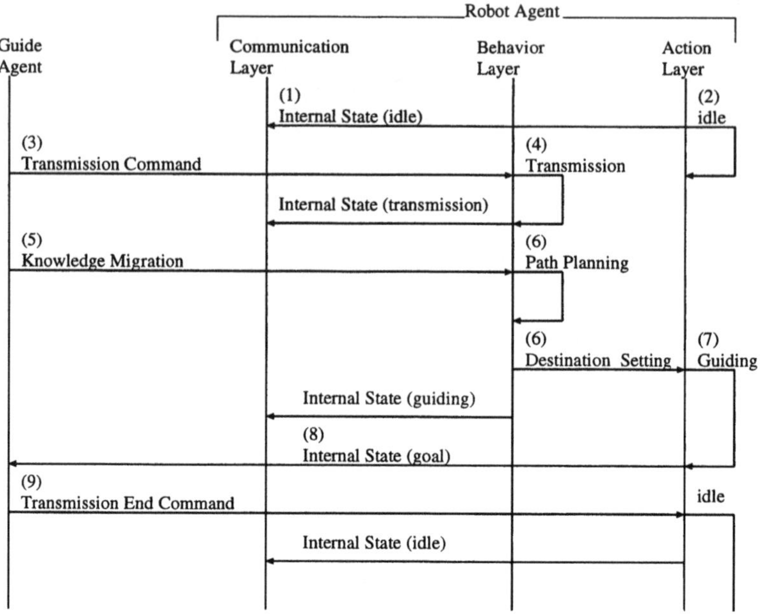

Figure 11. Protocol trace to transit agent's behavior state.

tion layer to the behavior layer. If the agent cannot receive this command within 60 ms, it will default to the Idle state in the behavior layer (back to above 2.).

4.3 Path Planning

Figure 12 and Figure 13 indicate the division of a mobile robot's running space in the east and west directions. The robotic software agent goes back and forth between a starting point of a site number "2" and an ending point of a site number "7" which exist of rooms and exhibitions available for guiding. A visitor can walk between a site number "1" and "8" with the hand-held mobile computer. In the guiding state, an agent calculates the distance from the present location to a destination's location by executing path planning in the behavior layer and orders a running command for this distance to the action layer. In this path planning, one path is defined as the section between site numbers indicating changed points. The agent runs until it receives a site number of a destination by checking an order of receiving a site number, and it runs from the entrance of a final site and the destination location by checking its encoder value. The robotic software agent runs until a site number includes a destination, and when it enters a destination site area, it runs the distance between the entrance of a site point and the destination by using the encoder. The following procedure is taken.

1. checking a running direction
   If a running direction is different from that of the destination from the present

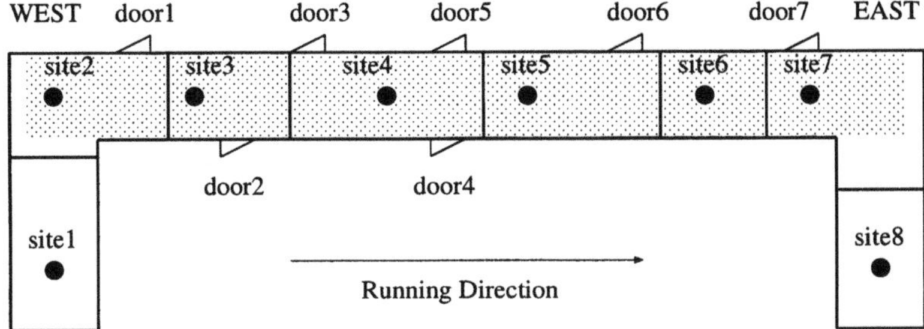

Figure 12. Space division in east direction where robotic software agent is running.

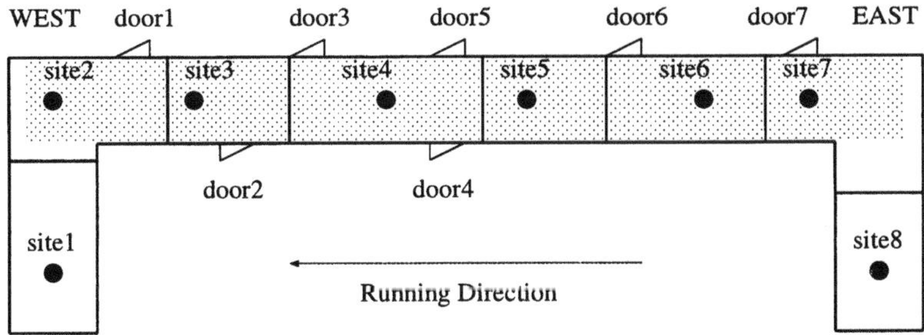

Figure 13. Space division in west direction where robotic software agent is running.

location, the robotic software agent rotates 180 degrees.

2. running until a site number includes a destination
3. calculating the distance from the entrance of a site including the destination to the destination. Goal-Distance means this distance. For example, when a running direction is east and the destination is door4, the entrance of a site indicates the first point to detect site4 closely connected with site3.
   When Goal-Site is detected,

$$GoalDistance \leftarrow (D1, ..., Dn),$$

   with
   D: the distance in mm between an entrance point in a site including the destination and the destination
   n: a door number including the destination
   Goal-Site: a site number including the destination
4. arranging running direction by using the location system
   In updating a site number, a heading degree is calculated by a theta value and a compass value in order to minimize the gap of the encoder value with respect to straight running.

5 Conclusion

The primary issue has been how to develop a new model of a distributed knowledge management. To accomplish this mission, we designed, formalized and analyzed its computational model. Our first approach provided the model among agents. Our social agency model's design and formalization on agents cooperation with knowledge migration are introduced into a prototype system of guide activities in a laboratory. Our goal was cooperation to achieve a common goal that is guiding a visitor to his destination, adaptability to manage both its knowledge and other agents' migrated knowledge to execute its behavior in knowledge-level, mobility to guide a visitor in a laboratory by knowledge migration between mobile computer and autonomous mobile robot, and transparency to construct transparent knowledge boundaries between real space and virtual space which computer generates in its display. Our second approach provided an autonomous agent's architecture with knowledge migration aimed at communicating with other agents in knowledge-level. This shows the model within one agent. In our autonomous agent's architecture, independent software components are developed, and two kinds of protocols are coordinated among those components. Our proposal of modeling a distributed knowledge management for cooperative agents is proved by our prototype system.

6 Future Research

Throughout this chapter we have a number of possible directions for future research into the design of computational models of agency. To conclude, we now briefly expand on a number of these ideas.

6.1 Introduction

This section studies the scaling problem in different knowledge representation schemes between people and robots with sensors in a dynamic, unpredictable environment. A robot has sensors to gather data about the environment and its robot has ontology-based knowledge to communicate between people and agents in guide activities [3], [4], [5], [6]. A agent on a robot has less knowledge about its task and th world. This knowledge will guide its behavior selection to achieve its goal with planning. It is important for a agent to choose a reasonable knowledge representation scheme in order to decide the scale of a task. In ontology-based knowledge granularity, agents including a robot and a mobile computer can communicate with people and they will decide their behavior to achieve a goal which people request. In bit-strings knowledge granularity which is sensors data, a robot can manage its control coped with the environment and one robot can communicate with another robot to enhance adaptability for cooperation among agents and learning against a dynamic, unpredictable environment.

Here, this chapter focuses on the bit-strings knowledge granularity to enhance adaptability in multi-agent system. This chapter proposes the model of a coevolutionary architecture for solving decomposable problems and apply it to the evolution of multi-agents, although this work is an preliminary step. The coevolutionary approaches utilizes a technique in which agents representing simpler subtasks are evolved in separate instances of learning classifier system. Collaboration among agents are formed representing complete solutions. Agents are created dynamically as needed. Results are presented in which the coevolutionary architecture produces higher quality solutions in fewer evolutionary trials on the problem of evolving agents in a grid world.

In this approach, each GA (a genetic algorithm) instance evolves a agent of individuals representing competing solutions to a subtask. Rather than evaluating solutions to these subtasks independently, the GA instances communicate with each other for the purpose of forming collaborations. This is accomplished by selecting representatives from each of the GA populations, and combining them into a single composite structures flows back to the individual subcomponents reflecting how well they collaborate with the other subcomponents to achieve the top level goal. This credit is then used by the local GAs to evolve better subcomponents.

This coevolutionary architecture is tested in the domain of learning rule sets for multi-agents under the problem of evolving agents in a grid world. This problem is a type of cooperative learning in a dynamic, unpredictable environment.

6.2 Basic Framework

The hypothesis underlying the idea presented here is that, in order to evolve solutions to more and more complex problems, explicit notions of modularity need to be introduced in order to provide reasonable opportunities for complex solutions to evolve in the form of interacting co-adaptive subcomponents. The difficulty comes in finding reasonable computational extensions to our current evolutionary paradigms in which such subcomponents emerge. At issue here is how to represent such subcomponents and how to apportion credit to them for their contributions to the problem solving activity such that the evolution of a solution to the top level goal proceeds without a human in the loop.

As shown in Figure 14, a cooperative coevolutionary architecture consists of a collection of GA, each attempting to evolve subcomponents (agents) which are useful as modules for achieving higher level goals. Complete solutions are obtained by assembling representative members of each of the agents present. Credit assignment at the agents level is defined in terms of the fitness of the complete solutions in which the agents members participate. This provides evolutionary pressure for agents to cooperate rather than compete. However, competition still exists among individuals within the same subpopulation. In the system used in this chapter, this model of a cooperative coevolutionary architecture is implemented with learning classifier, and the evolution of each agent was handled by a standard GA.

The use of multiple interacting subpopulations has also been explored as an alternate mechanism for coevolving using island model [18]. In the island model, a fixed number of subpopulations evolve competing rather than cooperating solutions. In ad-

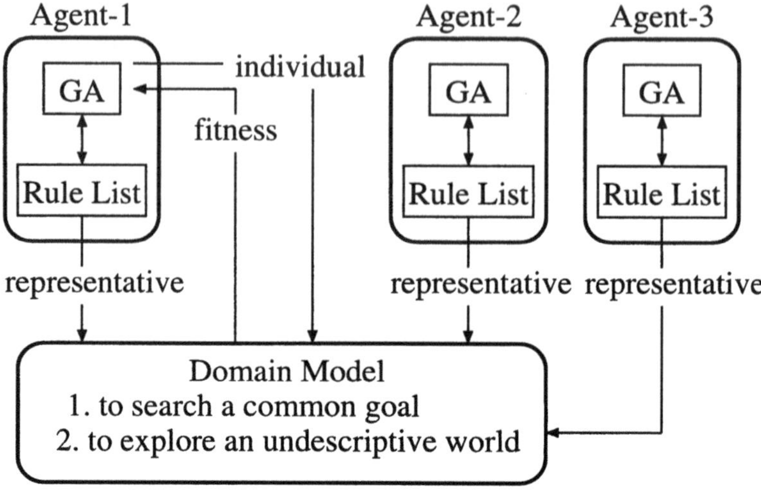

Figure 14. Model of cooperative evolutionary architecture.

dition, individuals occasionally migrate from one subpopulation (island) to another, so there is a mixing of genetic materials. Based on this idea, each agent evolves one subpopulation and each agent shares each subpopulation.

6.3 Future Work

A basic framework for agent-based coevolution has been presented in which a collection of GA running in evolving subcomponents which are combined into a composite structure capable of being evaluated on a top level goal. Because credit assignment at the agent level is defined in terms of the fitness of the complete solutions in which the agent members participate, there is evolutionary pressure for individuals to collaborate rather than compete with other individuals in coevolving agents.

This model of agent-based coevolution is applied to multi-agents. Results have been presented in which the coevolutionary architecture will produce higher quality solutions in fewer evolutionary trials. Although we have achieved considerable performance improvements, our primary motivation has been a better understanding of issues related to the evolution of interacting cooperative, adapted subcomponents.

Future work will focus on agent-based coevolution including people. There are two main difficulties introduced when one attempts this type of coevolution against people for more efficiency:

(1) Interactions with humans are poor.
(2) When opponents are random, known techniques for coevolution become impossible.

The first problem is common to multi-agent system that will learn from a real environment: interactions are slow and costly. We address this problem by nesting an extra loop of coevolution. While the system is waiting for human opponents, it runs more and more generations of coevolution among agents. The second problem led to develop a new evaluation strategy. It decides when each agent shares each sub-population in trials. With it, we have been able to prove that the system has been learning through interaction with people. This strategy also gives us the possibility for a fitness function that could solve the first problem.

In this future research, we will investigate the efficiency of modeling and design in agency model.

Acknowledgments

This chapter is based upon work supported in part by Graduate School of Information Science, Nara Institute of Science and Technology under Doctor's Thesis Research. The author thanks ATR Media Integration and Communications Research Laboratories for use of the Pioneer1 Mobile Robot, for graphical character design by Jun Kurumizawa, for useful discussions and technical support, and for the CMAP and CMAP-II projects.

References

1. J. Bradshaw, editor. (1997) Software Agents. MIT Press.
2. Noriko Etani. (1998) Using A Classifier System to Learn Adaptive Strategies for Collision Avoidance. Master's Thesis. Department of Information Processing, Graduate School of Information Science, Nara Institute of Science and Technology. NAIST-IS-MT9651016.
3. Noriko Etani. (1999) Robot Media Communication: An Interactive Real-World Guide Agent. Proceedings of First International Symposium on Agent Systems and Applications (ASA '99), Third International Symposium on Mobile Agents (MA '99). the IEEE Computer Society. ISBN 0-7695-0340-3. 234–241.
4. Noriko Etani. (1999) Robot Media Communication: A Real-world Guide Agent to Construct Transparent Knowledge Boundaries Between Real and Virtual Spaces. In Jiming Liu and Ning Zhong, editors. Intelligent Agent Technology: Systems, Methodologies and Tools (Proceedings of First Asia-Pacific Conference on Intelligent Agent Technology, 14-17 December, 1999). WORLD SCIENTIFIC PUBLISHING COMPANY PTE LTD. ISBN981-02-4054-6. 53–57.
5. Noriko Etani. (2002) Modeling Autonomous Agent's Architecture with Knowledge Migration in Social Agency. Proceedings of First international NAISO Congress on Autonomous Intelligent Systems (ICAIS '2002). ICSC Academic Press, Canada/The Netherlands.
6. Noriko Etani. (2002) Modeling a Distributed Knowledge Management for Cooperative Agents. In A. Abraham and M. Koeppen, editors. Hybrid Information Systems. Physica Verlag, Heidelberg. 513–526.
7. S. Fels, S. Sumi, T. Etani, N. Simonet, K. Kobayashi and K. Mase. Progress of

C-Map: A Context-Aware Mobile Assistant. Proceedings of the AAAI Spring Symposium on Intelligent Environments. March. 1998.
8. T. Finin, Y. Labrou and J. Mayfield. (1997) KQML as an agent communication language. In J. Bradshaw, editor. Software Agents. MIT Press.
9. FRIEND21: Future Personalized Information Environment Development. (1994) Human interface Architecture Rules Document. PIE, MITI.
10. L. Gasser. (1991) Social conceptions of knowledge and action: DAI foundations and open systems semantics. Artificial Intelligence **47**, 107–138.
11. Kurt G. Konolige. (1998) Saphira Software Manual. Version 6.1e.
12. Dejan S. Milojicic, William LaForge and Deepika Chauhan. (1998) Mobile Objects and Agents (MOA). Proceedings of the Fourth USENIX Conference on Object-Oriented Technologies and Systems (COOTS98). Santa Fe, New Mexico.
13. K. Nagao and J. Rekimoto. (1996) Agent augmented reality: A software agent meets the real world. Proceedings of the Second International Conference on Systems (ICMAS-96). AAAI Press. 228–235.
14. Jim Odell and William Tozier. (1999) Agents and Complex Systems. ASA/MA Tutorial. The Joint Symposium ASA/MA99, First International Symposium on Agent Systems and Applications (ASA'99), Third International Symposium on Mobile Agents (MA'99).
15. T. Ono, M. Imai, T. Etani and R. Nakatsu. (2000) Construction of Relationship between Humans and Robots. Transactions of Information Processing Society of Japan. Vol. 41. Number 1. 158–166.
16. A. S. Rao and M. P. Georgeff. (1991) Modeling Agents within a BDI-Architecture. In International Conference on Principles of Knowledge Representation and Reasoning (KR). Cambridge. Massachusetts. April. Morgan Kaufmann. 473–484.
17. Munindar P. Singh. (1994) Multiagent Systems: A Theoretical Framework for Intentions, Know-How, and Communications. Springer-Verlag. Heidelberg, Germany.
18. D. Whitley and T. Starkweather. (1990) Genitor II: a distributed genetic algorithm. Journal of Experimental and Theoretical Artificial Intelligence 2. 189–214.

Chapter 11

Intelligent Information Systems Based on Paraconsistent Logic Programs

Kazumi Nakamatsu

Summary. This chapter provides two intelligent system frameworks, an action control framework and a safety verification framework, based on a paraconsistent logic program called EVALPSN. Two examples for the EVALPSN based intelligent frameworks, an intelligent robot action control system and an automated safety verification system for railway interlocking, are introduced.

Keywords: paraconsistent logic, annotated logic program, defeasible deontic reasoning, intelligent robot action control, safety verification, railway interlocking.

1 Introduction

Annotated logics are a family of paraconsistent logics that were proposed by da Costa et al. [4, 22]. They were developed from the viewpoint of logic programming and applied to the semantics for knowledge bases by Subrahmanian et al. [3, 8, 21]. The annotated logic program [3] was extended to have strong (ontological) negation and named Annotated Logic Program with Strong Negation (ALPSN for short) by Nakamatsu and Suzuki [11]. The main purpose of the introduction of ALPSN was to deal with non-monotonic reasoning in a framework of annotated logic programming, and it was shown that ALPSN can provide a declarative semantics for default reasoning and a non-monotonic ATMS based on the stable model semantics for ALPSN. However, ALPSN is not so appropriate for dealing with defeasible reasoning or decision making, although defeasible reasoning is known as one of formalizations for non-monotonic reasoning [18, 19]. Therefore, in order to deal with defeasible reasoning in a framework of annotated logic programming, a new version of ALPSN called Vector Annotated Logic Program with Strong Negation (VALPSN for short) was also introduced by Nakamatsu et al. [13, 14, 15]. Moreover, VALPSN was extended to Extended VALPSN (EVALPSN for short) for dealing with defeasible deontic reasoning by Nakamatsu et al. [16, 17].

There are various situations in which agents have to choose one thing among some conflicting things in our world. Such decision making process is called defeasible reasoning. Defeasible logic is a formalization for defeasible reasoning. There are also many cases in which agents have to decide their actions based on norms such as law, policy, regulation etc., then some deontic notions such as obligation, permission, forbiddance etc. are used in the reasoning process. Such reasoning is formalized in various deontic logics. The combination of the two kinds of reasoning is called defeasible deontic reasoning and studied by Nute et al. [20]. Let us show a casual example for defeasible deontic reasoning. Suppose that you are wondering if you should go back to your home or drink beer in front of a pub. You are forbidden from drinking by your doctor, however you love to drink. Then you have a conflict between the forbiddance and permission in your mind. But you have to make a decision with some reasons such as "payday" or "no money" after all. Such reasoning is very common in our everyday life. EVALPSN is an inference tool for defeasible deontic reasoning.

In this chapter, two general frameworks for EVALPSN based defeasible deontic reasoning systems are introduced, and two intelligent information systems, an intelligent action control system for a virtual robot and an automated safety verification system for railway interlocking are presented as simple examples, although there are many other applicable areas of the EVALPSN based intelligent frameworks. The key idea of these EVALPSN based intelligent systems is as follows: each intelligent system has norms such as policy, guidelines, regulation etc. for its behavior, and the next state of the system is computed by defeasible deontic reasoning based on the norms; therefore, if the norms can be formalized in EVALPSN, the intelligent control for the system can be computed by EVALPSN programming.

In this chapter, two examples for EVALPSN based intelligent systems are introduced briefly.

Defeasible Deontic Action Control for Autonomous Robots

A virtual beetle robot traveling through a maze with obstacles, the wall of the maze, pitfalls and alcohol (the robot is forbidden from drinking), to the goal is supposed. The robot has three different kinds of sensors to detect the obstacles and the policy to act. According to the sensor values, some conflicting forbiddances with different levels of strength are created, and roughly speaking, its next action is decided by defeasible deontic reasoning between the forbidances. This intelligent robot action control system can be formulated in an EVALPSN such that the inputs to the EVALPSN are different kinds of sensor values and the output is the robot's next action.

Automated Safety Verification System for Railway Interlocking

Safety verification for railway interlocking is to verify the safety when securing or releasing railway routes, and it is carried out by checking whether route interlocking requests or sub-route release requests by signal operators contradict the safety properties that must be satisfied or not. Morley proposed a safety verification method for railway interlocking in station yards in his Ph.d thesis [10], which is a logical method based on a higher order logic language HOL [7]. An EVALPSN based safety verification method for railway interlocking is introduced. The basic idea of the EVALPSN based safety verification is that: the safety properties, route interlocking and sub-route release requests can be expressed deontically in an EVALPSN with no strong negation (EVALP for short); therefore, the EVALP based interlocking safety verification can be executed as usual logic programming inquiry.

Generally, if an EVALPSN contains strong negations, it has stable model semantics [6, 11] that is not so tractable, and the computation of the stable models takes long time. Therefore, it is not appropriate for real time processing. However, if an EVALPSN is a stratified program, it has a well-founded model [5], and the strong negation in the EVALPSN can be treated as the Negation as Failure in usual logic programming. Fortunately, since the EVALPSNs used for the intelligent robot action control and the automated safety verification are stratified, it can be appropriate for real time processing.

This chapter is organized as follows: first, an annotated logic program EVALPSN is reviewed; next, two EVALPSN based intelligent information system frameworks are provided; and last, two examples, an EVALPSN based robot action control system and an EVALPSN based automated safety verification system for railway interlocking, are introduced.

2 From VALPSN to EVALPSN

In this section, VALPSN is extended to EVALPSN after reviewing VALPSN briefly. The reader is assumed to be familiar with the usual notions of ordinary first order logics and logic programming in Lloyd [9].

2.1 VALPSN

Generally, in annotated logic programs, a truth value called an *annotation* is explicitly attached to each literal. For example, let p be a literal, μ an annotation, then $p:\mu$ is called an *annotated* literal. A partially ordered relation is defined on the set of annotations that has a complete lattice structure. An annotation in VALPSN called a *vector annotation* is a 2-dimensional vector such that each component is a non-negative integer. Thus the complete lattice of vector annotations is defined as: for a non-negative integer m,

$$\mathcal{T}_v = \{(i,j) \mid 0 \leq i \leq m,\ 0 \leq j \leq m,\ \ i \text{ and } j \text{ are integers }\}.$$

The ordering of the lattice $\mathcal{T}_v$ is denoted in the usual fashion by a symbol $\preceq_v$ and defined as: let $\vec{v_1} = (x_1, y_1)$ and $\vec{v_2} = (x_2, y_2)$ be vector annotations,

$$\vec{v_1} \preceq_v \vec{v_2} \quad \texttt{iff} \quad x_1 \leq x_2 \text{ and } y_1 \leq y_2. \tag{1}$$

In a vector annotated literal $p:(i,j)$, the first component i of the vector annotation (i,j) indicates the degree of positive information (true) to support the literal p and the second one j indicates the degree of negative information (false) as well as. Usually, vector annotated literals are interpreted epistemically. Thus, for example, the vector annotated literal $p:(3,2)$ is intuitively interpreted as "the literal p is known to be true of strength 3 and false of strength 2", and the vector annotated literal $q:(0,0)$ is interpreted as "the literal q is known to be neither true nor false".

Originally, VALPSN was introduced to provide the annotated semantics for the defeasible logics [1, 2], and the following correspondence between the satisfiability of VALPSN and the derivabilities of the defeasible logics was shown in [13, 14, 15]: let p be a literal,

$\models p:(3,0)$	corresponds to	p is definitely derivable.
$\models p:(2,0)$	corresponds to	p is defeasibly derivable.
$\models p:(1,0)$	corresponds to	p is defeasibly underivable.
$\models p:(0,0)$	corresponds to	p is unknown to be derivable.

Therefore, the integer m appearing in the lattice $\mathcal{T}_v$ of vector annotations is assumed to be 3 throughout this chapter.

Generally, annotated logics have two kinds of negations, an epistemic negation ($\neg$) and an ontological negation ($\sim$). The epistemic negation followed by an annotated literal is interpreted as a mapping between annotations, and the ontological negation is as a strong negation that appears in classical logics. The epistemic negation of vector annotated logic is defined as the following exchange between the components of vector annotations: let p be a literal,

$$\neg(p:(i,j)) = p:\neg(i,j) = p:(j,i). \tag{2}$$

Therefore, the epistemic negation followed by a vector annotated literal can be eliminated by the above syntactic operation (2). On the other hand, the ontological negation is defined by the epistemic negation [4].

Definition 1 (Strong Negation, $\sim$). Let A be an arbitrary formula in annotated logic.

$$\sim A =_{def} A \rightarrow (\neg(A \rightarrow A) \wedge (A \rightarrow A)). \tag{3}$$

Therefore, the epistemic negation followed by a non-literal formula is interpreted as the strong negation.

Definition 2 (well vector annotated literal). Let p be a literal.

$$p:(i,0) \quad \text{or} \quad p:(0,j)$$

are called *well vector annotated literals*, where i and j are non-negative integers such that $1 \leq i,j \leq 3$.

Definition 3 (VALPSN). If $L_0, \cdots, L_n$ are well vector annotated literals,

$$L_1 \wedge \cdots \wedge L_i \wedge \sim L_{i+1} \wedge \cdots \wedge \sim L_n \rightarrow L_0 \tag{4}$$

is called a *vector annotated logic program clause with strong negation* (VALPSN clause for short). A *Vector Annotated Logic Program with Strong Negation* (VALPSN for short) is a finite set of VALPSN clauses.

Now the semantics for VALPSN (the basic interpretation of VALPSN) is introduced briefly. Since the set $\mathcal{T}_v$ is a complete lattice, the Herbrand interpretation I of a VALPSN P over $\mathcal{T}_v$ can be considered to be a mapping from the Herbrand base B_P into the lattice $\mathcal{T}_v$. Usually the interpretation I is denoted by the set:

$$\{\, p:\sqcup\vec{\mu_i} | I \models (p:\vec{\mu_1} \wedge \cdots \wedge p:\vec{\mu_n}) \,\}, \tag{5}$$

where p is a literal and $\sqcup\vec{\mu_i}$ is the least upper bound of $\{\vec{\mu_1}, \cdots, \vec{\mu_n}\}$. The ordering $\preceq_v$ over $\mathcal{T}_v$ is extended to interpretations in a natural way and the notions of satisfaction are defined. In the rest of this chapter, a VALPSN P is assumed to be a set of ground VALPSN clauses.

Definition 4. Let I_1 and I_2 be any interpretations and A a vector annotated atom.

$$I_1 \preceq_I I_2 =_{def} (\forall A \in B_P)(I_1(A) \preceq_v I_2(A)), \tag{6}$$

where $I_1(A), I_2(A) \in \mathcal{T}_v$.

Definition 5 (Satisfaction). Let A be an atom. An interpretation I is said to satisfy
[1] a ground vector annotated atom $A:\vec{\mu}$ *i.e.*, $(I \models A:\vec{\mu})$ *iff* $\vec{\mu} \preceq_v I(A)$,
[2] a formula F *iff* I satisfies every closed instance of F,
[3] a formula $\sim F$ *iff* I does not satisfy F.
The satisfaction of other formulas are defined as same as usual logic.

We omit the details of the particular semantics for VALPSN such as the fixed point one, the stable model one and the well-founded one.

2.2 EVALPSN

The main difference between VALPSN and EVALPSN is the difference between their annotations. An *extended vector annotation* in EVALPSN has a form of $[(i, j), \mu]$ in which the first component (i, j) is a 2-dimentional vector annotation as same as a vector annotation in VALPSN and the second one μ is an index that denotes concepts such as fact (α), obligation (β), non-obligation (γ), and so on. The complete lattice $\mathcal{T}$ of extended vector annotations is defined as follows: $\mathcal{T} = \mathcal{T}_v \times \mathcal{T}_d$,

$$\mathcal{T}_v = \{(i,j) |\, 0 \le i \le 3,\ 0 \le j \le 3\} \quad \text{and} \quad \mathcal{T}_d = \{\bot, \alpha, \beta, \gamma, *_1, *_2, *_3, \top\}, \tag{7}$$

where i and j are non-negative integers. The ordering of $\mathcal{T}_d$ is denoted by a symbol $\preceq_d$ and the complete lattices $\mathcal{T}_v$ and $\mathcal{T}_d$ are described by the Hasse's diagrams in Figure 1. The intuitive meanings of the members of $\mathcal{T}_d$ are: $\bot$ (unknown), α (fact), β (obligation), γ (non-obligation), $*_1$ (both fact and obligation), $*_2$ (both obligation and non-obligation), $*_3$ (both fact and non-obligation) and $\top$ (inconsistent). The Hasse's diagram (cube) shows that the lattice $\mathcal{T}_d$ is a tri-lattice in which the direction $\overrightarrow{\gamma\beta}$ indicates *deontic truth*, the direction $\overrightarrow{\bot *_2}$ indicates the amount of *deontic knowledge* and the direction $\overrightarrow{\bot\alpha}$ indicates *factuality*. Therefore, for example, the annotation β can be intuitively interpreted to be deontically truer than the annotation γ and the annotations $\bot$ and $*_2$ are deontically neutral, i.e., neither obligation

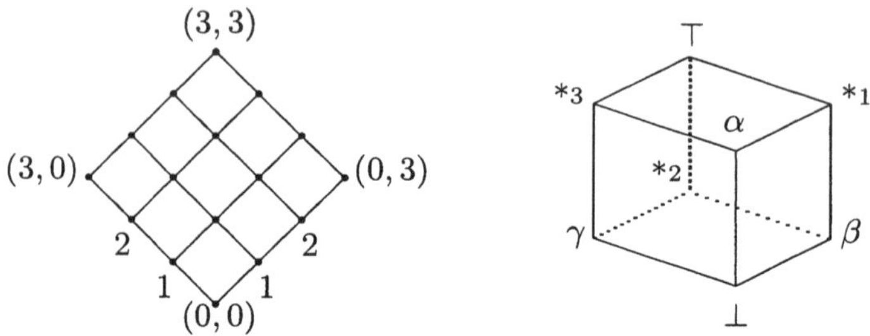

Figure 1. Lattice $\mathcal{T}_v$ (*left*) and Lattice $\mathcal{T}_d$ (*right*).

nor not-obligation The ordering of $\mathcal{T}$ is denoted by a symbol $\preceq$ and defined as: let $[(i_1, j_1), \mu_1]$ and $[(i_2, j_2), \mu_2]$ be extended vector annotations,

$$[(i_1, j_1), \mu_1] \preceq [(i_2, j_2), \mu_2] \quad \texttt{iff} \quad (i_1, j_1) \preceq_v (i_2, j_2) \text{ and } \mu_1 \preceq_d \mu_2.$$

There are two kinds of epistemic negations, $\neg_1$ and $\neg_2$, in EVALPSN, which are defined as mappings over $\mathcal{T}_v$ and $\mathcal{T}_d$, respectively.

Definition 6 (Epistemic Negations of EVALPSN, $\neg_1$ and $\neg_2$).

$$\begin{aligned}
&\neg_1([(i,j),\mu]) = [(j,i),\mu], \quad \forall \mu \in \mathcal{T}_d \\
&\neg_2([(i,j),\bot]) = [(i,j),\bot], \ \neg_2([(i,j),\alpha]) = [(i,j),\alpha], \\
&\neg_2([(i,j),\beta]) = [(i,j),\gamma], \ \neg_2([(i,j),\gamma]) = [(i,j),\beta], \\
&\neg_2([(i,j),*_1]) = [(i,j),*_3], \ \neg_2([(i,j),*_2]) = [(i,j),*_2], \\
&\neg_2([(i,j),*_3]) = [(i,j),*_1], \ \neg_2([(i,j),\top]) = [(i,j),\top].
\end{aligned}$$

The epistemic negations, $\neg_1$ and $\neg_2$, followed by extended vector annotated literals can be eliminated by the syntactic operations in the above definition, and the strong negation ($\sim$) in EVALPSN is defined by one of the epistemic negations.

Deontic notions such as obligation, and fact can be represented by extended vector annotations as follows:

fact is represented by an extended vector annotation, $[(m, 0), \alpha]$
obligation is represented by an extended vector annotation, $[(m, 0), \beta]$
forbiddance is represented by an extended vector annotation, $[(0, m), \beta]$
permission is represented by an extended vector annotation, $[(0, m), \gamma]$,

where $m(1 \leq m \leq 3)$ is a non-negative integer. For example, an extended vector annotated literal $p\!:[(3,0),\alpha]$ can be intuitively interpreted as "the literal p is known to be a fact of strength 3", and $q:[(0,2),\beta]$ can be also as "the literal q is known to be forbidden of strength 2".

Definition 7 (well extended vector annotated literal). Let p be a literal.

$$p : [(i, 0), \mu] \quad \text{or} \quad p : [(0, j), \mu]$$

are called *well extended vector annotated literals*, where i and j are non-negative integers ($1 \leq i, j \leq 3$) and $\mu \in \{ \alpha, \beta, \gamma \}$.

Definition 8 (EVALPSN). If $L_0, \cdots, L_n$ are well extended vector annotated literals,

$$L_1 \wedge \cdots \wedge L_i \wedge \sim L_{i+1} \wedge \cdots \wedge \sim L_n \rightarrow L_0 \tag{8}$$

is called an *extended vector annotated logic program clause with strong negation* (EVALPSN clause for short). An *Extended Vector Annotated Logic Program with Strong Negation* is a finite set of EVALPSN clauses.

Due to the strong negation, it cannot be said that EVALPSN has the tractable fixed point semantics [9] and generally they have the stable model semantics introduced by Gelfond and Lifschitz [6]. In the case of the stable model semantics, some EVALPSNs may have more than two stable models or no stable model. Anyway the treatment of the stable model semantics is not so appropriate for practical use. However, fortunately if an EVALPSN is a stratified program, it has a tractable well-founded model introduced by Gelder et al. [5], then the strong negation in the stratified EVALPSN can be processed as the Negation as Failure used in usual logic programming. Since all EVALPSNs appearing in this chapter are stratified, the stable model semantics does not have to be taken into account. Therefore, we do not pay our attention to the stable model semantics anymore. The following simple example shows extended vector annotated logic programming without the stable models semantics.

Example 1. Suppose an EVALPSN

$$P = \{ \ p:[(1,0),\alpha], \tag{9}$$

$$\sim p:[(2,0),\alpha] \rightarrow q:[(0,2),\beta], \tag{10}$$

$$\sim q:[(0,2),\gamma] \rightarrow r:[(3,0),\beta] \quad \}. \tag{11}$$

The stable model semantics does not have to be considered, as the EVALPSN P is stratified. Since $[(1,0),\alpha] \preceq [(2,0),\alpha]$, from (9) and (10), the extended vector annotated literal

$$q:[(0,2),\beta] \tag{12}$$

is derived. Furthermore, since $[(0,2),\beta] \not\preceq [(0,2),\gamma]$, from (11) and (12), the vector annotated literal $r:[(3,0),\beta]$ is derived.

3 Frameworks for EVALPSN Based Intelligent Systems

In this section, two EVALPSN based intelligent system frameworks, an intelligent action control framework and an automated safety verification framework are provided.

3.1 Action Control Framework

Usually, an autonomous agent has more than two kinds of actions such as going forward, left turn and right turn, and has to decide the next action according to the input information and some regulations. An EVALPSN based framework for intelligent action control systems is presented.

The following information in a form of EVALPSN clauses are supposed to be input to the intelligent action control system:

- sensor values,
- anticipatory values for system output, and
- previous system output stored in the memory.

Basically, the EVALPSN defeasible deontic action control consists of the following two phases:

Phase I (Forbiddance Derivation);
according to the intelligent action control system input, some forbiddances from actions are derived and those forbiddances have levels of strength;

Phase II (Obligatory Action Derivation);
only the weakest forbiddance from an action derived in the previous phase is chosen and changed to permission, and the obligation to carry out the action is derived as the action control output.

These phases are formulated in EVALPSN based on some norms, such as action control policy, regulation, constraint and so on, for the intelligent action control. As the feedback step, the action control output in a form of EVALPSN clauses is stored into the system memory and may be feedbacked to the following action control process. This control flow is described in Figure 2. There seem to be various autonomous machines to which EVALPSN defeasible deontic action control can be applied. The following section introduces a beetle robot traveling through the maze with three ob-

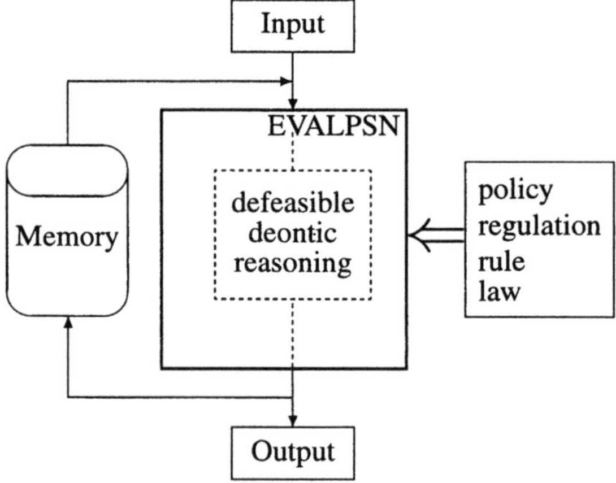

Figure 2. Framework for action control.

stacles, the wall of the maze, pitfalls and alcohol to the goal. He has three different kinds of sensors to detect those obstacles and three kinds of actions, going forward, left turn and right turn. This robot action control can be implemented as EVALPSN programming such that the inputs of the action control system are different kinds of sensor values and the output is an action which should be carried out. For example, suppose that a pitfall is detected in front of the robot and alcohol is detected in left front of it, and there is no way to avoid those obstacles, then, the following action control is carried out: the stronger forbiddance by a pitfall and the weaker one by alcohol are derived in the *Phase I*; then the weakerly forbidden action is chosen and the obligation to turn to the left is derived in the *Phase II*.

3.2 Safety Verification Framework

Safety verification is one of crucial issues, especially, in transportation control systems such as railway signal control, traffic light control, air traffic control and so on. An EVALPSN based intelligent framework for automated safety verification systems is presented. The following information in a form of EVALPSN clauses is supposed to be input to the EVALPSN based intelligent safety verification system:

1. detected information such as the state of railway interlocking or radar information for air traffic control, and
2. requests or instructions by system operators such as signal operators or air traffic controllers.

Generally, safety verification systems have safety properties to be kept and the safety properties are represented in an EVALPSN. This EVALPSN works as the inference engine of the safety verification system. Basically, the safety verification is carried out by verifying whether the system operators' requests or instructions contradict the safety properties or not based on EVALPSN programming as follows: suppose that the information 2 is input to the EVALPSN representing the safety properties as inquiry; if the answer *yes* is replied, there is no contradiction between the operators' request and the safety properties (Figure 3).

4 Intelligent Robot Action Control

In this section, EVALPSN based defeasible deontic action control is introduced with a virtual beetle robot example. The robot is supposed to travel through a maze with obstacles, the wall of the maze, pitfalls and alcohol to the goal. The robot has three different kinds of sensors and antennas to detect those obstacles and the goal of the maze. It can do three kinds of actions, going forward, left turn and right turn. The robot's actions are computed as the results of EVALPSN based defeasible deontic reasoning.

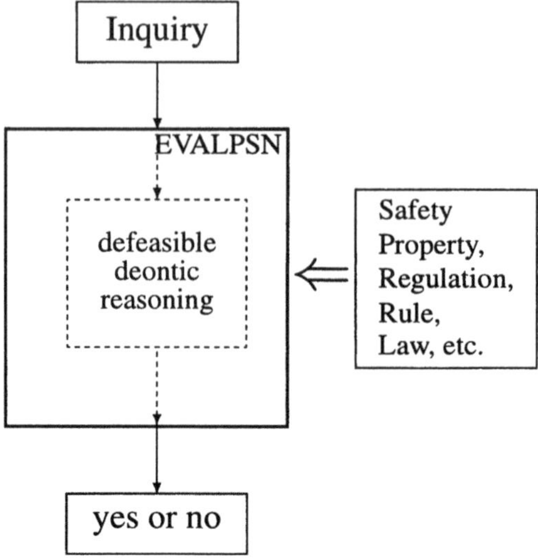

Figure 3. Framework for safety verification.

4.1 Beetle Robot Mr.A

The beetle robot is called Mr.A and assumed to be forbidden from drinking alcohol. Thus, alcohol can be regarded as an obstacle from him. He has four pairs (left and right) of sensors and antennas:

- ultra-sonic sensors represented by a predicate ob for detecting the wall of the maze;
- floor sensors represented by a predicate fl for detecting pitfalls;
- alcohol sensors represented by a predicate al for detecting alcohol; and
- antennas represented by a predicate an for detecting the goal of the maze.

We also assume that:

- the maze can be divided into many square blocks, the maze pass is 2 block wide, and each obstacle and Mr.A himself can be held in just one square block;
- Mr.A can do just three kinds of actions, going forward, left turn and right turn; and
- Mr.A's action is carried out at each time point of a discrete time sequence $\{ 0, 1, 2, \cdots \}$.

Sensor values are represented by vector annotations that indicate the locations of the obstacles. For example,

$$ob(t) : [(2, 2), \alpha] \tag{13}$$

expresses that it is a fact that the wall of the maze has been detected in front of Mr.A at the time t, and the first component of the vector annotation $(2, 2)$ represents the strength degree of the left ultra-sonic sensor value and the second component does that of the right one value. Each vector annotation (i, j) attached to the block border line in Figure 4 expresses that if an obstacle is detected in the block, the sensor value is (i, j). The sensory areas of those sensors are also described in Figure 4. The sensor values $\{\ (3, 0),\ (0, 3),\ (3, 3)\ \}$ out of the sensory areas are used only for the antennas.

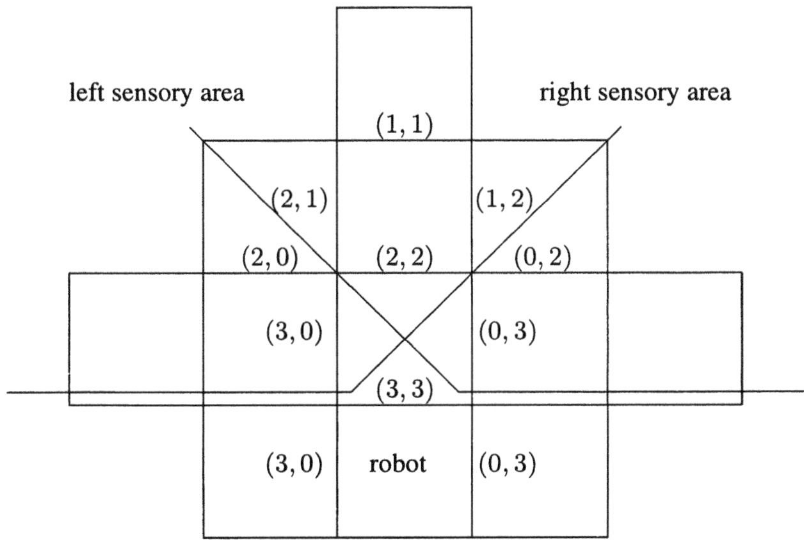

Figure 4. Sensory areas.

We use some abbreviations to express the sensor values in EVALPSN clauses shown in Table 1, which indicate the locations of the obstacles, where the symbols `l` and `r` are used only for the sensor values by the antennas.

Table 1. Abbreviations for vector annotations.

front (`f`) : $(3, 3)$	left (`l`) : $(3, 0)$
right (`r`) : $(0, 3)$	far front (`ff`) : $(2, 2)$
left front (`lf`) : $(3, 0)$	right front (`rf`) : $(0, 3)$
left far front (`lff`) : $(2, 1)$	right far front (`rff`) : $(1, 2)$
left left front (`llf`) : $(2, 0)$	right right front (`rrf`) : $(0, 2)$

The robot Mr.A has to decide whether avoiding or climbing over the obstacle when he detects it. We assume that Mr.A can not climb over the wall of the maze; he might jump over a pitfall in the maze, however, he does not want to do that, as he has the risk of injury; and he loves drinking alcohol, however, he is forbidden from drinking. He has the following policies and superiority relations between them:

Policies for Mr.A's Behavior
1. He must not crash the wall.
2. He should not fall into a pitfall.
3. He should advance with following the right.
4. It might be better not to drink alcohol.
***Superiority Relations*:** *4 < 3 < 2 < 1* .

The superiority relations say that policy *1* is superior to policy *2*, policy *2* is superior to policy *3*, and so on.

The intelligent robot action control based on the policies and the superiority relations is formalized in EVALPSN as shown later.

4.2 EVALPSN Formalization for Action Control

Basically, the intelligent action control system based on EVALPSN consists of two phases, *Forbiddance Derivation* and *Obligatory Action Derivation* as described before. For example, suppose that there are a pitfall in front of the robot Mr.A and two pubs (alcohol) in left and right front of him, and he has no way to avoid all those obstacles. Then, one stronger forbiddance by the pitfall and two weaker forbiddances are derived. Furthermore, he has to determine which action he should take based on his behavior policies. Since he has to obey the policy *3* "he should advance with following the right", then the weaker forbiddance from right turn is ignored and right turn is taken as his next obligatory action. Such defeasible deontic reasoning is formalized in EVALPSN. The strength levels (1 to 3) of forbiddance are determined based on the superiority relations between the policies. For example, the detection of the wall can be recognized as the strongest forbiddance, because Mr.A can not climb over it anyway, which is represented by an extended vector annotation $[(0,3),\beta]$.

Note 1. The sensor values of the antennas are not treated in the following EVALPSN formalization for simplicity.

4.2.1 Forbiddance Derivation

If it is a fact that an obstacle has been detected, one forbidden action is derived and the strength level of the forbiddance depends on the kind of the obstacle. Such forbiddance derivation can be formalized in EVALPSN. Mr.A's three actions, going forward, left turn and right turn, are represented by predicates, *forward, lturn* and *rturn*, respectively.

- If the wall is detected in far front (`ff`) of Mr.A, then, he is forbidden from going forward. This forbiddance is strongest (level 3), never defeated by other forbiddances and formulated as an EVALPSN clause,

$$ob(t) : [\mathtt{ff},\alpha] \rightarrow \mathit{forward}(t) : [(0,3),\beta], \tag{14}$$

  where the expression $ob(t) : [\mathtt{ff},\alpha]$ denotes that it is a fact that the wall of the maze has been detected in far front of Mr.A at the time t, and is transformed into an EVALPSN clause $ob(t) : [(2,0),\alpha] \wedge ob(t) : [(0,2),\alpha]$.

Note 2. Abbreviations such as ff and llf are used in EVALPSN clauses instead of vector annotations such as $(2,2)$ and $(2,1)$ when expressing sensor values for easy understanding.

- If the wall is detected in left front (lf) of Mr.A, then he is forbidden at the strength level 3 from turning to the left.

$$ob(t) : [\texttt{lf}, \alpha] \to lturn(t) : [(0,3), \beta]. \tag{15}$$

- If the wall is detected in right front (rf) of Mr.A, then he is forbidden at the strength level 3 from turning to the right.

$$ob(t) : [\texttt{rf}, \alpha] \to rturn(t) : [(0,3), \beta]. \tag{16}$$

- If a pitfall is detected in far front (ff) of Mr.A, then he is forbidden at the strength level 2 from going forward.

$$fl(t) : [\texttt{ff}, \alpha] \to forward(t) : [(0,2), \beta]. \tag{17}$$

- If a pitfall is detected in left front (lf) of Mr.A, then he is forbidden at the strength level 2 from turning to the left.

$$fl(t) : [\texttt{lf}, \alpha] \to lturn(t) : [(0,2), \beta]. \tag{18}$$

- If a pitfall is detected in right front (rf) of Mr.A, then he is forbidden at the strength level 2 from turning to the right.

$$fl(t) : [\texttt{rf}, \alpha] \to rturn(t) : [(0,2), \beta]. \tag{19}$$

- If alcohol is detected in far front (ff) of Mr.A, and other kinds of obstacles are detected in left front (lf) and right front (rf) of him, then he is forbidden at the strength level 1 from drinking alcohol (going forward). This case can be formulated in EVALPSN as follows:

$$al(t):[\texttt{ff},\alpha] \wedge ob(t):[\texttt{lf},\alpha] \wedge ob(t):[\texttt{rf},\alpha] \to forward(t):[(0,1),\beta], \tag{20}$$
$$al(t):[\texttt{ff},\alpha] \wedge ob(t):[\texttt{lf},\alpha] \wedge fl(t):[\texttt{rf},\alpha] \to forward(t):[(0,1),\beta], \tag{21}$$
$$al(t):[\texttt{ff},\alpha] \wedge ob(t):[\texttt{lf},\alpha] \wedge al(t):[\texttt{rf},\alpha] \to forward(t):[(0,1),\beta], \tag{22}$$
$$al(t):[\texttt{ff},\alpha] \wedge fl(t):[\texttt{lf},\alpha] \wedge ob(t):[\texttt{rf},\alpha] \to forward(t):[(0,1),\beta], \tag{23}$$
$$al(t):[\texttt{ff},\alpha] \wedge fl(t):[\texttt{lf},\alpha] \wedge fl(t):[\texttt{rf},\alpha] \to forward(t):[(0,1),\beta], \tag{24}$$
$$al(t):[\texttt{ff},\alpha] \wedge fl(t):[\texttt{lf},\alpha] \wedge al(t):[\texttt{rf},\alpha] \to forward(t):[(0,1),\beta], \tag{25}$$
$$al(t):[\texttt{ff},\alpha] \wedge al(t):[\texttt{lf},\alpha] \wedge ob(t):[\texttt{rf},\alpha] \to forward(t):[(0,1),\beta], \tag{26}$$
$$al(t):[\texttt{ff},\alpha] \wedge al(t):[\texttt{lf},\alpha] \wedge fl(t):[\texttt{rf},\alpha] \to forward(t):[(0,1),\beta], \tag{27}$$
$$al(t):[\texttt{ff},\alpha] \wedge al(t):[\texttt{lf},\alpha] \wedge al(t):[\texttt{rf},\alpha] \to forward(t):[(0,1),\beta], \tag{28}$$

- If alcohol is detected in left front (lf) of Mr.A, and other kinds of obstacles are detected in right front (rf) and far front (ff) of him, then he is forbidden at the strength level 1 from drinking alcohol (turning to the left). This case can be formulated in EVALPSN as follows:

$$al(t):[\mathtt{lf},\alpha] \wedge ob(t):[\mathtt{rf},\alpha] \wedge ob(t):[\mathtt{ff},\alpha] \rightarrow \mathit{lturn}(t):[(0,1),\beta], \quad (29)$$
$$al(t):[\mathtt{lf},\alpha] \wedge ob(t):[\mathtt{rf},\alpha] \wedge fl(t):[\mathtt{ff},\alpha] \rightarrow \mathit{lturn}(t):[(0,1),\beta], \quad (30)$$
$$al(t):[\mathtt{lf},\alpha] \wedge ob(t):[\mathtt{rf},\alpha] \wedge al(t):[\mathtt{ff},\alpha] \rightarrow \mathit{lturn}(t):[(0,1),\beta], \quad (31)$$
$$al(t):[\mathtt{lf},\alpha] \wedge fl(t):[\mathtt{rf},\alpha] \wedge ob(t):[\mathtt{ff},\alpha] \rightarrow \mathit{lturn}(t):[(0,1),\beta], \quad (32)$$
$$al(t):[\mathtt{lf},\alpha] \wedge fl(t):[\mathtt{rf},\alpha] \wedge fl(t):[\mathtt{ff},\alpha] \rightarrow \mathit{lturn}(t):[(0,1),\beta], \quad (33)$$
$$al(t):[\mathtt{lf},\alpha] \wedge fl(t):[\mathtt{rf},\alpha] \wedge al(t):[\mathtt{ff},\alpha] \rightarrow \mathit{lturn}(t):[(0,1),\beta], \quad (34)$$
$$al(t):[\mathtt{lf},\alpha] \wedge al(t):[\mathtt{rf},\alpha] \wedge ob(t):[\mathtt{ff},\alpha] \rightarrow \mathit{lturn}(t):[(0,1),\beta], \quad (35)$$
$$al(t):[\mathtt{lf},\alpha] \wedge al(t):[\mathtt{rf},\alpha] \wedge fl(t):[\mathtt{ff},\alpha] \rightarrow \mathit{lturn}(t):[(0,1),\beta], \quad (36)$$
$$al(t):[\mathtt{lf},\alpha] \wedge al(t):[\mathtt{rf},\alpha] \wedge al(t):[\mathtt{ff},\alpha] \rightarrow \mathit{lturn}(t):[(0,1),\beta]. \quad (37)$$

- If alcohol is detected in right front (rf) of Mr.A, and other kinds of obstacles are detected in far front (ff) and left front (ff) of him, then he is forbidden at the strength level 1 from drinking alcohol (turning to the right). This case can be formulated in EVALPSN as follows:

$$al(t):[\mathtt{rf},\alpha] \wedge ob(t):[\mathtt{ff},\alpha] \wedge ob(t):[\mathtt{lf},\alpha] \rightarrow \mathit{rturn}(t):[(0,1),\beta], \quad (38)$$
$$al(t):[\mathtt{rf},\alpha] \wedge ob(t):[\mathtt{ff},\alpha] \wedge fl(t):[\mathtt{lf},\alpha] \rightarrow \mathit{rturn}(t):[(0,1),\beta], \quad (39)$$
$$al(t):[\mathtt{rf},\alpha] \wedge ob(t):[\mathtt{ff},\alpha] \wedge al(t):[\mathtt{lf},\alpha] \rightarrow \mathit{rturn}(t):[(0,1),\beta], \quad (40)$$
$$al(t):[\mathtt{rf},\alpha] \wedge fl(t):[\mathtt{ff},\alpha] \wedge ob(t):[\mathtt{lf},\alpha] \rightarrow \mathit{rturn}(t):[(0,1),\beta], \quad (41)$$
$$al(t):[\mathtt{rf},\alpha] \wedge fl(t):[\mathtt{ff},\alpha] \wedge fl(t):[\mathtt{lf},\alpha] \rightarrow \mathit{rturn}(t):[(0,1),\beta], \quad (42)$$
$$al(t):[\mathtt{rf},\alpha] \wedge fl(t):[\mathtt{ff},\alpha] \wedge al(t):[\mathtt{lf},\alpha] \rightarrow \mathit{rturn}(t):[(0,1),\beta], \quad (43)$$
$$al(t):[\mathtt{rf},\alpha] \wedge al(t):[\mathtt{ff},\alpha] \wedge ob(t):[\mathtt{lf},\alpha] \rightarrow \mathit{rturn}(t):[(0,1),\beta], \quad (44)$$
$$al(t):[\mathtt{rf},\alpha] \wedge al(t):[\mathtt{ff},\alpha] \wedge fl(t):[\mathtt{lf},\alpha] \rightarrow \mathit{rturn}(t):[(0,1),\beta], \quad (45)$$
$$al(t):[\mathtt{rf},\alpha] \wedge al(t):[\mathtt{ff},\alpha] \wedge al(t):[\mathtt{lf},\alpha] \rightarrow \mathit{rturn}(t):[(0,1),\beta]. \quad (46)$$

- If alcohol is detected in far front (ff), left front (lf) or right front (rf) of Mr.A, and there is at least one direction with no obstacle, then he is permitted to advance toward the direction and forbidden at the strength level 2 not 1 from drinking alcohol. In this case, the strength level of abstinence is upgraded, as there is at least one direction with no obstacle. This case can be formulated in EVALPSN as follows:

$$al(t):[\mathtt{ff},\alpha] \wedge \sim al(t):[\mathtt{lf},\alpha] \wedge \sim fl(t):[\mathtt{lf},\alpha] \wedge \sim ob(t):[\mathtt{lf},\alpha] \rightarrow \mathit{forward}(t):[(0,2),\beta], \quad (47)$$
$$al(t):[\mathtt{ff},\alpha] \wedge \sim al(t):[\mathtt{rf},\alpha] \wedge \sim fl(t):[\mathtt{rf},\alpha] \wedge \sim ob(t):[\mathtt{rf},\alpha] \rightarrow \mathit{forward}(t):[(0,2),\beta], \quad (48)$$
$$al(t):[\mathtt{lf},\alpha] \wedge \sim al(t):[\mathtt{ff},\alpha] \wedge \sim fl(t):[\mathtt{ff},\alpha] \wedge \sim ob(t):[\mathtt{ff},\alpha] \rightarrow \mathit{lturn}(t):[(0,2),\beta], \quad (49)$$

$$al(t):[\texttt{lf},\alpha] \wedge \sim al(t):[\texttt{rf},\alpha] \wedge \sim fl(t):[\texttt{rf},\alpha] \wedge \sim ob(t):[\texttt{rf},\alpha] \rightarrow \mathit{lturn}(t):[(0,2),\beta], \quad (50)$$

$$al(t):[\texttt{rf},\alpha] \wedge \sim al(t):[\texttt{ff},\alpha] \wedge \sim fl(t):[\texttt{ff},\alpha] \wedge \sim ob(t):[\texttt{ff},\alpha] \rightarrow \mathit{rturn}(t):[(0,2),\beta], \quad (51)$$

$$al(t):[\texttt{rf},\alpha] \wedge \sim al(t):[\texttt{lf},\alpha] \wedge \sim fl(t):[\texttt{lf},\alpha] \wedge \sim ob(t):[\texttt{lf},\alpha] \rightarrow \mathit{rturn}(t):[(0,2),\beta]. \quad (52)$$

4.2.2 Obligatory Action Derivation

If there are some forbidden actions, only one action should be chosen among those ones based on Mr.A's behavior policies. We formalize such defeasible reasoning in EVALPSN. If Mr.A travels through the maze with following the right, then there are superiority relations between his actions,

$$\text{left turn} < \text{going forward} < \text{right turn},$$

between his actions by policy *3*. Taking the superiority relation into account, we obtain the following EVALPSN to derive the next action from the forbidden actions.

- If left turn is not forbidden, and both going forward and right turn are forbidden, then Mr.A should turn to the left. This derivation is formulated in EVALPSN as follows:

$$\sim \mathit{lturn}(t):[(0,2),\beta] \wedge \mathit{forward}(t):[(0,2),\beta] \wedge \mathit{rturn}(t):[(0,2),\beta] \rightarrow \mathit{lturn}(t+1):[(3,0),\beta]. \quad (53)$$

- If going forward is not forbidden, and both left turn and right turn are forbidden, then Mr.A should go forward. This derivation is formulated in EVALPSN as follows:

$$\mathit{lturn}(t):[(0,2),\beta] \wedge \sim \mathit{forward}(t):[(0,2),\beta] \wedge \mathit{rturn}(t):[(0,2),\beta] \rightarrow \mathit{forward}(t+1):[(3,0),\beta]. \quad (54)$$

- If right turn is not forbidden, and both going forward and right turn are forbidden, then Mr.A should turn to the right. This derivation is formulated in EVALPSN as follows:

$$\mathit{lturn}(t):[(0,2),\beta] \wedge \mathit{forward}(t):[(0,2),\beta] \wedge \sim \mathit{rturn}(t):[(0,2),\beta] \rightarrow \mathit{rturn}(t+1):[(3,0),\beta]. \quad (55)$$

- If neither left turn nor going forward is forbidden, and right turn is forbidden, then Mr.A should go forward. This derivation is formulated in EVALPSN as follows:

$$\sim \mathit{lturn}(t):[(0,2),\beta] \wedge \sim \mathit{forward}(t):[(0,2),\beta] \wedge \mathit{rturn}(t):[(0,2),\beta] \rightarrow \mathit{forward}(t+1):[(3,0),\beta]. \quad (56)$$

- If neither going forward nor right turn is forbidden, and left turn is forbidden, then Mr.A should turn to the right. This derivation is formulated in EVALPSN as follows:

$$\begin{aligned} &\mathit{lturn}(t):[(0,2),\beta]\wedge \sim \mathit{forward}(t):[(0,2),\beta]\wedge \sim \mathit{rturn}(t):[(0,2),\beta] \\ &\qquad\qquad\qquad\qquad\qquad\qquad \rightarrow \mathit{rturn}(t+1):[(3,0),\beta]. \end{aligned} \quad (57)$$

- If neither left turn nor right turn is forbidden, and going forward is forbidden, then Mr.A should turn to the right. This derivation is formulated in EVALPSN as follows:

$$\begin{aligned} &\sim \mathit{lturn}(t):[(0,2),\beta] \wedge \mathit{forward}(t):[(0,2),\beta]\wedge \sim \mathit{rturn}(t):[(0,2),\beta] \\ &\qquad\qquad\qquad\qquad\qquad\qquad \rightarrow \mathit{rturn}(t+1):[(3,0),\beta]. \end{aligned} \quad (58)$$

- If there is no forbidden action, then Mr.A should turn to the right. This derivation is formulated in EVALPSN as follows:

$$\begin{aligned} &\sim \mathit{lturn}(t):[(0,2),\beta]\wedge \sim \mathit{forward}(t):[(0,2),\beta]\wedge \sim \mathit{rturn}(t):[(0,2),\beta] \\ &\qquad\qquad\qquad\qquad\qquad\qquad \rightarrow \mathit{rturn}(t+1):[(3,0),\beta]. \end{aligned} \quad (59)$$

The EVALPSN $P_B = \{\ (15), \cdots, (59)\ \}$ provides the basic action control for Mr.A. A simple example for Mr.A's action control is presented.

Example 2. Suppose that there are a pub (alcohol) in left front, a pitfall in far front and the wall in right front of Mr.A, respectively. First of all, sensor values are input to the EVALPSN P_B as the following EVALPSN clauses:

$$al(0) : [\mathtt{lf}, \alpha], \quad (60)$$
$$fl(0) : [\mathtt{ff}, \alpha], \quad (61)$$
$$ob(0) : [\mathtt{rf}, \alpha]. \quad (62)$$

Then, three forbiddances,

$$\mathit{lturn}(0) : [(0,1),\beta], \quad (63)$$
$$\mathit{forward}(0) : [(0,2),\beta], \quad (64)$$
$$\mathit{rturn}(0) : [(0,3),\beta], \quad (65)$$

are derived by the EVALPSN clauses $\{(16),(17),(30),(60),(61),(62)\}$. The weakest forbiddance (63) is chosen and the obligation to turn to the left,

$$\mathit{lturn}(1) : [(3,0),\beta], \quad (66)$$

is derived by the EVALPSN clause (53).

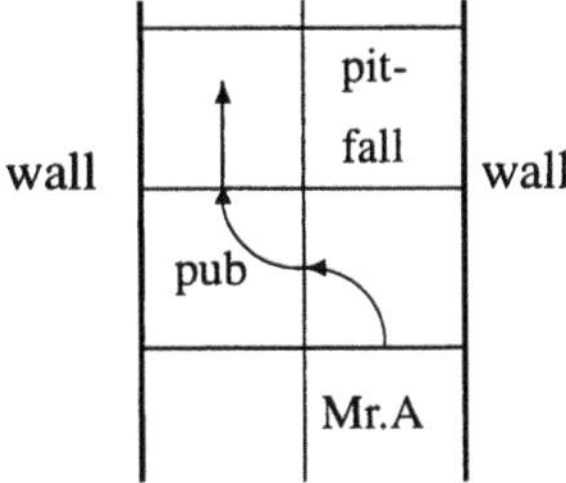

Figure 5. Example for robot action control.

5 Intelligent Safety Verification for Railway Interlocking

In this section, an automated safety verification system for railway interlocking based on EVALPSN programming is introduced.

Safety Verification for railway interlocking is a crucial issue to avoid railway accidents. One logical verification method for railway interlocking safety was introduced in Morley's Ph.D. thesis [10] with some examples of British Railways. In his thesis, an automated safety verification system based on a higher order logic language called HOL [7] is proposed. Our safety verification system is based on EVALPSN programming and does not have the incompleteness problem due to higher order logic.

The EVALPSN based safety verification is carried out as the following three steps:

1. the safety properties, which is proposed in [10] and must be kept when interlocking, are translated into an EVALPSN;
2. requests that should be verified and issued by signal operators, route security requests called Panel Route Requests and route release requests called Sub-Route Release, are translated into EVALPSN clauses;
3. since all EVALPSN clauses in 1 and 2 contain no strong negation, the EVALPSN clauses in 2 are inquired from the EVALPSN in 1 as usual logic programming, then if *yes* is returned, the request is assured, otherwise, not assured.

The details of these steps will be shown later.

5.1 Basic Terminology in GLD

The basic terminologies given in [10] are represented in EVALPSN. In Morley [10], Geographic Data Language (GDL) in which the interlocking functions are encoded and its semantics are introduced.

First of all, Geographic Data Language is reviewed with a concrete example of signaling schema (Figure 6) from [10]. The physical entities declared in the network are:

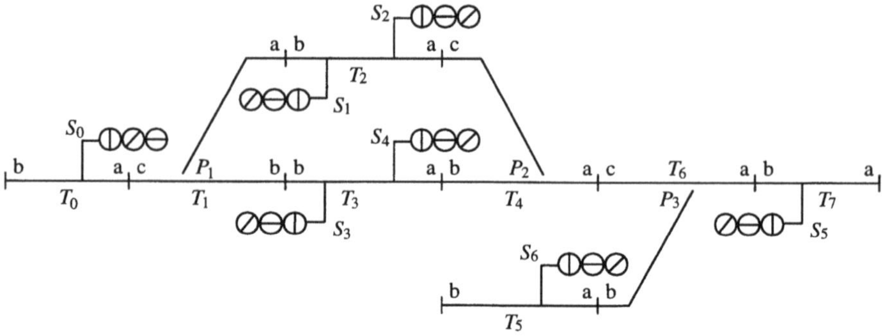

Figure 6. Signaling schema for WEST.

- track sections $\mathcal{T} = \{T_0, T_1, \ldots, T_7\}$,
- points $\mathcal{P} = \{P_1, P_2, P_3\}$,
- signals $\mathcal{S} = \{S_0, S_1, \ldots, S_6\}$,

and the logical control entities are:

- routes $\mathcal{R} = \{R_{02}, R_{04}, R_1, R_2, R_3, R_4, R_5, R_{51}, R_{53}, R_6\}$,
- sub-routes $\mathcal{U} = \{T_0^{ab}, T_0^{ba}, T_1^{ca}, \ldots, T_7^{ba}\}$.

For example, the sub-route T_0^{ab} denotes the railway from a to b in the track section T_0, the route R_{02} denotes the route from the signal S_0 to the signal S_2, and the route R_{02} consists of the sub-routes T_0^{ba}, T_1^{ca} and T_2^{ba}. Each entity has logical or physical states.

Sub-route has two states *locked* (`l`) and *free* (`f`).

"The sub-route is locked" means that the sub-route is supposed to be occupied by a train and "free" means unlocked. eg. T_0^{ba} `l` denotes the sub-route T_0^{ba} is scheduled to be occupied by a train.

Route has two states *set* (`s`) and *unset* (`xs`).

"The route is set" means that all sub-routes in the route are locked and "unset" means not set.

Track section has two states *occupied* (`o`) and *cleared* (`c`).

"The track section is occupied" means that the train is passing through the track section and "cleared" means that the train has already passed through the track section.

Point has four states:

controlled normal (`cn`) eg. P_1 `cn` denotes the point P_1 is controlled normal direction (ca or ac directions in the track section T_1);

controlled reverse (`cr`) eg. P_1 `cr` denotes the point P_1 is controlled reverse direction (cb or bc directions in the track section T_1);

controlled normal or *free to move* (`cnf`) eg. P_1 `cnf` denotes the point P_1 is controlled normal, or if it is not, P_1 can be moved to reverse position if the normal sub-routes are free; and *controlled reverse* or *free to move* (crf) eg. P_1 crf denotes the point P_1 is controlled reverse, or if it is not, P_1 can be moved to normal position if the reverse sub-routes are free.

The interlocking safety verification is carried out by checking whether route interlocking requests called a Panel Route Request or a route release request called a Sub-Route Release contradict the safety properties for interlocking or not. The set $\mathcal{Q}_{PRR}$ of Panel Route Requests is declared:

$$\mathcal{Q}_{PRR} = \{\mathtt{Q02}, \mathtt{Q04}, \ldots, \mathtt{Q6}, \ldots\}. \tag{67}$$

Panel Route Request is a request to secure the route issued by signal operators. For example,

$$\begin{aligned} \mathtt{Q02} \quad \text{if} \quad & P_1\ \mathtt{crf},\ T_1^{ac}\ \mathtt{f},\ T_2^{ab}\ \mathtt{f} \\ \text{then}\ & R_{02}\ \mathtt{s},\ P_1\ \mathtt{cr},\ T_1^{ca}\ \mathtt{l},\ T_2^{ba}\ \mathtt{l} \end{aligned} \tag{68}$$

is the Panel Route Request for the route R_{02} from the signal S_0 to the signal S_2. The set $\mathcal{Q}_{SRR}$ of Sub-Route Releases is also declared:

$$\mathcal{Q}_{SRR} = \{\mathtt{SR02}, \mathtt{SR04}, \ldots, \mathtt{SR6}, \ldots\}. \tag{69}$$

Sub-Route Release is a request to release all sub-routes included in the route that has been set and cleared. For example,

$$\begin{aligned} \mathtt{SR02} \quad \text{if} \quad & R_{02}\ \mathtt{xs},\ T_1\ \mathtt{c},\ T_2\ \mathtt{c} \\ \text{then}\ & T_1^{ca}\ \mathtt{f},\ T_2^{ba}\ \mathtt{f} \end{aligned} \tag{70}$$

is the Sub-Route Release request for the route R_{02}. The details of the safety verification for the requests SR02 and Q02 will be described as examples later.

5.2 Safety Properties in EVALPSN

The safety properties *MX, RT, PT* are represented in EVALPSN. The interlocking safety verification is carried out by checking the contradiction with the safety properties that consist of the following conditions:

MX It is never the case that two or more of the sub-routes over a given track section are simultaneously locked;

RT Whenever a route is set, all its component sub-routes are locked;

PT Whenever a sub-route over a track section containing points is locked, the points are controlled in alignment with that sub-route.

The above safety properties *MX, RT, PT* are represented in EVALPSN, then the symbols $\{\mathtt{l}, \mathtt{f}, \mathtt{s}, \mathtt{xs}, \mathtt{cn}, \mathtt{cnf}, \mathtt{cr}, \mathtt{crf}, \mathtt{o}, \mathtt{c}\}$ that represent the states of entities are used as the first components in extended vector annotations instead of usual vector annotations (2-dimensional vectors). The following mappings (the epistemic negations) between those annotations are defined:

$$\neg_1([\mathtt{l},\mu]) = [\mathtt{f},\mu], \qquad \neg_1([\mathtt{f},\mu]) = [\mathtt{l},\mu], \tag{71}$$
$$\neg_1([\mathtt{s},\mu]) = [\mathtt{xs},\mu], \qquad \neg_1([\mathtt{xs},\mu]) = [\mathtt{s},\mu], \tag{72}$$
$$\neg_1([\mathtt{cn},\mu]) = [\mathtt{cr},\mu], \qquad \neg_1([\mathtt{cr},\mu]) = [\mathtt{cn},\mu], \tag{73}$$
$$\neg_1([\mathtt{cnf},\mu]) = [\mathtt{crf},\mu], \qquad \neg_1([\mathtt{crf},\mu]) = [\mathtt{cnf},\mu], \tag{74}$$
$$\neg_1([\mathtt{o},\mu]) = [\mathtt{c},\mu], \qquad \neg_1([\mathtt{c},\mu]) = [\mathtt{o},\mu], \tag{75}$$

where $\mu \in \{\alpha,\beta,\gamma\}$. For example, an EVALPSN clause

$$T(0,ab):[\mathtt{f},\alpha] \to T(0,ba):[\mathtt{f},\gamma] \tag{76}$$

is intuitively interpreted as "if it is a fact that the sub-route T_0^{ab} is free, then the sub-route T_0^{ba} is permitted to be locked".

First the safety property *MX* for sub-routes are translated into EVALPSN clauses.

- *MX*

  Generally, the safety property *MX* represents that it is forbidden that two or more of the sub-routes over a given track section are simultaneously locked.

  The condition

$$MX[T_0^{ab},T_0^{ba}]$$

  can be interpreted as "if one of the sub-routes T_0^{ab}, T_0^{ba} is free, the other sub-route is permitted to be locked", which is translated into the following EVALPSN clauses:

$$T(0,ab):[\mathtt{f},\alpha] \to T(0,ba):[\mathtt{f},\gamma], \tag{77}$$
$$T(0,ba):[\mathtt{f},\alpha] \to T(0,ab):[\mathtt{f},\gamma]. \tag{78}$$

  Similarly, the conditions

$$MX[T_2^{ab},T_2^{ba}],$$
$$MX[T_3^{ab},T_3^{ba}],$$
$$MX[T_5^{ab},T_5^{ba}] \quad \text{and}$$
$$MX[T_7^{ab},T_7^{ba}]$$

  are also translated into the following EVALPSN clauses:

$$T(2,ab):[\mathtt{f},\alpha] \to T(2,ba):[\mathtt{f},\gamma], \tag{79}$$
$$T(2,ba):[\mathtt{f},\alpha] \to T(2,ab):[\mathtt{f},\gamma], \tag{80}$$
$$T(3,ab):[\mathtt{f},\alpha] \to T(3,ba):[\mathtt{f},\gamma], \tag{81}$$
$$T(3,ba):[\mathtt{f},\alpha] \to T(3,ab):[\mathtt{f},\gamma], \tag{82}$$
$$T(5,ab):[\mathtt{f},\alpha] \to T(5,ba):[\mathtt{f},\gamma], \tag{83}$$
$$T(5,ba):[\mathtt{f},\alpha] \to T(5,ab):[\mathtt{f},\gamma], \tag{84}$$
$$T(7,ab):[\mathtt{f},\alpha] \to T(7,ba):[\mathtt{f},\gamma], \tag{85}$$
$$T(7,ba):[\mathtt{f},\alpha] \to T(7,ab):[\mathtt{f},\gamma]. \tag{86}$$

The track section T_1 contains the point P_1, and the condition

$$MX[T_1^{ac}, T_1^{ca}, T_1^{bc}, T_1^{cb}]$$

can be interpreted as "if one of the normal (resp. reverse) side sub-routes T_1^{bc}, T_1^{cb} (T_1^{ac}, T_1^{ca}) is free and the point P_1 is permitted to be controlled normal (resp. reverse), the rest of the normal (resp. reverse) side sub-routes is permitted to be locked" so that the safety property *PT* can be considered. Therefore, the condition is translated into the following EVALPSN clauses:

$$T(1,cb):[\mathtt{f},\alpha] \wedge P(1):[\mathtt{cr},\gamma] \rightarrow T(1,bc):[\mathtt{f},\gamma], \tag{87}$$
$$T(1,bc):[\mathtt{f},\alpha] \wedge P(1):[\mathtt{cr},\gamma] \rightarrow T(1,cb):[\mathtt{f},\gamma], \tag{88}$$
$$T(1,ca):[\mathtt{f},\alpha] \wedge P(1):[\mathtt{cn},\gamma] \rightarrow T(1,ac):[\mathtt{f},\gamma], \tag{89}$$
$$T(1,ac):[\mathtt{f},\alpha] \wedge P(1):[\mathtt{cn},\gamma] \rightarrow T(1,ca):[\mathtt{f},\gamma]. \tag{90}$$

Similarly, the conditions

$$MX[T_4^{ac}, T_4^{ca}, T_4^{ab}, T_4^{ba}] \quad \text{and}$$
$$MX[T_6^{ac}, T_6^{ca}, T_6^{ab}, T_6^{ba}]$$

are also translated into the following EVALPSN clauses:

$$T(4,ba):[\mathtt{f},\alpha] \wedge P(2):[\mathtt{cr},\gamma] \rightarrow T(4,ab):[\mathtt{f},\gamma], \tag{91}$$
$$T(4,ab):[\mathtt{f},\alpha] \wedge P(2):[\mathtt{cr},\gamma] \rightarrow T(4,ba):[\mathtt{f},\gamma], \tag{92}$$
$$T(4,ca):[\mathtt{f},\alpha] \wedge P(2):[\mathtt{cn},\gamma] \rightarrow T(4,ac):[\mathtt{f},\gamma], \tag{93}$$
$$T(4,ac):[\mathtt{f},\alpha] \wedge P(2):[\mathtt{cn},\gamma] \rightarrow T(4,ca):[\mathtt{f},\gamma], \tag{94}$$
$$T(6,ca):[\mathtt{f},\alpha] \wedge P(3):[\mathtt{cr},\gamma] \rightarrow T(6,ac):[\mathtt{f},\gamma], \tag{95}$$
$$T(6,ac):[\mathtt{f},\alpha] \wedge P(3):[\mathtt{cr},\gamma] \rightarrow T(6,ca):[\mathtt{f},\gamma], \tag{96}$$
$$T(6,ba):[\mathtt{f},\alpha] \wedge P(3):[\mathtt{cn},\gamma] \rightarrow T(6,ab):[\mathtt{f},\gamma], \tag{97}$$
$$T(6,ab):[\mathtt{f},\alpha] \wedge P(3):[\mathtt{cn},\gamma] \rightarrow T(6,ba):[\mathtt{f},\gamma]. \tag{98}$$

Next, the safety property *RT* for routes are translated into EVALPSN clauses.

- *RT*
  The safety property *RT* represents that if all the sub-routes contained in one route are permitted to be locked, the route is permitted to be set.
  The condition
  $$RT(R_{02}, [T_1^{ca}, T_2^{ba}])$$
  can be interpreted as "if both the sub-routes T_1^{ca} and T_2^{ba} are permitted to be locked, the route R_{02} is permitted to be set", which is translated into the following EVALPSN clause:
  $$T(1,ca):[\mathtt{f},\gamma] \wedge T(2,ba):[\mathtt{f},\gamma] \rightarrow R(02):[\mathtt{xs},\gamma]. \tag{99}$$

Although there are many routes in the network, only the routes R_1, R_3, R_{51}, R_{53} are translated into EVALPSN clauses as examples. The conditions

$$RT(R_1, [T_1^{ac}, T_0^{ab}]),$$
$$RT(R_3, [T_1^{bc}, T_0^{ab}]),$$
$$RT(R_{51}, [T_6^{ac}, T_4^{ac}, T_2^{ab}]) \text{ and}$$
$$RT(R_{53}, [T_6^{ac}, T_4^{ab}, T_3^{ab}])$$

are translated into the following EVALPSN clauses:

$$T(1, ac):[\mathtt{f}, \gamma] \wedge T(0, ab):[\mathtt{f}, \gamma] \rightarrow R(1):[\mathtt{xs}, \gamma], \quad (100)$$
$$T(1, bc):[\mathtt{f}, \gamma] \wedge T(0, ab):[\mathtt{f}, \gamma] \rightarrow R(3):[\mathtt{xs}, \gamma], \quad (101)$$
$$T(6, ac):[\mathtt{f}, \gamma] \wedge T(4, ac):[\mathtt{f}, \gamma] \wedge T(2, ab):[\mathtt{f}, \gamma] \rightarrow R(51):[\mathtt{xs}, \gamma], \quad (102)$$
$$T(6, ac):[\mathtt{f}, \gamma] \wedge T(4, ab):[\mathtt{f}, \gamma] \wedge T(3, ab):[\mathtt{f}, \gamma] \rightarrow R(53):[\mathtt{xs}, \gamma]. \quad (103)$$

Last, the safety property *PT* for points are translated into EVALPSN clauses.

- *PT*
  The safety property *PT* represents the relation between point control and sub-route interlocking that should be kept.
  The conditions

$$PT\ \mathtt{cn}(P_1, [T_1^{bc}, T_1^{cb}]) \text{ and}$$
$$PT\ \mathtt{cr}(P_1, [T_1^{ac}, T_1^{ca}])$$

can be interpreted as "if one of the normal (resp. reverse) side sub-routes T_1^{bc}, T_1^{cb} (T_1^{ac}, T_1^{ca}) is free and P_1 is controlled normal (resp. reverse) or free to move, then the point P_1 is permitted to be controlled normal (resp. reverse)", which are translated into the following EVALPSN clauses:

$$T(1, bc):[\mathtt{f}, \alpha] \wedge P(1):[\mathtt{cnf}, \alpha] \rightarrow P(1):[\mathtt{cr}, \gamma], \quad (104)$$
$$T(1, cb):[\mathtt{f}, \alpha] \wedge P(1):[\mathtt{cnf}, \alpha] \rightarrow P(1):[\mathtt{cr}, \gamma], \quad (105)$$
$$T(1, ac):[\mathtt{f}, \alpha] \wedge P(1):[\mathtt{crf}, \alpha] \rightarrow P(1):[\mathtt{cn}, \gamma], \quad (106)$$
$$T(1, ca):[\mathtt{f}, \alpha] \wedge P(1):[\mathtt{crf}, \alpha] \rightarrow P(1):[\mathtt{cn}, \gamma]. \quad (107)$$

The conditions

$$PT\ \mathtt{cn}(P_2, [T_4^{ab}, T_4^{ba}]),$$
$$PT\ \mathtt{cr}(P_2, [T_4^{ac}, T_4^{ca}]),$$
$$PT\ \mathtt{cn}(P_3, [T_6^{ac}, T_6^{ca}]) \text{ and}$$
$$PT\ \mathtt{cr}(P_3, [T_6^{ab}, T_6^{ba}])$$

for the points P_2 and P_3 are also translated into the following EVALPSN clauses:

$$T(4,ab):[\mathtt{f},\alpha] \wedge P(2):[\mathtt{cnf},\alpha] \rightarrow P(2):[\mathtt{cr},\gamma], \quad (108)$$
$$T(4,ba):[\mathtt{f},\alpha] \wedge P(2):[\mathtt{cnf},\alpha] \rightarrow P(2):[\mathtt{cr},\gamma], \quad (109)$$
$$T(4,ac):[\mathtt{f},\alpha] \wedge P(2):[\mathtt{crf},\alpha] \rightarrow P(2):[\mathtt{cn},\gamma], \quad (110)$$
$$T(4,ca):[\mathtt{f},\alpha] \wedge P(2):[\mathtt{crf},\alpha] \rightarrow P(2):[\mathtt{cn},\gamma], \quad (111)$$
$$T(6,ac):[\mathtt{f},\alpha] \wedge P(3):[\mathtt{cnf},\alpha] \rightarrow P(3):[\mathtt{cr},\gamma], \quad (112)$$
$$T(6,ca):[\mathtt{f},\alpha] \wedge P(3):[\mathtt{cnf},\alpha] \rightarrow P(3):[\mathtt{cr},\gamma], \quad (113)$$
$$T(6,ab):[\mathtt{f},\alpha] \wedge P(3):[\mathtt{crf},\alpha] \rightarrow P(3):[\mathtt{cn},\gamma], \quad (114)$$
$$T(6,ba):[\mathtt{f},\alpha] \wedge P(3):[\mathtt{crf},\alpha] \rightarrow P(3):[\mathtt{cn},\gamma], \quad (115)$$

Here we consider Sub-Route Release in terms of the route unset conditions for routes. Sub-Route Release is issued sequentially along to the route to be unset. For example, suppose that the sub-route T_0^{ab} is being released. If the sub-routes T_1^{ac} or T_1^{bc} are permitted to be free and the track section T_0 is cleared, then T_0^{ab} is permitted to be free. Moreover, suppose that some different routes have sub-routes in common. If all the routes must be unset and the track section is cleared, then the common sub-routes is permitted to be free, or if all the precedent sub-routes are permitted to be free and the track section is cleared, then the sub-route is permitted to be free. For example, if both the routes R_{53} and R_{51} are unset, and the track section T_6 is cleared, the sub-route T_6^{ac} is permitted to be free. We need the following EVALPSN to represent such sub-route release information:

$$T(1,ac):[\mathtt{1},\gamma] \wedge T(1,bc):[\mathtt{1},\gamma] \wedge T(1):[\mathtt{c},\alpha] \rightarrow T(0,ab):[\mathtt{1},\gamma], \quad (116)$$
$$T(6,ab):[\mathtt{1},\gamma] \wedge T(5):[\mathtt{c},\alpha] \rightarrow T(5,ab):[\mathtt{1},\gamma], \quad (117)$$
$$T(4,ab):[\mathtt{1},\gamma] \wedge T(3):[\mathtt{c},\alpha] \rightarrow T(3,ab):[\mathtt{1},\gamma], \quad (118)$$
$$T(1,cb):[\mathtt{1},\gamma] \wedge T(3):[\mathtt{c},\alpha] \rightarrow T(3,ba):[\mathtt{1},\gamma], \quad (119)$$
$$T(4,ac):[\mathtt{1},\gamma] \wedge T(2):[\mathtt{c},\alpha] \rightarrow T(2,ab):[\mathtt{1},\gamma], \quad (120)$$
$$T(1,ca):[\mathtt{1},\gamma] \wedge T(2):[\mathtt{c},\alpha] \rightarrow T(2,ba):[\mathtt{1},\gamma], \quad (121)$$
$$R(1):[\mathtt{xs},\alpha] \wedge T(1):[\mathtt{c},\alpha] \rightarrow T(1,ac):[\mathtt{1},\gamma], \quad (122)$$
$$R(02):[\mathtt{xs},\alpha] \wedge T(1):[\mathtt{c},\alpha] \rightarrow T(1,ca):[\mathtt{1},\gamma], \quad (123)$$
$$R(04):[\mathtt{xs},\alpha] \wedge T(1):[\mathtt{c},\alpha] \rightarrow T(1,cb):[\mathtt{1},\gamma], \quad (124)$$
$$R(3):[\mathtt{xs},\alpha] \wedge T(1):[\mathtt{c},\alpha] \rightarrow T(1,bc):[\mathtt{1},\gamma], \quad (125)$$
$$T(6,ac):[\mathtt{1},\gamma] \wedge T(4):[\mathtt{c},\alpha] \rightarrow T(4,ab):[\mathtt{1},\gamma], \quad (126)$$
$$R(4):[\mathtt{xs},\alpha] \wedge T(4):[\mathtt{c},\alpha] \rightarrow T(4,ba):[\mathtt{1},\gamma], \quad (127)$$
$$T(6,ac):[\mathtt{1},\gamma] \wedge T(4):[\mathtt{c},\alpha] \rightarrow T(4,ac):[\mathtt{1},\gamma], \quad (128)$$
$$R(2):[\mathtt{xs},\alpha] \wedge T(4):[\mathtt{c},\alpha] \rightarrow T(4,ca):[\mathtt{1},\gamma], \quad (129)$$
$$R(5):[\mathtt{xs},\alpha] \wedge T(6):[\mathtt{c},\alpha] \rightarrow T(6,ab):[\mathtt{1},\gamma], \quad (130)$$
$$R(6):[\mathtt{xs},\alpha] \wedge T(6):[\mathtt{c},\alpha] \rightarrow T(6,ba):[\mathtt{1},\gamma], \quad (131)$$
$$R(53):[\mathtt{xs},\alpha] \wedge R(51):[\mathtt{xs},\alpha] \wedge T(6):[\mathtt{c},\alpha] \rightarrow T(6,ac):[\mathtt{1},\gamma], \quad (132)$$
$$T(4,ca):[\mathtt{1},\gamma] \wedge T(4,ba):[\mathtt{1},\gamma] \wedge T(6):[\mathtt{c},\alpha] \rightarrow T(6,ca):[\mathtt{1},\gamma], \quad (133)$$

$$T(6, ca):[1,\gamma] \wedge T(6, ba):[1,\gamma] \wedge T(7):[\mathrm{c},\alpha] \rightarrow T(7, ba):[1,\gamma]. \tag{134}$$

The EVALPSN provided in this section does not contain the strong negation. Actually, there is no strong negation in the EVALPSN representing the safety properties and the sub-route release information. Therefore, the EVALPSN based safety verification can be treated as well as usual logic programs.

5.3 Safety Verification Examples

In this subsection, some examples of safety verification for Panel Route Request and Sub-Route Release are presented. The basic idea to verify the interlocking safety based on EVALPSN is:

- since those safety properties can be regarded as regulations that imply deontic notions, obligation, forbiddance and permission, they are interpreted deontically and represented in EVALPSN;
- both the requests consist of if-part and then-part, and if the conditional part is assumed and the conclusion part does not contradict the safety properties, then the safety of the request is guaranteed; we interpret this verification process as that if the conditional part holds as obligation and the conclusion part is permitted against the safety properties, then the safety of the request is guaranteed; therefore, those requests are also represented as an EVALP and checked the safety in an EVALP programming system.

Generally, a Panel Route Request $\mathtt{Q}x$ has a form of

$$\mathtt{Q}x \quad \text{if } A_1, \cdots, A_m \text{ then } B_1, \cdots, B_n\backslash,$$

and a Sub-Route Release $\mathtt{SR}y$ also has a form of

$$\mathtt{SR}y \quad \text{if } C_1, \cdots, C_s \text{ then } D_1, \cdots, D_t\backslash.$$

The route requests, Panel Route Request and Sub-Route Release, are checked their safety by consulting the safety properties as follows:

(I) let an EVALPSN EP be the set

$$\{\ (77), \ldots, (134)\ \} \tag{135}$$

of EVALPSN clauses that represents the safety properties and the sub-route release conditions;

(II) translate each $A_i (1 \leq i \leq m)$ into an EVALPSN clause in a form of fact, and add all the EVALPSN clauses to the EVALPSN EP; and

(III) translate each $B_j (1 \leq j \leq n)$ into an EVALPSN clause in a form of permission, and inquire it of the EVALPSN EP obtained at (II), then if *yes* is returned, the safety for the route request is verified, and if *no* is returned, not verified.

Example 3. Now we take the Panel Route Request Q02 as an example of the Panel Route Request safety verification:

$$\mathtt{Q02} \quad \text{if} \quad P_1\ \mathtt{crf},\ T_1^{ac}\ \mathtt{f},\ T_2^{ab}\ \mathtt{f} \quad \text{then} \quad R_{02}\ \mathtt{s},\ P_1\ \mathtt{cr},\ T_1^{ca}\ \mathtt{l},\ T_2^{ba}\ \mathtt{l}\backslash.$$

The if-part of the Panel Route Request Q02 is translated into the following EVALPSN clauses:

$$P(1):[\texttt{crf},\alpha], \tag{136}$$
$$T(1,ac):[\texttt{f},\alpha], \tag{137}$$
$$T(2,ab):[\texttt{f},\alpha], \tag{138}$$

which are added to the EVALPSN EP. Then, the then-part of the Panel Route Request Q02 is verified as follows:

- the EVALPSN clause
$$T(2,ba):[\texttt{f},\gamma] \tag{139}$$
is derived by the EVALPSN clauses $\{\ (79),(138)\ \}$;
- the EVALPSN clause
$$P(1):[\texttt{cn},\gamma] \tag{140}$$
is derived by the EVALPSN clauses $\{\ (106),(137)\ \}$;
- the EVALPSN clause
$$T(1,ca):[\texttt{f},\gamma] \tag{141}$$
is derived by the EVALPSN clauses $\{\ (90),(137),(140)\ \}$;
- the EVALPSN clause
$$R(02):[\texttt{xs},\gamma] \tag{142}$$
is derived by the EVALPSN clauses $\{\ (99),(141),(139)\ \}$.

Therefore, the answer *yes* is returned and the safety for the Panel Route Request Q02 is assured.

Next, we verify the safety for the Sub-Route Release

$$\texttt{SR02} \quad \text{if} \quad R_{02}\,\texttt{xs},\ T_1\,\texttt{c},\ T_2\,\texttt{c} \quad \text{then} \quad T_1^{ca}\,\texttt{f},\ T_2^{ba}\,\texttt{f}\backslash.$$

The if-part of the Sub-Route Release SR02 is translated into the following EVALPSN clauses:

$$R(02):[\texttt{xs},\alpha], \tag{143}$$
$$T(1):[\texttt{c},\alpha], \tag{144}$$
$$T(2):[\texttt{c},\alpha], \tag{145}$$

which are added to the EVALPSN EP. Then, the then-part of the Sub-Route Release SR02 is verified as follows:

- the EVALPSN clause
$$T(1,ca):[\texttt{l},\gamma] \tag{146}$$
is derived by the EVALPSN clauses $\{\ (143),(144),(123)\ \}$;
- the EVALPSN clause
$$T(2,ba):[\texttt{l},\gamma] \tag{147}$$
is derived by the EVALPSN clauses $\{\ (145),(146),(121)\ \}$.

Therefore, the answer *yes* is returned and the safety for the Sub-Route Release SR02 is assured.

6 Remarks and Future Works

Two types of frameworks for EVALPSN based intelligent information systems and their examples, an autonomous robot action control system and an automated safety verification system for railway interlocking, are introduced.

Although the action control is for a virtual robot and not so real, it shows the guidelines for more practical applications of EVALPSN based intelligent action control. Since the EVALPSN used in the intelligent robot action control system is a stratified logic program, it is easily implemented as software programs. We are considering to translate stratified EVALPSN or EVALP into electronic circuits on a microchip, though we have not addressed about it. In fact, we have already finished the circuit design for the EVALPSN based robot action control system as a prototype of faster real time processing systems.

On the other hand, the EVALPSN based safety verification system for railway interlocking seems to be a very useful application and expected to be applied to actual systems. Then, temporal reasoning function may be implemented in the safety verification system for dealing with not only safety verification but also train driving scheduling, moreover, distributed systems should be expected if efficient processing is required. There seem to be various applications of the EVALPSN based intelligent safety verification. Some of them require faster real time processing, eg. safety verification for air traffic control. Here what we want to remark again is that stratified EVALPSN or EVALP can be easily implemented on a microchip as electronic circuits, and the existence of such a microchip must extend the applicable area of paraconsistent logic programming.

References

1. Billington, D. (1993): Defeasible Logic is Stable. J. Logic and Computation **3**, 379-400
2. Billington, D. (1997): Conflicting Literals and Defeasible Logic. Nayak, A. and Pagnucco, M. (eds.) Proc. 2nd Australian Workshop on Commonsense Reasoning, The Australian Computer Society, 1–15
3. Blair, H.A. and Subrahmanian, V.S. (1989): Paraconsistent Logic Programming. Theoretical Computer Science **68**, 135–154
4. da Costa, N.C.A., Subrahmanian, V.S., and Vago, C. (1989): The Paraconsistent Logics $P\mathcal{T}$. Zeitschrift für Mathematische Logic und Grundlangen der Mathematik **37**, 139–148
5. Gelder, A.V., Ross, K.A. and Schlipf, J.S. (1991): The Well-Founded Semantics for General Logic Programs. J. the Association for Computing Machinery **38**, 620–650
6. Gelfond, M. and Lifschitz, V. (1989): The Stable Model Semantics for Logic Programming. Proc. 5th International Conference and Symposium on Logic Programming, MIT Press, 1070–1080

7. Gordon, M.J.C. and Melham, T.F. (1993), *Introduction to HOL*, Cambridge Univ. Press
8. Kifer, M. and Subrahmanian, V.S. (1992): Theory of Generalized Annotated Logic Programming and its Applications. J.Logic Programming **12**, 335–368
9. Lloyd, J.W. (1987): *Foundations of Logic Programming* 2nd edition. Springer-Verlag
10. Morley, J.M. (1996): Safety Assurance in Interlocking Design. Ph.D Thesis, University of Edinburgh
11. Nakamatsu, K. and Suzuki, A. (1994): Annotated Semantics for Default Reasoning. Dai, R. (ed.) Proc. 3rd Pacific Rim International Conference on Artificial Intelligence, International Academic Publishers, 180–186
12. Nakamatsu, K. and Suzuki, A. (1998): A Nonmonotonic ATMS Based on Annotated Logic Programs. Agents and Multi-Agents Systems, LNAI **1441**, Springer-Verlag, 79–93
13. Nakamatsu, K. and Abe, J.M. (1999): Reasonings Based on Vector Annotated Logic Programs. Computational Intelligence for Modelling, Control & Automation, Concurrent Systems Engineering Series **55**, IOS Press, 396–403
14. Nakamatsu, K., Abe, J.M., and Suzuki, A. (1999): Defeasible Reasoning Between Conflicting Agents Based on VALPSN. Proc. AAAI Workshop Agents' Conflicts, AAAI Press, 20–27
15. Nakamatsu, K., Abe, J.M., and Suzuki, A. (1999): Defeasible Reasoning Based on VALPSN and its Application. Proc. The Third Australian Commonsense Reasoning Workshop, 114–130
16. Nakamatsu, K., Abe, J.M., and Suzuki, A. (2001): A Defeasible Deontic Reasoning System Based on Annotated Logic Programming. Proc. the Fourth International Conference on Computing Anticipatory Systems, American Institute of Physics, AIP Conference Proceedings 573, 609–620
17. Nakamatsu, K., Abe, J.M., and Suzuki, A. (2001): Annotated Semantics for Defeasible Deontic Reasoning. Proc. the Second International Conference on Rough Sets and Current Trends in Computing, LNAI 2005, Springer-Verlag, 432–440
18. Nute, D. (1987): Defeasible Reasoning. Proc. Hawaii International Conference on System Science, 470-477
19. Nute, D. (1992): Basic Defeasible Logics. *Intensional Logics for Programming*. Oxford University Press, 125–154
20. Nute, D. (ed.) (1997): *Defeasible Deontic Logic*. Kluwer Academic Publisher, 287–316
21. Subrahmanian, V.S. (1994): Amalgamating Knowledge Bases. ACM Transactions on Database Systems **19**, 291–331
22. Subrahmanian, V.S. (1987): On the Semantics of Qualitative Logic Programs. Proc. 4th IEEE Symposium on Logic Programming, 178-182

Chapter 12

Neuro-Fuzzy Paradigms for Intelligent Energy Management

Ajith Abraham and Muhammad Riaz Khan

Summary. Intelligent energy management has become one of the major research fields in electrical engineering. It constitutes an important tool for efficient planning and operation of power systems and its significance has been intensifying particularly, because of the recent movement towards open energy markets and the need to assure high standards on reliability. Hybrid neuro-fuzzy paradigms have recently gained a lot of interest in research and application. In this chapter, we discuss two neuro-fuzzy paradigms for intelligent energy management. In the first approach, a neural network learning algorithm is used to fine tune the parameters of a Mamdani and Takagi-Sugeno Fuzzy Inference System (FIS). Mamdani FIS is used to predict the energy demand and the Takagi-Sugeno FIS is used to predict the reactive power flow. In the second approach, fuzzy *if-then* rules were embedded into an Artificial Neural Network (ANN) learning algorithm (fuzzy-neural network) to achieve improved performance for short-term load forecast. The performance of the different neuro-fuzzy paradigms were tested using real world data and compared with a direct neural network and FIS approach. The different performance results obtained clearly demonstrates the importance of the proposed techniques for intelligent energy management.

Keywords: neuro-fuzzy, computational intelligence, hybrid systems, neural network, fuzzy system.

1 Introduction

Accurate load forecasting is of great importance for power system operation. It is the basis of economic dispatch, hydrothermal coordination, unit commitment, and system security analysis among other functions [23]. Short-term load forecasts have become increasingly important since the rise of the competitive energy markets [24], [25], [27], [30]. Many countries have recently privatized and deregulated their power systems, and electricity has been turned into a commodity to be sold and bought at market prices. Since the load forecasts play a crucial role in the composition of these prices, they have become vital for the supply industry.

Load forecasting is however a difficult task. First, because the load series is complex and exhibits several levels of seasonality: the load at a given hour is dependent not only on the load at the previous hour, but also on the load at the same hour on the previous day, and on the load at the same hour on the day with the same denomination in the previous week. Secondly, because there are many important exogenous variables that must be considered, specially weather-related variables.

We consider two different ways of integrating neuro-fuzzy paradigms [4]. The first approach is to apply a learning algorithm to a FIS [9], which is represented in a special ANN like architecture [26]. However the conventional ANN learning algorithms (gradient descent) cannot be applied directly to such a system as the functions used in the inference process are usually non differentiable. This problem can be tackled by using differentiable functions in the inference system or by not using the standard neural learning algorithm. The performance of the algorithms are validated by practical energy data [3], [6], [8].

In the second approach, the input parameters consisting of load patterns and weather parameters are fuzzified and used to train a neural network. We applied the backpropagation algorithm to train neural network to find a preliminary forecast load. In addition, the rule base of the fuzzy inference machine contains linguistic importance attached to them in terms of membership functions with knowledge in the form of fuzzy "*if-then*" rules. It makes the load correction inference from historical information and past forecast load errors to calculate the forecast load error. Adding the current forecast load error to the preliminary forecast load, we obtained the final forecast load. The effectiveness of the proposed approach to the short-term load-forecasting problem is demonstrated by the practical data collected from the Czech Electric Power Company (CEZ), Czech Republic [19]-[22].

This paper is organized as follows. In Sections 2 and 3, we present the different neuro-fuzzy paradigms followed by the different experimentation results in Section 4. Some conclusions are also provided towards the end.

2 Integrating Neural Networks and Fuzzy Inference System

A conventional fuzzy controller makes use of a model of the expert who is in a position to specify the most important properties of the process. Expert knowledge is often the main source to design the fuzzy inference systems. Figure 1 shows the architecture of the fuzzy inference system controlling a process. According to the performance measure of the problem environment, the MFs, rule bases and the inference mechanism are to be adapted [7].

Several research works are going on exploring the adaptation of fuzzy inference systems [1], [2], [11], [13], [15], [26], [29], [36]. These include the adaptation of membership functions, rule bases, aggregation operator's, etc. These techniques include but are not limited to:

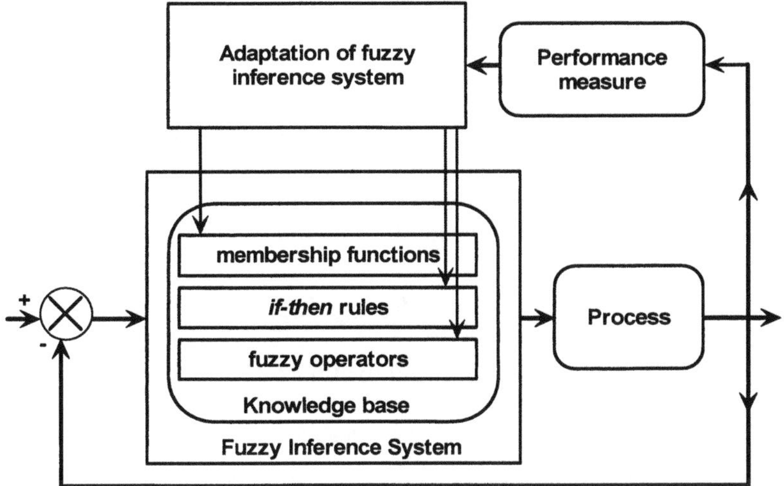

Figure 1. Architecture of adaptive fuzzy inference systems.

- Self-organizing process controller by Procyk et al. [34], which considered the issue of rule generation and adaptation.
- Evolutionary algorithms to optimize the fuzzy parameters, rule base, etc. [5], [10], [31].
- Gradient descent and its variants have been applied to fine-tune the parameters of the input and output membership functions [38].
- Pruning the quantity and adapting the shape of input/output membership functions [39].
- Fuzzy discretization and clustering techniques [40].

In most cases the inference of the fuzzy rules is done using the '*min*' and '*max*' operators for fuzzy intersection and union. If the T-norm and T-conorm operators are parameterized then gradient descent technique could be used in a supervised learning environment to fine-tune the fuzzy operators.

In an integrated model, neural network learning algorithms are used to determine the parameters of fuzzy inference systems. Integrated neuro-fuzzy systems share data structures and knowledge representations. A fuzzy inference system can utilize human expertise by storing its essential components in rule base and database, and perform fuzzy reasoning to infer the overall output value. The derivation of *if-then* rules and corresponding membership functions depends heavily on the *a priori* knowledge about the system under consideration. However there is no systematic way to transform experiences of knowledge of human experts to the knowledge base of a fuzzy inference system. There is also a need for adaptability or some learning algorithms to produce outputs within the required error rate. On the other hand, neural network learning mechanism does not rely on human expertise. Due to the homogenous structure of neural network, it is hard to extract structured knowledge from either the weights or the configuration of the network. The weights of the neural network represent the coefficients of the hyper-plane that partition the input space into two regions with different output values. If we can visualize this hyper-plane structure from the training data then the subsequent learning procedures in a neural network can be reduced. However, in reality, the a priori knowledge is usually obtained from human experts and it is most appropriate to express the knowledge as a set of fuzzy if-then rules and it is very difficult to encode into an neural network. Modeling integrated neuro-fuzzy systems implementing Mamdani and Takagi-Sugeno FIS is presented in Sections 2.1.1 and 2.1.2.

2.1 Adaptive Network Based Fuzzy Inference System (ANFIS)

ANFIS [15] is perhaps the first integrated hybrid neuro-fuzzy model. ANFIS structure as shown in Figure 2 is capable of implementing the Takagi and Sugeno FIS [14]. The detailed functioning of each layer is as follows:

Layer-1 (*fuzzification layer*): Every node in this layer has a node function

$$O_i^1 = \mu_{A_i}(x)\text{, for } i=1,2 \tag{1}$$

O_i^1 is the membership grade of a fuzzy set A ($= A_1$, A_2, or B_1 or B_2) and it specifies the degree to which the given input x (or y) satisfies the quantifier A. Usually the node function can be any parameterized function.. A Gaussian membership function is specified by two parameters c (membership function center) and σ (membership function width).

$$\text{Gaussian } (x, c, \sigma) = e^{-\frac{1}{2}\left(\frac{x-c}{\sigma}\right)^2} \tag{2}$$

Parameters in this layer are referred to as premise parameters.

Layer-2 (*rule firing strength layer*): Every node in this layer multiplies the incoming signals and sends the product out. Each node output represents the firing strength of a rule.

$$O_i^2 = w_i = \mu_{A_i}(x) \times \mu_{B_i}(y), i = 1,2...... . \tag{3}$$

In general any T-norm operators that perform fuzzy AND can be used as the node function in this layer.

Layer-3 Every i-th node in this layer calculates the ratio of the i-th rule's firing strength to the sum of all rules firing strength.

$$O_i^3 = \overline{w_i} = \frac{w_i}{w_1 + w_2}, i = 1,2.... . \tag{4}$$

Layer 4 (*rule strength normalization*): Every node in this layer calculates the ratio of the i-th rule's firing strength to the sum of all rules firing strength

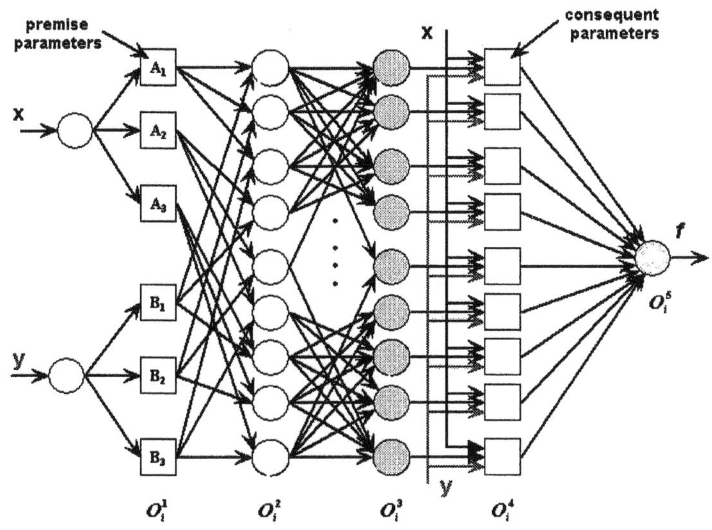

Figure 2. Architecture of the ANFIS.

$$\overline{w_i} = \frac{w_i}{w_1 + w_2}, \ i = 1,2.... \quad (5)$$

Layer-5 (*rule consequent layer*): Every node i in this layer is with a node function

$$\overline{w_i f_i} = \overline{w_i}\,(p_i x_1 + q_i x_2 + r_i) \quad (6)$$

where $\overline{w_i}$ is the output of layer 4, and $\{p_i, q_i, r_i\}$ is the parameter set. A well-established way is to determine the consequent parameters using the least means squares algorithm.

Layer-6 (*rule inference layer*): The single node in this layer computes the overall output as the summation of all incoming signals:

$$Overall\ output = \sum_i \overline{w_i f_i} = \frac{\sum_i w_i f_i}{\sum_i w_i} \quad (7)$$

Takagi-Sugeno neuro-fuzzy systems make use of a mixture of back propagation to learn the membership functions and least mean square estimation to determine the coefficients of the linear combinations in the rule's conclusions [14]. A step in the learning procedure got two parts: In the first part the input patterns are propagated, and the optimal conclusion parameters are estimated by an iterative least mean square procedure, while the antecedent parameters (membership functions) are assumed to be fixed for the current cycle through the training set. In the second part the patterns are propagated again, and in this epoch, back propagation is used to modify the antecedent parameters, while the conclusion parameters remain fixed. This procedure is then iterated. Assuming a single output ANFIS represented by

$$output = F(\vec{I}, S) \qquad (8)$$

where I is the set of input variables and S is the set of parameters, if there exist a function H such that the composite function $H \circ F$ is linear in some of the elements of S, then these elements can be identified by the least squares method. More formally the parameter set S can be decomposed into two sets:

$$S = S_1 \oplus S_2 \text{ (where } \oplus \text{ represents direct sum),} \qquad (9)$$

such that $H \circ F$ is linear in the elements of S_2. Then upon applying H to (8), we have:

$$H(output) = H \circ F(\vec{I}, S) \qquad (10)$$

which is linear in the elements of S_2. Now the given values of elements of S_1, we can plug P training data into (10), and obtain a matrix equation:

$$AX = B \text{ (}X = \text{unknown vector whose elements are parameters in } S_2\text{)} \qquad (11)$$

If $|S_2| = M$, (M = number of linear parameters) then the dimensions of A, X and B are $P \times M$, $M \times 1$ and $P \times 1$ respectively. Since P is always greater than M, there is no exact solution to (11). Instead, a Least Square Estimate (LSE) of X, X^*, is sought to minimize the squared error $\|AX - B\|^2$. X^* is computed using the pseudo-inverse of X:

$$X^* = (A^T A)^{-1} A^T B \qquad (12)$$

where A^T is the transpose of A and $(A^T A)^{-1} A^T$ is the pseudo-inverse of A where $A^T A$ is non-singular. Due to computational complexity, in ANFIS a sequential method is deployed as follows:

Let the *i-th* row vector of matrix A defined in (11) be a_i^T and *i-th* element of matrix B defined be b_i^T, then X can be calculated iteratively using the following sequential formulae:

$$\begin{aligned} X_{i+1} &= X_i + S_{i+1} a_{i+1} (b_{i+1}^T - a_{i+1}^T X_i) \\ S_{i+1} &= S_i - \frac{S_i a_{i+1} a_{i+1}^T S_i}{1 + a_{i+1}^T S_i a_{i+1}}, \quad i = 0,1,\ldots\ldots, P-1 \end{aligned} \qquad (13)$$

where S_i is often called the covariance matrix and the least squares estimate X^* is equal to X_P. The initial condition to bootstrap (13) are $X_O = 0$ and $S_O = \gamma I$, where γ is a positive large number and I is the identity matrix of dimension $M \times M$. For a multi-output ANFIS, (13) is still applicable except the $output = F(\vec{I}, S)$ will become a column vector. Each epoch of this hybrid learning procedure is composed of a forward pass and a backward pass. In the forward pass, we have to supply the input data and functional signals go forward to calculate each node output until the matrices A and B in (11) are obtained, and the parameters in S_2 are identified by the sequential least squares formulae given in (13). After identifying parameters in

S_2, the functional signals keep going forward till the error measure is calculated. In the backward pass, the error rates propagate from the output layer to the input layers, and the parameters in S_1 are updated by the gradient method given by

$$\Delta\alpha = -\eta \frac{\partial E}{\partial \alpha} \tag{14}$$

where α is the generic parameter, η is a learning rate and E the error measure. For given fixed values of parameters in S_1, the parameters in S_2 thus found are guaranteed to be the global optimum point in the S_2 parameter space due to the choice of the squared error measure[14].

The procedure mentioned above is mainly for offline learning version. However the procedure can be modified for an online version by formulating the squared error measure as a weighted version that gives higher weighting factors to more recent data pairs. This amounts to the addition of a forgetting factor λ to (13).

$$\begin{aligned} X_{i+1} &= X_i + S_{i+1}a_{i+1}(b_{i+1}^T - a_{i+1}^T X_i) \\ S_{i+1} &= \frac{1}{\lambda}\left[S_i - \frac{S_i a_{i+1} a_{i+1}^T S_i}{\lambda + a_{i+1}^T S_i a_{i+1}}\right] \quad i = 0,1,\ldots\ldots, P-1 \end{aligned} \tag{15}$$

The value of λ is between 0 and 1. The smaller the λ is, faster the effects of old data decay. But a smaller λ sometimes causes numerical instability and should be avoided.

2.2 Evolving Fuzzy Neural Networks

Evolving Fuzzy Neural Network (EFuNN) (Figure 3) implements a Mamdani type FIS and all nodes are created during learning [16]. The nodes representing membership functions (MFs) can be modified during learning. Each input variable is represented here by a group of spatially arranged neurons to represent a fuzzy quantization of this variable. For example, three neurons can be used to represent "small", "medium" and "large" fuzzy values of the variable. Different membership functions can be attached to these neurons (triangular, Gaussian, etc.). New neurons can evolve in this layer if, for a given input vector, the corresponding variable value does not belong to any of the existing MF to a degree greater than a membership threshold.

The third layer contains rule nodes that evolve through hybrid supervised/unsupervised learning. The rule nodes represent prototypes of input-output data associations, graphically represented as an association of hyper-spheres from the fuzzy input and fuzzy output spaces. Each rule node, e.g., r_j, represents an association between a hyper-sphere from the fuzzy input space and a hyper-sphere from the fuzzy output space; $W_1(r_j)$ connection weights representing the coordinates of the center of the sphere in the fuzzy input space, and $W_2(r_j)$ – the coordinates in the fuzzy output space. The radius of an input hyper-sphere of a rule node is defined as $(1 - Sthr)$, where $Sthr$ is the sensitivity threshold parameter defining the minimum activation of a rule node (e.g., r_1, previously evolved to represent a data point (X_{d1}, Y_{d1})) to an input vector (e.g., (X_{d2}, Y_{d2})) in order for the new input vector to be associated with this rule node. Two pairs of fuzzy input-output data vectors $d_1 = (X_{d1}, Y_{d1})$ and $d_2 = (X_{d2}, Y_{d2})$ will be allocated to the first rule node r_1 if they fall into the r_1 input sphere and in the r_1 output sphere, i.e. the local normalised fuzzy difference between X_{d1} and X_{d2} is smaller than the radius r and the local normalised fuzzy difference between Y_{d1} and Y_{d2} is smaller than an error threshold $Errthr$. The local normalised fuzzy difference between two fuzzy membership vectors d_{1f} and d_{2f} that represent the membership degrees to which two real values d_1 and d_2 data belong to the predefined MF, are calculated as $D(d_{1f}, d_{2f}) = sum(abs(d_{1f} - d_{2f})) / sum(d_{1f} + d_{2f})$.

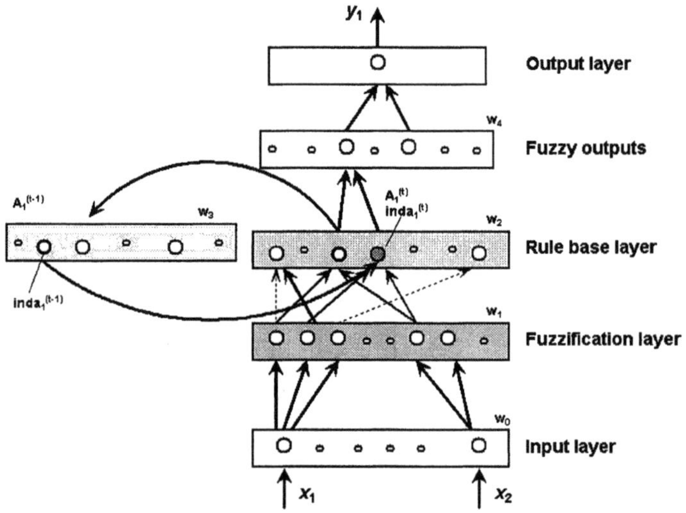

Figure 3. Architecture of EFuNN.

If data example $d_1 = (X_{d1}, Y_{d1})$, where X_{d1} and X_{d2} are correspondingly the input and the output fuzzy membership degree vectors, and the data example is associated with a rule node r_1 with a center r_1^1, then a new data point $d_2 = (X_{d2}, Y_{d2})$, will also be associated with this rule node through the process of associating (learning) new data points to a rule node. The centers of this node hyper-spheres adjust in the fuzzy input space depending on a learning rate lr_1, and in the fuzzy output space depending on a learning rate lr_2, on the two data points d_1 and d_2. The adjustment of the center r_1^1 to its new position r_1^2 can be represented mathematically by the change in the connection weights of the rule node r_1 from $W_1(r_1^1)$ and $W_2(r_1^1)$ to $W_1(r_1^2)$ and $W_2(r_1^2)$ according to the following vector operations:

$$W_2(r_1^2) = W_2(r_1^1) + lr_2 \,.\, Err(Y_{d1}, Y_{d2}) \,.\, A_1(r_1^1) \quad (16)$$

$$W_1(r_1^2) = W1\,(r_1^1) + lr_1 \,.\, Ds\,(X_{d1}, X_{d2}) \quad (17)$$

where $Err(Y_{d1}, Y_{d2}) = Ds(Y_{d1}, Y_{d2}) = Y_{d1} - Y_{d2}$ is the signed value rather than the absolute value of the fuzzy difference vector; and $A_1(r_j^1)$ is the activation of the rule node r_j^1 for the input vector X_{d2}.

While the connection weights from W_1 and W_2 capture spatial characteristics of the learned data (centers of hyper-spheres), the temporal layer of connection weights W_3 captures temporal dependencies between consecutive data examples. If the winning rule node at the moment $(t-1)$ (to which the input data vector at the moment $(t-1)$ was associated) was $r_1 = inda_1(t-1)$, and the winning node at the moment t is $r_2 = inda_1(t)$, then a link between the two nodes is established as follows:

$$W_3(r_1,r_2)^{(t)} = W_3(r_1,r_2)^{(t-1)} + lr_3.\ A_1(r_1)^{(t-1)}\ A_1(r_2))^{(t)}, \tag{18}$$

where $A_1(r)^{(t)}$ denotes the activation of a rule node r at a time moment (t); lr_3 defines the degree to which the EFuNN associates links between rules (clusters, prototypes) that include consecutive data examples (if $lr_3 = 0$, no temporal associations are learned in an EFuNN structure).

The learned temporal associations can be used to support the activation of rule nodes based on temporal, pattern similarity. Here, temporal dependencies are learned through establishing structural links. The ratio spatial-similarity/temporal-correlation can be balanced for different applications through two parameters S_s and T_c such that the activation of a rule node r for a new data example d_{new} is defined as the following vector operations:

$$A_1(r) = f(S_s\ .\ D(r,\ d_{new}) + T_c\ .\ W_3(r^{(t-1)},\ r)) \tag{19}$$

where f is the activation function of the rule node r, $D(r,\ d_{new})$ is the normalised fuzzy distance value and $r^{(t-1)}$ is the winning neuron at the previous time moment. The fourth layer of neurons represents fuzzy quantification for the output variables. The fifth layer represents the real values for the output variables.

EFuNN evolving algorithm is given as a procedure of consecutive steps [16]:

1. Initialize an EFuNN structure with a maximum number of neurons and zero value connections. If initially there are no rule nodes connected to the fuzzy input and fuzzy output neurons, then create the first node r_j=1 to represent the first data example EX= $(X_{d1},\ Y_{d1})$ and set its input $W_1\ (r_j)$ and output $W_2\ (r_j)$ connection weights as follows:
   <Create a new rule node r_j> to represent a data sample EX: $W_1(r_j) = EX$: $W_2(r_j)$ $= TE$, where TE is the fuzzy output vector for the (fuzzy) example EX.
2. *While <there are data examples> Do*
   Enter the current, example $(X_{di},\ Y_{di})$, EX being the fuzzy input vector (the vector of the degrees to which the input values belong to the input membership functions). If there are new variables that appear in this example and have not been used in previous examples, create new input and/or output nodes with their corresponding membership functions.
3. Find the normalized fuzzy similarity between the new example EX (fuzzy input vector) and the already stored patterns in the case nodes $r_j= r_1,\ r_2, \ldots, r_n$
   $D(EX,r_j) = sum\ (abs\ (EX - W_1(r_j)))\ /\ sum\ (W_1(r_j) + EX)$.

4. Find the activation $A_1(r_j)$ of the rule nodes $r_j= r_1, r_2, \ldots, r_n$. Here radial basis activation (*radbas*) function, or a saturated linear (*satlin*) one, can be used, i.e., $A_1(r_j) = radbas(S_s\ D(EX, r_j - T_c\ W_3)$, or $A_1(r_j) = satlin(1 - S_s\ D(EX, r_j + T_c\ W_3))$.
5. Update the pruning parameter values for the rule nodes, e.g., age, average activation as predefined.
6. Find m case nodes r_j with an activation value $A_1(r_j)$ above a predefined sensitivity threshold *Sthr*.
7. From the m case nodes, find one rule node $inda_1$ that has the maximum activation value $maxa_1$.
8. If $maxa_1 < Sthr$, then <*create a new rule node*> using the procedure from step 1.

   Else
9. Propagate the activation of the chosen set of m rule nodes $(r_{j1},\ldots,r_{jm})$ to the fuzzy output neurons: $A_2 = satlin(A_1(r_{j1},\ldots,r_{jm}) \cdot W_2)$.
10. Calculate the fuzzy output error vector $Err = A_2 - TE$.
11. If $(D(A_2,TE) > Errthr)$, <*create a new rule node*> using the procedure from step 1.
12. Update (a) the input, and (b) the output of the $m-1$ rule nodes $k = 2 : j_m$ in case of a new node was created, or m rule nodes $k=j_1 : j_m$, in case of no new rule was created:
    $Ds(EX - W_1(r_k)) = EX - W_1(r_k)$; $W_1(r_k) = W_1(r_k) + lr_1 . Ds(EX - W_1(r_k))$, where lr_1 is the learning rate for the first layer;
    $A_2(r_k) = satlin(W_2(r_k) \cdot A_1(r_k))$; $Err(rk) = TE - A_2(r_k)$;
    $W_2(r_k) = W_2(r_k) + lr_2 \cdot Err(r_k) \cdot A_1(r_k)$, where lr_2 is the learning rate for the second layer.
13. Prune rule nodes r_j and their connections that satisfy the following fuzzy pruning rule to a pre-defined level representing the current need of pruning:
    IF (a rule node r_j is OLD) and (average activation $A_1av(r_j)$ is LOW) and (the density of the neighboring area of neurons is HIGH or MODERATE) (i.e. there are other prototypical nodes that overlap with j in the input-output space; this condition apply only for some strategies of inserting rule nodes as explained below) THEN the probability of pruning node (r_j) is HIGH. The above pruning rule is fuzzy and it requires that the fuzzy concepts as OLD, HIGH, etc. are predefined.
14. Aggregate rule nodes, if necessary, into a smaller number of nodes. A C-means clustering algorithm can be used for this purpose.
15. End of the *while* loop and the algorithm

The rules that represent the rule nodes need to be aggregated in clusters of rules. The degree of aggregation can vary depending on the level of granularity needed. At any time (phase) of the evolving (learning) process, fuzzy, or exact rules can be inserted and extracted [17]. Insertion of fuzzy rules is achieved through setting a new rule node for each new rule, such as the connection weights W_1 and W_2 of the rule node represent the fuzzy or the exact rule. The process of rule extraction can be performed as aggregation of several rule nodes into larger hyper-spheres. For the aggregation of two-rule nodes r_1 and r_2, the following aggregation rule is used

$$If\,(D(W_1(r_1), W_1(r_2)) <= Thr_1)\ and\ (D(W_2(r_1), W_2(r_2)) <= Thr_2) \quad (20)$$

then aggregate r_1 and r_2 into r_{agg} and calculate the centers of the new rule node as

$$W_1(r_{agg}) = average(W_1(r_1),\ W_1(r_2)),\ W_2(r_{agg}) = average(W_2(r_1),\ W_2(r_2)) \quad (21)$$

Here the geometrical center between two points in a fuzzy problem space is calculated with the use of an average vector operation over the two fuzzy vectors. This is based on a presumed piece-wise linear function between two points from the defined through the parameters *Sthr* and *Errthr* input and output fuzzy hyperspheres.

2.3 Hybrid Fuzzy Neural Network (FNN)

In a hybrid model, the neural network is used to learn and classify the patterns, automatically creating the fuzzy rules and the fuzzy logic is used to infer the defuzzified output. The structure of the fuzzy-neural network used is shown in Figure 4 [21].

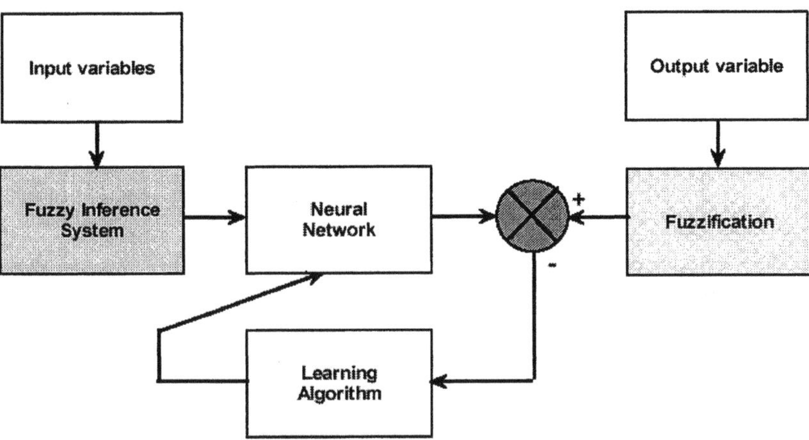

Figure 4. Structure of the fuzzy neural network.

The main features of FNN are:

- It provides a general method for combining available numerical information and human linguistic information in a common framework.
- It requires much less construction time than a comparable neural network.
- Significant accuracy is achieved in predicting chaotic time-series models.
- The mean absolute error is reduced in case of FNN as compared to ANN.
- The advantage of the FNN is its efficient adaptive tracking capability that results in the development of a robust and accurate forecasting technique, which gives an accurate forecast even under diverse weather conditions.

This hybrid approach utilizes the inherent properties of artificial neural networks, such as generalization and adaptability, self-organization, ability to solve nonlinear problems, retrieval from partial information, learning from well-defined patterns; and properties of fuzzy logic, such as abstract reasoning and human-like responses in cases involving uncertainty and contradictory data.

A fuzzy processor has been utilized for preprocessing the inputs with the application of fuzzy rules. The fuzzy processor effectively handles the numeric data and produces a fuzzy output vector, which is then fed to 3-layered feedforward neural network trained using backpropagation algorithm. The inputs to the neural network are processed by the ANN, which generates response at its output layer. This response is compared with the desired output and their difference is propagated backwards through the network connections to reduce this error. The learning data is presented repeatedly until the errors are reduced to an acceptable level. Once trained, the outputs of the neural network, interpreted as fuzzy membership functions of the desired output, are defuzzified to get the required output. In defuzzification, all significant fuzzy outputs are combined into a specific, comprehensive result for that output variable. In this process, all the fuzzy output values effectively modify their respective output membership functions. The ANN is allowed to train until it maps the input-output relationship with the desired accuracy [19], [20], [22].

3 Modern Energy Management

Energy generation, distribution and management is changing dramatically. In a few years, the concepts we have all grown up will might be gone forever. The traditional gas and electric companies as we know them today will be in a very different business, competing on the open market and possibly combining services. Deregulated electricity supply markets can mean substantial electricity cost savings for consumers willing to take the time to study the opportunities. In a deregulated electricity supply market, there are theoretically numerous merchant power companies with generation assets standing by to provide consumers with the power they need to operate the homes and businesses at a fair price. Consumers are theoretically able to negotiate with the merchant power companies to agree to a price for power, which will be added to the charges imposed by the incumbent transmission and distribution grid owners to deliver power to the consumer. However, not all customers will be able to get the same good deal as every other customer, economies of scale aside.

In a perfect world, every customer would use a constant amount of power at all times of the day and every day of the year. This would make it easy for the companies saddled with the responsibility of maintaining and operating the electricity generation and distribution systems to keep everything running smoothly, and at an economical price. Unfortunately for everybody, the world doesn not work that way. People use more power at peak hours during the day when they are operating power-hungry machines under bright fluorescent lights in air-conditioned offices, than they use at night when they at home in bed. This means that utility companies must make allowances for mid-day peaks in power consumption as they provide for the generation, transmission and distribution of power to customers in the city.

The prediction of electricity demand has been of much interest to the electricity supply industry for some years, both to aid long term planning strategies, involving the forecasting of seasonal peak demands, and for use in the short term (up to 24 hours) operation of generating plant. The nature of electricity market is changing very rapidly with a widespread international movement towards competitiveness. Traditionally, the energy sector, and particularly the electricity sector, has been dominated by monopoly or near monopoly enterprises, typically either owned or regulated by government. The recent privatization of the electricity supply industry has brought a renewed interest in this subject.

Some countries, such as Norway, Chile, Japan, UK and the United States have commonly been supplied electricity by a large number of different regional Generators and have developed a variety of mechanisms to allow some form of trade between them. In 1994 Victoria started the process of privatization and restructuring electricity industry to generate competition. The objective was to promote a more flexible, cost-effective and efficient electricity industry with the aim of delivering cheaper electricity to business and the general community. Following success of this operation, Australia started the process of implementing a unified National Electricity Market in December 1998 [8].

3.1 Modeling Electricity Demand Prediction in Victoria (Australia)

To meet the electricity market demands a highly reliable supply and delivery system is required. Additionally, in order to gain a competitive advantage in this market through the competitive spot-market pricing an accurate forecast of electricity demand at regular time intervals is essential. Until 1996, Victorian Power Exchange (VPX) the body responsible for the secure operations of the power system, generated electricity demand forecasts based on weather forecasts and historical demand patterns [28]. Our research is focused on developing more accurate and reliable forecasting models that improves current forecasting methods. Our approach is to develop reliable and accurate prediction models predicting 96 half-hourly (two days ahead) demands for electricity, and compares their performance with forecasts used by VPX. We considered an integrated neuro-fuzzy system and a feedforward artificial neural network trained using the scaled gradient conjugate algorithm and backpropagation algorithm. For developing the forecasting models we used the energy demand data for ten months period from 27th January to 30th November 1995 in the State of Victoria. We also made use of the associated data stating the minimum and maximum temperature of the day, time of day, season and the day of week. The forecasting models were trained using 3 randomly selected samples containing 20% of the data during the period 27th January 1995 to

28th November 1995. To ascertain the forecasting accuracy the developed models were tested to predict the demand for the period (29th – 30th) November 1995 [8].

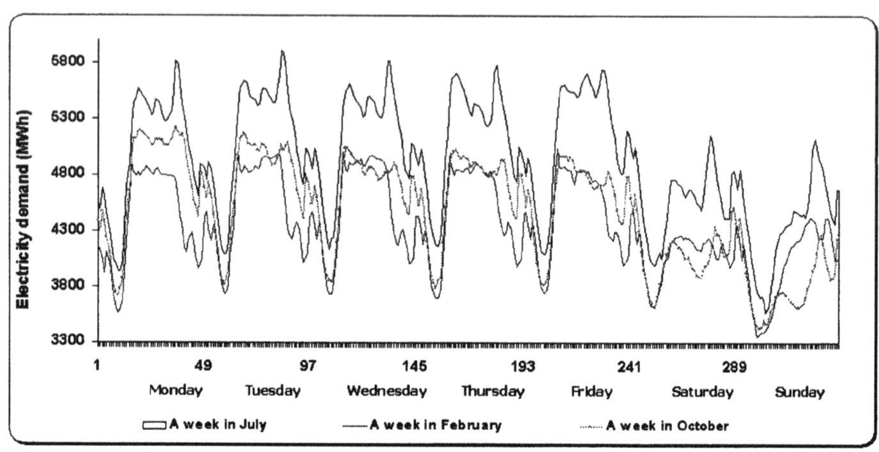

Figure 5. Typical weekly demand variations.

The data for our study were the recorded half-hourly actual electricity demand for the ten months period from January to November 1995 in the State of Victoria. Figure 5 shows a typical weekly cycle of electricity demand during three different months of the year. Fluctuations in daily demand are prevalent with peaks occurring around midday. Extreme weather conditions in winter and summer months accentuate peaks in electricity demand due to the widespread use of electricity for heating and cooling. Other times, electricity demand is dominated primarily by ambient temperature, time of day, working or non-working day and the day of the week.

The experimental system consists of two stages: modeling the prediction systems (training in the case of soft computing models) and performance evaluation. For network training, the six selected input descriptor variables were: the *minimum* and *maximum recorded temperatures*, *previous day's demand*, a value expressing the *half-hour period of the day*, *season*, and the *day of the week*. To evaluate the learning capability of the soft computing models, the network was trained only on 20% of the randomly selected data. We created 3 different samples of training data to study the effect of random sampling and periodicity. Each training sample consisted of 2937 data sets representing 20% random data. Our objective is to develop an efficient forecasting model capable of producing a short-term forecast of demand for electricity. The required time-resolution of the forecast is half-hourly, and the required time-span of the forecast is 2 days. This means that the system should be able to produce a forecast of electricity demand for the next 96 time periods. The training was replicated three times using three different samples of training data and different combinations of network parameters.

3.1.1 Neuro-Fuzzy Training

We used 4 Gaussian membership functions for each input variable and the following evolving parameters: sensitivity threshold *Sthr* = 0.99, error threshold *Errthr* = 0.001 and learning rates for first and second layer = 0.05. EFuNN uses a one pass training approach. The network parameters were determined using a trial and error approach. The training was repeated three times after reinitializing the network and the worst errors were reported. Online learning in EFuNN resulted in creating 2122 rule nodes. Training results and test results are summarized in Table 1.

Table 1. Test results and performance comparison of demand forecasting [8].

	EFuNN	ANN (BP)	ANN (SCGA)	ARIMA
Learning epochs	1	2500	2500	-
Training error (RMSE)	0.0013	0.116	0.0304	-
Testing error (RMSE)	0.0092	0.118	0.0323	0.0423
Computational load (in billion flops)	0.536	87.2	175.0	-

3.1.2 Neural Network Training

Our preliminary experiments helped us to formulate a feedforward neural network with 1 input layer, 2 hidden layers and an output layer [6-40-40-1]. Input layer consists of 6 neurons corresponding to the input variables. The first and second hidden layers consist of 40 neurons respectively using tanh-sigmoidal activation functions. To illustrate the convergence feature of Scaled Conjugate Gradient Algorithm (SCGA), we also trained a neural network (with same architecture) using backpropagation (BP) algorithm. To evaluate the neural network performance, training was terminated after 2500 epochs. Training and testing errors are summarized in Table 1. Figure 6 shows the convergence of SCGA with respect to BP algorithm. Figure 7 depicts the test results for prediction models considered. To have a performance evaluation the actual energy demand and the forecasts used by VHP and Box–Jenkins ARIMA model are also plotted in Figure 7. Compared to neural networks, an important advantage of neuro-fuzzy systems is its reasoning ability (*if-then* rules) of any particular state. A fully trained EFuNN could be replaced by a set of *if-then* rules [17]. A simple example of a learned EFuNN learned rule is illustrated below.

> "If the *maximum temperature* of the day is HIGH and *minimum temperature* of the day is LOW and *previous days demand* is MEDIUM and it is *summer* (HIGH) and *9.00 AM* (HIGH) and a *Monday* (HIGH) then the *electricity demand* is MEDIUM."

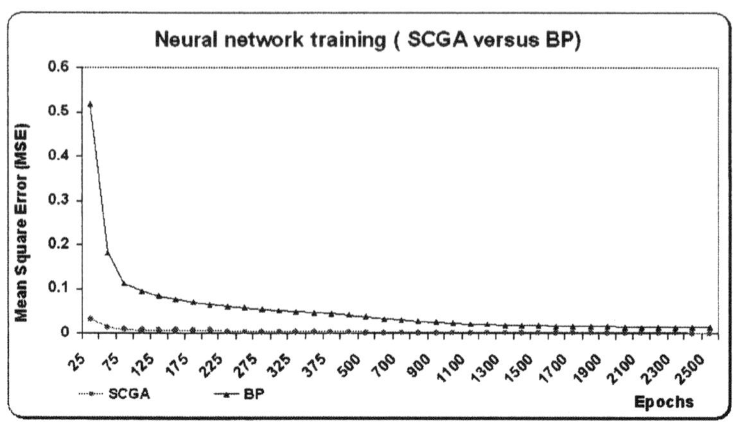

Figure 6. Convergence of neural network training.

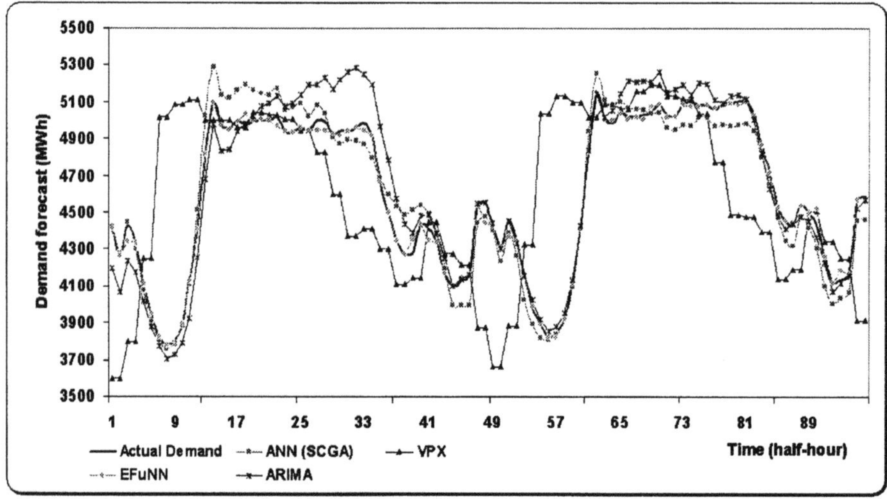

Figure 7. Test results and performance comparison of demand forecasts (2 days).

EFuNN uses a hybrid learning technique (a mixture of unsupervised and supervised learning) to fine-tune the parameters of the fuzzy inference system. As EFuNN adopts a single pass training (1 epoch) it is more adaptable and easy for further online training which might be highly useful for online forecasting and bidding. Another important feature of EFuNN is that the user has the flexibility to construct the network (by selecting the parameters). Hence for applications where speed is more important than the accuracy a faster network can be selected. However an important disadvantage of EFuNN is the determination of the network parameters like number and type of membership functions for each input variable, sensitivity threshold, error threshold and the learning rates. Even though a trial and error approach is practical, when the problem becomes complicated (large number of input variables) determining the optimal parameters will be a tedious task.

Our experiments on three separate data samples reveal that the results are not dependent on the data sample. We used only 20% of the total data to evaluate the learning capability of the soft computing models. Network performance could have been further improved by providing more training data. Another interesting fact about the considered soft computing models are their robustness and capability to handle noisy and approximate data that are typical in power systems, and therefore should be more reliable in worst situations.

3.2 Automation of Reactive Power Control

The ratio of active power (P) measured in watts to the apparent power (S) in volt-amperes is termed the power factor:

$$\text{Power factor} = \cos(\varphi) = \frac{P}{S} = \frac{\text{resistance } R}{\text{impedance } Z} \tag{22}$$

It has become a normal practice to say that the power factor is lagging when the current lags the supply voltage and leading when the current leads the supply voltage. This means that the supply voltage is regarded as the reference quantity. A majority of loads served by a power utility draw current at a lagging power factor. When the power factor of the load is unity, active power equals apparent power ($P = S$). But, when the power factor of the load is less than unity, say 0.6, the power utilized is only 60%. This means that 40% of the apparent power is being utilized to supply the reactive power, VAR, demand of the system. It is therefore clear that the higher the power factor of the load, the greater the utilization of the apparent power. For the generating and transmission stations, lower the power factor the larger must be the size of the source to generate that power, and greater must be the cross-sectional area of the conductor to transmit it. In other words, the greater is the cost of generation and transmission of the power. Moreover, lower power factor will also increase the I^2R (I denotes current) losses in lines/equipment as well as result in poor voltage regulation.

Most of the utility companies use a complex set of formulas, rewards/penalties etc. to receive an adequate return for their considerable investment in the larger capacity generators, transformers, cables and switchgear required to provide necessary KVA service to their customers. These formulas are generally referred to as power factor adjustments or KVAR reactive demand charges. In recent years, increased attention has been given to plant automation to reduce operational costs. Many manufacturing industries use human operators or timer controlled switching relays to turn on the power capacitors to compensate the reactive power requirement. Operational costs could be reduced and utilization efficiency improved if the power capacitor switching on/off process is automated using some intelligent techniques. We proposed a neuro-fuzzy approach to predict the reactive power trend (at time $t+1$) just by knowing the load current (at time t). Efficient usage of the VA loading will not only improve the overall grid condition but also reduce the consumer's industrial tariffs. Depending on the predicted reactive power demand, power factor corrective measures could be turned on or off to control the VA inflow into the plant. The developed model could be extremely useful for automated control of power inflow, especially in the countries where there are limitations on the usage of consumers' peak VA maximum demand [6].

We considered a heavy automobile manufacturing industry that works on 3 shifts of 8 hours duration for studying the load demand patterns. Observed data for a 24 hour period shows that the maximum and minimum VAR requirements are 2.96 MVAR and 0.014 MVAR, respectively [3], [6]. The variation of the reactive power flow during the peak hours of the three 8-hour shifts is depicted in Figure 8. If suitable power factor compensation was made when the reactive power demand was increasing, the plant might not have drawn much apparent power from the grid. The task is to predict the upward and downward trend of the reactive power demand and provide required reactive power compensation. Load flow analysis of the captioned plant reveals that the demand patterns are very similar every day (as long as the production of automobiles remain fixed).

The proposed neuro-fuzzy models and neural network were trained on the data taken at every minute for a 24-hour period to predict the reactive power demand, and tested to evaluate the prediction accuracy. To evaluate the efficiency of the prediction models, three different training and testing data sets were extracted and the experiments were performed three times.

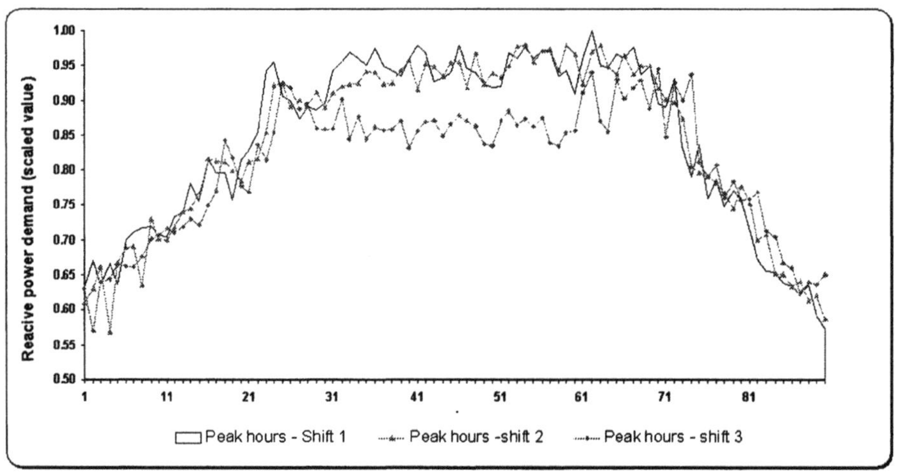

Figure 8. Reactive power demand variations during peak hours.

3.2.1 Experimentation Setup and Test Results

24-hour load flow patterns were used to train the neuro-fuzzy models and neural network. The training data comprises of 1440 data sets representing the 24-hour period. The input parameters considered are the phase voltage (*V*) and current (*I*).

The normal value of input parameter voltage (*V*) was fluctuated with +/- 2.5% of the normal value. All the data sets were scaled from 0 to 1. The input voltage was fluctuated to test the learning capability and robustness of the considered connectionist models. Training and testing data sets were extracted randomly from the complete dataset. 60% of data was used for training and remaining 40% for testing. To ensure that the data sample does not have any bias, we created 3 sets of data for training and testing (random extraction). Experiments with all 3 data sets were repeated 3 times for all the connectionist models.

3.2.2 Neural Network Training

We used a feedforward neural network with 2 hidden layers and trained using the backpropagation algorithm. The 2 input neurons correspond to the input variables and 1 output neuron for predicting reactive power. Initial weights, learning rate and momentum used were 0.3, 0.1 and 0.1, respectively. The training was terminated after 700 epochs.

3.2.3 ANFIS Training

In the ANFIS network, we used 3 Gaussian membership functions for each input parameter variable for predicting the reactive power demand. Nine rules were learned based on the training data. The training was terminated after 50 epochs.

3.2.4 EFuNN Training

We used 3 Gaussian membership functions and the following evolving parameters: sensitivity threshold *Sthr* = 0.95, error threshold *Errthr* = 0.05 and 544 rule nodes were created during training.

3.2.5 Performance and Results Achieved

Test data (input parameters) is passed through the trained connectionist models and the predicted output value is compared with the observed reactive power value to calculate the RMSE. Figure 9 illustrates the test results for predicted outputs using ANFIS, EFuNN and ANN. Table 2 shows an empirical comparative performance of the different connectionist models for the reactive power prediction problem. The empirical values shown in Table 2 are the worst values of the three trials with the three data sets for each model.

Among all the connectionist models neuro-fuzzy systems performed better than artificial neural network in terms of performance error achieved and training time. ANFIS performed marginally better than EFuNN in terms of low error. However ANFIS took more training time than EFuNN. Hence there is a compromise between performance error and training time. An important advantage of the EFuNN network is its online learning capability. Hence future training would be much easier. The predicted RMSE values are within acceptable rates and hence the developed models are reliable. The prediction accuracy could have been improved if we had not used the noisy input parameter (voltage) or if the actual voltage values were used.

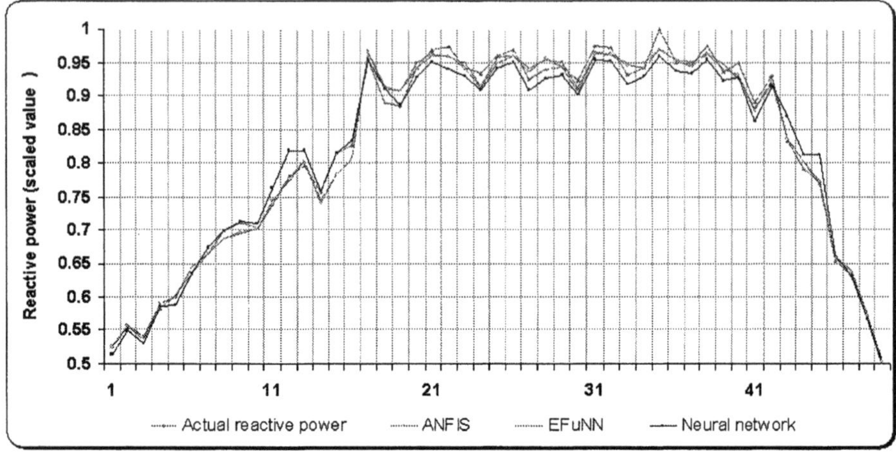

Figure 9. Test results showing the predicted reactive power using different models during the peak hours of shift 1.

Table 2. Reactive power prediction –comparative performance.

	ANFIS	EFuNN	ANN
Learning epochs	50	1	700
Training time (seconds)	36	25	188
Training error (RMSE)	0.0103	0.0116	0.0142
Testing error (RMSE)	0.0102	0.0120	0.0130

3.3 Load Forecasting in Czech Republic

The hourly load in the Czech Republic for the year 2000 is shown in Figure 10. The humidity is sorted into seven categories and labeled as extremely low (ExL), very low (VL), low (L), medium (M), high (H), very high (VH) and extremely high (ExH). The wind speed is labeled as zero (Z), positive very small (PVS), positive small (PS), medium (M), positive medium (PM), big (B) and positive big (PB). Similarly, wind chill is labeled as zero (Z), very very low (VVL), very low (VL), low (L), high (H), very high (VH) and extremely high (ExH) as shown in Figure 11. Inputs to the FNN are previous load and weather parameters i.e. temperature, humidity, wind-speed, and wind-chill and the output of the FNN is the load forecast for a given day [21], [22].

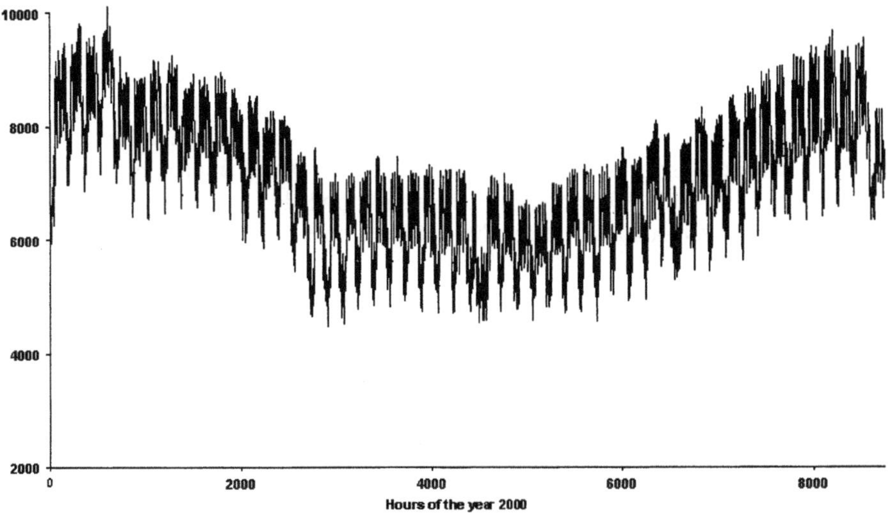

Figure 10. Hourly load of the year 2000.

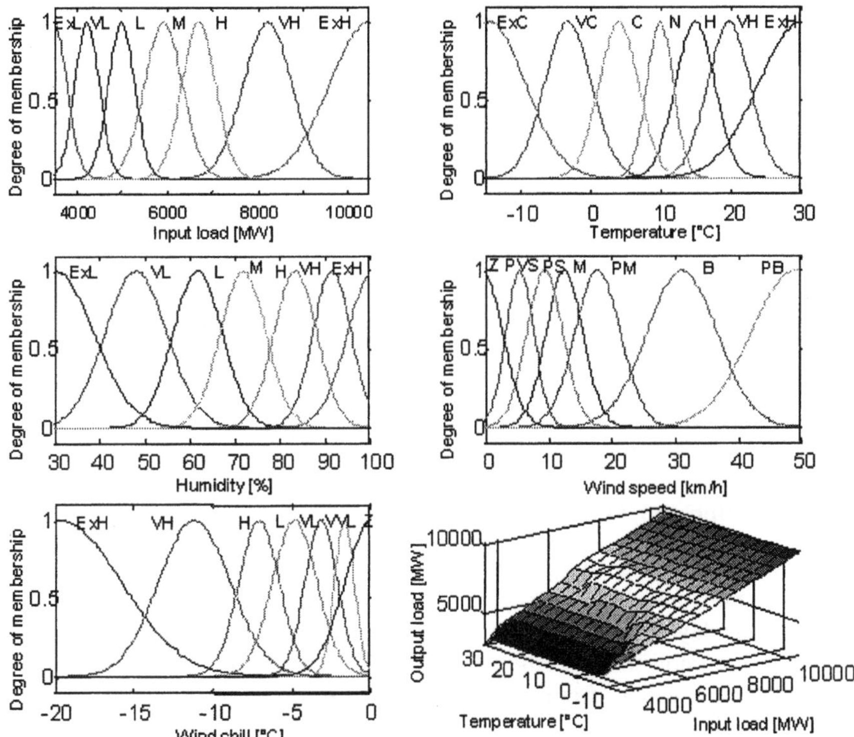

Figure 11. Input parameters using Gaussian-curve membership function.

The proposed fuzzy processor develops a linguistic model that describes dynamic behavior of the system by using a nonlinear mapping between input and output. Application of fuzzy inference rules allows amalgamation of different kinds of knowledge, reduces the dimensionality of the input data and effectively deals with nonprecise (fuzzy) information. The potentials of this technique come from its powerful hybrid methodology and extraordinary approximation power. Furthermore, the ability to incorporate human knowledge makes it a more beneficial candidate over others. The results obtained outperform the connectionist approaches.

The fuzzy processor increases the robustness of the forecast model in tackling uncertainties and ambiguity in the inputs. It also avoids the use of large chunk of historical data, and frequent retraining in response to changing input conditions. The fuzzy-neural network learns the training set near perfect and shows precise prediction. Further improvements to the current implementation could be be made using more complete historical load/meteorological data, improving the rules and refining the training strategies to incorporate incremental learning.

In our experiments, two years of historical load and weather data were used, one year (1999) for designing the fuzzy rule base design and the following year (2000) for testing the model performance. We used a Mamdani fuzzy inference system for predicting the 24-hour ahead (weekdays and weekends) load demand. To ensure prediction accuracy, the number of fuzzy membership functions and shape of the fuzzy membership functions were changed and new fuzzy rule base was obtained. The iterative process of designing the rule base, choosing a defuzzification algorithm, and testing the system performance was repeated several times with a different shapes number of fuzzy membership functions. The fuzzy rule base that provided the minimum error measure for the test set was selected for real-time forecast. We used various Membership Functions (MF) such as triangular, trapezoidal, Gaussian-curve and bell-shaped. The performance of triangular and Gaussian-curve membership functions were better as compared to bell-shaped and trapezoidal. Using different MF, the mean absolute percentage error (MAPE) and maximum absolute percentage error (MAP) for working days of the week and weekend are depicted in Tables 3 and 4 respectively. Empirical values from Tables 3 and 4 illustrate that the selection of different MFs, e.g., triangular, Gaussian, trapezoidal, etc., significantly affect the prediction performance.

3.3.1 Training and Test Data

The basic training set using 24 values of load data and 6 weather parameters (i.e. minimum, maximum and average temperatures, humidity, wind speed and wind chill) and the day of the week was used for training and testing as shown in Figure 12. To ensure good training, the network was trained using a large data set using data containing values from 1995 to 1999. Besides, the utilization of a large amount of training patterns is an effective antidote to fight against over fitting.

Table 3. MAPE and MAP for working days of a week using various MFs.

Working Days	Membership Functions					
	Triangular		Gaussian Curve		Trapezoidal	
	MAPE (%)	MAP (%)	MAPE (%)	MAP (%)	MAPE (%)	MAP (%)
Monday	2.58	6.46	2.84	7.79	2.73	5.84
Tuesday	1.67	4.89	1.18	4.97	2.59	6.82
Wednesday	0.99	3.23	1.87	4.64	1.91	4.74
Thursday	0.97	4.89	1.53	5.52	2.67	5.74
Friday	1.34	3.89	1.69	4.96	2.94	4.85

Table 4. MAPE and MAP for one weekend using various membership functions.

Weekend Days	Membership Functions					
	Triangular		Gaussian Curve		Trapezoidal	
	MAPE (%)	MAP (%)	MAPE (%)	MAP (%)	MAPE (%)	MAP (%)
Saturday	3.52	8.08	4.82	7.74	5.03	8.66
Sunday	3.76	7.47	4.68	8.71	5.06	9.84

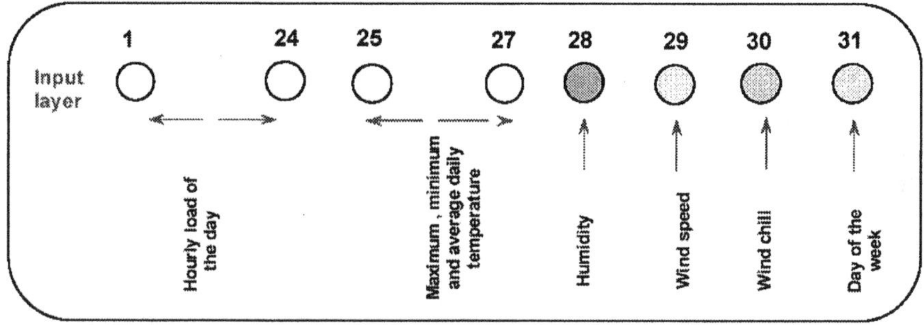

Figure 12. Input data scheme to load forecasting models.

On the other hand, the whole process is slower and morose, due to the computation time required. The load demand patters are different for working days, weekends and holidays as shown in Figure 13. The data for holidays were eliminated. Thus the training data was distributed into two sets for working days and weekends, respectively. The data for the year 2000 is used for testing and evaluation of generalization performance. For empirical comparison purposes, using the same training data, we also trained an artificial neural network using backpropagation algorithm and also developed a Takagi-Sugeno fuzzy inference system. Test data is passed through the trained networks and the performances are depicted in Tables 5 and 6. Figures 14 and 15 illustrate the forecast load versus the actual load along with the forecast error for one working day (Tuesday) and one weekend day (Sunday), respectively. Our training results also reveals that the FNN converged much faster when compared to a pure neural network approach.

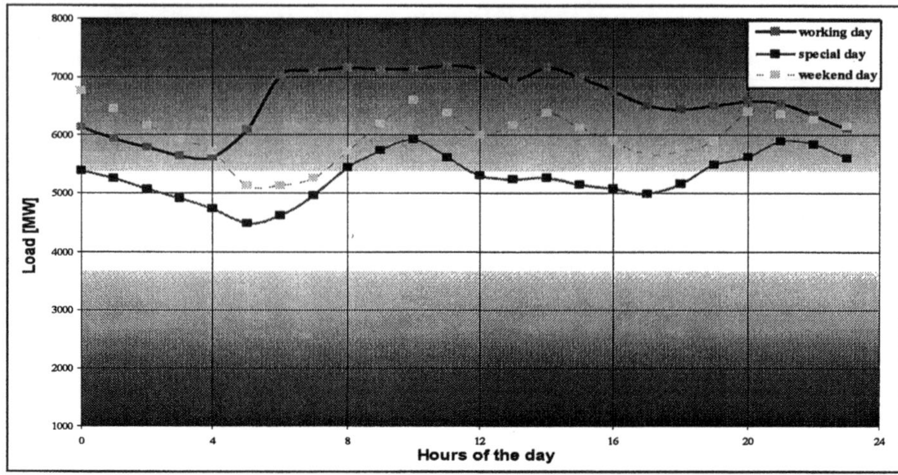

Figure 13. Load trend for a working day, weekend and holiday (special day) of a week in the year 2000.

Table 5. MAPE and MAP of 24-hour forecast during weekends.

Models	Saturday		Sunday	
	*MAPE (%)	*MAP (%)	MAPE (%)	MAP (%)
ANN	2.361	5.22	2.614	5.76
FIS	2.882	6.15	2.764	6.42
FNN	1.893	3.81	2.008	4.23

Table 6. MAPE and MAP of 24-hour forecast during weekdays (working days).

Models	Monday		Tuesday		Wednesday	
	MAPE	MAP	MAPE	MAP	MAPE	MAP
ANN	2.416	5.123	2.321	5.446	2.272	6.142
FIS	2.540	4.678	2.524	5.120	1.987	5.378
FNN	2.000	3.210	1.729	3.151	1.672	3.870

Models	Thursday		Friday	
	MAPE	MAP	MAPE	MAP
ANN	2.614	6.113	2.583	5.893
FIS	2.786	5.820	2.862	5.416
FNN	1.340	3.672	1.435	3.163

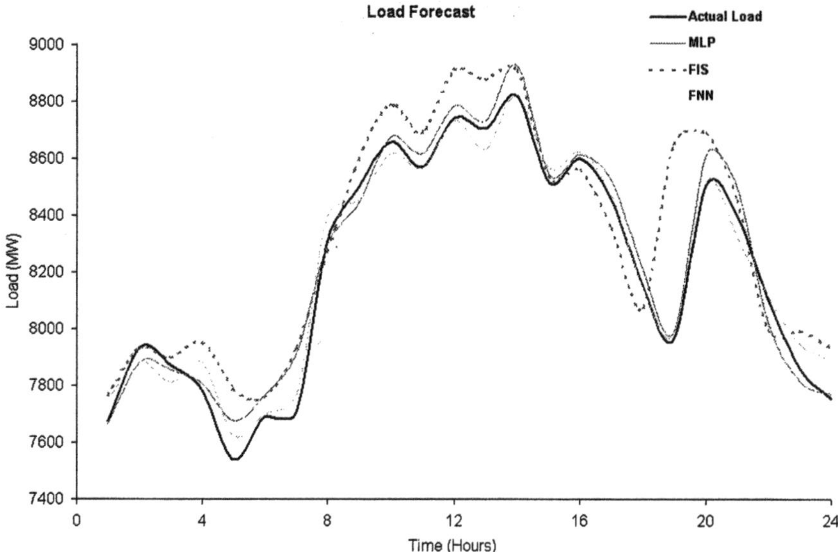

Figure 14. Comparison of 24 hours ahead load forecast for working day (Tuesday) using FNN, FIS and ANN.

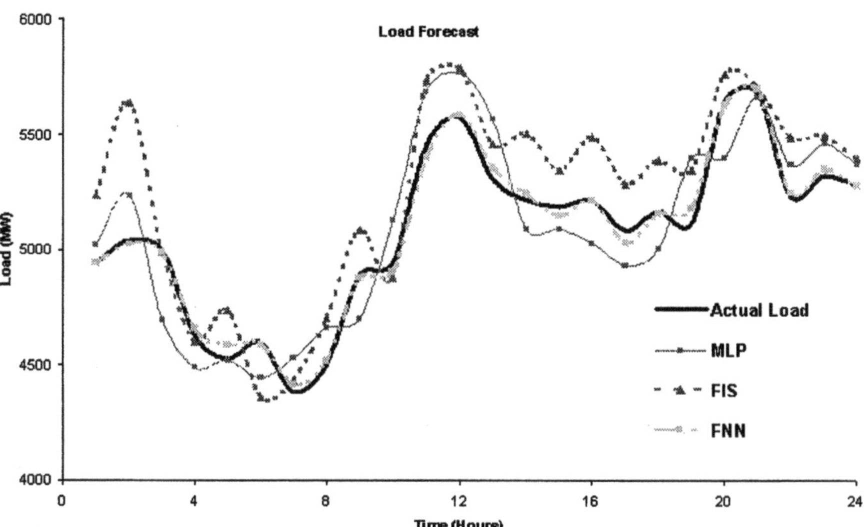

Figure 15. Comparison of 24 hours ahead load forecast for weekend day (Sunday) using FNN, FIS and ANN.

4 Conclusions

In this chapter, we attempted to model the integration of neural networks and fuzzy inference systems for energy management problems. For the demand energy demand and reactive power forecast problem, the proposed neuro-fuzzy approach performed better than the neural network model in terms of low RMSE error and less computational load (less performance time). As depicted in Figure 6 our experimentation results reveal that ANN trained by SCGA converged much faster than BP algorithm. Alternatively, BP training needs more epochs (longer training time) to achieve better performance. The neuro-fuzzy models considered on the other hand are easy to implement and produces desirable mapping function by training on the given data set. All the neuro-fuzzy models require information only about the input variables for generating forecasts thereby reducing the tedious analysis and trail and error methods as in the case of ARIMA model.

EFuNN makes use of the linguistic knowledge of FIS and the learning capability of neural networks. Hence, the neuro-fuzzy system is able to precisely model the uncertainty and imprecision within the data as well as to incorporate the learning ability of neural networks. Even though the performance of neuro-fuzzy systems is dependent on the problem domain, very often the results are better while compared to pure neural network approach. Compared to neural networks, an important advantage of neuro-fuzzy systems is its reasoning ability (*if-then* rules) of any particular state. A fully trained EFuNN could be replaced by a set of *if-then* rules. A simple example of a learned EFuNN learned rule is illustrated below.

> "If the *maximum temperature* of the day is HIGH and *minimum temperature* of the day is LOW and *previous days demand* is MEDIUM and it is *summer* (HIGH) and *9.00 a.m.* (HIGH) and a *Monday* (HIGH) then the *electricity demand* is MEDIUM."

As EFuNN adopts a single pass training (1 epoch) it is more adaptable and easy for further on-line training which might be highly useful for on-line forecasting and bidding. Another important feature of EFuNN is that the user has the flexibility to construct the network (by selecting the parameters). Hence, for applications where speed is more important than the accuracy a faster network can be selected. However, an important disadvantage of EFuNN is the determination of the network parameters like number and type of MF for each input variable, sensitivity threshold, error threshold and the learning rates. Even though a trial and error approach is practical, when the problem becomes complicated (large number of input variables) determining the optimal parameters will be a tedious task.

The hybrid fuzzy neural network combine the advantages of fuzzy systems, which deal with explicit knowledge, which can be explained and understood, and neural networks, which deal with implicit knowledge, that could be acquired by learning. Neural network learning provides a good way to adjust the expert's knowledge and automatically generate additional fuzzy rules and membership functions, to meet certain specifications and reduce design time and costs. The proposed FNN approach is found to be very powerful and robust for short-term load predictions. The simulated results using the FNN network are quite significant compared with the simple ANN and FIS based techniques. By using this hybrid approach, the load forecasting errors are reduced considerably. The advantage of the FNN is its adaptive tracking capability that has resulted in the development of a robust and effective forecasting technique. On the other hand, fuzzy logic enhances the generalization capability of a neural network system by providing more reliable output when extrapolation is needed beyond the limits of the training data. Furthermore, modified and new rules can be extracted from a properly trained hybrid network model, to explain how the results are derived.

Another interesting fact about the considered soft computing models are their robustness and capability to handle noisy and approximate data that are typical in power systems, and therefore, should be more reliable in worst situations.

Acknowledgments

Authors are grateful to the anonymous reviewers for the valuable comments which improved the clarity of the chapter.

References

[1] Abe A. and Lan M. S. (1995), A Method for Fuzzy Rules Extraction Directly from Numerical Data and its Application to Pattern Classification, IEEE Transactions on Fuzzy Systems, 3(1): pp. 18-28.

[2] Abe S. and Lan M. S. (1995), Fuzzy Rule Extraction Directly from Numerical Data for Function Approximation, IEEE Trans. Systems, Man & Cybernetics, 25: pp. 119-129.

[3] Abraham A. (2000), An Evolving Fuzzy Neural Network Model Based Reactive Power Control, In Proceedings of The Second International Conference on Computers in Industry, Bahrain, pp. 247-253.

[4] Abraham A. (2001), Neuro-Fuzzy Systems: State-of-the-Art Modeling Techniques, Connectionist Models of Neurons, Learning Processes, and Artificial Intelligence, Jose Mira and Alberto Prieto (Eds.), Lecture Notes in Computer Science 2084, Springer-Verlag Germany, pp. 269-276.

[5] Abraham A. (2002), EvoNF: A Framework for Optimization of Fuzzy Inference Systems Using Neural Network Learning and Evolutionary Computation, The 17th IEEE International Symposium on Intelligent Control, ISIC'02 Canada, IEEE Press, Canada.

[6] Abraham A. and Nath B. (1999), Artificial Neural Networks for Intelligent Real Time Power Quality Monitoring Systems, First International Power & Energy Conference, INT-PEC'99, CD ROM Proceeding, Isreb M. (Editor), ISBN 0732 620 945, Australia,

[7] Abraham A. and Nath B. (2000), Evolutionary Design of Fuzzy Control Sys-

tems – an Hybrid Approach, The Sixth International Conference on Control, Automation, Robotics and Vision, (ICARCV 2000), December 2000.

[8] Abraham A. and Nath B. (2001), A Neuro-Fuzzy Approach for Forecasting Electricity Demand in Victoria, Applied Soft Computing Journal, Elsevier Science, Volume 1/ 2, pp.127-138.

[9] Cherkassky V. (1998), Fuzzy Inference Systems: A Critical Review, Computational Intelligence: Soft Computing and Fuzzy-Neuro Integration with Applications, Kayak O, Zadeh LA et al. (Eds.), Springer, pp.177-197.

[10] Cordón O., Herrera F., Hoffmann F. and Magdalena L. (2001), Genetic Fuzzy Systems: Evolutionary Tuning and Learning of Fuzzy Knowledge Bases, World Scientific Publishing Company, Singapore.

[11] Furuhashi T. (1997), Development of If-Then Rules with the Use of DNA Coding, Fuzzy Evolutionary Computation, Pedrycz W (Ed.), Kluwer Academic Publishers, pp.107-125.

[12] Hippert, H.S. Pedreira, C.E. and Souza R.C. (2001), Neural networks for short-term load forecasting: A review and Evaluation, IEEE Transactions on Power Systems, Vol. 16, No. 1, pp. 44-55, February 2001.

[13] Jager R. (1995), Fuzzy Logic in Control, PhD Thesis, Technische Universiteit Delft, Netherlands.

[14] Jang J.S.R. (1992), Neuro-Fuzzy Modeling: Architectures, Analyses and Applications, PhD Thesis, University of California, Berkeley.

[15] Jang J.S.R., Sun C T and Mizutani E (1997), Neuro-Fuzzy and Soft Computing : A Computational Approach to Learning and Machine Intelligence, Prentice Hall Inc, USA.

[16] Kasabov N. (1998), Evolving Fuzzy Neural Networks - Algorithms, Applications and Biological Motivation, In Yamakawa T and Matsumoto G (Eds), Methodologies for the Conception, Design and Application of Soft Computing, World Scientific, pp. 271-274.

[17] Kasabov, N. and Woodford B. (1999), Rule Insertion and Rule Extraction from Evolving Fuzzy Neural Networks: Algorithms and Applications for Building Adaptive, Intelligent Expert Systems, In Proceedings of the FUZZ-IEEE'99 International Conference on Fuzzy Systems, Seoul, Korea, pp. 1406-1411.

[18] Kaufmann A. (1975), Introduction to the Theory of Fuzzy Subsets, New York, Academic Press.

[19] Khan M. R., Zak L., and Ondrusek C. (2001), Forecasting Weekly Load Using a Hybrid Fuzzy-Neural Network Approach, International Journal of Engineering Mechanics, pp. 327-336, No. 5, ISSN 1210-2717, Czech Republic.

[20] Khan M.R. (2001), Short-term Load Forecasting for Large Distribution Systems Using Artificial Neural Networks and Fuzzy Logic, Ph.D. Thesis, UVEE, FEI, VUT Brno, Czech Republic.

[21] Khan M.R. and Abraham A. (2003), Short Term Load Forecasting Models in Czech Republic Using Soft Computing Techniques, International Journal of Knowledge-Based Intelligent Engineering Systems, United Kingdom, (forth

coming).
[22] Khan M.R., Abraham A. and Ondrusek C. (2002), Soft Computing Models for Short-Term Load Forecasting in Czech Republic, 1st International Workshop on Hybrid Intelligent Systems, Physica Verlag, Germany, pp. 207-222.
[23] Khotanzad, A. Afkhami-Rohani, R. and Maratukulam, D. (1998), ANNSTLF-Artificial Neural Network Short-term Load Forecaster-Generation Tree, IEEE Transactions on Power Systems, Vol. 13, No. 4, pp. 1413-1422.
[24] Khotanzad, A. Hwang R. C. Abaye, A. and Maratukulam, D. (1995), An Adaptive Modular Artificial Neural Network Hourly Load Forecaster and its Implementation at Electric Utilities, IEEE Transactions on Power Systems, Vol. 10, No. 3, pp. 1716-1722.
[25] Kiartzis, S.J. Zoumas, C.E. Theocharis, J.B. Bakirtzis, A.G. and Petridis V. (1997), Short-term Load Forecasting in an Autonomous Power System Using Artificial Neural Networks, IEEE Transactions on Power Systems, Vol. 12, No. 4, pp. 1591-1596.
[26] Lin C.T. and Lee C.S.G. (1996), Neural Fuzzy Systems: A Neuro-Fuzzy Synergism to Intelligent Systems, Prentice Hall Inc, USA.
[27] Mori H, Hidenori K. (1996), Optimal Fuzzy Inference for Short-term Load Forecasting, IEEE Transactions on Power Systems, Vol. 11, No. 1, pp. 390-396.
[28] Nath B. and Nath M. (2000), Using Neural Networks and Statistical Methods for forecasting Electricity Demand in Victoria, International Journal of Management and Systems (IJOMAS), Special issue on Mathematics for Industry, Volume 16, No. 1, pp. 105-112.
[29] Nauk D., Klawonn F. and Kruse R. (1997), Foundations of Neuro-Fuzzy Systems, John Wiley & Sons Ltd, United Kingdom.
[30] Papalexopoulos A. D. Hao, S. Peng, T. M. (1994), An Implementation of a Neural Network Based Load Forecasting Model for the EMS, IEEE Transactions on Power Systems, Vol. 9, No. 4, pp. 1956-1962.
[31] Pedrycz W (Editor) (1997), Fuzzy Evolutionary Computation, Kluwer Academic Publishers, USA.
[32] Peng, T.M. Hubele, N.F. and Karady, G.G. (1992), Advancement in the Application of Neural Networks for Short-term Load Forecasting, IEEE Transactions on Power Systems, Vol. 7, No. 1, pp. 250-257.
[33] Piras, A. Germond, A. and Buchenel, B. Imhof, K. and Jaccard, Y. (1996), Heterogeneous Artificial Neural Network for Short-term Electrical Load Forecasting, IEEE Transactions on Power Systems, Vol. 11, No. 1, pp. 397-402.
[34] Procyk T J and Mamdani E H (1979), A Linguistic Self Organizing Process Controller, Automatica, Volume 15, pp. 15-30.
[35] Ranaweera D.K., Hubele N.F., Karady G.G. (1996), Fuzzy Logic for Short-Term Load Forecasting, Electrical Power and Energy Systems, Vol. 18, No. 4, pp. 215-222.
[36] Takagi T. and Sugeno M. (1983), Derivation of Fuzzy Control Rules from

Human Operators Control Actions, Proceedings of the IAFC Symposium on Fuzzy Information, Knowledge Representation and Decision Analysis, pp 55-60.

[37] Tsukamoto Y. (1979), An Approach to Fuzzy Reasoning Method, Gupta MM et al. (Eds.), Advances in Fuzzy Set Theory and Applications, pp. 137-149.

[38] Wang L.X. and Mendel J.M. (1992), Backpropagation fuzzy system as Nonlinear Dynamic System Identifiers, In Proceedings of the First IEEE International conference on Fuzzy Systems, San Diego, USA, pp. 1409-1418.

[39] Wang L.X. and Mendel J.M. (1992), Generating Fuzzy Rules by Learning from Examples, IEEE Transactions on Systems, Man and Cybernetics, Volume 22, No 6., pp. 1414-1427.

[40] Yoshinari Y., Pedrycz W., Hirota K. (1993), Construction of Fuzzy Models Through Clustering Techniques, Fuzzy Sets and Systems, Volume 54, pp. 157-165.

[41] Zadeh L.A. (1965), Fuzzy Sets, Information and Control, Volume 8: pp. 338-353.

[42] Zadeh L.A. (1973), Outline of a New Approach to the Analysis of Complex Systems and Decision Processes, IEEE Transactions on Systems, Man and Cybernetics, 3(1): pp. 28-44.

Chapter 13

Information Space Optimization for Inductive Learning

Ryohei Orihara, Tomoko Murakami, Naomichi Sueda, and Shigeaki Sakurai

Summary. New feature construction methods are presented. The methods are based on the idea that a *smooth* feature space facilitates inductive learning thus it is desirable for data mining. The methods, Category-guided Adaptive Modeling (CAM) and Smoothness-driven Adaptive Modeling (SAM), are originally developed to model human perception of still images, where an image is perceived in a space of index colors. CAM is tested for a classification problem and SAM is tested for a Kansei scale value (the amount of the impression) prediction problem. Both algorithms have been proved to be useful as preprocess steps for inductive learning through the experiments. We also evaluate CAM and SAM using datasets from the UCI repository and the result has been promising.

Keywords: human perception, Kansei engineering, data mining, television commercial, real-coded genetic algorithm.

1 Introduction

Human perception is personal and adaptive. Impressions evoked by stimuli are heavily influenced by memories and situations. There appears to be agreement in cognitive psychology that our perception is the result of the interaction of two distinct processes, i.e., a bottom-up process that is driven by the stimuli and a top-down process that is controlled by memory [1, 2]. Therefore, what a person perceives from particular stimuli depends on who the person is. The fact is especially important when the stimuli come from an artifact created by another human to convey some messages, e.g., a work of art such as a drawing or an advertisement such as a television commercial. One person might consider a particular television commercial persuasive and another might not. A producer of television commercial needs to know how a particular commercial will be perceived by the target audience.

We aim to build a software architecture that enables adaptive cognition of situations by simulating human perception in a computer model. We have chosen analysis of television commercials as a problem area. Possible applications include a system to support a user in identifying common characteristics of successful (or failed) past television commercials, and a system to support a user in identifying components of a television commercial that are particularly persuasive to a certain class of viewers.

Thus it is not only interesting but also commercially important to simulate human perception in a computer model.

We recreate a human's mental space with a problem solver that uses perceived stimuli in the space as inputs. Because we cannot decide adequateness of the simulated mental space by comparing it with a real one, the adequateness must be judged on the basis of outputs of the problem solver.

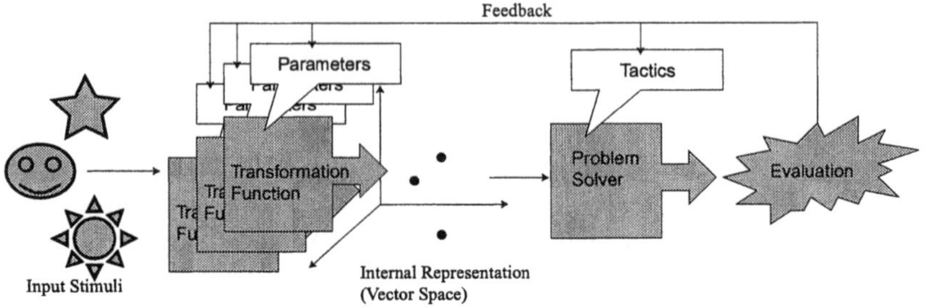

Figure 1. Perception and problem solving.

In Figure 1, our view of perception and problem solving is summarized. Input stimuli are translated into internal representations, namely, points in a vector space that represents the artificial mental space, by means of transformation functions. The points become an input for a problem solver. An output of the problem solver is evaluated with the environment and feedback is given to the transformation functions and the problem solver. Conventional machine learning only considers adaptation of the problem solver. We think adaptation of the transformation functions is more important and we investigate the possibility of achieving that. We employ a vector space as a simulated mental space because of the following advantages.

- Similarity between situations in the software's mental space can be easily measured by means of Euclidean distance between corresponding points in the vector space.
- It is easy to manipulate the points by something other than the input stimuli, if they are related to the axes composing the vector space.
- A set of points in the vector space can be seen as a private concept of the software. This view allows us to give it a simple means of forming a new concept, i.e., by standard clustering techniques.

Regarding the transformation, behavior of a transformation function should be systematically controllable to allow the software to gradually improve its perception through the execution of tasks. This can be done by parameterizing the function.

Creating a mental model in the artificial mental space can be thought as change of representation. Representation has always been an issue in Artificial Intelligence (AI). In traditional AI, the same formalization is often employed for the input stimuli and the internal representation. This approach is prone to be criticized on the ground that is a *single representation trick* [3], namely the software's programmer can design the formalization with knowledge of a task to be done so that the intelligent software can easily perform the task.

In neural network approaches, weights between neurons are trained by input-output pairs of data. A set of the weights can be seen as the neural network's internal representation of the situation defined by the training data. Although there is an indirect way to know the meaning of the set [4], it cannot be probed in a direct manner. Manipulating the set by means of factors other than the training data, for example, by memory, has yet to be investigated.

The issue of representation has been most actively studied in cases where the problem solver's task is inductive concept learning. The relationship between constructive induction, in which a learner creates new features to facilitate the learning, and our approach will be discussed in Section 6.2. However, the authors think that Rendell's work [5] should be mentioned here because the work has greatly influenced our method.

Rendell pointed out that an inductive learner performs poorly when an instance space has a lot of peaks, namely when the space is not regular in terms of the class membership function. Such space should be converted to a smoother space with fewer peaks by means of new features. He proposed a generate-and-test method to create the new features. Although the *smoothness* played a major role in explaining the difficulty of certain learning problems, it did little for the method proposed by him.

Our approach is, in a nutshell, to adjust the parameters of the transformation functions in order to maximize the smoothness of the vector space. Regarding the model shown in Figure 1, we adjust the parameters based on the property of the instance space that is closely related to behavior of the problem solver, rather than the behavior itself. In the next section we explain the objective of the project where we are conducting the study and data we are using. In Section 3 we describe the implementation of our method for still images. As mentioned above, we are dealing with analysis of television commercials. Among the data, (treatment of) the still image is particularly interesting. Because information such as what or who is in the scene is explicitly provided, conventional image understanding is not required here. Rather, it is important to evaluate emotional effects of the image on a viewer. This is one of research areas of *Kansei engineering*, an emerging technology domain where aspects of information processing related to *Kansei* — a Japanese word that implies human reaction under various stimuli ranging from sensory to mental state, that is sensitivity, sense, sensibility, feeling, esthetics, emotion, affection and intuition — are studied [6]. Later in Section 4 we will employ some methodologies from Kansei engineering to experimentally show the effectiveness of the method.

In a data mining context, the creation of a mental model can be seen as preprocessing. Although in general the transformation functions should be chosen to reflect the given task's characteristics, it is interesting if the method is effective for well-tested datasets such as ones in the UCI repository [7], with pre-selected some *general* transformation functions. We show the result in Section 5.

In Section 6, we compare our method with previous works, including various constructive induction approaches, as mentioned above. Finally, we summarize the chapter and show the future direction of the study in Section 7.

2 Project and Data

We aim to build a software architecture that enables adaptive cognition of situations. As a problem area we have chosen analysis of television commercials. Possible applications include:

- a system to support a user in identifying common characteristics of successful (or failed) past television commercials
- a system to support a user in identifying components of a television commercial that are particularly persuasive to a certain class of viewers
- and a system to suggest possible improvements for a given tentative commercial to a producer

Television commercial data consists of the following.

- Data on television commercials themselves. A product advertised in a commercial, scripts used in a commercial, composition description of a commercial, personalities featured in a commercial, representative still images of a commercial and so on. Although we are sure that video and audio are important components of a television commercial, we do not yet have data on them.
- Data on the ratings
- Data on commercial scheduling
- Data on the program scheduling
- Data on the viewers participating in the ratings research
- Data from market research on some familiar products

So far we have implemented our methods for text [8, 9], image [6] and general numercal data [10], such as attributes of personalities. The methods are capable of optimizing data spaces according to either categorized or ordered evaluation function. For the text data we employ the vector space model, a practice commonly used in the area of information retrieval (IR). Each document is converted to a vector based on frequencies of occurences of words. We employ the conversion as a transformation function. The method is evaluated in the IR context and proved to be useful. In the television commercial project, it is used for data such as dialogues and scripts.

In this chapter we focus on the methods for images and numerical data. The former is actually a special case of the latter, where an index color extraction algorithm is used as a transformation function. The method will be explained in detail in the next section.

3 Still Image Perception

A human obtains millions pieces of information from vision and simulating all of them exceeds this study's scope. In this chapter we focus on color information. This is consistent with our interest in the still images among the television commercial data because color information obviously has a great effect on the emotional aspects of image perception.

Usually a lot of colors are included in input stimuli. However, people do not care most of them. We *selectively* see colors, possibly based on our purposes. A previous study on art suggests that an index color need not occupy a large area in an image. For example, when we see a red rose on a white background, we feel the red is the index color [11].

The previous study also proposes an algorithm to extract index colors from an image [11]. The algorithm has several parameters to be tuned for a purpose. Our approach employs the algorithm as a transformation function for color information. The parameters of the algorithm are then adjusted to make a space of index colors useful for a task.

3.1 Index Color Extraction

The algorithm consists of the following three steps.

Reducing the Number of Colors. Usually a number of colors are used in an image and it is computationally prohibitive to treat all of them. The algorithm first reduces the number of colors in an image by mixing similar colors each other, using the color resolution threshold p as a parameter. The following shows an overview of the process.

1. All colors in an image are converted from RGB space into $L^*a^*b^*$ space. The distance of two colors is closer to human intuition in the $L^*a^*b^*$ space than in the RGB space [12]. Appendix A shows the details of the conversion.
2. The color list, which consists of a color and its frequency, is created as follows. When a new color is added to the list, if the distance between the color and one of the colors already in the list is less than a threshold p, then the new color and the color in the list are mixed with each other regarding their frequencies.
3. Less frequent colors in the color list are likely to be noise. If a color's frequency is less than 0.5% of all the pixels in the image, then the color is mixed with the nearest color in the list regarding their frequencies.
4. The number of colors in the list is reduced by mixing the two nearest colors regarding their frequencies. This step is iterated until the number of colors becomes m.
5. A histogram is created for each of L^*, a^*, b^* by partitioning the axis into $2m$ regions and summing the frequencies in the region. The maximum number of convexes m^* in the histograms represents the *true* number of colors, namely, the number of different colors that the designer intended to use. The number of colors in the list is reduced by mixing the two nearest colors regarding their frequencies until it becomes m^*. Then the histograms are recreated and m^* is recalculated. The procedure is iterated until m^* is not changed by the recalculation.

Calculating Conspicuousness of Colors. An index color must be conspicuous. Conspicuousness L of a color is defined as follows:

$$L \stackrel{def}{=} \sqrt{(e \cdot C_1)^2 + (f \cdot C_2)^2 + (g \cdot C_3)^2} \tag{1}$$

where C_1 denotes frequency of the color, C_2 denotes temptation of the color and C_3 denotes contrast of the color, and e, f and g are parameters. C_1, C_2 and C_3 are defined as follows:

- Frequency

$$C_1 \stackrel{def}{=} \frac{\text{Frequency of the color}}{\text{The number of pixels in the image}} . \tag{2}$$

- Temptation
  First, the color is converted into HSV space [12]. HSV stands for Hue, Saturation and Value. Appendix B shows the details of the conversion. A value H^* is defined based on H coordinate of the HSV space, using a predefined table [13], as shown in Appendix C. Then

$$C_2 \stackrel{def}{=} \frac{s \cdot H^* + t \cdot S + u \cdot V}{s + t + u} \tag{3}$$

  where s, t and u are parameters.
- Contrast
  Let V^* denote the V value of the color created by mixing all the colors in the list but the color under consideration. Then

$$C_3 \stackrel{def}{=} |V - V^*| . \tag{4}$$

Specifying Index Colors. The L is calculated for all the colors in the list. Index colors are N colors with larger L values in the list. Let us emphasize that we have introduced controllable parameters p, e, f, g, s, t, u and m for flexible modeling of images.

3.2 Algorithms for Color Space Optimization

By applying the index color extraction algorithm to an image, N colors are extracted from the image. That is $3N$ real values an image, since a color is represented by H, S and V. We regard this $3N$-dimensional space as an instance space for images.

We would like to analyze the images of a television commercial using the space from various viewpoints such as product category, brand separation and popularity of the commercial, as mentioned earlier. Classification by an inductive learner may be used as an engine of the analysis. However, as Rendell pointed out [5], a smooth instance space — smooth in terms of an evaluation function representing a viewpoint — is necessary for successful analysis. In this subsection we present two algorithms for that purpose: one for the case when the evaluation function is categorical and the other for the case when the evaluation function is continuous.

3.2.1 Category-Guided Adaptive Modeling

Suppose raw data, which consist of any-type-valued vectors and categories to which they belong, are given. Category-guided Adaptive Modeling, or CAM for short, defines functions $\{F_j\}$ so that different categories become separated and each category becomes compact in a space spanned by the functions. The space is called a *CAM space* and F_j is called a *CAM transformation function.*

Allowing a system to create arbitrary functions often results in too much complexity and generation of a lot of meaningless functions [14]. We assume that an outline of F_j is given as a parameterized function. The final F_j is determined by optimizing the parameters with respect to properties of categories in a CAM space.

Suppose we have data $\{\vec{r}_1, \ldots, \vec{r}_n\}$, the categories $\{L_1, \ldots, L_N\}$ that the raw data belong to and CAM transformation functions $\{F_1, \ldots, F_M\}$. Table 1 shows correspondences between values of the CAM transformation functions and the categories $\{L_1, \ldots, L_N\}$ that the raw data belong to. In Table 1, $\vec{r}_1$ and $\vec{r}_2$ belong to the same category L_1.

Table 1. Values of CAM transformation functions and corresponding categories.

$$\begin{array}{cccc|c} F_1(\vec{r}_1) & F_2(\vec{r}_1) & \ldots & F_M(\vec{r}_1) & L_1 \\ F_1(\vec{r}_2) & F_2(\vec{r}_2) & \ldots & F_M(\vec{r}_2) & L_1 \\ \vdots & \vdots & \vdots & \vdots & \vdots \\ F_1(\vec{r}_n) & F_2(\vec{r}_n) & \ldots & F_M(\vec{r}_n) & L_N \end{array}$$

Let $L(\vec{r}_i)$ denote the category that $\vec{r}_i$ belongs to. Let us define the following vector for each category L_k.

$$\vec{C}(L_k) = (C_1(L_k), \ldots, C_M(L_k)) \quad (5)$$

where

$$C_j(L_k) \stackrel{def}{=} \frac{\sum_{\vec{r}_i \in \{\vec{r} | L(\vec{r}) = L_k\}} F_j(\vec{r}_i)}{|\{\vec{r} | L(\vec{r}) = L_k\}|} \,. \quad (6)$$

Intuitively, we want F_j with the following property.

- The distance between mean values of F_j of different categories is large,
- The variance of values of F_j within a category is small.

It is desirable that the distance between mean values of F_j of different categories is large, and that the variance of values of F_j within a category is small. Let us assume that we give priority to the first criterion. The policy can be approximately implemented by minimizing the following function.

$$E(D_j, V_j) \stackrel{def}{=} \lceil 10000 e^{-D_j/1000} \rceil - e^{-V_j/1000} + 1 \quad (7)$$

where $\lceil \cdot \rceil$ represents the Gaussian symbol, i.e., the largest integer that does not exceed the value inside the symbol, and

$$D_j \stackrel{def}{=} \sum_{k_1,k_2 \in \{1,\ldots,N\} \wedge k_1 \neq k_2} |C_j(L_{k_1}) - C_j(L_{k_2})| \ , \tag{8}$$

$$V_j \stackrel{def}{=} max(V_{j,1}, \ldots, V_{j,N}) \ , \tag{9}$$

$$V_{j,k} \stackrel{def}{=} \sum_{\vec{r}_i \in \{\vec{r} | L(\vec{r}) = L_k\}} (F_j(\vec{r}_i) - C_j(L_k))^2 \ . \tag{10}$$

Each F_j is optimized separately. We employ a kind of Genetic Algorithm (GA), which is explained in 3.3, to implement the optimization.

Although it may not be obvious, the basic idea of CAM was drawn from Rendell's argument on *smoothness* [5]. When the class is categorical, Rendell considered a space as smooth if data with the same class occupy a compact region in the instance space, rather than spread all over the space. Equation (7) tries to implement the policy.

3.2.2 Smoothness-Driven Adaptive Modeling

Suppose raw data, which consist of any-type-valued vectors and values of an evaluation function E of them, are given. For clarity, E is called an *evaluation function in the narrow sense*, because we will have another evaluation function for optimization later. Smoothness-driven Adaptive Modeling, or SAM for short, defines functions $\{F_j\}$ so that a curved surface of E becomes smooth over a hyperplane spanned by the functions.

Intuitively, E represents a function to be predicted by learning and serves as a base of the smoothness of the instance space. A value of E of i-th data is denoted as E_i. The hyperplane is called a *SAM plane*, a space defined by a SAM plane and E is called a *SAM space* and F_j is called a *SAM transformation function.*

Again, we assume that an outline of F_j is given as a parameterized function. The final F_j is determined by optimizing the parameters with respect to smoothness of E's surface in a SAM space. The smoothness is measured using a heuristics described in the following.

Suppose we have data $\{\vec{r}_1, \ldots, \vec{r}_n\}$ and SAM transformation functions $\{F_1, \ldots, F_M\}$. Table 2 shows correspondences between values of the SAM transformation functions and values of E.

Table 2. Values of SAM transformation functions and corresponding values of E

$$\begin{array}{cccc|c} F_1(\vec{r}_1) & F_2(\vec{r}_1) & \ldots & F_M(\vec{r}_1) & E_1 \\ F_1(\vec{r}_2) & F_2(\vec{r}_2) & \ldots & F_M(\vec{r}_2) & E_2 \\ \vdots & \vdots & \vdots & \vdots & \vdots \\ F_1(\vec{r}_n) & F_2(\vec{r}_n) & \ldots & F_M(\vec{r}_n) & E_n \end{array}$$

We can assume that the data is sorted such that $E_1 \geq E_2 \geq \ldots \geq E_n$ without loss of generality because the order of the given data has no particular meaning. Let us consider the smoothness of E's surface in the SAM space. It becomes smoothest when $F_j(\vec{r_i})$ is monotonous with respect to E_i for all j. Now we try to define a function $H_j(F_j(\vec{r_1}), \ldots, F_j(\vec{r_n}))$ that represents the monotonousness.

Let us consider the following value.

$$\hat{d}_{j,i} \stackrel{def}{=} \frac{F_j(\vec{r}_{i+1}) - F_j(\vec{r_i})}{E_{i+1} - E_i} \ . \tag{11}$$

The basic idea is "if $\hat{d}_{j,i}$ takes a non-zero constant over i then the space is smooth with respect to F_j." The non-zero condition is necessary because otherwise the SAM plane is reduced to a point. Let the constant be 1. We can argue on desirability of $\hat{d}_{j,i}$ that is not equal to 1 as follows.

- $\hat{d}_{j,i} > 1$: Less desirable than $\hat{d}_{j,i} = 1$ but acceptable because this does not immediately lead to a peak in the SAM space. Desirability gradually decreases as $\hat{d}_{j,i}$ increases.
- $\hat{d}_{j,i} < 1$: Very undesirable because this immediately implies a gap in the SAM space. Desirability abruptly decreases as $\hat{d}_{j,i}$ decreases.

The following definition of G is consistent with the above discussion.

$$G(\hat{d}_{j,i}) = (\hat{d}_{j,i} - p)^2 + \frac{-A(\hat{d}_{j,i} - p)}{1 + e^{\hat{d}_{j,i} + B - p}} \quad \text{if } E_{i+1} - E_i \neq 0 \ , \tag{12a}$$

$$G(\hat{d}_{j,i}) = (\hat{d}'_{j,i} + 1 - p)^2 + \frac{-A(\hat{d}'_{j,i} + 1 - p)}{1 + e^{\hat{d}'_{j,i} + B + 1 - p}} \quad \text{otherwise} \ , \tag{12b}$$

$$\text{where } \hat{d}'_{j,i} \stackrel{def}{=} F_j(\vec{r}_{i+1}) - F_j(\vec{r_i}) \ .$$

A and B defines the shape of G. In the following experiments, we use $A = 20$ and $B = 2$. p is a constant to be set so that G takes the smallest value at $\hat{d} = 1$ in equation (12a).

With G, the evaluation function for F_j is defined as the following. Here, the lower the value of H_j, the smoother the space is.

$$H_j = \sum_{i=1}^{n-1} G(\hat{d}_{j,i}) \ . \tag{13}$$

Each F_j is optimized separately. We again employ a kind of GA, as in CAM, to implement the optimization. Now we show details of the GA.

3.3 GA with Simplex Crossover

We employ a kind of Genetic Algorithm (GA) to optimize the parameters of the transformation functions. There have been several approaches to apply GA to function optimization. We use one of them, an algorithm presented in [15], because the algorithm was reported to be superior than standard bit-string coding [15]. The algorithm employs a special crossover method called *simplex crossover*. The following is the outline of the algorithm.

- A chromosome is a vector of parameters of the function to be optimized. Suppose $a_1, a_2, \ldots, a_l$ are the parameters. A chromosome is a l-dimensional real-valued vector.
- Crossover is carried out as follows:
  1. Choose $l+1$ parents $\vec{x}_0, \vec{x}_1, \ldots, \vec{x}_l$ randomly.
  2. Let $\vec{g}$ denote the center of gravity of the parents. Namely,

$$g_k \stackrel{def}{=} \frac{1}{l+1} \sum_{i=0}^{l} x_{ik} \tag{14}$$

     where $\vec{g} = (g_1, ..., g_l), \vec{x}_i = (x_{i1}, ..., x_{il})$.
  3. Suppose $\vec{c}^0 \stackrel{def}{=} (0, \ldots, 0), \vec{p}^0 \stackrel{def}{=} \vec{g} + \alpha(\vec{x}_0 - \vec{g})$.
  4. Compute the following equations for $k = 1, \ldots, l$.

$$\vec{p}^k = \vec{g} + \alpha(\vec{x}_k - \vec{g}) \tag{15}$$
$$\vec{c}^k = R_{k-1}(\vec{p}^{k-1} - \vec{p}^k + \vec{c}^{k-1}) \tag{16}$$

     where

$$R_k \stackrel{def}{=} u(0,1)^{\frac{1}{k+1}} \tag{17}$$

     while $u(0,1)$ denotes the uniform random number in $[0,1]$.
  5. A child $\vec{x}^c$ is defined as the following.

$$\vec{x}^c \stackrel{def}{=} \vec{p}^l + \vec{c}^l \ . \tag{18}$$

- The crossover parameter α is defined as the following.

$$\alpha \stackrel{def}{=} \sqrt{l+2} \ . \tag{19}$$

- There is no mutation.
- Generation renewal is carried out as follows:
  1. Choose $l+1$ individuals as parents from the population randomly.
  2. Create a child by means of the crossover.
  3. Choose two individuals randomly from the parent.
  4. Replace the two individuals chosen in 3 with (1) an individual with the highest fitness value among the two and the child, and (2) an individual chosen by roulette selection from the three.

In order to use this algorithm to optimize the parameters of the CAM/SAM transformation functions, we define a chromosome as a sequence of the parameters. The fitness function is defined by means of equations (6),(7),(8),(9) and (10) for CAM and (11),(12a) and (12b) for SAM, respectively.

4 Experiments with Still Images

We have tested CAM and SAM using images of television commercials. Each image size is $128 \times 88 \times 16$ (bits).

4.1 CAM Experiment

In this experiment, the task is to classify images according to product category. CAM is used as a preprocessing step for inductive learning and results of learning with and without it are compared. Although CAM itself can be seen as a classifier and compared with linear discriminant analysis (LDA) or its relatives [16], we leave it for future work.

First, we divided all the images into two categories: the images taken from television commercials advertising computer games, and the other images. The task is to learn a rule for the computer game in a space of index colors. The following three methods were compared.

1. Using index colors specified by the index color extraction algorithm with default parameters, $p = 7.0, m = 10, s : t : u = 2 : 1 : 1$ and $e : f : g = 3 : 1 : 2$, which were determined by a color designer (a domain expert) [11], a $3N$-dimensional space is created. The rule is learned in the space by C4.5 [17].
2. Before applying C4.5 in method 1, a feature selection algorithm Relief [18] is used to select important features. Relief is known to be effective for a number of learning problems. C4.5 is applied to the features selected by Relief. Relief's parameters, the number of iteration and the threshold, are set to 30,000 and 0.0058 respectively, to be consistent with the guideline shown in [18].
3. An optimized color space by CAM is used, namely, the index color extraction algorithm is used as a CAM transformation function. Here CAM is used to optimize $e : f : g$ only. The default values are used for the rest of the parameters. The rule is learned in the space by C4.5.

Each method was tested for $N = 2, 3, 4, 5, 6$ and 7. In method 3, CAM's results were to optimize $e : f : g$ as $2 : 1 : 1$ for all the index colors for all N. We evaluated performance of a decision tree created by each method by performing 4-fold cross validation (CV) using 268 instances.

Figure 2 and 3 show the results regarding accuracy and size of the learned trees, respectively. In both figures the horizontal axis shows N. In Figure 2, the vertical axis shows prediction accuracy, average of 4-CV trials. In Figure 3, the vertical axis shows the number of nodes in the learned tree, including both internal nodes and leaves, also average of 4-CV trials. In this experiment, Relief served as a conventional preprocessing method for inductive learning, to make a comparison with CAM. In every case, CAM (method 3) yielded the best result among the three methods. Relief's struggling performance (method 2) indicates that there is insufficient information for the task even in the 7×3-dimensional color space. Nevertheless, CAM provided improvements, especially in terms of compactness.

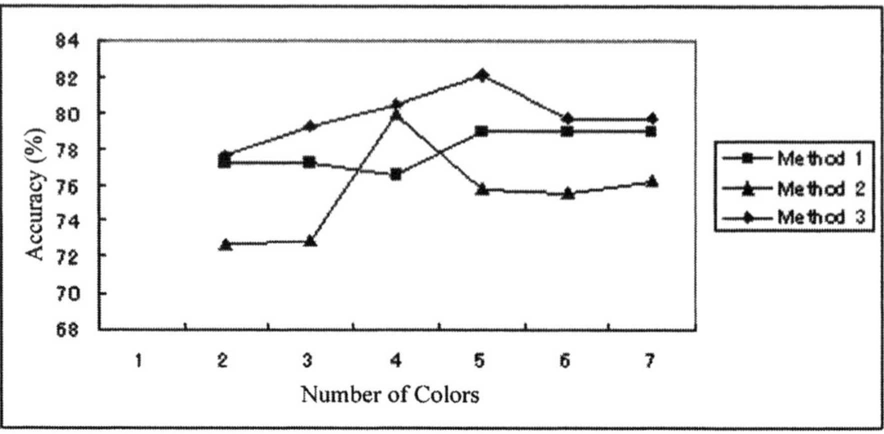

Figure 2. Decision tree accuracy for the CAM experiment.

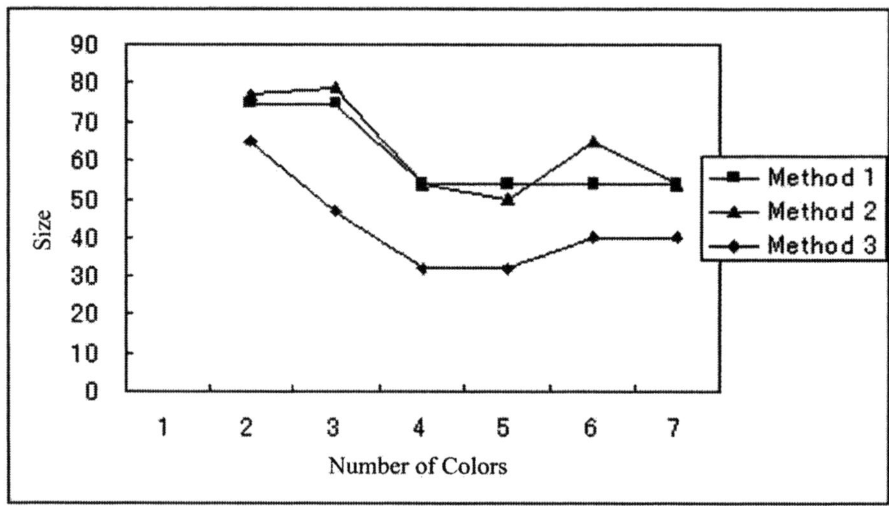

Figure 3. Decision tree compactness for the CAM experiment.

4.2 SAM Experiment

In this experiment, the task is to predict each image's *Kansei scale* value. The Kansei scale quantitatively represents the intuitive impression that a human feels when viewing an image, by means of a pair of Kansei image words and a range of real numbers [13]. For example, when we consider a Kansei scale consisting of a pair *dark⇔bright* and a range $[-3, 3]$, a value 2 for an image means that the image looks rather bright. The prediction itself is done by inductive learning. SAM is used as a preprocessing step for the inductive learning and results of learning with and without it are compared.

We conducted a cognitive experiment to gather Kansei scale data. 60 examinees are shown 100 images and asked to score brightness (or darkness) of each image using integers $\{-3, -2, -1, 0, 1, 2, 3\}$, -3 being the darkest and 3 being the brightest. By averaging values for each image, we have the Kansei scale value from a human's viewpoint. Then we extract 42 images whose absolute value for the scale is greater than or equal to 1.5, from the 100 images. The 42 images consist of 35 *bright* images and 7 *dark* images. The task is to learn a model of brightness in a space of index colors, using the 42 examples. The following three methods were compared. This time the number of index colors is fixed to 2.

1. Using index colors specified by the index color extraction algorithm with default parameters shown in the CAM experiment, a 6-dimensional space is created. The model is learned in the space by C4.5 and a back propagation-based neural network KINOsuite-PR [19].
2. Before applying the inductive learners in method 1, Relief is used to select important features. Relief's parameters are set as in the CAM experiment. C4.5 and KINOsuite-PR are applied to the features selected by Relief.
3. A color space optimized by SAM is used, namely, the index color extraction algorithm is used as a SAM transformation function. Here SAM is used to optimize $e : f : g$ only. The default values are used for the rest of the parameters. The model is learned in the space by C4.5 and KINOsuite-PR.

In method 3, SAM's results were to optimize $e : f : g$ as $3 : 2 : 2$ for both index colors. We evaluated the performance of a model created by each method by performing 4-fold CV.

Table 3 shows the result. Each score is the average of 4-CV trials.

Table 3. Results for Kansei scale prediction experiment.

	C4.5		Neural net
	Accuracy (%)	# of Nodes	Accuracy (%)
method 1	57.35	5.25	64.14
method 2	69.85	9.25	67.43
method 3	74.91	8.25	71.71

Comparing with the CAM experiment, the Relief (method 2) yielded good results. However, SAM (method 3) is better. Furthermore, it should be noted that both SAM and Relief improve the results regardless of the inductive learner. As we will see in Section 6.2, there is a family of preprocess algorithms that is closely connected with an inductive learner. Such algorithms inherently limit freedom of selection of an inductive learner, which is sometimes an undesirable property in data mining applications.

Most research on the Kansei scale consists of attempts to establish a common medium for exchanging human impressions. Our objective is different. We believe that each person has their own Kansei scales and it should be beneficial to them to make the scales explicit. Such individual scales can still facilitate the exchange of impressions, if they are made explicit.

5 Experiments with UCI Datasets

In this section the authors test CAM and SAM against the UCI datasets [7]. Usually the transformation functions should be carefully determined considering the task's property, like we did in the last section. In this section, we examine the methods as a general preprocessing method. Namely, we consider cases where each of the following two functions is used as the transformation function.

$$T_s(X) = \frac{1}{1+e^{-a(X-b)}} , \tag{20}$$

$$T_t(X) = sin(a(X-b)) . \tag{21}$$

In the both cases a and b are parameters to be adjusted by CAM or SAM.

T_s is a sigmoid function. Applying it to a continuous feature has a similar effect to clustering, with *crispness* of the boundary of the clusters depending on a. If we use a learning algorithm that partitions a continuous attribute, like C4.5, it likely helps the algorithm to find a better cut point to create a value range, in terms of generalization.

T_t is chosen because it is expected to merge peaks in the original feature space, just like what **mod 2** function did for the concept *even* [5].

To use GA, we have to set ranges for a and b for each attribute. This is rather cumbersome because the reasonable range depends on the range of the original features. Therefore continuous features other than the class are normalized into $[0, 1]$. The ranges of a and b are shown below.

- For T_s, a ranges $[1, 100]$ and b ranges $[0, 1]$.
- For T_t, a ranges $[1, 10]$ and b ranges $[0, 1]$.

We use C4.5 as a learning algorithm and compare the following three cases in terms of error rates and size of decision trees after the pruning.

1. C4.5 is directly used for learning problems.
2. Each continuous attribute is altered with CAM/SAM, with T_s as the transformation function, then C4.5 is applied.
3. Each continuous attribute is altered with CAM/SAM, with T_t as the transformation function, then C4.5 is applied.

5.1 CAM Experiment

To evaluate CAM, the authors chose datasets based on the following policy:

- The class must be categorical and there is no order among the class values.
- At least one attribute must be continuous, to be adjusted by CAM.
- No missing values present, because the experiment is not meant to evaluate robustness of the algorithm against the missing values.

As a result, the following datasets were chosen:

- **Ecoli**
- **Glass**
- **Image**
- **Ionosphere**
- **Iris**
- **Letter-recognition**
- **Liver-disorders**
- **New-thyroid**
- **Optdigits**
- **Pendigits**
- **Pima-indians-diabetes**
- **Sonar**
- **Wisconsin(wdbc)**
- **Wine**
- **Yeast**

Tables 4 and 5 show the result of the experiment through 10-fold cross validation for the error rates and the sizes of trees (the number of internal nodes plus leaves) respectively. The average values of the 10-CV trials are shown. Because CAM employs GA, it can yield different results run by run. Thus we make 10 iterations for each CV trial for CAM to capture the nondeterministic nature of the algorithm, and the average and the standard deviation of the 10-CV averages are shown. For easier comparison between the methods, Figures 4 and 5 relatively show the result for the error rates and the sizes of trees respectively, using the case with C4.5 only as the base (100%).

As for the error rates, CAM with T_s gave more than 10% improvements in 6 datasets, and more than 50% improvements in 2 datasets, out of 15. It yielded more than 10% degradations in 3 datasets, and more than 50% degradation in a dataset. On the other hand, CAM with T_t also gave more than 10% improvements in 6 datasets, and more than 50% improvement in a dataset. But it resulted more than 10% degradations in 9 datasets.

Although there are no clear common properties of datasets for which CAM with T_s did not perform well (**Image** and **Pima...**), 6 out of 9 datasets where CAM with T_t struggled share a common property; their number of class values are relatively large (more than 5). This might indicate a shortcoming of the CAM's evaluation function (7). Because it almost solely relies on the distance between mean values of F_j of different categories as for the class separation, it can favor such F_j that the values are very sparse and overlapped between classes.

As for the sizes, the result is not as clear as the error rates. CAM with T_s gave more than 10% improvements in 3 datasets, and more than 20% improvements in 1 dataset, out of 15. It yielded more than 10% degradations in 2 datasets, and there were no datasets where it resulted more than 20%. Although CAM with T_t gave more than 10% improvements in 2 datasets, it resulted more than 10% degradations in 6 datasets.

Table 4. CAM Result with UCI Datasets: Error Rates (%).

Datasets		Ecoli	Glass	Image	Ionosphere	Iris	Letter-recognition	Liver-disorders
C4.5 only		14.8	30.9	3.7	9.1	4.7	11.8	35.4
CAM (T_s)	ave.	8.9	23.4	6.5	8.6	0	11.39	36.14
	std.dev.	0	0	0	0	0	0.03	0.12
CAM (T_t)	ave.	31.62	24.07	4.48	11.39	2.68	14.44	39.12
	std.dev.	0.632	2.46	0.271	1.23	1.24	0.206	1.38

Datasets		New-thyroid	Optdigits	Pendigits	Pima...	Sonar	wdbc	Wine	Yeast
C4.5 only		7.8	10.1	3.6	28.7	38.5	5.4	6.2	43.3
CAM (T_s)	ave.	4.6	10.9	4	36.8	32.6	0.4	5.7	44.8
	std.dev.	0	0	0	0	0	0	0	0
CAM (T_t)	ave.	4.6	11.76	4.51	32.03	24.46	3.7	0	48.6
	std.dev.	0	0.233	0.07	0.76	1.57	0.467	0	1.01

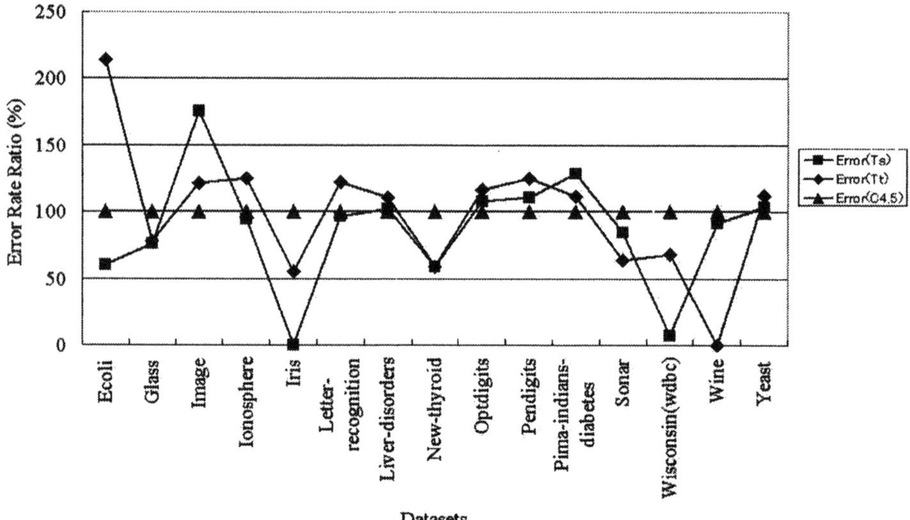

Figure 4. CAM: Relative Error Rates.

From the experiment, the authors can say that using CAM with T_s as preprocessing improves the error rates in most cases. It sometimes improves the size of the trees too, but the effect is not as significant as in the case of the error rates. The effect of CAM with T_t is not robust. This observation can be also read from the standard deviations in Tables 4 and 5.

5.2 SAM Experiment

To evaluate SAM, the authors chose datasets based on the following policy:

Table 5. CAM Result with UCI Datasets: Tree Sizes.

Datasets		Ecoli	Glass	Image	Ionosphere	Iris	Letter-recognition	Liver-disorders
C4.5 only		34	49.4	85.8	24.2	9.4	2365.4	82.6
CAM (T_s)	ave.	33	47	75	18	11	2436.44	86
	std.dev.	0	0	0	0	0	0.512	0
CAM (T_t)	ave.	44.14	49.6	88.26	25.38	8.26	2621.18	70.6
	std.dev.	1.81	1.16	1.41	1.42	0.31	305	4.94

Datasets		New-thyroid	Optdigits	Pendigits	Pima...	Sonar	wdbc	Wine	Yeast
C4.5 only		15.4	421	361	120	31	20.8	10.2	379.8
CAM (T_s)	ave.	17	427	362.56	120.6	29.86	21	9	384.2
	std.dev.	0	0	0.52	0	0.09	0	0	0
CAM (T_t)	ave.	17.22	411.94	416.42	117.8	33.04	26.3	9	448.48
	std.dev.	0.189	2.63	3.66	5.08	1	1.49	0	7.95

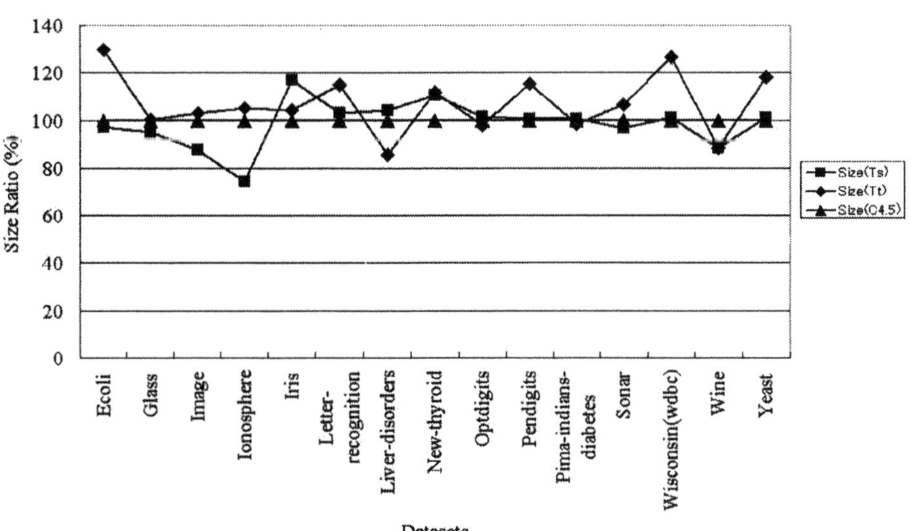

Figure 5. CAM: Relative Tree Sizes.

- The class must be continuous. To use SAM, we have to be able to measure differences between class values.
- At least an attribute must be continuous, to be adjusted by SAM.
- No missing values present, because the experiment is not meant to evaluate robustness of the algorithm against the missing values.

As a result, the following datasets were chosen. Because all the datasets have a continuous class and the authors use C4.5 as a learning algorithm, how we define learning problems using the data is also explained.

- **Abalone**: The integer class is used as the discrete class without any change (28 class values).
- **Auto-Mpg**: The continuous class is rounded to integer to create the discrete class (36 class values).
- **Automobile**: The integer class is used as the discrete class without any change (5 class values).
- **Computer**: The integer class "PRP" is translated into the new class as the following (8 class values).
  - $PRP < 21 : 1$
  - $20 < PRP < 101$: 2
  - $100 < PRP < 201$: 3
  - $200 < PRP < 301$: 4
  - $300 < PRP < 401$: 5
  - $400 < PRP < 501$: 6
  - $500 < PRP < 601$: 7
  - $600 < PRP$: 8
- **Housing**: The continuous class is rounded to integer to create the discrete class (43 class values).
- **Servo**: The integer class is multiplied by 10 then rounded to integer to create the discrete class (29 class values).
- **Wisconsin(wpbc)**: First, only the records of recurrent patients are extracted. Then integer attribute of "Time To Recur" is divided by 10 then rounded to integer to create the discrete class (7 class values).

Tables 6 and 7 show the result of the experiment through 10-fold cross validation for the error rates and the sizes of trees (the number of internal nodes plus leaves) respectively. The average values of the 10-CV trials are shown. Like CAM, it can yield different results run by run. Thus we make 10 iterations for each CV trial for SAM to capture the nondeterministic nature of the algorithm, and the average and the standard deviation of the 10-CV averages are shown.

For easier comparison between the methods, Figures 6 and 7 relatively show the result for the error rates and the sizes of trees respectively, using the case with C4.5 only as the base (100%).

The error rates are generally very poor except for **Automobile** and **Computer**. This is likely due to the fact that the number of class values is too large. As for the comparison of the methods, the differences in the error rates are largely insignificant, except:

- In **Auto-Mpg**, SAM with T_t yielded the slight improvements.
- In **Automobile**, the both SAM methods yielded the significant improvements.
- In **Servo**, SAM with T_t yielded the slight degradation.
- In **Wisconsin(wpbc)**, SAM with T_t yielded the slight improvement.

The number of class values is relatively small in **Automobile** and **Wisconsin(wpbc)**, and large in **Auto-Mpg** and **Housing**. It is possible that the degree of the improvement is related to this property. However, we could not prove the hypothesis from

Table 6. SAM Result with UCI Datasets: Error Rates (%).

Datasets		Abalone	Auto-Mpg	Automobile	Computer	Housing	Servo	wpbc
C4.5 only		78.6	80.4	22.7	30	88.4	78.5	72
SAM	ave.	77.03	78.16	13.3	29.76	90.52	75.95	71.75
(T_s)	std.dev.	0.303	1.27	3.53	0.50	1.10	0.15	4.70
SAM	ave.	80.33	74.32	12.8	25.62	89.13	83.58	66.9
(T_t)	std.dev.	0.35	0.58	2.30	1.18	1.30	2.92	4.25

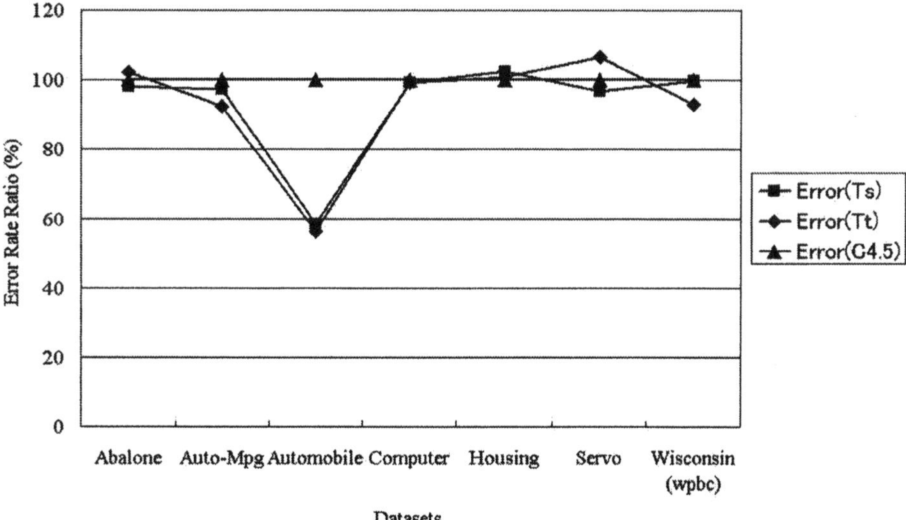

Figure 6. SAM: Relative Error Rates.

this experiment.

For the sizes, SAM with T_s yielded the improvements in most cases. The only exception is **Automobile**, where it gave the improvement in terms of the error rate.

From the experiment, the authors can say that using SAM with T_s as preprocessing does not harm, at least. It often gives the improvement in the error rate or the size. On the other hand, SAM with T_t is not safe to use as a general preprocessing method.

It is interesting that while in the CAM experiment the error rates are generally improved and the sizes are unchanged, the effects of the algorithm are reversed for SAM.

6 Related Work

In this section, we review some previous works related ours in Kansei engineering and machine learning areas.

Table 7. SAM Result with UCI Datasets: Tree Sizes.

Datasets		Abalone	Auto-Mpg	Automobile	Computer	Housing	Servo	wpbc
C4.5 only		2163.1	215.9	53.8	35.3	296.4	64.9	21.2
SAM	ave.	441.05	178.34	77.6	25.22	273.93	62.88	22.16
(T_s)	std.dev.	84.55	13.40	3.37	2.05	14.03	0.24	0.62
SAM	ave.	2156.61	216.88	66.6	30.54	299.45	67.95	21.24
(T_t)	std.dev.	12.46	1.54	3.51	1.75	3.52	2.55	1.05

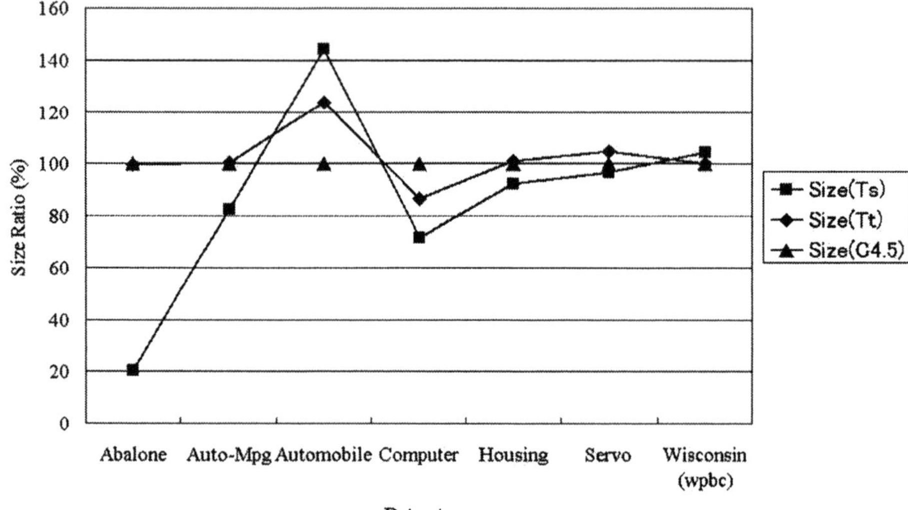

Figure 7. SAM: Relative Tree Sizes.

6.1 Kansei Engineering

From the Kansei modeling point of view, representing subjective perceptions is necessary because each person has his or her own viewpoints. One approach is to construct a model that implements a mechanism for the subjective interpretation based on visual perception [20]. In that approach, a visual perception model is composed of a physical level and a mental feeling level and the subjective interpretations are represented by means of a ratio for each features in the physical level, such as color, direction and position in relation to stimuli. These ratio is calculated by statistics [21] or neural network [20]. Although the neural net approach has a merit to simulate a nonlinear nature of the human perception, it is difficult to grasp what a perceptual model created by the neural net means. An information space created by CAM/SAM is easier to be understood if meanings of the transformation functions are clear, like in the still image experiments.

6.2 Change of Representation

Because of its potential, change of representation has been an active issue in machine learning for a long time. Most work has been done in the area of constructive induction, feature creation and selection. Relief, which is used in Section 3, is one of most successful fruits of research into feature selection [22, 23, 24].

In this subsection the authors review some of the past work in the area of constructive induction and feature creation, and explain their relationships to this study.

6.2.1 Analysis of the Problem

In this subsection we review past studies that dealt with constructive induction as a general problem, rather than proposing a working system.

As already discussed, Rendell [25, 5] pointed out that smoothness of the instance space is important for learning and feature creation is potentially powerful because it can provide means to merge peaks in the original instance space. The view inspired the authors to directly measure the smoothness, as in CAM and SAM.

However, a solution provided by Rendell himself [25] was not entirely satisfactory. He proposed *Variable Bias Management System*, where an instance space is elegantly connected to a bias space through a 3-layer model. However, in real applications the system required a problem space that basically did all the bias management jobs in it.

An elegant solution were proposed to the problem. Mehra [26] made an intriguing proposal, an inverse space, where a feature becomes a point and instances become axes. The space was useful to manipulate features, but there was a fundamental problem of dimensionality caused by a number of instances.

Using a problem of identifying coding regions of a DNA, Craven and Shavlik [27] discussed inductive learner's performance on different problem representations — nucleotides representation and codons representation. After showing the codons representation's superiority over the competitor, they tested typical feature creation algorithms (ones shown in the next subsection) to find out if they can generate the codons representation from the nucleotides representation. Although the codons representation looked relatively easy to create and was within the creation capability of the algorithms, they failed. Based on the result they pointed out that feature creation by operator application was not strong enough. In CAM/SAM, we provide an outline of transformation functions. This is a kind of background knowledge and constrains types of features to be generated. As a vague analogy, the parameterized function is to SAM/CAM as the insight to focus on triplets of nucleotides — but exactly which three nucleotides should be focused on is unknown, e.g., they might be AAA, AGA, or AGT — is to the DNA problem.

6.2.2 Feature Creation by Operator Application

CN2-MCI [28] creates a new feature by means of clustering upon the Cartesian product of the domains of base features. Then the new features are filtered by a kind of information theoretic criterion. On the other hand, Piater's method [29] for a Bayesian network classifier for image classification uses Kolmogorov-Smirnoff distance to determine a threshold to create a binary feature from two base features. They are similar to CAM in the sense that they use a discriminating power-based measure to create a new feature. However CAM is much more restrictive than they are because CAM does not allow combination of arbitrary base features.

AQ-HCI [30] creates a feature with a special form, M-of-N, which is a generalization of XOR. It is similar to CAM/SAM in the sense that an outline of a new feature is given.

While most of the systems in this category, whose well-known examples include FRINGE [31] and CITRE [32], use the output of an inductive learner such as a decision tree to guide the feature construction, GALA [33] is similar to CAM/SAM in the sense that it is not connected with any particular inductive learners. Hu argued that performance of a hypothesis-based constructive induction system depends on quality of the original hypothesis. CAM/SAM also consider the feature creation as a preprocess step. However, as John pointed out in the context of feature selection [34], there are trade-offs; a feature selection (creation) system that has access to the inductive learner, which is later used to create hypotheses, has an advantage because in selecting (creating) features it can take the learner's bias into account.

If features with very strong discriminating power are created by a feature creation system, a new space spanned by the new features does not require separate inductive learners to perform classification. The classical instance of this type of feature creation is linear discriminant analysis (LDA). Torkkola [16] proposed the use of mutual information to create features. It can be considered to be an extension of LDA. CAM can also be considered to be an extension of LDA.

6.2.3 Inductive Logic Programming

DUCE [14] introduces new predicates in the course of execution with the inter-construction, intra-construction and dichotomization operators. When it creates a tentative new predicate, DUCE asks an oracle if the predicate is meaningful in the domain. If the oracle affirmatively answers then the predicate is added to the theory. As Muggleton pointed out, typical constructive induction systems tend to create a lot of meaningless features. CAM/SAM avoid this by sticking to given parameterized transformation functions.

Flach and Lavrač [35] pointed out that a first-order learning problem can be systematically converted to a propositional learning problem through a special (and simple) type of predicate invention: first-order feature construction. In the sense that it constrains the type of feature creation, it is related to CAM/SAM and AQ-HCI [30].

7 Conclusion and Future Work

We pointed out the importance of good perceptions in problem solving for intelligent software and formalized a framework of adaptive perception using vector spaces. We also constructed a prototype system for still image perception, where an image is perceived in a space of index colors and the space is adapted with respect to categorical (CAM) or continuous (SAM) evaluation function. Both algorithms have been proved to be useful as preprocess steps for inductive learning in the Kansei engineering domain. Furthermore CAM and SAM have been tested as general preprocessing methods. CAM with a parameterized sigmoid function has yielded significant improvement in accuracy in many cases. SAM with the parameterized sigmoid function has generally resulted less complicated decision trees and sometimes yielded significant improvement in accuracy.

The adaptive perception is possibly very powerful. Inductive learning achieves an object-level generalization, meaning that it gives classification of unknown objects for the given concept. If an appropriate vector space is learned, it will achieve a concept-level generalization, meaning that it gives insight respecting unknown objects even for an unknown concept, by means of clustering.

Subjects for future work include the following.

- To build a commercial creator support system by combining the still image perception module with a text data perception module [8, 9], and to have it tested by domain experts
- to show the effectiveness of CAM as an extension of LDA
- to show CAM/SAM's superiority over typical feature creation algorithms

Acknowledgments

The authors are grateful to people who maintain UCI machine learning repository, for making their experiments possible.

References

1. Hofstadter, D.R., et al. (1995): Fluid Concepts and Creative Analogies: computer models of the fundamental mechanisms of thought. BasicBooks
2. Indurkhya, B. (1992): Metaphor and Cognition. Volume 13 of Studies in Cognitive Systems. Kluwer Academic
3. Russell, S. (1989): The use of knowledge in analogy and induction. Pitman Press
4. Tsukimoto, H. (1997): Extracting propositions from trained neural networks. In: Proc. of IJCAI '97. 1098–1105
5. Rendell, L. (1990): Feature construction for concept learning. In Benjamin, D., ed.: Change of representation and inductive bias. Kluwer Academic, 327–353
6. Murakami, T., et al. (2001): Specification on similar kansei in an adaptive perceptual space. In: Proc. of Knowledge-Based Intelligent Information Engineering Systems & Allied Technologies. 973–978
7. Blake, C., Merz, C. (1998): UCI repository of machine learning databases. http://www.ics.uci.edu/~mlearn/MLRepository.html
8. Murakami, T., et al. (2000): Friendly information retrieval through adaptive

restructuring of information space. In: Proc. of AIE/IEA 2000. 639–644
9. Murakami, T., Orihara, R. (2000): Friendly information retrieval through adaptive restructuring of information space. New Generation Computing **18**, 137–146
10. Orihara, R., et al. (2002): Information space optimization with real-coded genetic algorithm for inductive learning. In Abraham, A., Koeppen, M., eds.: Hybrid Information Systems. Heidelberg, Physica Verlag, 415–430
11. Morohara, Y., et al. (1995): Automatic picking of index colors in textile pictures for designers. Trans. Inf. Process. Soc. Jpn. **36**, 329–337, in Japanese.
12. Ikeda, M. (1980): Foundation of Color Engineering. Asakura Shoten, in Japanese.
13. Yamazaki, H., Kondo, K. (1998): A method of changing a color scheme with kansei scales. In: Proc. of 8th International Conference on Engineering Computer Graphics and Descriptive Geometry. 210–214
14. Muggleton, S. (1987): DUCE, an Oracle based Approach to Constructive Induction. In: Proc. of 10th IJCAI. 287–292
15. Higuchi, T., et al. (2001): Simplex crossover for real-coded genetic algorithms. Transactions of the JSAI **16**, 147–155 in Japanese.
16. Torkkola, K., Campbell, W. (2000): Mutual information in learning feature transformations. In: Proc. of ICML 2000. 1015–1022
17. Quinlan, J. (1993): C4.5: Programs for Machine Learning. Morgan Kaufmann
18. Kira, K., Rendell, L. (1992): The feature selection problem: Traditional method and a new algorithm. In: Proc. of AAAI'92. 129–134
19. Toshiba Corporation (1997): Data Mining Tool KINOsuite-PR. http://www2.toshiba.co.jp/datamining/index.htm
20. Kato, T. (1996): Cognitive user interface to cyber space database: human media technology for global information infrastructure. In: Proceedings of the International Symposium on Cooperative Database Systems for Advanced Applications. 184–190
21. Shibata, T., Kato, T. (1998): General model of subjective interpretation for street landscape image. In: DEXA. 501–510
22. Orihara, R. (1999): Explanation-based feature subset selection for decision tree induction. Journal of JSAI **14**, 296–306
23. Orihara, R. (1999): Knowledge Description and Semantics in Non-Deductive Reasoning. PhD thesis, University of Tsukuba
24. Liu, H., Motoda, H. (1998): Feature Extraction, Construction, and Selection. Kluwer Academic Publishers
25. Rendell, L., et al. (1987): Layered concept – learning and dynamically variable bias management. In: Proc. of 10th IJCAI. 308–314
26. Mehra, P., et al. (1989): Principled constructive induction. In: Proc. of 11th IJCAI. 651–656
27. Craven, M.W., Shavlik, J.W. (1994): Investigating the value of a good input representation. In: Proc. of ML-COLT '94 Workshop on Constructive Induction and Change of Representation. 15–28
28. Kramer, S. (1994): CN2-MCI: A Two-Step Method for Constructive Induction.

In: Proc. of ML-COLT '94 Workshop on Constructive Induction and Change of Representation. 35–41

29. Piater, J.H., Grupen, R.A. (2000): Constructive feature learning and the development of visual expertise. In: Proc. of ICML 2000. 751–758
30. Wnek, J., Michalski, R.S. (1994): Discovering representation space transformations for learning concept descriptions combining DNF and M-of-N rules. In: Proc. of ML-COLT '94 Workshop on Constructive Induction and Change of Representation. 61–68
31. Pagallo, G. (1989): Learning DNF by Decision Trees. In: Proc. of 11th IJCAI. 639–644
32. Matheus, C.J., Rendell, L.A. (1989): Constructive induction on decision trees. In: Proc. of 11th IJCAI. 645–650
33. Hu, Y., Kibler, D. (1996): Generation of attributes for learning algorithms. In: Proc. of AAAI96. 806–811
34. John, G., et al. (1994): Irrelevant features and the subset selection problem. In: Proc. of the 11th Machine Learning. 121–129
35. Flach, P.A., Lavrač, N. (2000): The role of feature construction in inductive rule learning. In Raedt, L.D., Kramer, S., eds.: Proceedings of the ICML2000 workshop on Attribute-Value and Relational Learning: crossing the boundaries, Stanford, USA, 17th International Conference on Machine Learning. 1–11

Appendix A

Conversion from RGB to $L^*a^*b^*$ [12]

Let R, G and B denote R, G and B coordinates of the color. Then X, Y, Z, X_0, Y_0 and Z_0 are defined as the following.

$$X \stackrel{def}{=} 2.7689R + 1.7517G + 1.1302B \tag{22}$$

$$Y \stackrel{def}{=} 1.0000R + 4.5907G + 0.0601B \tag{23}$$

$$Z \stackrel{def}{=} 0.0565G + 5.5943B \tag{24}$$

$$X_0 \stackrel{def}{=} 95.045 \tag{25}$$

$$Y_0 \stackrel{def}{=} 100 \tag{26}$$

$$Z_0 \stackrel{def}{=} 108.892 \tag{27}$$

The $L^*a^*b^*$ coordinates are defined using the values.

$$L^* \stackrel{def}{=} 116\left(\frac{Y}{Y_0}\right)^{\frac{1}{3}} - 16 \tag{28}$$

$$a^* \stackrel{def}{=} 500\left(\left(\frac{X}{X_0}\right)^{\frac{1}{3}} - \left(\frac{Y}{Y_0}\right)^{\frac{1}{3}}\right) \tag{29}$$

$$b^* \stackrel{def}{=} 200\left(\left(\frac{Y}{Y_0}\right)^{\frac{1}{3}} - \left(\frac{Z}{Z_0}\right)^{\frac{1}{3}}\right) \tag{30}$$

Appendix B

Conversion from RGB to HSV [12]

Let R, G and B denote R, G and B coordinates of the color. Then the H, S and V coordinates of the color, H, S and V respectively, are defined as the following, using $r \stackrel{def}{=} \frac{V-R}{V-X}$, $g \stackrel{def}{=} \frac{V-G}{V-X}$ and $b \stackrel{def}{=} \frac{V-B}{V-X}$.

- V is decided as the following.

$$V \stackrel{def}{=} max(R, G, B) \ . \tag{31}$$

- S is decided as the following.

$$S \stackrel{def}{=} \frac{V - X}{V} \tag{32}$$

  where $X \stackrel{def}{=} min(R, G, B) \ .$
- H is decided depending on how V is decided.
  - If $R = V$

$$\text{If } G = X \text{ then } H \stackrel{def}{=} \frac{5 + b}{6} \tag{33}$$

$$\text{Otherwise } H \stackrel{def}{=} \frac{1 - g}{6} \ . \tag{34}$$

  - If $G = V$

$$\text{If } B = X \text{ then } H \stackrel{def}{=} \frac{1 + r}{6} \tag{35}$$

$$\text{Otherwise } H \stackrel{def}{=} \frac{3 - b}{6} \ . \tag{36}$$

  - If $B = V$

$$\text{If } R = X \text{ then } H \stackrel{def}{=} \frac{3 + g}{6} \tag{37}$$

$$\text{Otherwise } H \stackrel{def}{=} \frac{5 - r}{6} \ . \tag{38}$$

Appendix C

Determining H^* [13]

H represents hue of the color. H^* is determined using Table 8 shown below, depending on the hue.

Table 8. Hue and H^*.

Hue	H^*
Red	1.00
Yellowish red	0.89
Yellow	0.93
Yellowish green	0.61
Green	0.28
Blue green	0.22
Blue	0.18
Blue violet	0.08
Violet	0.00
Red violet	0.35

Chapter 14

Detecting, Tracking, and Classifying Human Movement Using Active Contour Models and Neural Networks

Ken Tabb, Neil Davey, Rod Adams, and Stella George

Keywords: Active contour model, snake, neural network, axis crossover vector, pedestrian, tracking, shape recognition

1 Introduction

Detecting and tracking moving objects in the visual field is a task which has interested the computer vision discipline for some years [4, 7, 12, 13, 14]. A hybrid technique is described in this chapter for detecting and tracking moving objects in a sequence of images, and for identifying them as 'human' or 'non-human'.

We use modified Active Contour Models (ACMs) [3, 11, 18, 22] to detect and track moving objects in successive frames of a video. Neural networks have been developed to then classify the shapes obtained by the ACM as either 'human' or 'non-human'. In order for the neural network to analyze only the shape of the contour, several characteristics of the contour are removed before it is fed into the neural network. Scale-, location-, resolution- and rotation-independence are all achieved by re-representing the contour as an axis crossover vector (ACV) [19, 20].

Generalizing from a representative training set of axis crossover vectors, the neural network is able to accurately classify previously unseen instances of computer generated (CG) humans, dogs and horses, moving in a variety of directions along the ground plane. Real humans moving in real-world outdoor environments are also correctly classified by the neural network, despite the training set consisting solely of CG shapes. Examples of previously unseen object classes are also classified by the neural network in terms of how human they are, demonstrating the neural network's ability to generalize when presented with novel classes of shapes. All experiments used in this study are described, along with an analysis of their results.

The chapter ends with a general discussion of the technique, including potential applications of the system.

2 Active Contour Models

Active Contour Models [11], also referred to as 'snakes', provide a method of abstracting an object's shape from a single image or a sequences of images, resulting in a contour representation of the object's perceived boundary in each image (Figure 1).

Figure 1. Active Contour Models tracking a moving human in different frames of a video sequence. The ACM actually operates on a preprocessed copy of the video frame; the ACM has been displayed on the original video frames for illustrative purposes. Intermediate frames have been omitted from the figure.

An ACM consists of two parts. Firstly, the shape of the contour is defined by a sequence of control points, which are connected together to form a spline. This sequence of points can be initialized either manually [11] or automatically [18]. The control points each have an (x,y) coordinate position within the image's coordinate system. Secondly, an energy function is minimized for each control point on the contour, determining where each control point is to be positioned. The energy function therefore governs the shape that the contour will take and ultimately determines how accurately a specific active contour model will be able to lock onto a given object's shape in a given image.

The energy function usually involves component functions for different types of energy: internal, image, and external energy (discussed later). Each component energy function can be customized, allowing features pertinent to a specific problem domain to influence snake movement, and in so doing tailoring the ACM's behavior to specific applications.

The different types of energy can be weighted so that, for instance, image energy has a stronger influence than internal energy in positioning a control point. Quite how this precedence of energy types is defined depends upon the specific problem domain; an energy type which seems critical to an ACM's success in one task might appear inconsequential in other applications.

Each type of energy works by assigning high energy values to locations within the image which fail to meet that component energy function's criteria, and low energy values to locations which best meet the criteria (Figure 2). After each competing location for a given control point has been assigned energy values for each type of component energy, minimizing a given control point's energy involves comparing each location's total energy value, and selecting the location with the lowest overall energy. The control point is then moved to the new location. The ACM iterates through its control points, moving them one at a time. The energy minimization process is repeated until the ACM relaxes, that is, until a sufficient proportion of the control points no longer move between iterations.

Internal energy influences the contour's shape by measuring the geometrical relationships between adjacent control points on the contour. For instance, in order to obtain an accurate approximation of an object's outline, control points can be spread equally around the contour, to prevent them from becoming 'bunched up' and missing potentially important parts of the shape (Figure 3). Such energy is called continuity energy, and is well defined mathematically [22]. In order to achieve this, continuity energy can move control points such that they become more evenly spread along the contour.

Curvature energy is another type of internal energy which governs how smooth or sharp a contour is allowed to become, by imposing limits on how sharp an angle each control point is allowed to form with its neighbors (Figure 4). Curvature energy does not, however, prevent the ACM from forming complex shapes, provided a sufficient density of points exist on the contour (Figure 5).

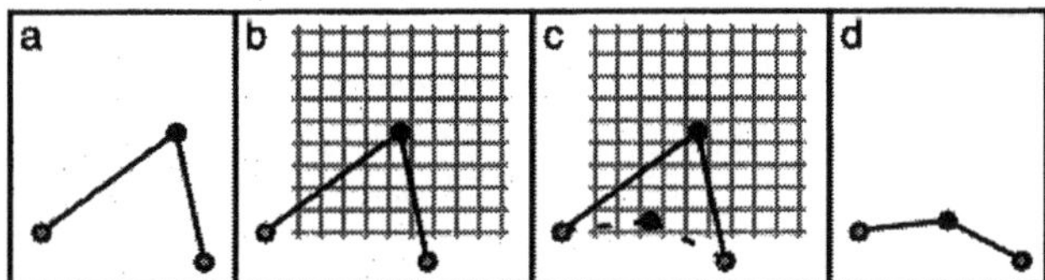

Figure 2. Energy minimization for a control point. [a] The original locations for three adjacent control points on the contour. [b] In order to move the middle of the three control points (black), each location in the control point's neighborhood has an energy value calculated as per the ACM's energy function. [c] For the sake of example, the position in the neighborhood with the lowest energy value is the one indicated towards the lower left of the neighborhood grid. [d] The control point is moved to the more suitable location and the process repeats for the next control point on the contour.

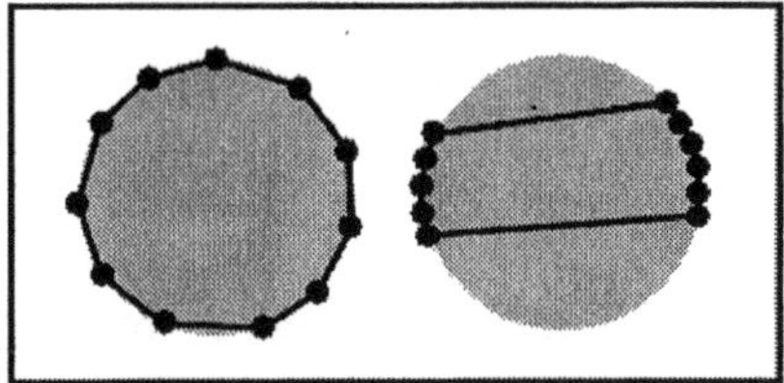

Figure 3. Control point continuity energy. [Left] Spacing control points evenly around the contour (black) results in a more accurate shape mapping of the contour to the object (gray). [Right] When control points bunch up along the contour, the contour may miss parts of the object's shape.

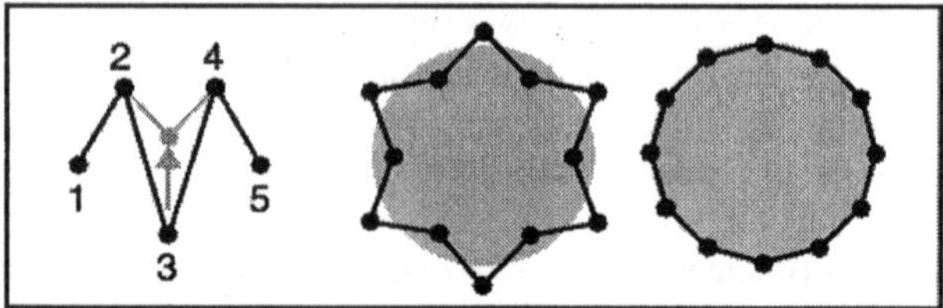

Figure 4. Control point curvature energy. Increasing the angle between neighboring control points [left], when applied to the entire contour [middle] removes corners from the contour, allowing the contour to lock onto objects with curved shapes [right]. Conversely if the target object contains many jagged edges, curvature energy should be disabled in the ACM energy function, so that the contour may form complex shapes.

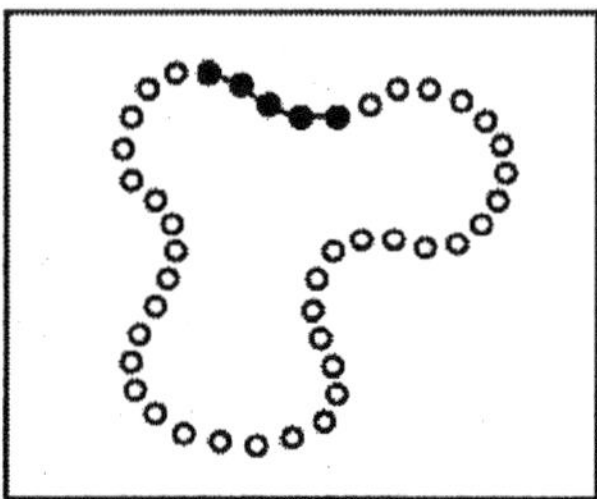

Figure 5. Complex shape formation despite the effects of curvature energy. With sufficient numbers of control points, an ACM can still form complex shapes, even when curvature energy prevents adjacent control points from forming corners.

Image energy influences control point movement based on desired features, such as colors and edges, in the image. As such image energy is independent of the geometrical relationships between control points. The image energy function is usually specific to an application. For instance if the target object in a scene is always red, and is the only red object in the scene, then capitalizing on the object's unique color would be a good basis for an image energy function. This can be implemented in the image energy function by assigning a low image energy value to any pixel in the image which contains a high concentration of red in the RGB colorspace. In this case, minimizing image energy attracts control points to areas in the image which contain red.

The image energy definition used in our system involves preprocessing the image by performing background subtraction [10, 17], edge enhancement via Sobel edge masks [17], and edge normalization [22], as can be seen in Figure 6. The image energy value for a given pixel in the image is then its normalized edge value.

External energy influences snake movement by responding to user interaction, such as the user pushing or pulling the snake around using the mouse. For instance, if there are several local minima in an image which the snake is getting caught on, the user can push the snake from one local minima to the next [11]. In our implementation we do not use external energy, as the snake must automatically track the moving human, rather than being manually controlled.

Figure 6. Image preprocessing used to calculate image energy. [Left] The original color video sequence's frames are differenced, providing motion detection [Middle]. Sobel edge masks are then passed over the image [Right], extracting the edges of moving objects in the video sequence. These edge values are then normalized, so that local differences in edge strength do not appear disproportionately large.

The organic nature of the objects being tracked in our work mean that the target shapes contain relatively flat edges and both curved and sharp corners. Curvature energy, however, necessitates that a decision needs to be made during the design of the ACM's energy function regarding how much curvature the ACM should be allowed to form. Consequently it is difficult to match a global, contour-wide curvature energy criteria which meets the local needs of a particular control point. In other words it may be the case that, generally speaking, the ACM should refrain from forming unnecessary corners, but that individual control points should be allowed to form corners in the contour where it makes sense to do so, for example on the toes of the human in Figure 1.

In order to allow certain control points to form corners where necessary, our ACM includes a control point locking mechanism, as proposed in [22]. The condition, detailed in [18], locks a given control point in place for one iteration of the contour, if that control point meets two criteria. Firstly, the control point must be already positioned in a location of low image energy. Typically this involves edges of objects in the image, although this depends upon the definition of the ACM's image energy function. Secondly, the control point must form a sufficiently small angle, or sharp corner, between itself and the two adjacent control points on the contour. Only when both of these criteria are met is the control point locked in place. The lock lasts for one energy minimization iteration of the contour. During the next iteration around the contour, a locked control point is reassessed as to whether the lock should remain. For instance if the two adjacent control points have since formed a straighter line with the current control point, such that the angle of the current control point is increased beyond the threshold value of the locking condition, then the lock is removed, even if the current control point still satisfies the condition requiring low image energy.

By implementing a control point locking device, which allows necessary corners to form in the contour, an ACM can successfully detect and track moving human shapes (Figures 1 and 7), as reported in [19, 20].

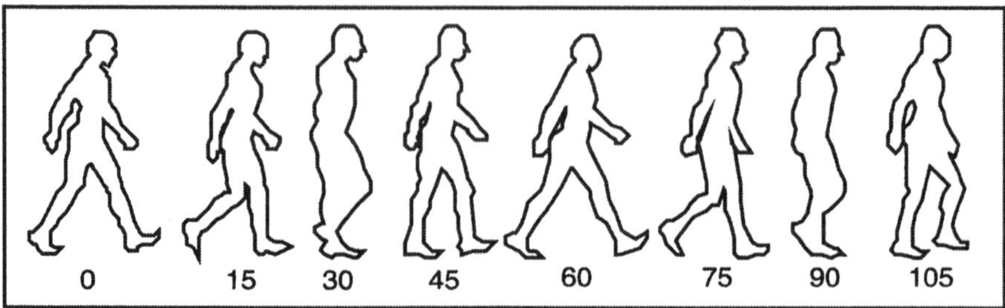

Figure 7. Tracking a human object using an ACM. If necessary localized corners can be formed without extraneous corners from forming, via a control point locking device, allowing the ACM to successfully lock onto objects which have infrequent curvature in their outline, and / or which are articulated. Shown underneath each contour in the figure is the frame number of the video sequence in which the shape was detected.

3 Active Contour Vector Translation

In this section we discuss the issues preventing an active contour from being used unaltered as an input pattern to a neural network. We propose a solution in the form of the axis crossover vector, and identify the most appropriate level of detail for encoding human shape information, following a series of systematic experiments with neural networks and axis crossover vectors.

3.1 Obtaining Generic Shape Descriptions from Snakes

Snakes themselves have no way of determining what type of object they are tracking, they merely track visible outlines in the image, irrespective of the object(s) that the outline belongs to. In order to determine whether or not the object being tracked by a snake is human, the shape of the object somehow needs to be analyzed. Due to the large variation of human poses and subsequent shapes, the wide variety of non-human objects which may be encountered in real world situations, and the general ill-defined nature of the problem, we have opted to use a feedforward error-backpropagation neural network as the means by which an object being tracked using a snake is classified human or non-human. More details of the neural network, and of experiments involving the network, are presented in the following section.

The reliance of active contour models upon features in an image brings both advantages and disadvantages to the task of object identification. By locking onto features in an image, an active contour is able to abstract the actual object in the image, rather than an approximation or a prediction. This provides the highest resolution shape possible from the image data alone. Conversely, the control points on the contour have absolute coordinates based in the image coordinate space, as the control points relate to pixel locations in the image. This dependence upon image coordinates means that the native snake representation exhibits many undesirable qualities when used as a generic shape description.

Such contours are not location-invariant; two equally shaped and sized contours residing in different parts of the image will result in different coordinate vectors. Additionally the vectors of contours which have the same shape but different sizes will also differ, thus the contours are not scale-invariant either. If one contour has more control points than another equally shaped contour, they will have different vector lengths, making the vector dependent upon the resolution of the contour, that is, upon the number of points defining the shape. Finally, if two contours are of the same shape, in the same image location, and have the same number of control points, but each contour's first control point is on a different part of the outline, their vectors will be in a different order and are thus rotation-variant. All of these factors make the comparison of contours difficult.

In order to use active contours' shapes as input data for a neural network performing a shape classification task, these qualities need to be removed. The re-representation therefore must provide for scale-, location-, resolution- and rotation-invariance if it is to be useful as a generic shape description. Furthermore the representation needs to remove the pairing of data, so that the neural network is not expected to have to group each control point's x and y locations together. This in itself presents a challenge when devising an input pattern representation for the neural network.

The solution which we have developed is the axis crossover vector [19]. The center of the contour is calculated, which in our implementation is simply the mean control point location. From that center, a pre-defined number of axes are projected outwards at specified angles, to the furthest edges of the contour (Figure 8). The distance from the contour's center to its furthest edge along that axis is then stored in a vector. The vector length is equal to the number of axes being projected. This allows the vector to be location-invariant, as no image coordinates are stored in the axis crossover vector, only distances from the center of the shape to its edges, along the axes. In addition the vector is resolution-invariant, that is, independent of the number of control points on the contour. Similarly, it does not matter where on the contour the first control point resides, and so is rotation-invariant.

Once all of the axes have been measured and their distances stored in the vector, the vector is normalized. This ensures that the vector is scale-invariant, as the largest value in the vector will at this point be 1.0, which will be the longest of the axes projected.

In our studies involving deformable objects' shapes, we do not know in advance what pose or orientation the human will take. For this reason, we project all axes evenly around 360°, as in Figure 8.

The resulting normalized vector can then be used as a training or test input pattern for a neural network, with each vector element being a different input neuron's input (Figure 9).

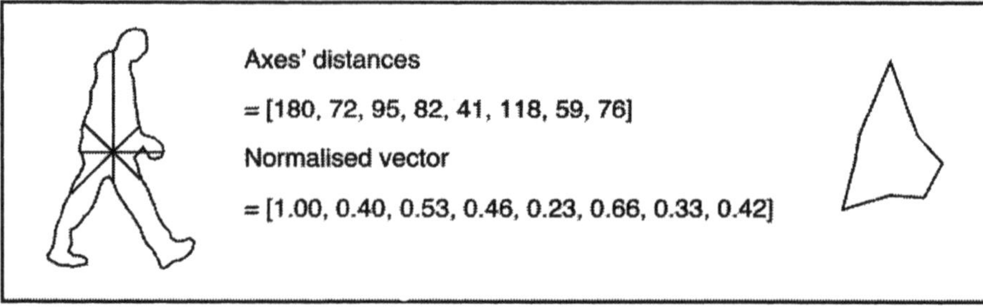

Figure 8. An 8-axis crossover vector. [Left] A predetermined number of axes are projected from the center of a contour to its furthest edges at specified angles. The distances of these axes are then stored in a vector and normalized. [Right] A polygonal visualization of the actual shape information encoded by the axis crossover vector.

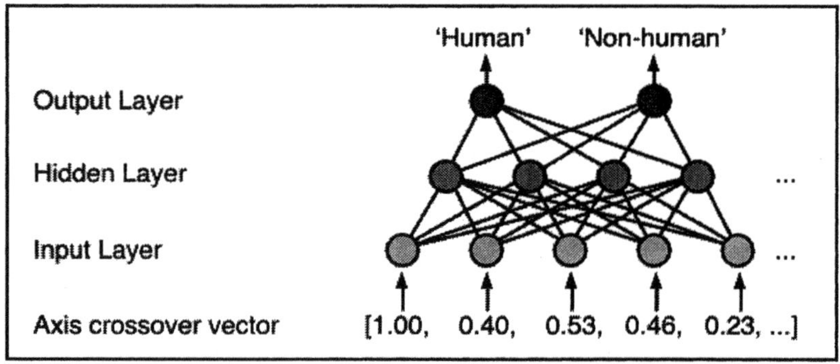

Figure 9. Axis crossover vectors as input patterns for neural networks. A given vector element maps onto a given input unit in the neural network, thus the size of input layer and axis crossover vector must equate.

There are two limitations to using axis crossover representations as input to neural networks. Firstly, a given axis must be projected in the same direction between shapes. For instance the first axis in all of our shapes is projected at 0° (vertically). If axes are projected differently between shapes then the task of comparing axis crossover vectors is complicated.

Secondly, the granularity of shape description can be tailored to a specific task by using a larger or smaller number of axes. However, because vector elements map onto input neurons, the number of axes used must be equal in all vectors used to train or test a given neural network.

In order to ascertain the most appropriate number of axes to use for representing a given object class' shapes, it was necessary to develop several neural networks, each with a different input layer size, and to determine which network gave the best results.

3.2 Verification of Axis Crossover Vectors as Generic Shape Descriptors for Neural Networks

It was necessary to test whether or not a neural network could distinguish one group of crossover vectors (pedestrians) from other groups of crossover vectors (non-pedestrians).

In order to classify shapes as human or non-human using a supervised neural network, it was necessary to obtain a training and test set of examples and counter examples. All objects in the training and test sets of experiments reported in this section were computer generated using 3D modeling and animation software, which allowed for tighter control over objects, and meant in the case of animate objects such as humans, that their gender, height, weight, age, direction of movement, and gait could be finely controlled.

The axis crossover representation allows for different numbers of axes to be used in the contour representation. Having fewer axes simplifies the neural network's task, however enough axes need to be used that the pedestrian qualities of the contours are encapsulated in the vectors, so that they can be differentiated from the non-pedestrian vectors. It was decided to test several different numbers of axes used in the representations, which in turn meant developing and testing several neural networks, each with as many input units as there were axes in the representations. It was hoped that these experiments would identify the optimal number of axes to use in representing the particular class of contours relevant to this project.

The same set of CG shapes were used in each neural network's training set, encoded with axis crossover vectors containing the appropriate number of axes, for example 4 axes for the neural network containing 4 input units. The training set contained 150 pedestrian shapes, and 150 non-pedestrian shapes. The non-pedestrian shapes consisted of inanimate outdoor objects, such as cars, trees and lampposts.

The test set contained the same set of 10 CG pedestrian and 10 CG non-pedestrian shapes across all neural networks, again encoded using appropriately sized axis crossover vectors for each neural network. All shapes in the test set were unseen, but were of object classes included in the training set.

The networks were trained to within 0.05 error, and when tested, network output was allowed some lenience; an output of 0 - 0.2 was classed '0', and an output of 0.8 - 1 was classed '1'.

Each shape used as a training or test pattern had a snake locked onto it, which was then translated into an axis crossover vector. Axis crossover vectors containing 4, 8, 12, 16, 20 and 24 axes were used, each on 10 identical neural networks containing the corresponding number of input units. Each of the 10 identical neural networks were initialized with a different random weight matrix, to lessen the chances of a given network architecture becoming trapped in local minima in the weight space.

All networks contained 2 output units. All training patterns required an output of '1, 0' (human pattern) or '0, 1' (non-human pattern). Using 2 output units allowed the network's confidence value to be identified. The confidence value is simply the difference between the two output units' values. This provides a measure of how confident the network considers its classification of a given shape to be; a value of +1.0 denotes maximum confidence that the object is human, a value of –1.0 denotes maximum confidence that the object is non-human, and a value of 0 denotes minimum confidence (maximum uncertainty). Values for shapes which the neural network is not confident in classifying can be analyzed to determine why the network is not confident; a value of 0 could mean that the network output is '1, 1', '0, 0' or '0.5, 0.5', for instance. Each means something different; the network might consider the shape to be both human and non-human, neither human nor non-human, or partly human and partly non-human. If we had used only a single bipolar output unit a value of '0' would only tell us that the network was uncertain, and not why it was uncertain.

Shapes which the network is uncertain at classifying can be incorporated into the training set, so that the training set becomes more representative of the possible shapes which will be encountered, and improves the network's generalization abilities with these types of shape.

Experiments previously reported in [19] show that the networks which used 16-axis crossover representations, and therefore 16 input units, displayed the best performance. Of these 16 input unit networks, those containing 13 hidden units were best at distinguishing human from non-human shapes, being able to classify 90% of unseen human shapes and 60% of unseen non-human shapes correctly. Interestingly, using more than 16 axes gave no benefit in performance whilst requiring more epochs during training. The neural network with 16 input units was therefore used henceforth, together with 16-axis crossover vectors.

To test the chosen network's confidence in its categorizations, it was necessary to look at the average difference between the values output by both of its output units during the previous experiment to see how 'confident' it was that a pedestrian vector was pedestrian and that a non-pedestrian vector was not.

Figure 10 shows the average results for the 10 unseen CG human and 10 CG outdoor non-human vectors. The network's output units are split into a 'human' unit and a 'non-human' unit (labeled OU1 and OU2 respectively in Figure 10) which should ideally fire mutually exclusively of one another, as something cannot be both pedestrian and non-pedestrian. However, as was found in the previous experiment, the network is not 100% accurate, resulting in some uncertainties about the answers it gives. Nevertheless a clear divide can be seen between the values output when classifying a pedestrian vector correctly versus classifying a non-pedestrian vector correctly.

The average confidence value when classifying an unseen CG human shape correctly is 0.81 (the difference between the two output unit values), with 1.0 representing complete confidence that the shape is human, whereas the average confidence value when classifying an unseen CG non-human shape correctly is –0.65, with –1.0 representing complete confidence that the shape is non-human. A confidence value of 0 represents the least confident classification. The confidence values shown in Figure 10 reflect the trend exhibited in the previous experiment, that the network is more accurate at classifying human shapes than non-human shapes. In particular, the average output values when given non-human shapes only just fall within the 'classified correctly' zones of 0 - 0.2 and 0.8 - 1.

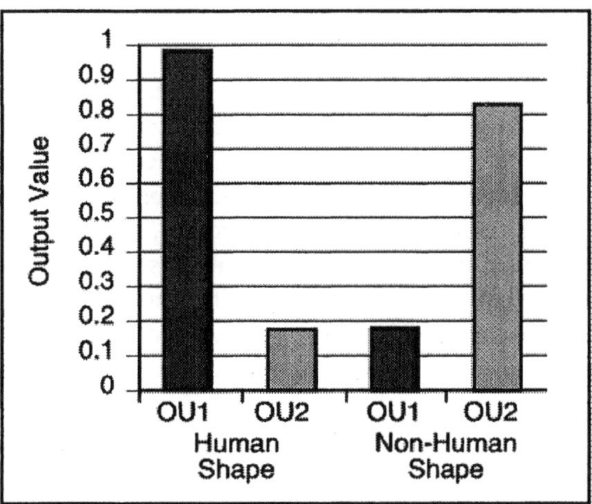

Figure 10. Mean test set classification results, from experiments on ten identical double output unit neural networks trained using 16-axis CG human and CG outdoor non-human shapes. Each network was initialized with a different initial weight matrix.

4 Shape Analysis and Classification

In this section we investigate how the shape representation developed in the previous sections can be used to produce neural networks that can distinguish between a variety of CG and real object classes. This section uses exclusively a network with 16 input units, 13 hidden units and 2 binary output units, where one output unit was trained to identify human shapes, and the other non-human shapes, as in the previous section.

4.1 CG and Real Object Classification – Lateral Object Movement

The neural network architecture developed in the previous section was retrained, this time with 400 examples each of CG human and CG non-human shapes in the training set. The non-human shapes consisted of 200 inanimate outdoor objects such as trees, cars and streetlights; and deformable animate objects, namely 100 shapes each of CG dogs and horses, again generated using 3D inverse kinematics software. All non-human shapes were trained to produce a 'non-human' classification. Each shape in the training set had a snake locked onto it, and the snake's contour was then re-represented as a 16-axis crossover vector, which was used as the input pattern. All animate objects used in the training set (the CG humans, horses and dogs) were moving laterally across the image, either from left to right or right to left. The network was trained to within an error margin of 0.05.

The test set for this categorization experiment contained 100 unseen CG human shapes, 100 unseen real human shapes, and 100 unseen CG non-human shapes. Within the non-human test patterns, a previously unseen object class was introduced to the network: CG velociraptors. Velociraptors were used as they constituted unseen non-human bipeds, and so could further determine how distinct the network considered human objects to be. The CG non-human objects consisted of 25 shapes each of CG inanimate objects, CG horses, CG dogs, and CG velociraptors. As with the training patterns, all animate objects were moving laterally across the image from either left to right or right to left. The real human shapes were obtained from complex outdoor scenes, as illustrated in Figure 1, and included male and female humans of varying heights, weights and ages.

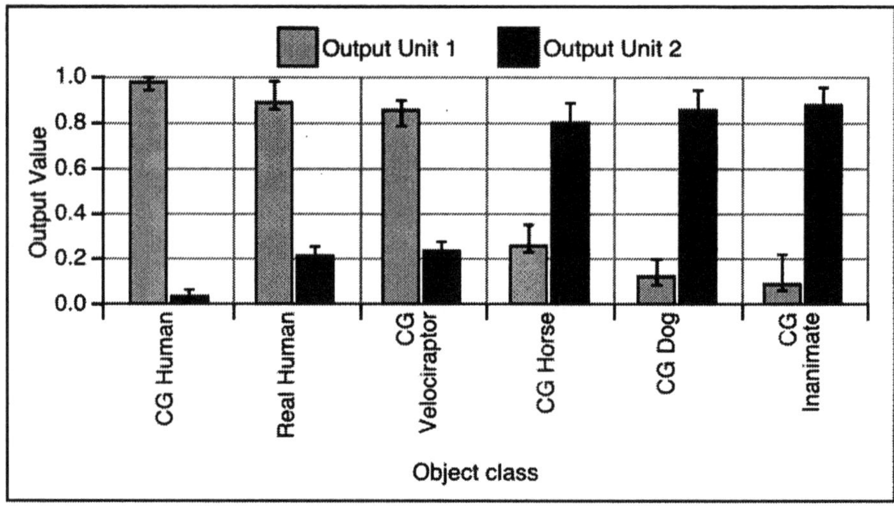

Figure 11. Mean neural network categorization values and standard deviations for different types of unseen objects. The network was trained and tested using shapes of objects moving laterally across the image.

The results in Figure 11 show that the network clearly identifies unseen CG humans as 'human' with a high level of confidence, and unseen CG dogs, horses, and inanimate objects as 'non-human' with a high, albeit lesser, level of confidence. The CG velociraptors are classified as more human than non-human, but the network still clearly distinguishes between the CG humans and the CG velociraptors, with mean confidence values of 0.94 and 0.63, respectively. Real humans were not classified as accurately as CG humans, but were nevertheless again categorized as 'human' with more confidence than the CG velociraptors, obtaining a mean confidence value of 0.69. It is highly significant that the network, trained only on CG human / non-human shapes, can then correctly classify real humans. In addition, it is impressive that the network can distinguish the real human shapes from a previously unseen object class.

4.2 CG and Real Object Classification – Omni-Directional Object Movement

Despite the results from the previous experiment showing some promise that the neural network was able to distinguish both real and CG humans from CG non-human objects, the only objects used so far, in both training and testing, had all been walking laterally to the camera. It was therefore important to test the same neural network, with no additional training, on objects moving omni-directionally in the ground plane; that is, not just from side to side but in arbitrary directions along the ground plane. Consequently, the test set for this experiment included 320 objects for each object class: CG humans, real humans, CG velociraptors, CG horses, CG dogs, and CG inanimate objects. Each class of object was tested from 16 different viewpoints around the compass: 0°, 22.5°, 45°, 67.5°, 90°... 337.5°. There were thus 20 CG humans moving at 0°, 20 CG humans moving at 22.5°, and so on.

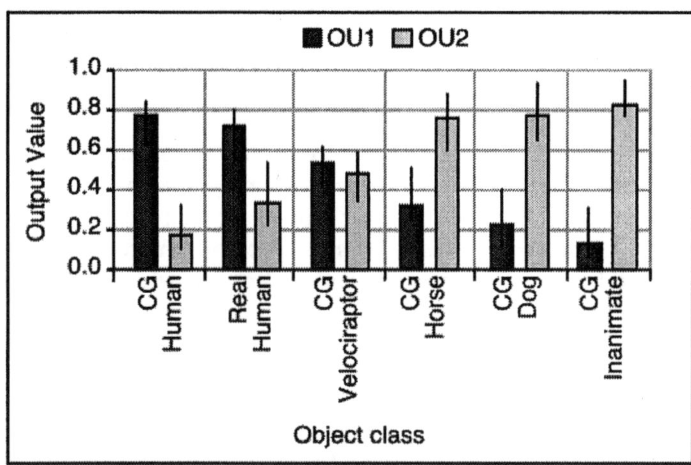

Figure 12. Mean neural network categorization values and standard deviations for different types of unseen objects. The network was trained using shapes of objects moving laterally across the image. The objects used in the test set were moving in all directions (0° - 360°) on the ground plane.

The standard deviations in Figure 12 show that the network is no longer as able to distinguish object types as well when test patterns involve objects moving in different directions. This is unsurprising, as the training set includes no allowance for objects which aren't moving laterally to the camera. In addition to the network's inaccuracy at correctly classifying objects which aren't moving laterally, the standard deviation for the network's classification of each object class, shown on the graph, has also increased (Figure 12). A breakdown of how the network performed with each object class at each angle of motion can be seen in Figure 13.

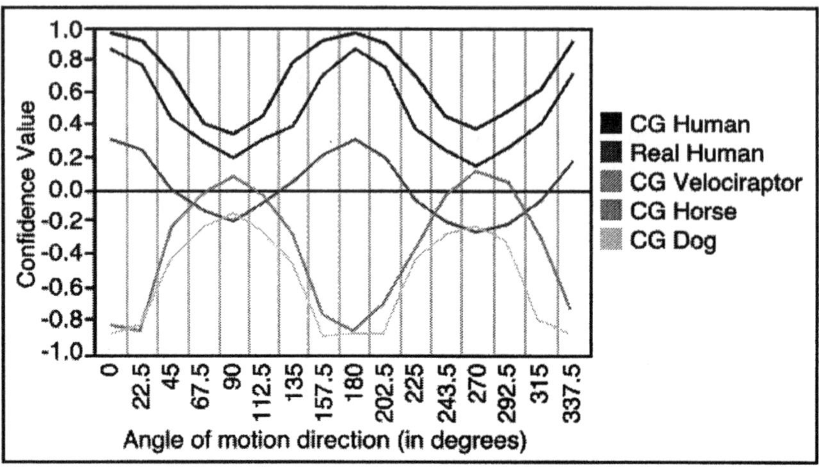

Figure 13. Mean neural network classification results when tested with objects moving in different directions along the ground plane. Each plot shows the mean classification of 20 examples of an object class moving in a given direction along the ground plane. The neural network was trained with only laterally moving objects.

Of note is that the network is still able to classify the object classes as successfully as before when those objects are moving laterally, that is, when their direction approaches 0° or 180°. However as their movement approaches 90° or 270°, the network's ability to correctly classify them is severely compromised. Nevertheless for the system to be applicable to complex outdoor scenarios, it was necessary to accommodate all movement along the ground plane. Thus it was decided to use a different training set, which incorporated objects moving in different directions along the ground plane.

4.3 CG and Real Object Classification – Omni-Directional Object Movement Following Retraining with Omni-directional Objects

The network used in the previous section was retrained to accommodate the same classes of object that were in the previous training set, but this time moving in arbitrary directions along the ground plane. Consequently the size of the training set increased; the training set included 600 CG humans, 300 CG horses, 300 CG dogs and 200 CG inanimate objects, totaling 1,400 objects, moving in different directions on the ground plane. Again, no velociraptors were used in the training process. The network shared the same architecture as per previous experiments, and was again trained to within 0.05 error.

The results of classifying unseen omni-directional objects, following training using the new training set, are shown in Figure 14. The same test set was used as for the previous experiment, and involved unseen omni-directional objects. Of note is that both the network's accuracy and standard deviation at classifying all object classes has improved, including for the unseen real human and CG velociraptor classes. A more detailed breakdown can be seen in Figure 15, where objects moving in arbitrary directions obtained a consistent classification, irrespective of their direction of movement.

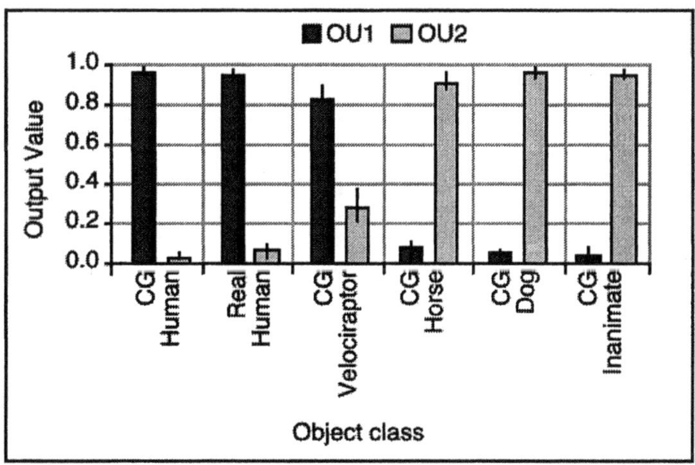

Figure 14. Mean neural network categorization values and standard deviations for different types of unseen objects. The network was trained and tested using shapes of objects moving in arbitrary directions (0° - 360°) on the ground plane.

5 Discussion

It has been clearly demonstrated that the combination of active contour models and neural network classifiers provides a useful tool for the identification of human and non-human forms.

Snakes based upon the Fast Snake model have been used to successfully track moving humans around the visual field. The snakes' contours have then been re-represented in a scale-, location-, resolution- and control point rotation-invariant axis crossover vector. We have found that using 16 axes when generating the axis crossover vector provides an effective level of detail for encoding human shape information. These vectors have been used to train a neural network to differentiate human shapes from non-human shapes. The resultant neural network achieves a high level of success when presented with both unseen CG shapes and unseen real shapes, having been trained using CG shapes only.

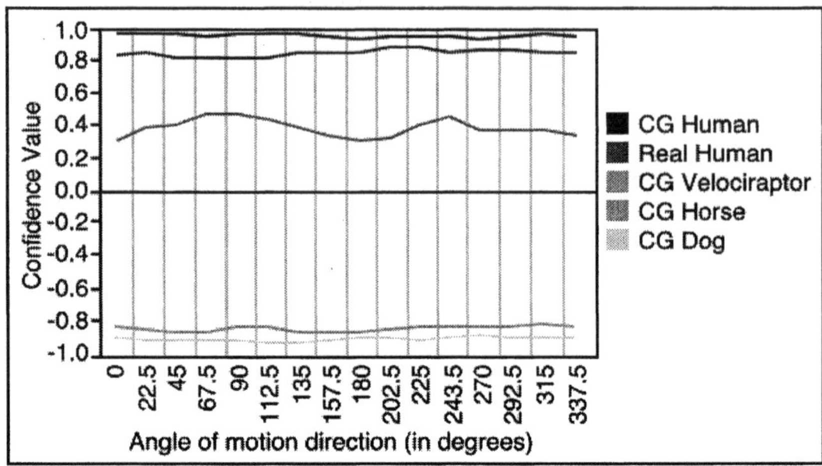

Figure 15. Mean neural network classification results when tested with objects moving in different directions along the ground plane. Each plot shows the mean classification of 20 examples of an object class moving in a given direction along the ground plane. The neural network was trained with objects moving in arbitrary directions along the ground plane. The network was tested with the same set of shapes as the network in Figure 13.

When retraining the network with a more representative training set, the network was able to correctly classify objects which were moving in arbitrary directions along the ground plane.

One consequence of the axis crossover vector's resolution- and rotation- independence is that axis crossover vectors can be used to represent contours which contain a different number of control points. The amount of detail of contours which are used to train the system can therefore be different to that of contours being classified, as it is the number of axes which define the shape representation used by the neural network. This makes the technique applicable to situations where the number of control points on contours is not known beforehand, or may change at runtime. This is in contrast to other methods of shape description such as the Point Distribution Model [1, 8], Active Shape Model [2, 5] and Active Appearance Model [6, 9, 21], which rely on there being a one to one correspondence between landmark points on the shape model and the current contour. That is not to say such techniques are superceded by our technique; in situations requiring individual recognition, for instance, an Active Appearance Model provides far more functionality than our technique. The technique we present is a method for determining which object class a given shape belongs to, rather than for determining which specific object it is.

Further work in this research area involves identifying if the neural network's accuracy in classifying real human objects could be improved by training it using real human shapes instead of, or as well as, CG human shapes. Clearly this would make the task of generating the training set more complex, as there would be less control over the variety and movement of the real human subjects than currently exists in the 3D modeling and animation software, making it more cumbersome to obtain a representative training set from which the neural network can generalize.

Potential applications for this technique would be military and CCTV surveillance systems, providing an indication that a certain class of object has entered, or is moving in, the image.

It may also be interesting to combine this technique with a History Space Classifier [15] to detect abnormalities in an individual's walking pattern, perhaps providing a tool for the medical domain.

References

1. Baumberg AM & Hogg DC (1993). Learning flexible models from image sequences. University of Leeds School of Computer Studies Research Report Series, Report 93.36.
2. Baumberg AM & Hogg DC (1994). An efficient method for contour tracking using active shape models. University of Leeds School of Computer Studies Research Report Series, Report 94.11.
3. Blake A & Isard M (1998). Active Contours. Springer-Verlag.
4. Broggi A, Bertozzi M, Fascioli A & Sechi M (2000). Shape-based pedestrian detection. In Proceedings IEEE IV-2000, Intelligent Vehicles Symposium, pp.215-220, Detroit, USA, 3-5 October 2000.
5. Cootes TF & Taylor CJ (1992). Active shape models – 'Smart Snakes'. In Proceedings of the British Machine Vision Conference 1992, pp.266-275.
6. Cootes TF, Edwards GJ & Taylor CJ (1998). Active Appearance Models. In Proceedings of the European Conference on Computer Vision 1998 (H. Burkhardt & B. Neumann Eds.). Vol. 2, pp. 484-498.
7. Cootes TF, Page GJ, Jackson CB & Taylor CJ (1995). Statistical grey-level models for object location and identification. In Proc. British Machine Vision Conference 1995, pp.533-542.
8. Cootes TF, Taylor CJ, Cooper DH & Graham J (1992). Training models of shape from sets of examples. In Proceedings of the British Machine Vision Conference 1992, pp. 9-18.
9. Edwards GJ, Cootes TF & Taylor CJ (1998). Face recognition using active appearance models. In Proceedings of the European Conference on Computer Vision 1998 (H.Burkhardt & B. Neumann Ed.s). Vol. 2, pp. 581-695.
10. Gordon G, Darrell T, Harville M & Woodfill J (1999). Background estimation and removal based on range and color. In Proceedings of the IEEE Computer Society Conference on Computer Vision and Pattern Recognition, June 1999. pp 459-464.
11. Kass M, Witkin A & Terzopoulos D (1988). Snakes: active contour models. In International Journal of Computer Vision (1988), pp. 321-331.
12. Koller D, Daniilidis K & Nagel HH (1993). Model-based object tracking in

monocular image sequences of road traffic scenes. In International Journal of Computer Vision 10:3 (1993) pp.257-281.
13. Kompatsiaris I, Tzovaras D & Strintzis MG (1998). Flexible 3D motion estimation and tracking for multiview image sequence coding. In Signal Processing: Image Communication 14 (1998) pp.95-110.
14. Lin MH (1999). Tracking articulated objects in real-time range image sequences. In Proceedings of the International Conference on Computer Vision, September 1999, pp.648-653.
15. Magee DR & Boyle RD (2000). Spatio-temporal modeling in the farmyard domain. In Proceedings of the First International Workshop, Articulated Motion and Deformable Objects, Mallorca, September 2000, pp. 83-95.
16. Shimida N, Shirai Y & Kuno Y [2000]. Model adaptation and posture estimation of moving articulated object using monocular camera. In Proceedings of the First International Workshop, Articulated Motion and Deformable Objects, Mallorca, September 2000, pp. 158-172.
17. Sonka M, Hlavac V & Boyle R (1994). Image Processing, Analysis and Machine Vision. Chapman & Hall.
18. Tabb K & George S (1998). Snakes and their influence on visual processing. Internal Technical Report, Department of Computer Science, University of Hertfordshire.
19. Tabb K, Davey N, George S & Adams R (1999). Detecting partial occlusion of humans using snakes and neural networks. In Proceedings of the 5th International Conference on Engineering Applications of Neural Networks, 13-15 September 1999, Warsaw, pp.34-39.
20. Tabb K, Davey N, Adams R & George S (2002). The recognition and analysis of animate objects using neural networks and active contour models. In Neurocomputing 43 (2002) pp.145-172.
21. Walker KN, Cootes TF & Taylor CJ (2002). Automatically building appearance models from image sequences using salient features. In Image and Vision Computing Vol.20, Issues 5-6, pp.435-440.
22. Williams DJ & Shah M (1992). A fast algorithm for active contours and curvature estimation. In CVGIP – Image Understanding 55, pp. 14-26.

Chapter 15

Fuzzy Sets in Investigation of Human Cognition Processes

Goran Trajkovski

Summary. This chapter discusses two applications of the theory of fuzzy sets in investigating and evaluating human learning abilities and cognitive processes. They are an integral part of the Interactivist-Expectative Theory on Agency and Learning (IETAL) and its multiagent expansion known as Multi-Agent Systems Interactive Virtual Environments (MASIVE).

In the first application presented, a fuzzy set is defined to ease and automate the process of detection of negative variation in filtered brain waves during the Dynamic Cognitive Negative Variation (CNV) experiment. The automatic detection of brain waveforms that are contingent of negative variation is a crucial part of the experiment that measures individual human learning parameters. By eliminating the direct influence of the human expert, a level of objectivity is being maintained over the duration of the whole experiment. The decision process is significantly shorter, which contributes to more accurate measuring, as is the case in numerous experiments involving human subjects and learning.

In the second application, fuzzy sets serve as tools in the process of grading, which is a highly cognitive, but ill-defined problem. The fuzzy evaluation framework that is given is very general, and straightforwardly applicable in any evaluation process when the evaluator is expected to quantize one or several aspects of a given artifact.

Keywords: fuzzy sets, cognitive processes, learning, evaluation, CNV brain waves

1 Introduction

Our work in the domain of the IETAL theory [10], during the past years has focused on exploring the concept of learning the environment by an autonomous agent, and the problems it encounters during its stay there [16]. It has been motivated explorations of human cognition and adaptation [11], especially in the domain of CNV waves.

The key concepts of IETAL are those of expectancy and learning through interactions with the environment, while building an intrinsic model of it. The agent uses a projection of its intrinsic model in order to navigate within the environment during its quest to satisfy a dynamic set of active drives. Every entry in the contingency table is attributed an emotional context, depending on the current active drives, and the likelihood of usefulness of the entries towards the satisfaction of the said drives.

The automation of subsystems of experimental setups that are heavily dependent on human experts is always a challenge. When successful the application, the results are very beneficiary. It eliminates the costly human experts, it cuts down the time of the experimental runs, and it does not allow time for the human subject to defocus from the experiment.

Applying tools from the fuzzy set theory is the natural way to go when dealing with processes that deal with ill-defined problems, problems that are strongly based on human implicit (and expertise) knowledge, lack of information, uncertainty, etc.

In this chapter we overview two applications of the fuzzy set theory in experiments that involve human cognition, learning and/or adaptation. Although seemingly unrelated, they concern with the same problems on two different levels. As a part of the first application (Fuzzy sets in the Dynamic CNV Experiment), we define and apply fuzzy sets as classifiers/decision makers in an experiment of human cognitive abilities on a very low, almost "hardware" level. The second application deals with the evaluation of the mental perception in the process of evaluation by a human agent, and is basically the same type of problem, but dealt on a much higher and complex level. Our investigation has proven both proposed applications applicable and successful.

Section 2 of the chapter discusses an application of fuzzy sets in the Dynamic CNV experiment. It compares the results of such an application with the classification results given by a human expert. Section 3 proposes a fuzzy evaluation environment. The problem of personal standardized fuzzy criterion, multi-evaluator criterions, and consistency of criteria are also discussed. This application is also briefly discussed in the light of the MASIVE theory. The last section overviews the chapter, and gives directions for further work in the domain of the two applications, and beyond.

2 Fuzzy Sets as CNV Classifiers

2.1 Brain Waves

The brain waves can be classified as spontaneous and event-related, according to the cause of appearance. The former are result of the regular brain activity, whereas the latter are due to inner or outer stimuli.

The evoked brain potentials can, on the other hand, be classified as anticipatory and evoked. The anticipatory waves appear in anticipation of an event, i.e., they represent expectation and/or motor preparation in the human brain in link with a corresponding event. The evoked response potentials (ERP) are changes in the electrical activity of the nervous system initiated by efferent or afferent physical stimuli or psycho-physiological events.

As an ERP subclass, the slow brain potentials feature slow shifts in cortex potential, ranging from 0.5 to 5 seconds (or even longer), and are said to represent special psycho-physiological states caused by efferent stimuli (somatosensory, mental, auditory, or visual).

The ERPs are buried in louder EEG signals. To extract it, we used weighted average procedure

$$ERP\ (n) = (1 - k)\ ERP\ (n - 1) + k\ EEG\ (n), \tag{1}$$

where n is the trial number, and the value $k = 0.1$ is used in our experiments.

2.2 CNV Brain Waves

The CNV brain potentials are the most illustrative example of slow ERP waves. A typical waveform is presented in Figure 1. These waves were first registered in an experimental procedure referred to as the CNV paradigm [19]. The experimental setup is shown in Figure 2.

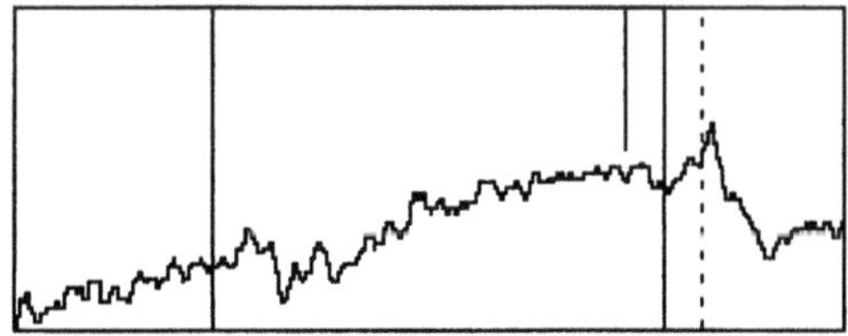

Figure 1. Typical CNV wave recorded from Cz electrode. By convention negative EEG values are up. The vertical lines denote the time of the warning and imperative stimuli (see text for explanation).

Figure 2. The classical CNV experimental setup.

The human subject is presented with two stimuli. The signal S_1 is the warning and S_2 is the imperative stimulus. The subject is told that after the first, warning stimulus, there would be a second one, which he/she is supposed to interrupt. The stimuli (visual, auditory or somatosensory) are usually chosen to cause either a brain reaction or a simple mental decision.

The CNV wave is the negative shift of the cortex potential between two stimuli. That is the reason why these potentials are also sometimes referred to as expectation waves and are believed to be a good measure for the subject's anticipation of the second signal. After a while, the subject learns the pattern S_1-S_2, which results in CNV in the filtered frequencies of the brain activity being recorded. The wave is contingent of negative variation, if the negative shift in the potential should appear in between the two stimuli, the potential increases monotonously; and the shift vanishes after the imperative stimulus (action that the human subject undertakes).

One of the simplest models for these potentials is the ramp model [9]. A schema with the characteristic parameters, mainly suggested by a CNV expert) is shown in Figure 3.

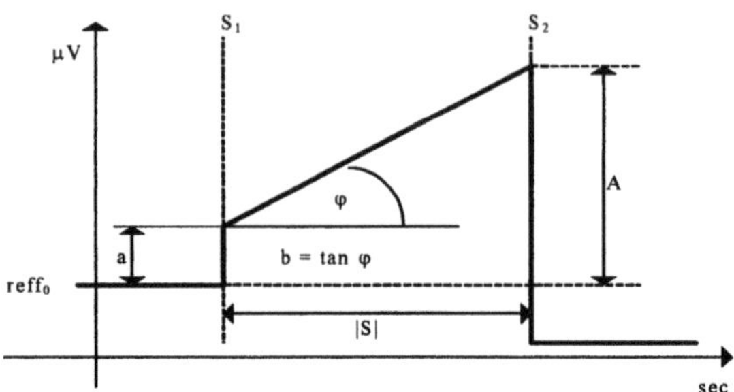

Figure 3. The ramp model for CNV brainwave forms.

The parameters *a* and *b* are results of linear regressive interpolation in the inter-stimulus interval. The parameter $reff_0$ (referent zero) represent the average level of the signal before the first stimulus. The value of the variable *A* (amplitude) is the difference between the signal in S_2, and the reference value $reff_0$.

In a human-generated brainwave, the amplitude *A* does not always equal $a+b|S|$, and therefore jump $a\text{-}reff_0$, the angle φ and the amplitude A are all considered a non-redundant set of (independent) parameters, characterizing the wave at hand.

2.3 Emulating Abstract Agents on Humans

Humans are the only linguistically competent agents in the sense of the IETAL theory. They are able to filter out stimuli and easily emulate an abstract bio-agent. In simplified terms, to have an agent in IETAL, it is initially necessary to have defined an action repertoire, stimuli set, a contingency schema, and a goal. Examples of defining an abstract agent for the purpose of the CNV experiment are given below.

We define the stimuli set S and action set A, sets of the agent by telling the subject what inputs should be paid attention to pay attention to and what possible motor actions are allowed to be undertaken. For example, to define the stimuli set S, the subject is told: "You will be hearing beeps like this (demonstration of the beeps)." To define expected actions repertoire A, the subject is told: "You are allowed to press or not to press these buttons (experimenter shows the buttons to the subject)." Schemas are defined similarly. To define contingency schema, the subject is told: "Whenever you here beep S_1, a beep S_2 will follow after 2 seconds (demonstration follows)".

The experimenter can now define the structure of the environment by defining, for example, the order of appearance of stimuli, the response of the environment on agent's actions, and the like, or implement a decision engine that would decide whether to fire the stimuli or not.

Having defined all the initial elements, the experimenter can now observe various aspects of the agent's sojourn in that particular environment. Of course, of particular interest are the structures that emerge during the agent-environment interaction. These are the networks of concept/behaviors. Because they are not directly observable the experimenter has to reconstruct them out of the agent's observable parameters, and define and compute various other parameters that may be relevant for the problem being observed.

Apart from parameters like the reaction time, introspective verbal reports, eye movements, and the like, we found Evoked Brain Potentials to be most helpful in reconstructing the conceptual networks in abstract agents [13].

The experimenter communicates the subject only the stimuli set S, the elementary actions set A (which coincides with the basic behaviors set) and the goal: S = {S_1, S_2, ...,S_n}; A = {push button, expect S_2, don't expect S_2}; Goal = {interrupt S_2 as soon as possible}.

This experiment is motivated by the following question: What will happen if during the experiment the experimenter changes the environment configuration? For example, what happens if after some trials in the above experiment, S_2 doesn't follow S_1 any more? What happens if then we return to the initial environment? The DCNV experiment (Section 2.4) is a variation of the classical CNV experiment with agents defined as above, and is designed to answer some of these questions.

2.4 DCNV Paradigm Experiment

In the Dynamic CNV (DCNV) experimental setup [4], there is a feedback that makes the occurrence of the imperative stimulus S_2 dependent on the current status of the CNV waveform (Figure 4). The idea of the experiment is to measure learning parameters in the human, in that manner that, once the S_1-S_2 pattern is learned (as per a criterion that establishes that a wave is CNV), the human is forced to forget the pattern (by introducting absence of S_2), and then the subject is being trained the pattern over again.

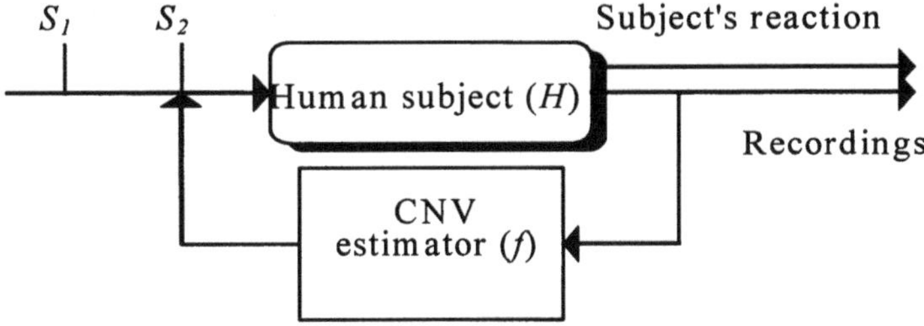

Figure 4. The setup for the DCNV experiment.

Figure 5 represents another view at the DCNV experimental setup. *H* is the human subject, *g* the extraction (filtering) function of the brain activity recordings, *h* is the subject's response to the stimuli, and f is the function that decides whether the extracted wave contains CNV or not. *S* is the signal producer. Based on the decision made by *f*, S produces S_1 or an S_1-S_2 sequence (Figure 6).

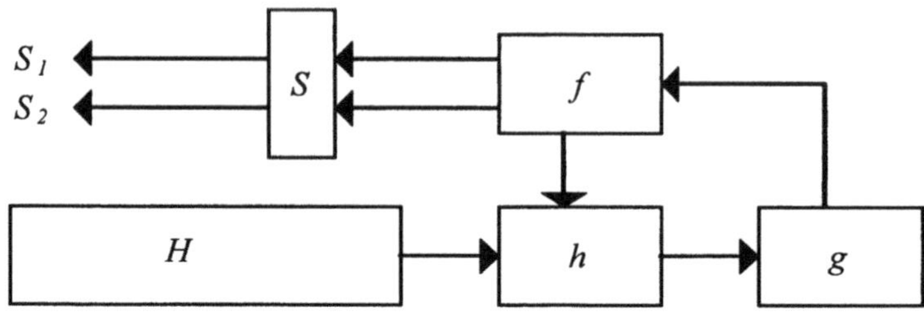

Figure 5. Definition of the DCNV experiment. See text for explanation.

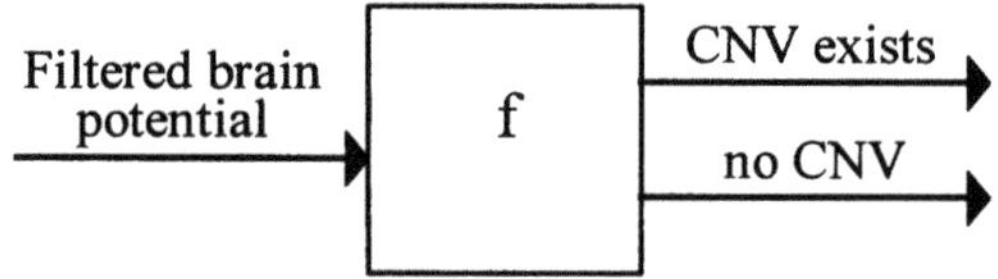

Figure 6. The function *f* as a classifier.

So, *f* is a classifying function, revealing whether the experimental output contains CNV or not.

The speed at which *f* performs is also crucial for the experiment, having into consideration that this is an experiment that involves human subjects, and is an effort in evaluation of human cognitive performance.

As an evaluator (*f*), a CNV human expert has been used in the initial experiments. Since there is no formal definition of a CNV wave in terms of value ranges of the parameters (or for that matter, in terms of anything but experts' intuition), the human expert in the experimental runs had had increased decision times, that is strongly believed that somewhat biased the experimental dynamics and values of the recordings. Eliminating the human expert from the process is expected to improve the experimental environment in the sense of the discussed above.

It is clear now that the classifying/deciding role of *f* is crucial for the DCNV experiment.

2.5 Fuzzy Set for CNV Detection

CNV detection in filtered potentials is a hard task due to noise introduced by the apparatus, human and other random factors. On top of that, there is no formal definition of CNV waveforms. In the process of interviewing CNV experts to "define" what a CNV wave is (or is not), all the statements that were collected were abundant in fuzzy terms [21], which was motivating enough to initiate an effort to designing a CNV evaluator using tools from the fuzzy paradigm.

For example, here is what a typical CNV expert's statement would read, in terms of the parameters of the ramp wave model (Figure 3): *"If the angle* φ*, the shift a-reff*$_0$ *and the amplitude A are* big enough, *then the wave contains* CNV.*"*

The efforts in that sense resulted in a CNV fuzzy set, that was proved successful for the problem of automatic detection of CNV in the brainwaves. We give definition of a fuzzy set [6, 20], with satisfactory results in detection of CNV waves, centered on values that were suggested as threshold values by some of the experts.

Let the domains of the filtered wave parameters φ, *a-reff*$_0$ and *A*,

$$U_\varphi = [-90^\circ, 90^\circ],\ U_a = [-30\ \mu V, 30\ \mu V],\ \text{and } U_A = [0\ \mu V, 30\ \mu V], \quad (2)$$

be universal sets. Let the membership functions of the fuzzy sets $\tilde{\varphi}$, $\tilde{a}$ and $\tilde{A}$ on U_φ, U_a, and U_A respectively be as shown in Figure 7.

Let

$$U = U_\varphi \times U_a \times U_A \quad (3)$$

and $\tilde{C}$ be the fuzzy set with membership function μ defined as follows:

$$\mu(x, y, z) = \max\{\min\{\mu_\varphi(x), \mu_a(y)\}, \min\{\mu_\varphi(x), \mu_A(z)\}, \min\{\mu_a(y), \mu_A(z)\}\} \quad (4)$$

Having defined the fuzzy set for CNV detection $\tilde{C}$, its membership function is used as function *f* in the classical DCNV experimental setup, as shown in Figure 8.

2.6 Experimental Results

The results of a human expert and of the fuzzy set approach are compared in this section. An experimental ensamble of fifty trials are presented in Table 1. The *φ*, *a-reff*$_0$, and *A* columns (second, third and fourth column) are the measured parameters of the filtered wave. Each row represents one experimental trial. The fifth, *f* column is the decision of the human expert [9]. The degrees of membership for each trial are given in the μ columns.

Table 2 surveys the incidence of the "findings" of the human expert and the fuzzy set $\tilde{C}$, when two different levels are used in the cuts in the process of defuzzification. As it can be seen that the proposed method gives satisfactory results compared to the results obtained from the human expert. The agreement rate (the human and the fuzzy set provided the same output) between the human and the fuzzy set method when 0.50 cut is used is 90%, whereas when 0.40 cut is used, the rate is 88%.

In Figure 9, it can be seen how the human expert agent and the fuzzy set classifier work during the DCNV experiment. Since the experiment needs a reference in order to decide whether to present the human subject with the imperative stimulus or not, the human agent was chosen to act as that reference. The horizontal lines in Figure 9 represent the cuts on the 0.4 and 0.5 levels.

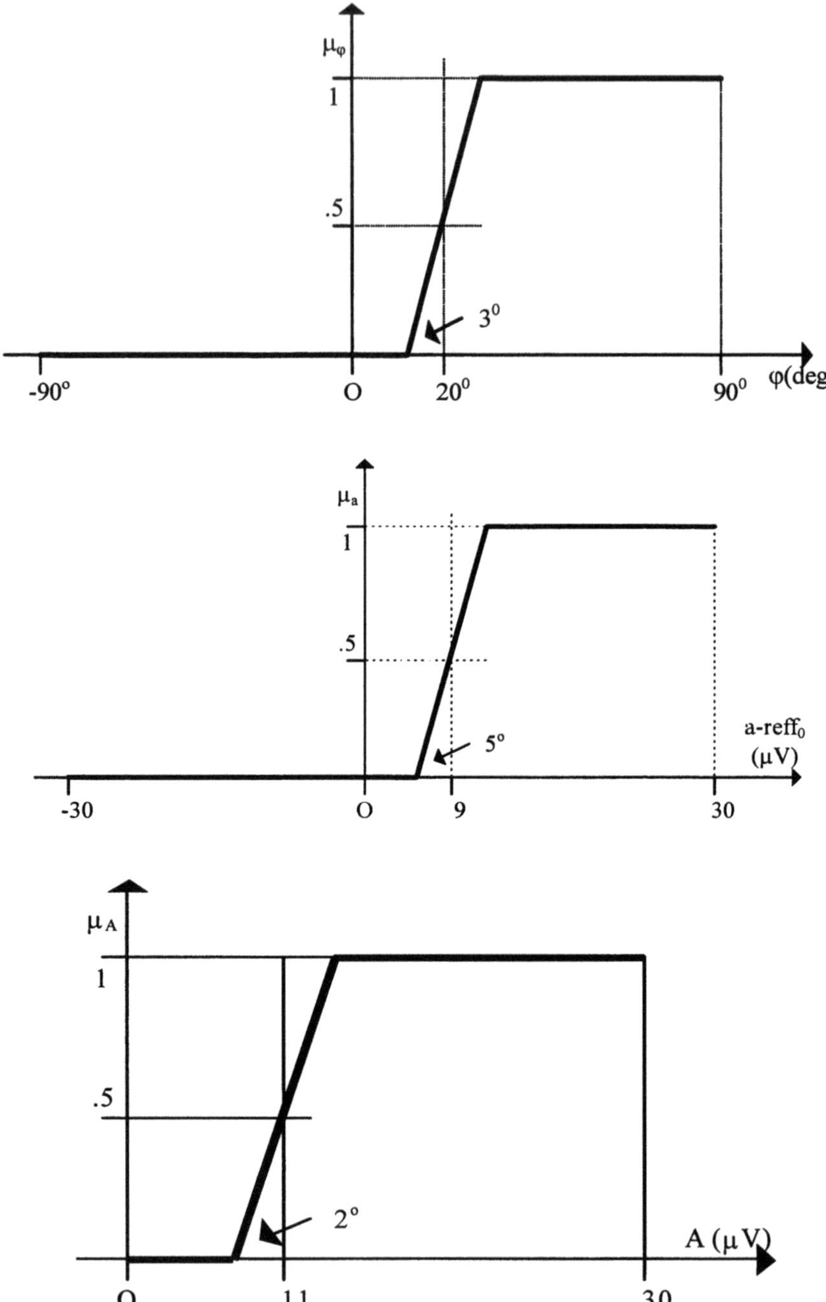

Figure 7. Membership functions of the fuzzy sets $\widetilde{\varphi}$, $\widetilde{a}$ and $\widetilde{A}$.

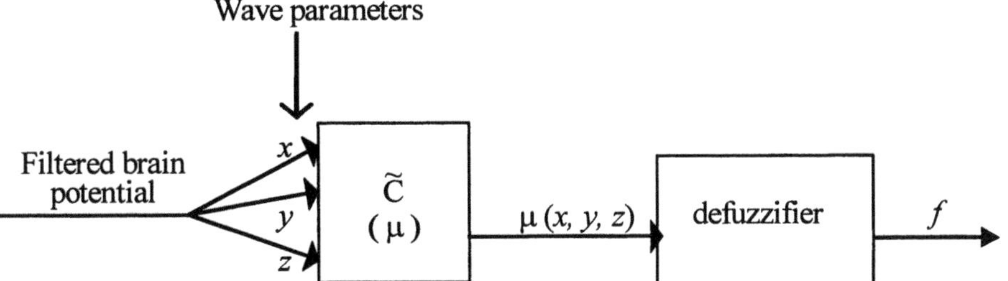

Figure 8. The membership function of $\widetilde{C}$ as f in the DCNV experiment.

Table 1. Experimental results (see text for explanation). The number in the first column is the order number of the experimental run.

	φ	a	A	f	μ
1.	9	−3	2	0	.00
2.	9	−3	2	0	.00
3.	22	−3	4	0	.00
4.	31	−2	6	0	.34
5.	33	−2	0	0	.38
6.	29	−2	7	0	.38
7.	33	−2	9	0	.45
8.	33	0	11	1	.52
9.	35	1	13	1	.59
10.	33	1	14	1	.62
11.	9	4	4	0	.06
12.	11	4	3	0	.06
13.	17	3	4	0	.27
14.	24	5	9	1	.45
15.	31	5	9	1	.45
16.	37	6	13	1	.59
17.	39	8	15	1	.66
18.	37	8	15	1	.66
19.	29	9	12	1	.55
20.	24	11	11	1	.65
21.	19	11	11	1	.52
22.	17	11	11	1	.52
23.	6	9	7	0	.38
24.	−9	12	5	0	.31
25.	−22	13	2	0	.20
26.	−11	13	4	0	.27
27.	−17	14	4	0	.27
28.	-6	13	6	0	.34
29.	14	7	7	0	.32
30.	19	6	9	0	.40
31.	9	8	7	0	.38
32.	9	7	5	0	.31
33.	9	7	4	0	.27
34.	9	6	4	0	.24
35.	9	4	2	0	.06
36.	11	4	3	0	.06
37.	17	5	7	0	.30
38.	22	5	9	1	.45
39.	24	5	10	1	.48
40.	22	6	12	1	.32
41.	17	7	13	0	.32
42.	11	7	12	0	.32
43.	−3	8	9	0	.41
44.	−9	9	9	0	.45
45.	−3	7	10	0	.32
46.	6	8	12	0	.41
47.	11	7	14	0	.32
48.	22	6	16	1	.60
49.	22	7	17	1	.60
50.	31	8	18	1	.76

Table 2. Number of agreements between the human expert and the fuzzy method for the 50 recordings in Table 1.

.50 cut defuzzification		
Human \ Fuzzy	No CNV	CNV
No CNV	32	0
CNV	5	13

.40 cut defuzzification		
Human \ Fuzzy	No CNV	CNV
No CNV	27	5
CNV	1	17

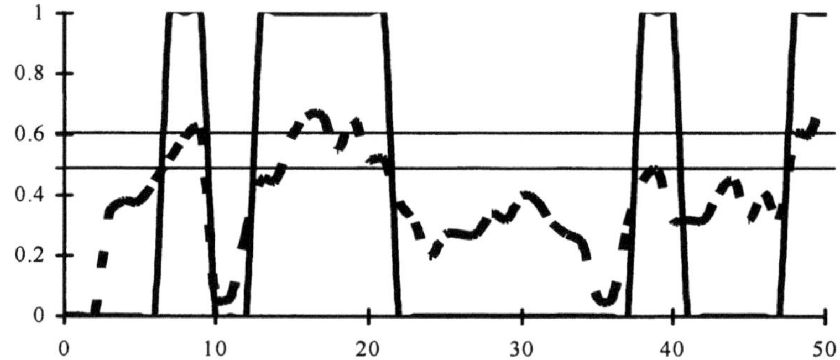

Figure 9. The human expert classification (solid line) and the fuzzy method estimation of CNV (dotted line) in a DCNV experimental series.The horizontal lines represent the 0.40 and 0.50 defuzzification cut levels used in the derivation of results in Table 2.

In this section it was shown how a fuzzy set can be used for CNV detection in the DCNV experiment. The membership value obtained for any triplet of wave parameters gives the membership of the wave in the fuzzy set $\tilde{C}$ of filtered waves containing CNV. Experimental results together with the comparison of the classical and this fuzzy detection method, have demonstrated the suitability of the proposed method.

3 Fuzzy Evaluation Framework

Human higher-level cognitive processes are hard to quantify and measure. Many times the process involves human experts, who themselves are subject to subjectivity, and include large amounts of uncertainty, due to either lack of information, contradictory information, or lack of expertise. In such cases, non-human methods are taken advantage of [1]. Complex problems usually lead to a waste number of possible solutions, each of which must be examined with care – an assessment task that is often well beyond normal human capability.

Expert systems are being developed commercially to solve nontraditional problems. During knowledge programming, planning establishes plausible antecedents and consequences of actions; these beliefs represent expectations [5].

Taking into account the non-formal nature of many problems that involve human arbitrage [11], the exploration of possibilities within the fuzzy set theory can help. The fuzzy logic is suitable for modeling human-like reasoning by the computing-with-words paradigm [6, 14].

This section proposes a fuzzy evaluating environment that eases the evaluation procedure, makes it more objective, and is founded upon a consistent criterion. The consistency of the criterion is congruent with the traditional evaluation, since the fuzzy evaluation system is eventually expected to provide with an evaluation in traditional terms.

Through the example of a fuzzy student evaluation environment, this section sets the foundation for fuzzification process in this domain, and discusses how problems that are encountered regularly within standard traditional evaluation environment can be approached in the new domain. The procedures discussed further in this section are directly implementable in virtually any situation when an evaluator is to evaluate a man-made artifacts (art, elegance of computer programs etc), when the evaluation process involves a lot of subjectivity, and cannot be explicitly defined beforehand, but is based on implicit knowledge and/or intuition of the evaluating expert. As it will become clear from the definitions below, the traditional approach will become a special, marginal case of the fuzzy evaluation procedure.

Prior to the evaluation process, the evaluation criteria is set in order to meet the personal expectation of the evaluator, as well as the possible expectations of an institution. We will illustrate such a process in this section. Maintaining a consistent criterion over time has always been a problem in student evaluation. This section also deals with the problem of establishing a standard evaluation criterion for an evaluator, and shows how it can generate grades/marks in the traditional sense (defuzzification part of the process).

3.1 Traditional Evaluation Environment

The traditional evaluation environments (TEE) depend on the institution that performs the evaluation, and on the historically established practices in evaluation. The usual approach is assigning grades or numerical values according to some criteria learned with practice. In this section we formalize the TEE and give a common example, form a perspective that would be congruent with the Fuzzy Evaluation Environment (FEE) definition.

Example 1. Let us assume that we are grading an essay-type of a question on a student's test in Sociology. As done in [8], we can suggest filling in work papers in the evaluation process. The evaluator is expected to provide a grade from a list, such as the one shown in Table 3.

Table 3. Evaluation grades and their notational values in a possible grading procedure.

Grades – notational	**Notional values (possible interpretation)**
4 (A)	Very adequate
3 (B)	Adequate
2 (C)	Moderately adequate
1 (D)	Inadequate
0 (F)	Nonsense

Definition 1. Let us denote by *G* the set of all n grades that can be given by an evaluator G = $\{G_1 > G_2 > \ldots > G_n\}$, with the corresponding grade points (weight function *w*, defined on G): $w_i = w(G_i)$, $I = 1, 2, \ldots, n$. The pair (G,w) is called TEE.

The standard evaluation procedure within the TEE consists generally of the following steps: (i) Thorough examination of the student answer; (ii) Marking the strengths and weaknesses to facilitate the overview, and (iii) Deciding which grade to awarded to the essay at hand.

The level of performance is decided in qualitative terms, which are fuzzy in nature. That can also be seen by examination of the notational values used in Table 3.

3.2 Fuzzy Evaluation Environment

Evaluation, as stated above, is a problem that involves the concepts of uncertainty and subjectivity. The knowledge-based decision and evaluation performed during the process are complex and hard to deal with, unless with fuzzy tools. The fuzzy paradigm is suitable for dealing with tasks where the observed parameters and phenomena are not easily expressed in terms of exact numerical values [12, 13, 18]. In case the evaluator want to distinguish different features of the essay he/she is grading (in a fuzzy manner), then he/she might give different fuzzy grades upon different features of the evaluated system.

Definition 2. The non-empty set

$$U = \{u_1 < u_2 < \ldots < u_m\} \subseteq N_{100} = \{1, 2, \ldots, 100\}, \qquad (5)$$

is called universal set for the FEE.

The elements are chosen in order to define the levels of evaluation on which the evaluator is to express his/her degrees of satisfaction by the aspect of the software product that is the subject of the evaluation.

Definition 3. A fuzzy set on U is a mapping M: U → [0, 1].

For a given element $u \in U$, the value M(u) is referred as the membership degree or level of membership of u to the set M.

Definition 4. The non-empty fuzzy set M defined on U is called fuzzy grade (f-grade) if its graph is of one of the following types: (i) Decreasing, (ii) Increasing, or (iii) At first increasing, then decreasing.

The f-grade is supposed to show the amount (weight) of belief $M_i = \mathrm{M}(u_i)$ of the evaluator that the essay covers u_i% of the evaluator's expectations.

The non-emptiness of the fuzzy set is required in order to prevent from division-by-zero complication when dealing with the similarity measures further in the test. However, that is not a very restrictive requirement, since empty fuzzy sets are the most likely to be assigned to a product by a completely ignorant evaluator, which is rarely the case.

Definition 5. Let M be an f-grade. Then the vector (m-tuple)

$$\mathbf{M} = (\mathrm{M}_1, \mathrm{M}_2, \ldots, \mathrm{M}_m) \tag{6}$$

is called f-grade vector of M.

Definition 6. The set F of all f-grades that can be awarded is called the f-grade set of the FEE.

Definition 7. The binary relation SS defined on F in the following manner

$$SS(\mathbf{A},\mathbf{B}) = \frac{\sum_{i=1}^{m} a_i b_i}{\max\left\{\sum_{i=1}^{m} a_i^2, \sum_{i=1}^{m} b_i^2\right\}}, \tag{7}$$

where $\mathbf{A}, \mathbf{B} \in F$, with $\mathbf{A} = (a_1, a_2, \ldots, a_m)$ and $\mathbf{B} = (b_1, b_2, \ldots, b_m)$, and "." denotes the dot product of vectors, is called the measure of similarity. The value $SS(\mathbf{A},\mathbf{B})$ is called the degree/coefficient of similarity between the f-grade vectors **A** and **B**, i.e., their corresponding f-grades. The fuzzy relation SS is a similarity (fuzzy equivalence) relation and is said to be a good choice as far as dilatation/concentration is concerned [3].

Definition 8. Let S_i, i=1,…,n be f-grades. The matrix **S**, defined as follows:

$$\mathbf{S} = [\, SS(\mathbf{S}_i, \mathbf{S}_j)\,]_{n\mathrm{x}n} \tag{8}$$

is called the similarity matrix of the sets $\mathbf{S}_i$.

Definition 9. The n-tuple $\mathbf{C}=(\mathbf{S}_1,\mathbf{S}_2,\ldots, \mathbf{S}_n)$ is called standard Fuzzy Criterion (SFC), if the values in the columns and the rows of the similarity matrix of the sets S_i, $i = 1,2,\ldots,n$; (i) Increase; (ii) Decrease or (iii) At first increase then decrease, as the row/column index increases/decreases. The sets S_i are then called Standard Fuzzy Sets (SFSs) or components of the criterion **C**.

Definition 9 ensures that for any f-grade **M**, one of the following inequalities will hold:

$$SS(\mathbf{M},\mathbf{S}_1) \bigcirc SS(\mathbf{M},\mathbf{S}_2) \bigcirc \ldots \bigcirc SS(\mathbf{M},\mathbf{S}_n), \tag{9}$$

where the n-1 circle signs (O) in (9) are substituted with either "≤" or "≥" signs, in that order that all the "≤" signs precede the "≥" signs in the inequality. Thus, it is ensured that the similarity between the f-grade **M** and one f-grades $\mathbf{S_i}$ is biggest, and the similarity grades decrease the further we go from the most similar $\mathbf{S_i}$ to **M**.

Example 2. Let G={**A, B, C, D, F**}, as in Example 1, let U={0,25,50,75,100}, and let $\mathbf{C_1}$={$\mathbf{S_A}$, $\mathbf{S_B}$, $\mathbf{S_C}$, $\mathbf{S_D}$, $\mathbf{S_F}$} where the membership degrees of the $\mathbf{S_i}$' s are given in Table 4.

Table 4. The membership functions of the components of C_1.

	0%	25%	50%	75%	100%
$\mathbf{S_A}$	0.0	0.1	0.5	0.8	1.0
$\mathbf{S_B}$	0.0	0.2	0.3	1.0	0.2
$\mathbf{S_C}$	0.0	0.3	0.6	1.0	0.2
$\mathbf{S_D}$	0.0	0.1	0.9	0.8	0.2
$\mathbf{S_F}$	0.0	0.2	0.9	0.1	0.1

A column of the similarity matrix **S** is shown in Table 5.

Table 5. The similarity coefficients of $\mathbf{S_A}$ from Table 4 with the other f- grades from $\mathbf{C_1}$.

SS	$\mathbf{S_A}$
$\mathbf{S_A}$	1.00
$\mathbf{S_B}$	0.62
$\mathbf{S_C}$	0.70
$\mathbf{S_D}$	0.68
$\mathbf{S_F}$	0.04

From Table 5 it can be seen that $\mathbf{C_1}$ cannot be chosen to represent a SFC, since $\mathbf{S_C}$ is more similar to $\mathbf{S_A}$ then to $\mathbf{S_B}$, which is not to be the case with a consistent evaluation process (according to Definition 9). The **n**-tuple $\mathbf{C_2}$, given in Table 6, however, is an example of a valid fuzzy criterion, which can be seen from Table 7.

Table 6. The membership functions of the components of $\mathbf{C_2}$.

	0%	25%	50%	75%	100%
$\mathbf{S_A}$	0.0	0.1	0.5	0.8	1.0
$\mathbf{S_B}$	0.0	0.2	0.3	1.0	0.8
$\mathbf{S_C}$	0.0	0.3	0.6	1.0	0.2
$\mathbf{S_D}$	0.0	0.4	0.8	0.8	0.2
$\mathbf{S_F}$	1.0	1.0	0.0	0.0	0.0

Table 7. The lower triangle of the C_2.similarity matrix.

	S_A	S_B	S_C	S_D	S_F
S_A	1.00				
S_B	0.93	1.00			
S_C	0.70	0.79	1.00		
S_D	0.67	0.72	0.97	1.00	
S_F	0.05	0.10	0.15	0.20	1.00

Definition 10. Let *M* be an f-grade and **C** a given SFC, with components $\mathbf{S}_i$, and

$$s_i = SS\,(\mathbf{M}, \mathbf{S}_i),\ i = 1, 2, \ldots, n. \tag{7}$$

Gj is said to be the correspondent traditional grade (CTG) to the f-grade *M* with respect to **C** if $s_j = \max\{s_i, i = 1, 2, \ldots, n\}$.

In case of having more than one maximal value between the coefficient s_i, it is left to the discretion of the evaluator to choose which grade to be awarded as the CTG to *M* wrt. C.

Example 3. Let $\mathbf{C}_2$ be the SFC for evaluation and $\mathbf{M} = (0.0, 0.1, 0.2, 0.3, 0.4)$. The similarity coefficient wrt. $\mathbf{S}_A$, $\mathbf{S}_B$, $\mathbf{S}_C$, $\mathbf{S}_D$, and $\mathbf{S}_F$ (actually these are the fuzzy **A, B, C, D,** and **F**), are 0.39, 0.40, 0.35, 0.35, and 0.05 respectively, so the CTG is **B** (excellent performance).

Definition 11. Let **G** be the grade set of TEE, **C** a SFC wrt. *SS* similarity measure, whose f-grades are defined on the universe U. Then, the quadruple (G, U, **C**, *SS)* is called a Fuzzy Evaluation Environment.

3.3 Multi-Evaluator Environments

Evaluators typically have different degree of appropriateness for a given task. The appropriateness of a particular evaluator is a function of how well the agent's skill matches the expertise to do the task, the extent to which its limited knowledge is adequate for the task, its current processing resources and the quality of its communication with the other agents. In this section we discuss the challenges of a multi-evaluator fuzzy evaluation environment.

In real life, like in situations when we ask for a second opinion, it is not unlikely that two different physicians are two different answers. The same kind of mismatch may happen when we ask two (or more) experts to evaluate an essay. The reason why that happens is *NOT* the incompetence of the expert. The reason lies in casually–defined terms and notions in the domain knowledge.

How to draw conclusions on the expert system features, when the traditional grades of the evaluators are different? The usual approach would be the use of voters. They can be majority, median, plurality and weighted voters [7], and we choose them so that they reflect the current situation with all the evaluating agents at hand with seemingly inevitable consequences on the reliability of the whole process [2].

3.4 Tailoring a Personal Evaluator's Standardized Criterion

In this subsection we give results of a practical approach towards a personalized SFC of an evaluator. It is a SFC obtained with the help of a backward propagation neural network (NN). Within this procedure we will try to reveal one of the four components of FEE, namely SFC, given the other three beforehand.

In order to standardize the criterion of an evaluator, we have chosen essays by 20 different students that have been graded in a TEE. Several weeks after the traditional process has taken place, we asked the same evaluator to assign f-grades to each of them, using the grade set from Table 3 and U = {0, 50, 100}, as a basis for the evaluation procedure.

Triplets were chosen in order to ease the evaluator's transition towards the FEE. On the other hand, we had left no possibility of subversion the evaluation process by just giving all-but-one 0s as membership degrees of the f-grades. We thought that the triplets, for the inexperienced fuzzy evaluator would be just the right transitional resolution in evaluation.

The evaluator was advised to use multiples of 0.1 as f-grade membership degrees, for we think that beliefs cannot be more effectively expressed using the whole set of real numbers, due to the presence of the human factor, and its inability to evaluate with an ideal precision and perfect resolution.

After the data set was collected, an artificial neural network with back propagation was trained. The training set is presented in Table 8.

After the neural network was trained and tested, it was given the task to evaluate all the possible f-grades in the given environment. We selected those that were evaluated as 4.00, 3.00, 2.00, 1.00 and 0.00, as "perfect" candidates for the SFC that we were to establish. They are given in Table 9.

In the next stage of the standardization procedure, we combined the x.00 f-grades to see which ones will give a SFC. This process resulted with the six SFC, one of which is given in Table 10.

Any one of these SFCs can be used in the fuzzy evaluation procedure.

In case the set of x.00 f-grades is too restricted, it can be enlarged with f-grades evaluated in any given ε-proximity ($\varepsilon \geq 0$) of the values 0.00, 1.00, 2.00, 3.00, and/or 4.00.

In this section we presented a fuzzy framework for evaluation of different aspects of students' work, especially those that are not easily quantifiable, and involve a lot of subjectivity in the grading process. The approach is actually very general and can be applied in any evaluation-type of activities, when there is any expert/human-like knowledge involved in the assessment. Moreover, we illustrated a neural network approach towards standardizing the evaluator's criteria, in order to be able to evaluate in a consistent manner, even in a multi-evaluator environment.

Table 8. The training set of the neural network.

f-grades			grade
M_1	M_2	M_3	
0.0	0.8	1.0	**A**
0.2	0.7	1.0	**A**
0.1	0.3	1.0	**A**
0.0	0.3	0.9	**A**
0.0	0.4	0.8	**A**
0.1	0.4	0.9	**A**
0.1	0.5	0.4	**B**
0.3	0.5	0.7	**B**
0.1	0.4	0.1	**C**
0.3	0.9	0.3	**C**
0.2	0.9	0.1	**C**
0.2	0.7	0.2	**C**
0.5	0.5	0.1	**D**
0.9	0.7	0.0	**D**
0.6	1.0	0.2	**D**
0.7	1.0	0.0	**D**
1.0	0.3	0.0	**D**
0.1	0.3	0.0	**D**
1.0	0.5	0.0	**F**
1.0	0.0	0.0	**F**

Table 10. Components of a SFC of the evaluator.

grade	f-grades		
	M_1	M_2	M_3
A	0.1	0.3	1.0
B	0.1	0.5	0.4
C	0.1	0.4	0.1
D	0.1	0.3	0.0
F	1.0	0.0	0.0

Table 9. F-grades evaluated as 4.00, 3.00, 2.00, 1.00 and 0.00 (x.00 f-grades).

NN output x.00	f-grades		
	M_1	M_2	M_3
4.00	0.2	0.7	1.0
	0.2	0.4	1.0
	0.1	0.6	0.9
	0.1	0.3	1.0
	0.0	0.8	1.0
	0.0	0.7	0.9
	0.0	0.5	0.8
	0.0	0.3	0.9
3.00	0.3	0.5	0.7
	0.1	0.5	0.4
	0.0	0.1	1.0
2.00	0.2	0.2	0.7
	0.1	0.4	0.1
1.00	1.0	0.3	0.0
	0.9	0.7	0.0
	0.8	1.0	0.0
	0.6	1.0	0.1
	0.6	0.7	0.5
	0.5	0.6	0.4
	0.5	0.5	0.1
	0.5	0.5	0.1
	0.4	0.4	0.1
	0.1	0.3	0.0
0.00	1.0	0.5	0.0
	1.0	0.6	0.5
	1.0	0.0	0.0

3.5 Fuzzy Evaluation in MASIVE

While working within the MASIVE Theory framework [15, 16], throughout the intra-agent interaction, a protolanguage of the environment emerges. The agents use similar proto-lexemes to build proto-concepts/metaphors that are later tokenized and become concepts/words. The process of evaluating the similarity and classifications of the proto-lexemes uses a fuzzy evaluation system. The criterion is built within the multi-agent society.

4 Conclusions

In this chapter two applications of the fuzzy set paradigm in problems that deal with human mental processes. The first one was the use of fuzzy sets as classifiers/decision makers in the Dynamic CNV experiment, and the second one in defining the fuzzy evaluation environment. The applications have been proven successful and applicable.

However, every question touched challenges to attack answering many more questions.

For the first application the following are a sample of the questions that will be dealt with in near future. What are the relevant derived parameters that would quantize the learning process in the human subject wrt. the DCNV experiment? How is the DCN experiment to be changed in order to be able to further extract valuable data on human abilities with respect to the low-level learning and adaptation abilities [16, 17]?

As far as the evaluation framework is concerned, it is challenging to work on devising a transition function between two standardized evaluation criteria. Thus, if several evaluators evaluate a given artifact, the evaluations of one can be transformed into an evaluation of the others.

References

1. Addis, T.R., *Designing knowledge-based systems*, Prentice Hall, Upper Saddle River, NJ, 1986.
2. Barbara, D., Garcia-Molina, H., "The reliability of voting mechanisms," *IEEE Trans. on Computers*, **C-36** (1987) 1197-1208.
3. Biswass, R., "An Application of fuzzy sets in students' evaluation," *Fuzzy Sets and Systems,* Elsevier, Amsterdam, Holland, **72** (1995) 187-194.
4. Bozinovska, L., Prevec, T., Stojanov, G., Bozinovski, S., "Dynamic CNV paradigm," *Proc. of the First European Psychology Conference*, Tilburg, Germany, (1991) 51-57.
5. Klahr, P., Waterman, D.A., *Expert systems: techniques, tools and applications*,

Adison-Wesley, Boston, MA, 1986.
6. Klir G.J., Juan B., *Fuzzy sets and fuzzy logic*, Prentice Hall, Upper Saddle River, NJ, 1995.
7. Lorczak, P.R., Caglayan, A.K., Eckhardt, D.E., "A theoretical investigation of generalized voters for redundant systems," *19th IEEE Int. Symposium on Fault-Tolerant Computing Digest of Papers*, IEEE Computer Society Press (1989) 444-451.
8. Perry, W., *Effective methods for software testing*, John Willey and Sons, Indianapolis, IN, 1995.
9. Stojanov, G., *Detection and extraction of evoked brain potentials*, MSc Thesis, University "SS. Cyril and Methodius"- Skopje, Macedonia, 1992 (in Macedonian).
10. Stojanov, G., Bozinovski, S., Trajkovski, G. (1997), "Interactionist-Expectative View on Agency and Learning", *IMACS Journal of Mathematics and Computers in Simulation*, Elsevier, Amsterdam, Holland, **44** (1997) 295-310.
11. Trajkovski, G., Cukic, B., Stojanov, G. (1999), "Fuzzy logic in neurophysiology: a case study," *Proc. Computational Intelligence: Methods and Applications (CIMA'99)*, Rochester, NY (1999) 21-25.
12. Trajkovski, G., *Fuzzy relations and fuzzy lattices*, M.Sc. Thesis, University "St. Cyril and Methodius, Skopje, Macedonia, 1997 (in Macedonian).
13. Trajkovski, G., Stojanov, G., Bozinovski, S., Bozinovska, L., Janeva, B., "Fuzzy sets and neural networks in CNV detection," *Proc. Interaction Technology Interfaces (ITI'97)*, Pula, Croatia (1997) 153-158.
14. Trajkovski, G., Janeva, B., "Towards a standardized personal fuzzy criterion for student evaluation", *Proc. 7th Int'l Fuzzy Systems Assoc. Congress*, Academia, Prague, Czech Republic, **III** (1997) 62-67.
15. Trajkovski, G., *Representation of Environments in Multiagent Systems*, PhD Thesis, Thesis, University "St. Cyril and Methodius, Skopje, Macedonia, 2002 (in Macedonian).
16. Trajkovski, G., "MASIVE: A Case Study in Multiagent Systems", *LCAI: Proc Third International Conference on Intelligent Data Engineering and Automated Learning IDEAL 2002*, Manchester, UK, (2002) (in print).
17. Trajkovski, G., Goode, M., Chapman, J., Swearingen, W., "Investigating learning in human agents: The POPSICLE experiment", submitted to: Knowledge-Based Intelligent Information & Engineering Systems (KES) 2002, Crema, Italy, (2002).
18. Turski, W.M., "Should/Could software be more reliable than the 'world' in which it is used? " *Proc. ISSRE '98*, Padeborn, Germany, (1998) 3-9.
19. Walter, G., Cooper, R., McCallum, W. (1964), "Contingent negative variation: an electric sign of sensory-motor association and expectancy in the human brain," *Nature* **203** (1964) 380-384.
20. Zadeh, L.A., "Fuzzy sets", *Information and Control*, **8** (1965) 295-303.
21. Zadeh, L.A., "Fuzzy logic = computing with words," *IEEE Trans. Fuzzy Systems*, **4** (1996) 103-111.

Chapter 16

A Full Explanation Facility for an MLP Network That Classifies Low-Back-Pain Patients and for Predicting MLP Reliability

M.L. Vaughn, S.J. Cavill, S.J. Taylor, M.A. Foy, and A.J.B. Fogg

Summary. This chapter presents a full explanation facility for any standard MLP network with binary input neurons that performs a classification task. The interpretation of any input case is represented by a non-linear ranked data relationship of key inputs. The knowledge that the MLP has learned is represented by ranked class profiles or as a set of rules. The explanation facility discovers the MLP knowledge bounds, enabling novelty detection to be implemented and the predictability of the MLP to be investigated. Results using the facility are presented for a 48-dimensional real world MLP that classifies low-back-pain patients.

Keywords: Explanation facility, interpretation, knowledge discovery, knowledge bounds, novelty detection, rules

1 Introduction

A full explanation facility has been developed [22-24] for interpreting the output on a case-by-case basis from any standard multi-layer perceptron (MLP) network that classifies binary input data in *n*-dimensional input space using sigmoidal activation functions and 1-in-c output layer neurons. The method represents a significant advance towards the goal of readily interpreting trained neural networks that solve real-world problems with a large number of input features [7].

The explanation facility is being developed for use by orthopedic surgeons at a hospital in the UK to assist in the diagnosis of low-back-pain patients by a MLP with 48 input neurons [25]. An interpretation is given, in both text and graphical forms, which shows the non-linear data relationship between the key patient symptoms, both absent and present, used by the MLP in making the classification. The interpretation method is presented in Section 2.1.

Using the full explanation facility, the complete knowledge learned by the MLP can be represented by class profiles of average ranked key inputs or as a set of automatically induced rules for the training set. This is discussed in Section 2.2. The rule extraction method is compared with other approaches in Section 5.4.

The MLP knowledge bounds have been defined [23] as the set of hidden layer decision regions in the *n*-dimensional input space which contain the correctly classified training examples. The discovery of the knowledge bounds from the hidden layer activations for all of these training data is discussed in Section 2.3.

For novelty detection, a new direct approach is taken [23], whereby the explanation facility warns the user that the classification is potentially unreliable when an input case is beyond the MLP network's knowledge bounds. This is discussed in Section 2.4.

Results from using the full explanation facility for the 48-dimensional operational low-back-pain MLP network are presented in Sections 3 to 6 of this chapter. The prediction of the reliability of the low-back-pain MLP is investigated in Section 7, where the training examples are discovered to occupy contiguous class threads of hidden layer decision regions across the 48-dimensional input space. Finally, the summary and conclusions are presented in Section 8.

2 The Interpretation and Knowledge Discovery Method

When an input vector from class k, $k = 1,c$ is presented to a trained MLP network that performs a classification task with c output neurons and sigmoidal activations, the activation of output neuron $k \rightarrow 1.0$ and all other output neuron activations $\rightarrow$ 0.0. The interpretation and knowledge discovery method [22, 23] defines output neuron k as the *classifying* output neuron for the input case. Hence, the classifying output neuron varies, depending on the class membership of the input case.

The interpretation method starts by examining the high activation at the classifying output neuron. This is shown schematically in Figure 1 where the classifying output neuron is the first output neuron and represents membership of class A with a high activation level of 0.9.

The activation level at a classifying neuron is the sigmoid of the combined input sum ($\Sigma h_{jk} w_{jk}$) from the hidden layer neurons H_j, $j = 1,m$. This sum is made up of a positive contribution from hidden neurons connected to the classifying neuron with positive weights and a negative contribution connected to the classifying neuron with negative weights.

Hence, the high activation level at any classifying output neuron is completely determined by the positive contribution to the combined input sum, and the negative part of the contribution only contributes a dampening effect. The hidden neurons that contribute to the positive combined input sum at the classifying neuron positively drive the classification of the input case. The method defines these neurons as the *hidden layer feature detectors*.

Hidden layer feature detectors are usually highly activated by the MLP so as to maximize the positive part of the combined input sum at the classifying output neuron. The method, similarly, finds those input neurons that contribute to the positive part of the combined input sum at each of the hidden layer feature detectors. This leads to the discovery of the *key positive inputs*; these are the relevant inputs that positively drive the classification of the input case.

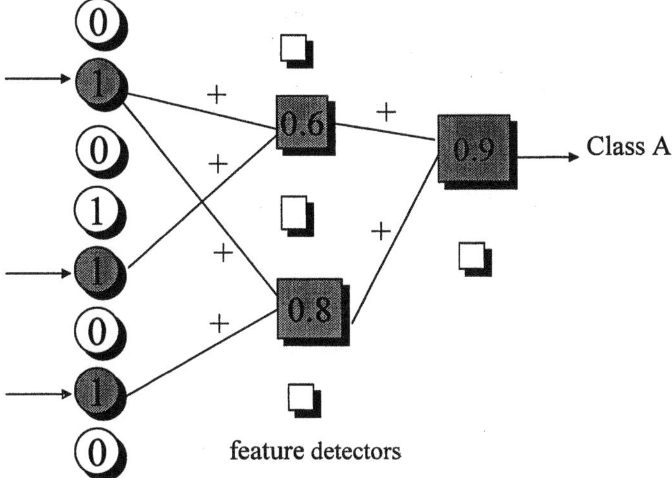

Figure 1. The discovery of the key positive inputs.

Following this, the *key negated inputs* are found. These are the relevant inputs that reduce the activation of the hidden neurons that are *not* feature detectors so as to minimize the negative contribution at the classifying output neuron [23].

2.1 Interpreting an Input Case – 'The Explanation'

The procedure for the interpretation of any input case presented to the MLP network is as follows.

2.1.1 Discovery of the Hidden Layer Feature Detectors

The method first finds the hidden layer feature detector neurons that collectively contribute all the positive input to the classifying output neuron. For sigmoidal activations, these are hidden neurons connected to the classifying neuron with *positive* weights. The hidden layer bias also makes a positive contribution when the connection weight to the classifying output neuron is positive.

2.1.2 Discovery of the Ranked Key Positive Inputs

The method next finds the key inputs which contribute all the positive input to each feature detector neuron. For binary inputs, these can only be *positive* inputs (value 1) connected to each feature detector with *positive* weights. The key positive inputs positively activate the feature detectors which, in turn, positively activate the classifying output neuron. The discovery of the feature detectors and the key positive inputs is shown schematically in Figure 1. Note that one of the positive inputs is not a key input.

The key positive inputs are ranked in order of the decrease in activation at the classifying output neuron when each is selectively switched *off* at the MLP input layer.

2.1.3 Discovery of the Ranked Key Negated Inputs

The role of the negated inputs (value 0) is to reduce the activation of the hidden layer neurons that are *not* feature detectors so as to minimize the negative contribution at the classifying output neuron. For binary inputs the key *negated* inputs are the zero-valued network inputs connected with *positive* weights to hidden neurons that are *not* feature detectors.

The key negated inputs are key positive inputs for another class which deactivate the feature detectors for that class when not active at the MLP input layer. The discovery of the *not* feature detectors and the key negated inputs is shown schematically in Figure 2. Note that two of the negated inputs are not key inputs.

The key negated inputs are ranked in order of the decrease in activation at the classifying output neuron when each is selectively switched *on* at the MLP input layer.

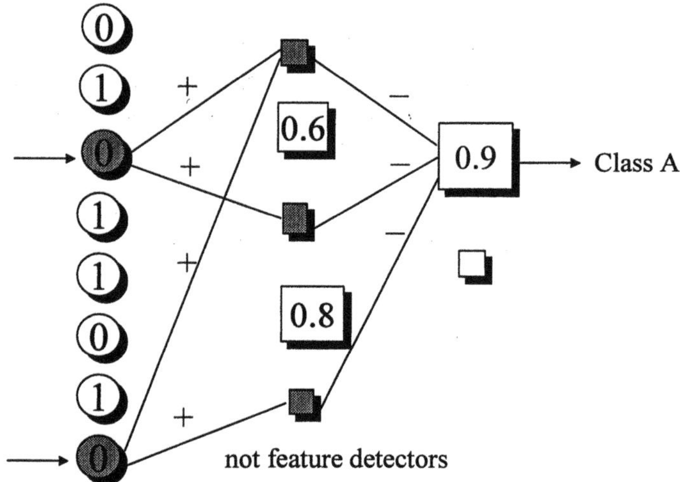

Figure 2. The discovery of the key negated inputs.

2.1.4 The Ranked Data Relationship – 'The Explanation'

The ranked key positive and negated inputs are first combined in order of decrease in activation at the classifying neuron. The ranked data relationship is then found by progressively switching off (and on) the key positive (and negated) inputs *together* until most (e.g., 95%) of the classifying output activation has been accounted for.

The data relationship is non-linear due to the effect of the sigmoidal activation functions and embodies the graceful degradation properties of the MLP. This is demonstrated in Section 4 in both graphical and text forms for three example low-back-pain cases.

2.2 Discovering the MLP Knowledge

The knowledge discovered from all training cases represents the knowledge that the MLP network has learned from the training set. The network knowledge can be represented as a set of ranked class profiles for all training cases or as a set of induced maximally general rules, which are valid for the training set [24]. Both representations of the MLP knowledge can be used by the domain experts to validate the MLP.

2.2.1 Ranked Class Profiles

The ranked positive inputs and negated inputs are first found for each successfully classified case in the MLP training set. The *average* of the ranking values of the key positive inputs and the key negated inputs is then taken separately by class resulting in a ranked positive input profile and a ranked negated input profile for each class. This is demonstrated for the operational low-back-pain network in Section 5.1.

2.2.2 Rule Induction

For a MLP input training case, a rule *which is valid for the training set* can be directly induced from the data relationship in order of the combined key input rankings. First, a candidate rule is built in combined ranked order starting with the highest ranked input. If the candidate rule is unique to successfully trained examples from the same class as the input case then the rule is valid, otherwise a new antecedent is added to the candidate rule in ranked input order. When a valid rule has been found, the induction of a new candidate rule is attempted *from the same input training case* starting with the *next highest* ranked key input.

Examples of rules induced from low-back-pain training cases are shown in Section 5.2.

2.2.3 Inducing Maximally General Rules

For each valid rule induced from a MLP training case, a maximally general rule can be induced by finding a valid rule *with the least number* of antecedents in the original rule. It is possible to induce more than one maximally general rule with the same minimum number of antecedents from the original rule. The set of maximally general rules induced from all training cases represents the generalized knowledge that the MLP network learns from the training set.

Examples of maximally general rules induced from low-back-pain training cases are shown in Section 5.2.

2.3 Discovering the MLP Knowledge Bounds

This section examines the part of the full explanation facility that discovers the position of the training examples in the *n*-dimensional input space, from a hidden layer decision region perspective. This determines the MLP knowledge bounds and enables the facility to implement novelty detection.

2.3.1 The Role of the MLP Hidden Neurons in 2-Dimensional Input Space

From Figure 3 it can be seen that the activation of a hidden neuron with a sigmoidal activation function and 2 input neurons has the exact value 0.5 when

$$w_1x_1 + w_2x_2 = T, \tag{1}$$

where T is the threshold of the hidden neuron. It can be seen from Figure 4 that the hidden neuron separates the 2-dimensional input space into two separate decision regions with the linear decision region boundary defined by (1). In one of the decision regions the activation of the hidden neuron is < 0.5 and in the other decision region the activation of the hidden neuron is ≥ 0.5 [13].

2.3.2 The Role of the MLP Hidden Neurons in *n*-Dimensional Input Space

In general, a hidden neuron H_j with n input neurons separates the n-dimensional input space into two separate decision regions with the $(n-1)$-dimensional hyperplane given by

$$w_{1j}x_1 + w_{2j}x_2 + \ldots + w_{ij}x_i + \ldots + w_{nj}x_n = T_j, \tag{2}$$

where T_j is the threshold at H_j. In one decision region $h_j \geq 0.5$, where h_j is the activation of H_j, and in the other decision region $h_j<0.5$.

In general, the MLP with m hidden neurons H_j, $j = 1,m$ creates a number of separate decision regions in n-dimensional input space from the intersection of m $(n-1)$-dimensional hyperplanes given by (2) for $j = 1,m$. So, the role of the MLP hidden layer neurons is to separate the training examples into the classes specified during supervised training [23].

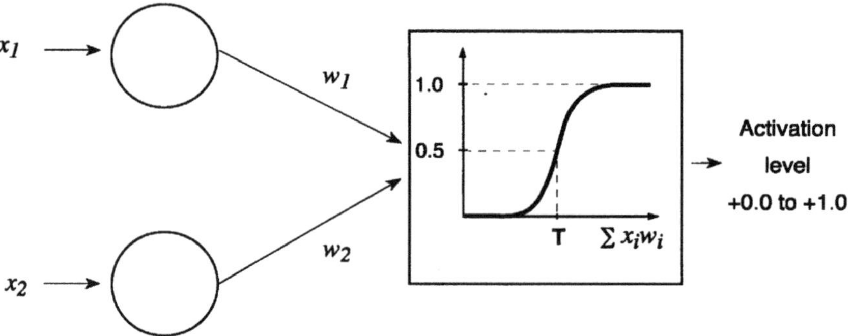

Figure 3. A hidden layer neuron in 2-dimensional input space.

This is illustrated in Figure 5 where the contours labeled A and B represent continuous training distributions for class A and class B in 2-dimensional input space. In Figure 5, an arbitrary separation of the classes into 10 decision regions by the hyper-planes of 4 MLP hidden neurons (with 2 input neurons) is shown.

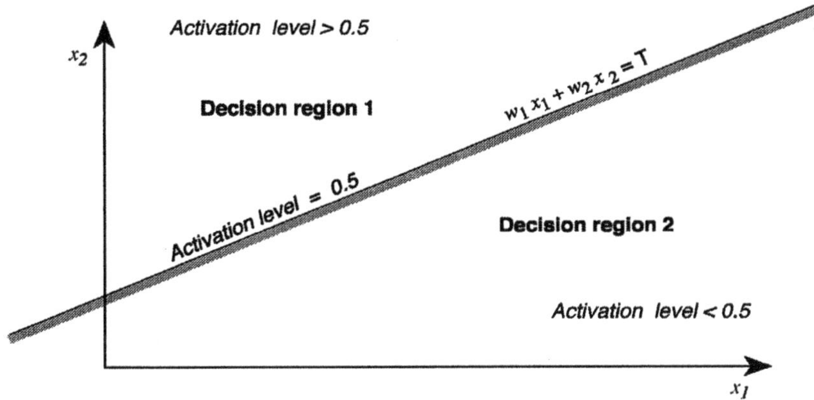

Figure 4. Separation of 2-dimensional input space into 2 decision regions by the decision boundary of a MLP hidden neuron.

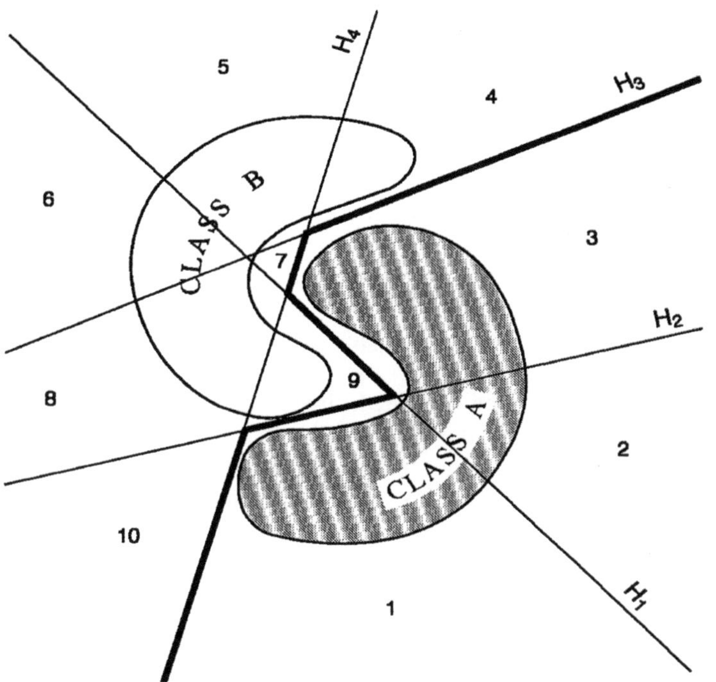

Figure 5. Separation of class A and B training distributions by the hyperplanes of 4 MLP hidden layer neurons in 2-dimensional input space.

Each decision region has a unique combination of hidden layer activations $h_j \geq 0.5$ or $h_j < 0.5$, for $j = 1,m$, and can be assigned a corresponding unique binary label, where 1 represents $h_j \geq 0.5$ and 0 represents $h_j < 0.5$ [23]. For example, the 10 decision regions shown in Figure 5 can be assigned the binary labels listed in Table 1, where it is assumed *for illustration purposes only* that $h_j \geq 0.5$ above the hyperplane of all H_j, $j = 1,4$.

Table 1. Hidden decision region labels corresponding to Figure 5.

region	h_1	h_2	h_3	h_4	region label
1	< 0.5	< 0.5	< 0.5	< 0.5	**0000**
2	≥ 0.5	< 0.5	< 0.5	< 0.5	**1000**
3	≥ 0.5	≥ 0.5	< 0.5	< 0.5	**1100**
4	≥ 0.5	≥ 0.5	≥ 0.5	< 0.5	**1110**
5	≥ 0.5	≥ 0.5	≥ 0.5	≥ 0.5	**1111**
6	< 0.5	≥ 0.5	≥ 0.5	≥ 0.5	**0111**
7	≥ 0.5	≥ 0.5	< 0.5	≥ 0.5	**1101**
8	< 0.5	≥ 0.5	< 0.5	≥ 0.5	**0101**
9	< 0.5	≥ 0.5	< 0.5	< 0.5	**0100**
10	< 0.5	< 0.5	< 0.5	≥ 0.5	**0001**
11	< 0.5	< 0.5	≥ 0.5	≥ 0.5	**0011**

2.3.3 The MLP Knowledge Bounds

The MLP knowledge bounds are defined by Vaughn [23] as the set of unique decision regions, identified by binary labels, as determined from h_j, $j = 1,m$, for *all correctly classified* training cases, after training is completed. For example, the knowledge bounds of the MLP shown in Figure 5 are regions 1, 2, and 3 for class A training examples and regions 5, 6, 7, 8, and 9 for class B training examples. The knowledge bounds for the operational low-back-pain MLP are presented in Section 6.

If the training data can be separated directly into the specified classes by the linear decision boundaries of the MLP hidden neurons the data is linearly separable. In general, training data is not linearly separable and the MLP hidden neurons separate the training examples into more than one decision region that contain training examples from the same class [23], as shown in Figure 5.

The role of the MLP output layer neurons is to map the training example hidden activations h_j, $j = 1,m$ into the output activations y_k, $k = 1,c$, that correspond to class membership at the c output neurons. The input/output mapping is continuous in the n-dimensional MLP input space since the sigmoidal functions at H_j, $j = 1,m$ and y_k, $k = 1,c$ are continuous.

2.4 MLP Novelty Detection

A new direct approach is taken for novelty detection [23], whereby the user is warned that the classification is potentially unreliable when an input case is beyond the MLP network's knowledge bounds. The position of *any* input case in the *n*-dimensional input space is found from the MLP hidden layer activations which are directly converted into a binary label, as discussed in Section 2.3. A novel case is detected when the decision region label for the input case is different from the decision region labels within the MLP knowledge bounds.

For example, the novel regions in the 2-dimensional MLP input space partly shown in Figure 5, are regions 10 and 11. (Region 11 is created by the intersection of the hyperplanes of H_2 and H_3.) The novel regions in the 48-dimensional input space of the operational low-back-pain MLP network are presented in Section 6.

3 The Low-Back-Pain MLP Network

Low-back-pain is one of the most common medical problems encountered in healthcare, with 60-80% of the population suffering some form of it within their life-span [11,26]. It has been estimated that only 15% of patients with low-back-pain obtain an accurate diagnosis of their problem with any degree of certainty, and many receive care that is less than optimal [2,11]. Low-back-pain is a difficult multi-factorial problem that includes physical, psychological and social aspects of illness [26].

For this study, low-back-pain is classified into three diagnostic classes: simple low back pain (SLBP) – mechanical low back pain, minor scoliosis and old spinal fractures; root pain (ROOTP) – nerve root compression due to either disc, bony entrapment or adhesions; and abnormal illness behavior (AIB) – mechanical low back pain, degenerative disc or bony changes with signs and symptoms magnified as a sign of distress in response to chronic pain.

Extensive data was collected from low-back-pain patients who agreed to participate in the study at the orthopedic clinic in the Princess Margaret Hospital, Swindon, UK, from 1995 to 1997. A data set of 198 actual cases was collected from patient questionnaires, physical findings and clinical findings [25].

3.1 Initial Low-Back-Pain MLP

The initial low-back-pain MLP network was a fully connected, feed-forward network with 92 binary encoded input neurons corresponding to 39 patient attributes, and 3 output layer neurons, each representing a diagnostic class [25].

Using the sigmoidal activation function and generalized delta learning rule the network was trained with 99 randomly selected patient cases. The MLP network architecture with the lowest test error was found with 10 hidden layer neurons at 1100 cycles when the training set had a 96% classification accuracy and a test set, with 99 patient cases, had a 67% classification accuracy.

In using the explanation facility to validate the knowledge that the initial network had learned from the training examples, it was discovered that two SLBP training cases had been incorrectly included in the training set [25]. These patients had been transferred to other clinicians in the same hospital.

In using the explanation facility to validate the 35 apparently misclassified testing examples, it was agreed by the domain experts that 16 of these cases were likely to have been *correctly* classified by the low-back-pain MLP. Most of these cases were correctly classified as AIB, which confirmed the difficulty experienced by clinicians in diagnosing AIB patients, as observed in other studies [27].

The initial low-back-pain network was re-trained with the corrected training and testing sets. The best test performance of the re-trained network was re-evaluated as 81.7%, a similar result to other researchers [4].

3.2 Operational Low-Back-Pain MLP

The operational low-back-pain MLP was developed from the initial 92-10-3 MLP for which the explanation facility was used to find the *least* relevant inputs. These were arbitrarily selected as the inputs ranked in 20th position or below in the positive input class profiles, which resulted in the removal of 44 inputs (representing 25 attributes) from the 92-10-3 MLP network input layer [25].

The reduced network was trained and tested with the same training set of 97 cases and test set of 99 cases as the re-trained initial network. The network architecture with the lowest test error was found with 5 hidden layer neurons at 600 cycles when the training set had a 99% classification accuracy and a test set, with 99 patient cases, had a 78.4% classification accuracy. This was a marginally reduced test performance compared with the 92-10-3 MLP result of 81.7% but with approximately half the number of inputs. This result appears to confirm the validity of the knowledge discovery method. The attributes and 48 value labels for the operational 48-5-3 MLP are shown in Table 2.

The 97 low-back-pain training cases consist of 35 SLBP cases, 45 ROOTP cases and 17 AIB cases. The 99 low-back-pain testing cases consist of 34 SLBP cases, 43 ROOTP cases and 22 AIB cases. Of these, 23 SLBP test cases, 33 ROOTP test cases and 20 AIB test cases were correctly classified with test performances of 67.6%, 76.7% and 90.9% respectively. The AIB class has a significantly higher test performance.

4 Interpreting Example Low-Back-Pain Training Cases

The interpretation of low-back-pain patient cases input to the operational 48-5-3 MLP is demonstrated for three example low-back-pain training cases: a class SLBP training case, a class ROOTP training case and a class AIB training case, which have classifying output activations of 0.94, 0.97 and 0.85, respectively. Similar interpretation results to those presented in this section have been found for all 196 low-back-pain training and testing cases, in real time.

Table 2. Operational 48-5-3 low-back-pain MLP input attributes.

Input attributes	Value labels
duration of pain	acute, recurring, chronic
pain brought on by :	fall, bending over, lifting
which is worse?	back pain, leg pain, equal
no back pain	yes, (no)
back pain aggravated by	coughing, standing, sitting, walking
no leg pain	yes, (no)
leg pain aggravated by	coughing, standing, walking
pain worse in evening?	yes, (no)
litigation	yes, (no)
claiming invalidity benefit	yes, (no)
use of walking aids	yes, (no)
lumbar flexion	> 45° , < 30°
lumbar extension	(5 – 15)°, < 5°
catch on extension	yes, (no)
straight leg raise (right)	> 70°, < 45°
raise (right) limited by	back pain, not limited
straight leg raise (left)	(45 – 70)°, < 45°
raise (left) limited by	leg pain, hamstrings
loss of reflexes	yes, (no)
nerve root	multiple, none
Wadell's inappropriate signs	(0 – 2), (3 – 5)
Oswestry Disability Index	(0 – 20), (80 – 100)
MSPQ<12	yes, (no)
Zung	< 17, > 33
DRAM	normal, distressed somatic, distressed depressive

4.1 An Example SLBP Training Case

The hidden layer feature detector neurons for the SLBP example case, as defined in Section 2.1, are hidden neurons H_2, H_4, and H_5. The top five combined ranked inputs which account for a total decrease in the classifying neuron activation of 95%, when switched off (and on) together at the MLP input layer, are shown in Table 3 where key negated inputs are preceded by '*not*'.

Table 3 represents the text form of the direct interpretation of the MLP classification of the SLBP example case. The corresponding graphical form of the interpretation is shown in Figure 6, which illustrates the non-linear nature of the data relationship with respect to the ranked key inputs.

Table 3. SLBP example case interpretation – the combined ranked data relationship.

rank	SLBP training case	Output activation
	Full example input case	0.943
1	back pain worse than leg pain	0.919
2	*not* straight left leg raise <45°	0.815
3	straight right leg raise unlimited	0.446
4	nerve roots (none)	0.133
5	straight right leg raise >70°	0.047

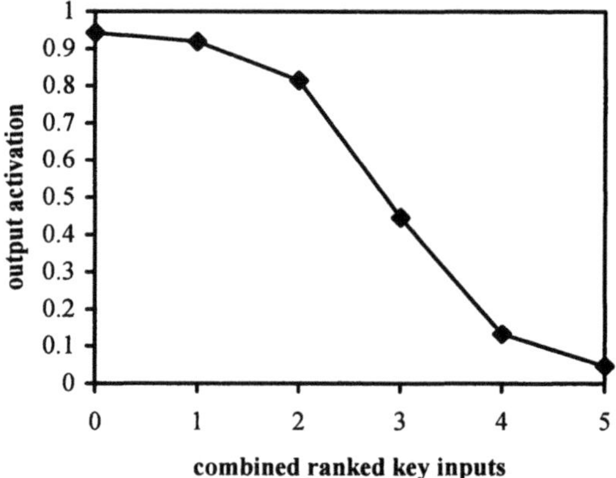

Figure 6. SLBP example case explanation.

4.2 An Example ROOTP Training Case

The hidden layer feature detector neurons for the ROOTP example case, as defined in Section 2.1, are hidden neurons H_3 and H_4. The top eight combined ranked inputs account for a total decrease in the classifying neuron activation of 96%, when switched off (and on) together at the MLP input layer, as shown in Table 4 where key negated inputs are preceded by '*not*'.

Table 4 represents the text form of the direct interpretation of the MLP classification of the ROOTP example case. The corresponding graphical form of the interpretation is shown in Figure 7, which illustrates the non-linear nature of the data relationship. The shallow descent for the first three combined inputs indicates that the class ROOTP output neuron is saturated for these inputs.

Table 4. ROOTP example case interpretation – the combined ranked data relationship.

rank	ROOTP training case	Output activation
	Full example input case	0.970
1	*not* back pain worse than leg pain	0.968
2	leg pain worse than back pain	0.960
3	back pain aggravated by coughing	0.929
4	*not* no leg pain	0.756
5	*not* straight right leg raise unlimited	0.333
6	straight left leg raise limited by leg pain	0.140
7	*not* straight right leg raise >70°	0.059
8	Waddell's inappropriate signs (0-2)	0.031

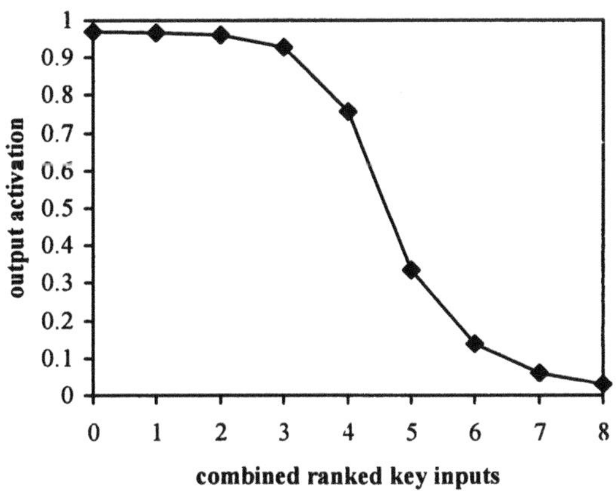

Figure 7. ROOTP example case explanation.

4.3 An Example AIB Training Case

The hidden layer feature detector neurons for the AIB example case, as defined in Section 2.1, are hidden neurons H_1 and H_2. The top five combined ranked inputs account for a total decrease in the classifying neuron activation of 95%, when switched off (and on) together at the MLP input layer, as shown in Table 5 where key negated inputs are preceded by '*not*'.

Table 5 represents the text form of the direct interpretation of the MLP classification of the AIB example case. The corresponding graphical form of the interpretation is shown in Figure 8, which illustrates the non-linear nature of the data relationship.

Table 5. AIB example case interpretation – the combined ranked data relationship.

rank	AIB training case	Output activation
	Full example input case	0.851
1	*not* Waddell's inappropriate signs (0-2)	0.758
2	*not* straight left leg raise limited by leg pain	0.347
3	*not* leg pain worse than back pain	0.146
4	pain brought on by bending	0.128
5	Waddell's inappropriate signs (3-5)	0.040

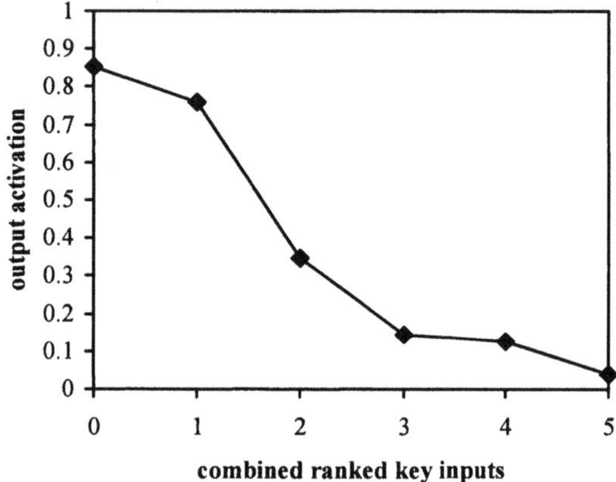

Figure 8. AIB example case explanation.

5 Discovering the Knowledge from the Low-Back-Pain MLP

The low-back-pain MLP knowledge can be represented as a set of ranked class profiles or as a set of induced maximally general rules which are valid for the training set, as discussed in Section 2.2.

5.1 Discovering the Low-Back-Pain MLP Knowledge as Ranked Class Profiles

The key ranked positive inputs and key ranked negated inputs were discovered separately for each of the 97 low-back-pain training cases. The *average* of the ranking values of the key positive inputs and the key negated inputs were then taken separately by class. This resulted in a ranked positive training profile and a ranked negated training profile for each diagnostic class, as shown in Table 6. Only the top ten ranked inputs are shown in each class because the class profiles decrease exponentially with respect to the key inputs.

Table 6. Positive and negated low-back-pain class training profiles.

rank	positive SLBP class profile	negated SLBP class profile
1	back pain worse than leg pain	straight left leg raise < 45°
2	no nerveroots	loss of reflexes
3	straight right leg raise > 70°	leg pain worse than back pain
4	right leg raise not limited	lumbar flexion < 30°
5	pain brought on by bending	worse(equal)
6	lumbar extension < 5°	back pain aggravated by coughing
7	straight left leg raise (45–70°)	pain brought on by lifting
8	lumbar flexion > 45°	straight right leg raise < 45°
9	no leg pain	claiming invalidity/ disability benefit
10	back pain aggravated by sitting	back pain aggravated by walking
rank	**positive ROOTP class profile**	**negated ROOTP class profile**
1	leg pain worse than back pain	back pain worse than leg pain
2	back pain aggravated by coughing	pain brought on by bending
3	left leg raise limited by leg pain	right leg raise not limited
4	low Waddell's inappropriate signs	lumbar flexion > 45°
5	lumbar flexion < 30°	pain brought on by falling
6	low MSPQ score	back pain aggravated by sitting
7	leg pain aggravated by walking	multiple nerveroots
8	input bias	straight right leg raise > 70°
9	loss of reflexes	no leg pain symptoms
10	pain brought on by lifting	distressed depressive
rank	**positive AIB class profile**	**negated AIB class profile**
1	back pain worse than leg pain	back pain aggravated by coughing
2	pain brought on by bending over	left leg raise limited by leg pain
3	multiple nerveroots	leg pain worse than back pain
4	pain brought on by falling over	low Waddell's inappropriate signs
5	back pain aggravated by sitting	low MSPQ score
6	claiming invalidity/ disability benefit	lumbar flexion < 30°
7	lumbar flexion > 45°	leg pain aggravated by walking
8	high Waddell's inappropriate signs	normal DRAM
9	distressed depressive	low Zung depression score
10	litigation	right leg raise limited by back pain

The positive input class profiles in Table 6 indicate that a typical SLBP patient presents with back pain worse than leg pain and a good range of lumbar flexion and leg raise, whereas a typical ROOTP patient presents with leg pain greater than back pain, limited leg raise and lumbar movements. A typical AIB patient has some indicators in common with the SLBP class, which is to be expected, since AIB patients are often SLBP patients that develop signs of illness behavior manifested by distress. Some of the positive AIB indicators indicate a distressed psychological profile.

The *negated* input profiles in Table 6 indicate that a typical SLBP patient does *not* present with many of the key positive inputs of class ROOTP patients and vice versa. Many of the key *negated* AIB inputs support the key positive inputs of class AIB patients.

The clinicians agreed that the low-back-pain MLP had determined largely relevant attributes as typical characteristics of patients belonging to the 3 diagnostic classes, as shown in the class profiles in Table 6

5.2 Discovering the Low-Back-Pain MLP Knowledge as a Set of Induced Rules

The rule induction method is demonstrated for the same example training cases that were interpreted in Section 4.

5.2.1 Induced Rule for SLBP Example Training Case

Using the method described in Section 2.2, the following valid rule for the training set was induced in combined ranked order from the SLBP data relationship shown in Table 3 and Figure 6:

IF back pain worse than leg pain
AND *not* straight left leg raise <45°
AND straight right leg raise unlimited
THEN class SLBP

This rule is valid for 7 out of the 35 SLBP training cases (and no ROOTP or AIB training cases). Using the method described in Section 2.2, the maximally general rule induced from the above rule, which is valid for the training set, is given by:

IF back pain worse than leg pain
AND straight right leg raise unlimited
THEN class SLBP

This rule is valid for the same 7 SLBP training cases and represents the most generalized form of the knowledge that the network has learned from these cases.

5.2.2 Induced Rule for ROOTP Example Training Case

The following valid rule for the training set was induced in combined ranked order from the ROOTP data relationship shown in Table 4 and Figure 7:

IF *not* back pain worse than leg pain
    AND leg pain worse than back pain
    AND back pain aggravated by coughing
    AND *not* no leg pain
    AND *not* straight right leg raise unlimited
    AND straight left leg raise limited by leg pain
THEN class ROOTP

This rule is valid for 4 out of the 45 ROOTP training cases (and no SLBP or AIB training cases). Using the method described in Section 2.2, the following maximally general rule can be induced from the above rule:

IF leg pain worse than back pain
    AND back pain aggravated by coughing
    AND straight left leg raise limited by leg pain
THEN class ROOTP

This rule is valid for 6 ROOTP training cases and represents the most generalized form of the knowledge that the network has learned from these training cases.

5.2.3 Induced Rule for AIB Example Training Case

The following valid rule for the training set was induced in combined ranked order from the AIB data relationship shown in Table 5 and Figure 8:

IF *not* Waddell's inappropriate signs (0-2)
    AND *not* straight left leg raise limited by leg pain
    AND *not* leg pain worse than back pain
    AND pain brought on by bending
    AND Waddell's inappropriate signs (3-5)
    AND multiple nerve roots
THEN class AIB

This rule is valid for 2 out of the 17 AIB training cases (and no SLBP or ROOTP training cases). However, it is of interest that the first 3 antecedents are common to 11 (out of 17) AIB training cases (and to 4 ROOTP cases). Note that this rule is valid only when the 6th ranked key input for the case is included, whereas the data relationship does not include this input.

Two maximally general rules can be induced from the above rule, which are valid for the low-back-pain training set. The following maximally general rule is valid for 9 of the 17 AIB training cases:

IF *not* Waddell's inappropriate signs (0-2)
    AND *not* straight left leg raise limited by leg pain
    AND multiple nerve roots
THEN class AIB

The following maximally general rule is valid for 3 of the 17 AIB training cases:

IF *not* Waddell's inappropriate signs (0-2)
AND pain brought on by bending
AND multiple nerve roots
THEN class AIB

Again, it is interesting to note that the maximally general rules also depend on the presence of the 6th ranked key input 'multiple nerve roots', whereas this attribute does not feature in the data relationship for the AIB example case.

5.3 Comparison of Data Relationships and Rules

The combined ranked data relationship, as shown in Tables 3 to 5 and Figures 6 to 8 for the example low-back-pain cases, embodies the graceful degradation properties of the MLP network. It reveals the non-linear relationship between the most important network inputs and the classifying output activation for the example training case. Furthermore, the combined ranked data relationship can be found for *any* input case that is presented to the MLP when the network is in general use. This represents the explanation of any input case to the user of the MLP.

In comparison the rules induced from the example training cases are brittle and do not indicate the relative importance of the attributes used by the MLP in classifying the case. The advantage of representing the knowledge as a rule, however, is that the rule is valid for the training set and the more general the rule, the easier it is for the domain expert to comprehend. Maximally general rules are a useful representation of the generalized MLP knowledge and are easily understood by the domain expert for validating the MLP.

However, the use of the positive and negated class profiles for the validation and development of the low-back-pain MLP [25] has been amply proven, as discussed in Section 3.1.

5.4 Comparison with Other Rule Extraction Methods

Many methods have been developed for extracting knowledge from neural networks [1,5,9,10,21,28] but such methods do not meet the goal of directly interpreting trained neural networks [7] as does the method presented in this chapter. Most approaches for extracting knowledge in the form of rules from trained neural networks use a search based approach [1]. Whereas rules are a useful form for representing the complete knowledge that the MLP has learned in the training process, a rule does not embody the degradation properties of the MLP network and a data relationship is considered a more appropriate representation for an explanation, as discussed in Section 5.3.

5.4.1 Search Based Methods

Many approaches for extracting knowledge in the form of rules from trained networks use a search based approach [1,9,10,21] where the MLP is regarded as a collection of two-layer perceptrons with activation functions which approximate step threshold units. A decompositional approach [1] is taken where rules are extracted for each hidden and output unit separately. The rule extraction algorithms search for combinations of input values which, when satisfied, guarantee that a given unit is maximally active.

A limitation of these methods is that obtaining all possible combinations of rules is NP-hard [17,20] and the methods are not universally applicable to arbitrary MLP networks [7]. Such methods are thus not feasible for extracting rules from real-world networks with a large number of input features, as in the current study.

5.4.2 Search Space Reduction Methods

To reduce the search space in the rule extraction process some approaches incorporate techniques such as specialized training procedures [6, 17, 20, 21] and network pruning [12, 14, 16, 17] or special network topologies [3, 15].

The Partial-RE method in [17] uses weight ordering to reduce the search space and can be used for larger size problems if a small number of premises per rule is sufficient, in which case the method is polynomial in n, where n is the number of input features.

5.4.3 Rules that Directly Map Inputs to Outputs

DEDEC [19] enables the extraction of rules that directly map inputs to outputs for feedforward networks which, as in the current study, also extracts rules by ranking the inputs of the MLP according to their importance. However, the ranking process is done by first examining the weight vectors of the network and then clustering the ranked inputs. Each cluster is used to generate a set of optimal binary rules that describes the functional dependencies between the attributes of a cluster and the network outputs [17].

Another approach that enables the extraction of rules that directly map inputs to outputs for arbitrary MLP networks is validity-interval analysis (VIA) [18] which uses linear programming to determine if a set of constraints on a network's activation values is consistent. However, the method is computationally expensive since it requires multiple runs of linear programming per rule. A further drawback is that the method assumes that activations of the hidden layer units are independent.

5.4.4 Rule Extraction Method in the Current Study

The rule extraction method described in the current study also enables the extraction of rules that directly map inputs to outputs for arbitrary MLP networks. However, the method in this study uses a direct, holistic approach which is not computationally expensive since the complexity of the method is linear in the number of hidden layer and output layer neurons. Unlike VIA, the approach makes use of the dependent hidden layer activations to interpret and discover knowledge from an input case.

The method aims to extract rules only from the training set because the network knowledge is expected to embody what it has learned from the training examples. As a result, the maximum number of rules that can be extracted is limited by the size of the training set. However, duplication of the most general rules is expected, especially for training examples in the same hidden layer decision region. Use of the method thus far, as indicated in this study and [24, 25] indicates good rule comprehensibility with a small number of premises per extracted rule due to the exponentially decreasing relationships learned by the network.

However, as discussed earlier in Section 5.3, the combined ranked data relationship is considered a more appropriate representation for the interpretation of an input case to the MLP network. Moreover, for the 48-5-3 low-back-pain MLP, this is achievable in real time.

6 Discovering the Low-Back-Pain MLP Knowledge Bounds

Using the full explanation facility, the knowledge bounds of the operational 48-5-3 network are found to consist of 5 hidden layer decision regions for each of the SLBP and ROOTP classes and 2 decision regions for the AIB class, as shown in Table 7. However, ROOTP decision region 11110 (30) is not considered part of the knowledge bounds since the single case within the region is not correctly classified.

Table 7. Knowledge bounds (training decision regions) of 48-5-3 low-back-pain MLP.

decision region	class	correct training cases	classification strength
01011 (11)	SLBP	20|20	strong
00011 (3)	SLBP	6|6	medium (some ROOTP)
11011 (27)	SLBP	4|4	medium (some AIB)
01010 (10)	SLBP	3|3	weak (some ROOTP)
11010 (26)	SLBP	1|1	mixed AIB
00110 (6)	ROOTP	37|37	very strong
10110 (22)	ROOTP	6|6	strong (some AIB)
10100 (20)	ROOTP	1|1	mixed AIB
00010 (2)	ROOTP	1|1	mixed SLBP
11110 (30)	ROOTP *– SLBP diagnosis*	0|1 *incorrect*	medium (some AIB)
11000 (24)	AIB	11|11	strong (some SLBP)
11100 (28)	AIB	6|6	medium (some ROOTP)

This result demonstrates that the low-back-pain training data are *not* linearly separable since there is more than one decision region per class, as discussed in Section 2.3.

A measure of the knowledge of the low-back-pain MLP has been defined [23] as the number of decision regions within the MLP knowledge bounds as a proportion of the maximum number of decision regions. For $n \geq m$ the maximum number of decision regions is 2^m [23], hence the knowledge bounds of the 48-5-3 low-back-pain network cover 11 out of a maximum of 32 regions, or 34%, of the 48-dimensional input space.

6.1 Novelty Detection for the Low-Back-Pain MLP

Using the full explanation facility, it is found that 94 of the 99 low-back-pain testing cases are positioned within the knowledge bounds of the low-back-pain network and three cases are each situated in new decision regions outside the knowledge bounds, as shown in Table 8. These three cases are considered novel, as defined in Section 2.4, and the full explanation facility currently warns the clinicians of the possibility of false positive diagnoses for such cases.

However, the classification is correct in two out of three of the novel regions, which indicates that the classification of novel low-back-pain patients is partly reliable in these regions.

Table 8. Testing decision regions of the 48-5-3 low-back-pain MLP.

decision region	knowledge bounds	class	correct test cases	classification strength
01011 (11)	YES	SLBP	12|15	mostly strong
00011 (3)	YES	SLBP	3|4	mixed ROOTP
11011 (27)	YES	SLBP	5|7	medium (some AIB)
01010 (10)	YES	SLBP	2|2	medium (some ROOTP)
11010 (26)	YES	SLBP	1|4	weak (mixed AIB)
00110 (6)	YES	ROOTP	28|34	mostly strong
10110 (22)	YES	ROOTP	1|3	medium (some AIB)
10100 (20)	YES	ROOTP	1|2	mixed AIB
00010 (2)	YES	ROOTP	1|2	mixed SLBP
11110 (30)	maybe	ROOTP	2|2	very weak to medium (some AIB and SLBP)
00111 (7)	*NO*	ROOTP	0|1	medium (some SLBP)
01110 (14)	*NO*	ROOTP	1|1	medium (some SLBP)
11000 (24)	YES	AIB	11|13	medium (some SLBP)
11100 (28)	YES	AIB	6|8	some to mixed ROOTP
11001 (25)	*NO*	AIB	1|1	mixed SLBP

The two remaining test cases outside of the knowledge bounds are situated in region 11110 (30). These two test cases are correctly classified, however, indicating that decision region (30) could be included in the MLP knowledge bounds.

7 Predicting the Reliability of the Low-Back-Pain MLP

Using the full explanation facility, the reliability of both the training data and the testing data classifications is investigated for each decision region within the knowledge bounds. Tables 7 and 8 also show the strength of the classification of the training and testing data respectively by decision region. (A strong classification is an activation > 0.85 at the class output neuron and < 0.10 at all others, a medium classification is ≥ 0.5 and < 0.25 respectively and a weak classification is < 0.5 at all output neurons.)

In several decision regions, as expected, there is some (low) activation for another class, as indicated in Tables 7 and 8. However, in some regions the classification is mixed, when the correct class has only a *marginally* higher activation, even though this activation can exceed 0.5. This was an unexpected result.

The training data reliability is investigated further in the following section and the testing data reliability is investigated further in Sections 7.2 and 7.3.

7.1 Low-Back-Pain Training Data Reliability Within the Knowledge Bounds

As shown in Table 7, the reliability of the training data classification depends on the position of the training data in the 48-dimensional input space since, for each class, the decision region with the most training cases is significantly more reliable. These are ROOTP region (6) with 37 (out of 37) very strongly classified training examples, SLBP region (11) with 20 (out of 20) and AIB region (24) with 11 (out of 11) strongly classified training examples. A theoretical explanation for this is presented in Section 7.4.

From a hamming distance analysis of the binary hidden layer decision region labels it is discovered that the MLP preserves the continuity of the three types of low-back-pain classification in separate contiguous threads of decision regions for each class across the 48-dimensional input space. Examples of the main threads for each class are shown in Figures 9 to 12. (Note: Each thread is analogous to taking a 'walk' in a forward direction across boundaries of neighboring hidden layer decision regions in the input space where a hidden layer boundary can only be crossed once. Hence, we can 'walk' across a maximum of 5 unique hidden layer decision region boundaries.)

The contiguous threads indicate that the training input distribution for each class is continuous in the input space and they also explain the mixed classifications in the training decision regions ROOTP/AIB (20), ROOTP/SLBP (2) and

SLBP/AIB (26). For example, in the ROOTP thread 1, shown in Figure 9, the strongest ROOTP region (6) is a neighbor to the next strongest ROOTP region (22) which adjoins the mixed ROOTP/AIB region (20) containing just a single case. However, the mixed classification in region(20) is consistent with its position between a ROOTP thread and AIB region (28), which has *some* ROOTP activation. Furthermore, Figure 9 indicates that the ROOTP training input distribution classification is continuous from the mixed SLBP/ROOTP region (2) to the mixed ROOTP/AIB (20) across the 48-dimensional input space.

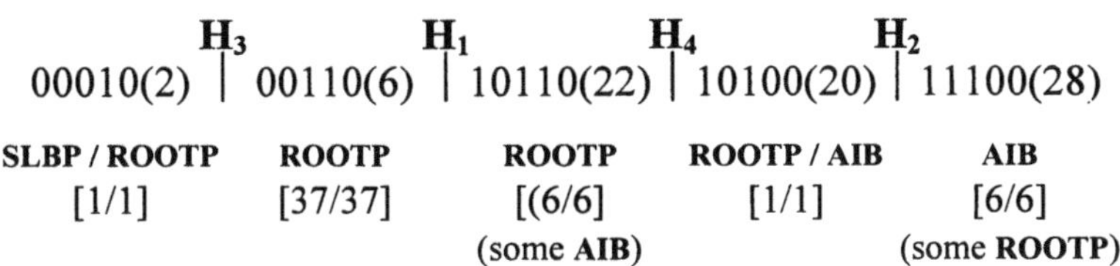

Figure 9. Training ROOTP thread 1.

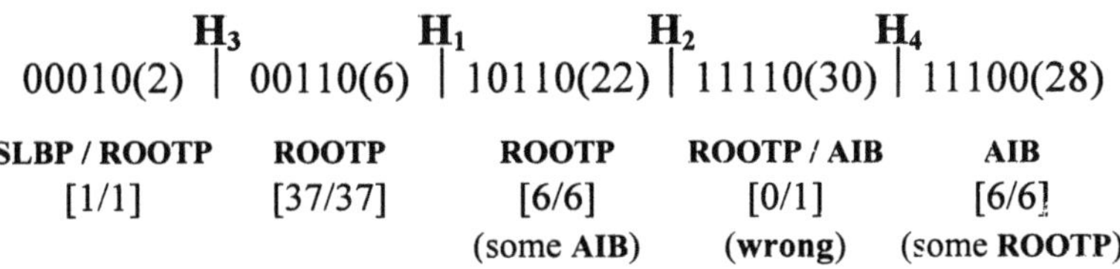

Figure 10. Training ROOTP thread 2.

H_2 H_5 H_1 H_3 H_4

00011(3) | 01011(11) | 01010(10) | 11010(26) | 11110(30) | 11100(28)

SLBP	**SLBP**	**SLBP**	**SLBP / AIB**	**ROOTP / AIB**	**AIB**
[6/6]	[20/20]	[3/3]	[1/1]	[0/1]	[6/6]
				(**wrong**)	(some **ROOTP**)

Figure 11. Training SLBP thread 1.

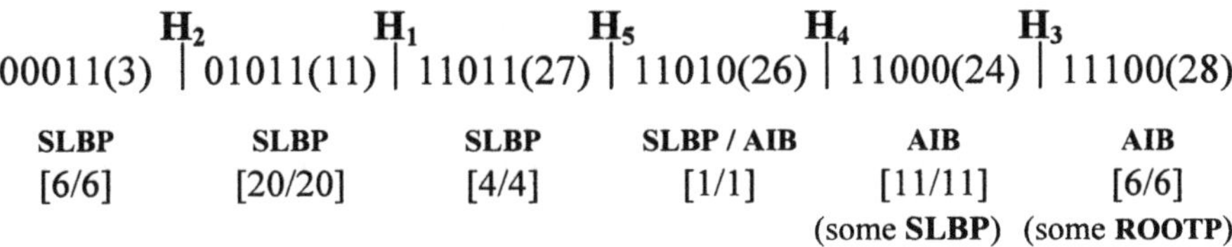

Figure 12. Training SLBP thread 2 and AIB thread.

In another view of the ROOTP thread, as shown in Figure 10, the incorrectly classified training case in the ROOTP region (30) (with some AIB activation) is seen to be consistent with its position between ROOTP (22) and AIB (28).

However, it can be seen from Figure 11 that region(30) is also positioned at the end of SLBP thread 1, adjoining AIB (28). This means that the incorrectly classified case in region (30) shares a boundary with ROOTP (22), SLBP/AIB (26) and AIB (28).

In another view across the 48-dimensional decision space, as shown in Figure 12, it can be seen that the AIB thread is contiguous to mixed SLBP/AIB region (26), the last region in SLBP thread 2.

7.2 Low-Back-Pain Testing Data Reliability Within the Knowledge Bounds

From Table 8 it can be seen that 96 of the 99 testing cases lie in the decision regions within the knowledge bounds with a similar distribution to the training cases. This confirms that the low-back-pain test cases have a similar distribution to the training cases in the 48-dimensional input space, as recommended in network training [8]. Of the 99 testing cases, 11 SLBP cases, 10 ROOTP cases and 2 AIB cases are miss-classified. From Table 7, it can be seen that the 23 miss-classified cases are distributed uniformly throughout the knowledge bounds. However, the most populated test regions within the knowledge bounds have a classification rate above 80% compared with the average test rate of 78%.

Comparing the training data classification strength, as shown in Table 7, with the testing data classification strength, as shown in Table 8, it is evident that the test classification strength is uniformly less than the training classification strength throughout the MLP knowledge bounds.

The classification of the test cases also shows a similar pattern of consistency within the contiguous threads of training decision regions but with a reduced reliability compared to the training cases, as already discussed.

The interpretation of each misclassified test case in terms of the ranked key positive and negated inputs used by the 48-5-3 MLP has been found to be consistent with the ranked class profiles as shown in Table 6, Section 5.1. The domain experts are currently re-assessing each miss-classified case and comparing their diagnoses with the explanations of the MLP classifications.

7.3 Low-Back-Pain Testing Data Reliability Beyond the Knowledge Bounds

As shown in Table 8, there are three new test data decision regions outside the 48-5-3 low-back-pain MLP knowledge bounds: regions(7), (14) and (25), each of which contains a single test case. Region(30), which contains the incorrectly classified training case, is also represented with two correctly classified test cases, albeit one case very weakly.

From a hamming distance analysis of the testing decision region labels, it was discovered that the three new test regions are each situated between training decision regions inside the knowledge bounds, as shown in Figures 13 to 15 for region

(14). The classification strengths in the novel testing regions are found for region (14). The classification strengths in the novel testing regions are found to be consistent with their position in the threads in the 48-dimensional input space. Since the diagnosis is correct in two out of three of these regions, this suggests that the classification of new low-back-pain patients is partly reliable in decision regions *between* neighboring regions in the knowledge bounds.

	H_4		H_1		H_5		H_3		H_2	
kb		kb		kb		**TEST**		kb		kb
10100(20)	|	10110(22)	|	00110(6)	|	**00111(7)**	|	00011(3)	|	01011(11)
ROOTP / AIB		**ROOTP**		**ROOTP**		**ROOTP**		**SLBP**		**SLBP**
[1/1]		[6/6]		[37/37]		[0/1] **(wrong)**		[6/6]		[20/20]

Figure 13. Position of novel test region (7) within the low-back-pain MLP knowledge bounds.

	H_4		H_1		H_2		H_3		H_5	
kb		kb		kb		**TEST**		kb		kb
10100(20)	|	10110(22)	|	00110(6)	|	**01110(14)**	|	01010(10)	|	01011(11)
ROOTP		**ROOTP**		**ROOTP**		**ROOTP**		**SLBP**		**SLBP**
[1/1]		[1/1]		[37/37]		[1/1]		[3/3]		[20/20]

Figure 14. Position of novel test region (14) within the low-back-pain MLP knowledge bounds.

	H_2		H_1		H_4		H_5		H_3	
kb		kb		kb		**TEST**		kb		kb
00011(3)	|	01011(11)	|	11011(27)	|	**11001(25)**	|	11000(24)	|	11100(28)
SLBP		**SLBP**		**SLBP**		**AIB / SLBP**		**AIB**		**AIB**
[6/6]		[20/20]		[4/4]		[1/1]		[11/11]		[6/6]

Figure 15. Position of novel test region (25) within the low-back-pain MLP knowledge bounds.

Since training region (30) contains two correctly classified test cases, albeit one very weakly, it suggests that this region could be considered as part of the low-back-pain MLP knowledge bounds. It is of interest that training region (30) also shares a boundary to test region (14), as shown in Figure 16.

Incorrectly classified test cases, both inside and outside the knowledge bounds, will be investigated in terms of key input data relationships used by the MLP in making the individual classifications, as provided on a case-by-case basis by the explanation facility. The reliability of classifications will be investigated in *all decision regions* in the entire 48-dimensional input space beyond the knowledge bounds.

	H_5		H_3		H_2		H_1		H_4	
kb		kb		kb		**TEST**		(kb)		kb
00011(3)	|	00010(2)	|	00110(6)	|	**01110(14)**	|	11110(30)	|	11100(28)
SLBP		**ROOTP**		**ROOTP**		**ROOTP**		**ROOTP**		**AIB**
[6/6]		[1/1]		[37/37]		[1/1]		[0/1]		[6/6]
								(wrong)		

Figure 16. Position of training region(30) with respect to novel test region (14).

7.4 Explaining the Low-Back-Pain MLP Training Data Classification Reliability

A possible explanation is now presented for the decision region with the most training cases for each class being significantly more reliable.

It has been shown [23] that, for a correctly classified input case, a MLP with binary inputs and sigmoidal activation functions must find sufficiently many H_j with activations $h_j \geq 0.5$ connected to the class output neuron with positive weights. These hidden neurons are defined as the hidden layer feature detector neurons [23]. (The MLP must also find sufficiently many H_j with activations $h_j \geq 0.5$ connected to the other output neurons with negative weights.)

In a strong output classification the hidden neurons connected to the class output neuron with negative weights have a low activation (<0.5) so as to minimize the negative part of the combined input sum to the class output neuron. This is the situation in the three training regions with the strongest classifications, SLBP region (11), ROOTP region (6) and AIB region (24), as shown in Table 7. For example, the binary label 01011 for SLBP (11) means that $h_2 \geq 0.5$, $h_4 \geq 0.5$ and $h_5 \geq 0.5$ in region (11).

Further investigation finds that H_2, H_4 and H_5 are all feature detector neurons since they are connected to the SLBP output neuron with *positive* weights. The binary label also means that $h_1 < 0.5$ and $h_3 < 0.5$. Further investigation finds that H_1 and H_3 are connected to the SLBP output neuron with *negative* weights and so are *not* feature detector neurons.

It is highly significant that the most populated training regions for each class have the strongest classifications and *are the only decision regions* for which all H_j with $h_j \geq 0.5$ are feature detector neurons and all H_j with $h_j < 0.5$ are not feature detector neurons. For example, the neighboring region SLBP 00011 (3) to region (11) has the same feature detectors H_2, H_4 and H_5 but H_2 is a feature detector with $h_l < 0.5$. Another neighboring region SLBP 11011 (27) to SLBP region (11) has the same feature detectors H_2, H_4 and H_5 but H_1 is not a feature detector with $h_1 >$ 0.5. Both of these situations are sub-optimal and they are possible reasons for the reduced classification strengths in these regions.

Similar effects are observed in ROOTP (22) and ROOTP (2) which are neighbors to strongly classified ROOTP (6). Also in neighboring region AIB (28) to strongly classified AIB (24) and neighboring region SLBP (10) to strongly classified SLBP (11). A combination of two of these sub-optimal effects is observed in ROOTP (20), ROOTP (30), and SLBP (26), where the classifications are mixed, potentially less reliable and each of which has only one training case.

It is concluded that the MLP is *constrained* to have a continuous classification distribution in contiguous threads of hidden decision regions, thereby limiting the classification strength in parts of the threads. This is possibly because the MLP only has 5 hyperplanes with which to partition the training data, one each for the 5 hidden neurons.

Furthermore, it has been shown [23] that the MLP classification is limited in bounded decision regions and unlimited in unbounded regions where the classification improves as input points move further from the hidden layer decision region boundaries. From the evidence in Table 7, it is considered that regions ROOTP (20), ROOTP (30) and SLBP (26) are almost certainly bounded and possibly also the regions with one sub-optimal effect. The three regions, SLBP region (11), ROOTP region (6) and AIB region (24), with a strong classification are almost certainly unbounded.

Future studies will compare the classification strength and reliability in the decision regions of other low-back-pain networks with more hidden neurons.

7.5 Predicting the Low-Back-Pain MLP Reliability

The contiguous threads in Figures 9 to 12 demonstrate the consistency of the classification strengths of the training data in each decision region within the knowledge bounds, as shown in Table 7 and discussed in Section 7.1. The classification of the testing cases also shows a similar pattern of consistency within the contiguous threads of training decision regions but with a uniformly reduced reliability compared to the training cases, as shown in Table 8.

The classification strength of the testing data shown in Table 8 is also uniformly less than the training data, throughout the MLP knowledge bounds, with an average test rate of 78%, and over 80% in the most populated decision regions.

In view of the consistency of the classification of the testing cases in the contiguous training threads within the knowledge bounds, it is concluded that the reliability of the low-back-pain MLP is predictable. It is expected that all network inputs within the network's knowledge bounds will have a similar performance to the test classifications in the corresponding decision regions, as shown in Table 8.

The reliability of classifications will be investigated in *all decision regions* in the entire 48-dimensional input space beyond the knowledge bounds. However, the novelty detection implemented for the operational network will continue to warn the clinicians of the potential unreliability of any input cases that arise beyond the MLP's knowledge bounds.

8 Summary and Conclusions

In summary, this chapter has presented a full explanation facility which directly interprets the output on a case-by-case basis from any standard MLP network in the form of a non-linear ranked data relationship of key inputs, in both text and graphical forms, as demonstrated in Section 4. It has shown how the knowledge that the MLP has learned can be represented by average ranked class profiles or as a set of rules induced from the training set cases, as demonstrated in Section 5 for the operational 48-5-3 low-back-pain MLP.

Since the interpretation and knowledge discovery method does not use combinatorial search or specialized training procedures, it potentially represents a significant advance towards the goal of readily interpreting trained neural networks that solve real-world problems with a large number of input features [7, 20], as in the current study.

The full explanation facility discovers the MLP knowledge bounds as the hidden layer decision regions containing correctly classified training examples and a novel input case is detected when the case is situated in a decision region outside the knowledge bounds. Results of using the full explanation facility have been presented in Section 6 for the operational 48-5-3 low-back-pain MLP.

An analysis of the reliability of the low-back-pain MLP in Section 7 shows that the MLP preserves the continuity of the three types of low-back-pain training classifications in separate contiguous threads of hidden layer decision regions for each class across the entire 48-dimensional input space. This demonstrates that the training input distribution for each class is continuous in the input space.

The reliability of the training data classification is found to depend on the position of the decision region within the class threads in the 48-dimensional input space. For each class, the decision region with the most training cases is significantly more reliable. Reasons for this are presented in Section 7.4 where it is concluded that the MLP is *constrained* to have a continuous classification distribution in contiguous threads of hidden decision regions, thereby limiting the classification strength in parts of the threads.

An analysis of the testing data decision regions confirms that test cases are positioned within the knowledge bounds with a similar distribution to the training cases. The classification of the test cases shows a similar pattern of consistency within the contiguous threads of training decision regions but with a uniformly reduced reliability (78.4%) compared to the training cases (99%).

It is concluded that the reliability of the low-back-pain MLP is predictable and that all possible network inputs within the knowledge bounds will have a similar performance to the test classifications in the corresponding training decision regions.

9 Future Work

The lower bound on the classification in the sub-optimal decision regions within the contiguous threads of hidden layer decision regions will be investigated, with a view to automatically determining, as part of the full explanation facility, when the output from the operational 48-5-3 low-back-pain network is unreliable.

The domain experts will investigate each miss-classified case and compare their diagnoses with the explanations of the operational low-back-pain network's classifications. For confirmed diagnoses, reasons will be sought for the MLP miss-classifications in terms of key factors used by the domain experts in making their diagnoses. The reliability of classifications will be investigated in the remaining 17 decision regions in the 48-dimensional input space beyond the knowledge bounds.

Future studies will compare the classification strength and reliability in the decision regions of the operational low-back-pain MLP with that of other architectures.

References

1. Andrews R, Diederich J, Tickle AB (1995) Survey and Critique of Techniques for Extracting Rules from Trained Artificial Neural Networks. Knowledge-Based Systems 8 (6):373-389
2. Bigos S, Bowyer O, Braen G (1994) Acute Low Back Problems in Adults. Clinical Practice Guideline No. 14. AHCPR Publication No. 95-0642. U.S. DHHS
3. Bologna, G (1996), Rule Extraction from the IMLP Neural Network: A Comparative Study. Proc. NIPS'96 Workshop of Rule Extraction from Trained Artificial Neural Networks. Queensland Univ. Technol
4. Bounds DG, Lloyd PJ, Mathew B (1990) A Comparison of Neural Network and Other Pattern Recognition Approaches to the Diagnosis of Low Back Disorders. Neural Networks Vol. 3. 583-591
5. Cloete I, Zurada JM (1999) Knowledge-Based Neurocomputing, MIT Press, Cambridge, MA.
6. Craven MW, Shavlik JW (1993) Learning symbolic rules using artificial neural networks. In: Proceedings Tenth International Conference on Machine Learning. Amherst MA, USA. Morgan Kauffman, pp 73-80.
7. Craven MW, Shavlik JW (1994) Using Sampling and Queries to Extract Rules from Trained Neural Networks. In: Machine Learning. Proceedings Eleventh International Conference on Machine Learning, Amherst, MA,USA. Morgan Kaufmann, pp 73-80
8. Denker J, Schwarz D, Wittner B, Solla S, Howard R, Jackel, L, Hopfield J (1987) Large Automatic Learning, Rule Extraction, and Generalisation. Complex Systems, 1, 877-922
9. Fu L (1994) Neural Networks in Computer Intelligence, McGraw-Hill, London

10. Gallant SI (1993) Neural Network Learning and Expert Systems. MIT Press, Cambridge, MA.
11. Jackson D, Llewelyn-Phillips H, Klaber-Moffett J (1996) Categorization of Back Pain Patients Using an Evidence Based Approach. Musculoskeletal Management 2:39-46
12. Krishnan R (1996) A Systematic Method for Decompositional Rule Extraction from Neural Networks. Proceedings NIPS'96 Workshop of Rule Extraction from Trained Artificial Neural Networks. Queensland University of Technology, pp 38-45
13. Looney CG (1997) Pattern Recognition Using Neural Networks. New York, Oxford University Press.
14. Maire F (1997) A Partial Order for the M-of-N Rule Extraction Algorithm. IEEE Trans. Neural Networks 8:1542-1544
15. Saito K, Nakano R (1996) Law Discovery Using Neural Networks. Proceedings NIPS'96 Workshop of Rule Extraction from Trained Artificial Neural Networks. Queensland University of Technology, pp 62-69
16. Setiono R (1997) Extracting Rules from Neural Networks by Pruning and Hidden Unit Splitting. Neural Computing l 9:205-225
17. Taha IA, Ghosh J (1999) Symbolic Interpretation of Artificial Neural Networks. IEEE Trans. Neural Networks 11(3):448-463
18. Thrun, SB (1995) Extracting rules from artificial neural networks with distributed representations. In: Tesauro G, Touretzky D, Leen T (eds) Advances in Neural Information Processing Systems. MIT Press, 7:505-512
19. Tickle AB, Orlowski M, Diederich J (1996) DEDEC: A Methodology for Extracting Rules from Trained Artificial Neural Networks. In: Proceedings NIPS'96 Workshop of Rule Extraction from Trained Artificial Neural Networks. Queensland University of Technology, pp 90-102
20. Tickle AB, Andrews R, Golea M, Diederich J (1998) The Truth Will Come to Light: Directions and Challenges in Extracting the Knowledge Embedded Within Trained Artificial Neural Networks. IEEE Trans. Neural Networks, 9(6):1057-1067
21. Towell, GG, Shavlik JW (1993) Extracting Refined Rules from Knowledge Based Neural Networks. Machine Learning 13:71-101
22. Vaughn ML (1996) Interpretation and Knowledge Discovery from the Multi Layer Perceptron Network: Opening the Black Box. Neural Computing & Applications Journal. 4(2):72-82
23. Vaughn ML, (1999) Derivation of the Weight Constraints for Direct Knowledge Discovery from the Multilayer Perceptron Network. Neural Networks 12:1259-1271
24. Vaughn ML, Cavill SJ, Taylor SJ, Foy MA, Fogg AJB (2000) Direct Explanations and Knowledge Extraction from a Multilayer Perceptron Network that Performs Low Back Pain Classification. In: Wermter S, Sun R (eds) Hybrid Neural Systems. Springer, pp 270-285
25. Vaughn ML, Cavill SJ, Taylor SJ, Foy MA, Fogg AJB (2001) Direct Explana-

tions for the Development and Use of a Multi-Layer Perceptron Network that Classifies Low Back Pain Patients. International Journal Neural Systems 11(4):335-347
26. Waddell G (1987) A New Clinical Model for the Treatment of Low-Back Pain. Spine 12(7):632-644
27. Waddell G, Bircher M, Finlayson D, Main C (1984) Symptoms and Signs: Physical Disease or Illness Behaviour. BMJ 289:739-741
28. Wermter S, Sun R (2000) Hybrid Neural Systems, Springer

Chapter 17

Automatic Translation to Controlled Medical Vocabularies

András Kornai and Lisa Stone

Summary. In the medical domain, over the centuries several *controlled vocabularies* have emerged with the goal of mapping semantically equivalent terms such as `fever`, `pyrexia`, `hyperthermia`, and `febrile` on the same (numerical) value. Translating unstructured natural language texts or *verbatims* produced by healthcare professionals to categories defined by a controlled vocabulary is a hard problem, mostly solved by employing human coders trained both in medicine and in the details of the classification system. In this chapter we survey the automatic translation or *autocoding* systems currently in use.

Keywords: medical text, controlled vocabulary, natural language processing

1 Introduction

The current widespread use of controlled medical vocabularies emerged in response to the data exchange and standardization needs of modern medical research and care. In this chapter we survey the methods used in translating unstructured natural language texts or *verbatims* produced by healthcare professionals to categories defined by a controlled vocabulary.

To this day, the primary method of translation is to use human coders, trained both in medicine and in the details of the classification system. In the past few decades, automated translation systems or *autocoders* of varying efficiency and reliability have emerged. Our focus will be on system that are *hybrid* not only in the contemporary sense of including both rule-based and statistical inference techniques but also in the older sense of having both human and automatic components.

In Section 2 we describe the historical origins and current status of controlled vocabularies in the medical domain. In Section 3 we present the general architecture of hybrid autocoding systems, and discuss their main components.

2 Controlled Medical Vocabularies

Statistical methods pervade every aspect of medicine. The early detection of epidemics, the research methods determining the efficacy of drugs and other treatments, the diagnosis and treatment protocols used in actual patient care, and even postmortem analysis are all based on careful statistical analysis of experimentally collected and naturally observed data points. But statistics can be meaningfully applied only if the source data is already properly classified. The need for controlled medical vocabularies to classify disease into general groups, and for detailed nomenclatures of signs, symptoms, diseases, and procedures has been recognized early on.

The City of London devised the *London Bills of Mortality* as an early warning system of the bubonic plague epidemics which periodically ravaged Europe. Beginning in 1603, these public postings provided London citizens with a detailed weekly mortality count, including cause of death and age of those fallen. After receiving a commission from King William III to study these records, John Graunt published his statistical analysis *Natural and Political Observations Made upon the Bills of Mortality* in 1662. This work is widely cited as the first known effort to classify human disease [23].

In the 18th century, as interest in natural history swept across Europe, physicians applied classification methods from botany and zoology to clinical medicine. They reasoned that diseases were entities like plants and animals and could likewise be arranged into taxomonic families, classes, species and genera on the basis of anatomical, clinical, and pathological criteria [28]. Francois B. de Sauvages, a professor of medicine at the University of Montpellier in France, published *Nosologia Methodica,* the first such systemic classification of disease in 1763. His system established 10 classes of disease, 44 orders, 315 genera, and approximately 2,400 separate entities. The scheme was enormously detailed but blemished by many inconsistencies and duplications [29]. Criticizing such cumbersome, all encompassing taxonomies, William Cullen of Edinburgh proposed a simpler arrangement in 1769. His *Synopsis Nosologiae Methodicae* was a didactic and practical index containing only 4 classes, 9 orders, and 151 genera [13].

The need for an internationally accepted classification system for statistical purposes and for public health control was recognized at the First International Statistical Conference in Brussels in 1853. The organization developed the *International List of Causes of Death.* This classification system, based on the work of the British statistician William Farr, had a two level hierarchy. The top level contained 5 general groups – epidemic diseases, constitutional diseases, local diseases according to their anatomical localization, diseases of the development, and diseases in direct consequence of a traumatism. At the lower level, 139 diseases were categorized. This classification was revised by the same organization in 1874, 1880, and 1886 [9].

In 1893, Dr. Jacques Bertillon, the chief statistician of Paris, resumed the work with the publication of *Nomenclatures de Maladies,* which became known as the Bertillon Classification of Causes of Death (WHO1) [54]. This system had a three level hierarchy. The highest level contained 44 groups, followed by a mid level containing 99 groups, and a lower level containing 161 disease entities. This system was adopted by many countries. The American Public Health Association recommended its use for vital record registries. Upon Bertillon's death in 1922, the Health Organization of the League of Nations took responsibility for the system, publishing the fourth revision in 1929 [55], and the fifth revision in 1938 [56].

After WWII, the newly formed World Health Organization of the United Nations accepted responsibility for the decennial revision published in 1946. In 1948, at the sixth revision conference, the final document was published as *International Statistical Classification of Disease, Injuries, and Causes of Death.* This revision extended the scope of the classification to non-fatal diseases and added causes of morbidity to the causes of mortality. Most developed nations agreed to use this system for classifying and reporting public health statistics [9]. A 7th version containing minor revisions was published in 1955 [57].

2.1 ICD

In 1969, WHO published the 8th edition, and revised the name to *Manual of the International Statistical Classification of Disease, Injuries, and Causes of Death* (ICD). This revision contained many modifications designed to help statisticians index information from medical and hospital records [58]. In the United States, groups of physicians and health records administrators recognized that the 8th revision still did not meet their needs. As a result, the US government published the ICD-A, or the *Eighth Revision International Classification of Diseases, Adapted for Use in the United States* [11].

In 1977, WHO published the ninth revision of ICD, which was immediately adapted in the United States for clinical use as ICD9-CM. Most insurance companies still use this revision as a method for controlling billing and payment, although Version 10 was published in 1992 [59].

ICD9-CM is a hierarchical system containing 4 levels. In the 9th revision, there are 17 chapters at the highest level. A fixed range of 3 digit 2nd level codes is assigned to each of the 17 high level groups. Within each 2nd level range, a fixed range of codes is assigned to a midlevel grouping. The individual 3 digit code defines a concept at the mid (3rd) level, and two more digits are added to the code to classify the concept more exactly (4th or lowest level).

Since ICD assigns a 5-digit numeric code to each entry at the lowest level, the hierarchical organization serves as an important mnemonic aid to the human coder. It would be next to impossible to remember that 162.3 is the code for `malignant neoplasm of the upper lobe, bronchus or lung`, but let us trace the levels of this code through the ICD9-CM system. At the first level we find

1. Infectious and parasitic diseases (001-139)
2. Neoplasms (140-239)
3. Endocrine, nutritional and metabolic diseases, and immunity disorders (240-279)
4. Diseases of the blood and blood-forming organs (280-289)
5. Mental disorders (290-319)
6. Diseases of the nervous system and sense organs (320-389)
7. Diseases of the circulatory system (390-459)
8. Diseases of the respiratory system (460-519)
9. Diseases of the digestive system (520-579)
10. Diseases of the genitourinary system (580-629)
11. Complications of pregnancy, childbirth, and the puerperium (630-676)

12. Diseases of the skin and subcutaneous tissue (680-709)
13. Diseases of the musculoskeletal system and connective tissue (710-739)
14. Congenital anomalies (740-759)
15. Certain conditions originating in the perinatal period (760-779)
16. Symptoms, signs, and ill-defined conditions (780-799)
17. Injury and poisoning (800-999)

Obviously, the coder will select `neoplasms`, providing the range 140-239, for which we now inspect the 2nd level:

- malignant neoplasm of lip, oral cavity, and pharynx (140-149)
- malignant neoplasm of digestive organs and peritoneum (150-159)
- malignant neoplasm of respiratory and intrathoracic organs (160-165)
- malignant neoplasm of bone, connective tissue, skin, and breast (170-176)
- malignant neoplasm of genitourinary organs (179-189)
- malignant neoplasm of other and unspecified sites (190-199)
- benign neoplasms (210-229)
- carcinoma in situ (230-234)
- neoplasms of uncertain behavior (235-238)
- neoplasms of unspecified nature (239)

Thus from the 2nd level `malignant neoplasm of respiratory and intrathoracic organs` can be easily selected, providing the range 160-165. In this range, we inspect the 3rd level:

161 Malignant neoplasm of larynx
162 Malignant neoplasm of trachea, bronchus, and lung
163 Malignant neoplasm of pleura
164 Malignant neoplasm of thymus, heart, and mediastinum
165 Malignant neoplasm of other sites within the respiratory system and intrathoracic organs

Thus the coder can easily find `malignant neoplasm of trachea, bronchus, and lung`, providing the code 162. Going into the lowest (4th) level we find

162.0 Trachea
162.2 Main bronchus
162.3 Upper lobe, bronchus or lung
162.4 Middle lobe, bronchus or lung
162.5 Lower lobe, bronchus or lung
162.8 Other parts of bronchus or lung
162.9 Bronchus and lung, unspecified

and this provides the final (in this case, four-digit) code 162.3. In 2.3 we will see how this method of successive approximation remains applicable when we move from human to machine coding.

The 5 digit ICD9-CM codes are highly indexical in the sense that the numerical code can be analyzed to determine the position of a term in the hierarchy. While providing a terse mechanism for communicating information, indexical codes introduce many problems. If a disease entity is changed to a new position in the hierarchy, it must be assigned a different code. Also, it may be necessary to reuse obsolete codes to maintain the numeric hierarchy.

Reliance on a base ten numerical code results in limited flexibility and the need for "other", "not otherwise specified" (NOS), and "not elsewhere classified" (NEC) terms. Furthermore, the upper bound on the number of available codes limits the size of the vocabulary. Another problem is that the lack of multiple axes (independent classification dimensions) results in duplicate terms [12]. This was addressed in subsequent classification systems, in particular SNOMED, to which we turn now.

2.2 SNOMED

In the 1960s another organizational effort, arising from the need for greater specificity, and with the express goal of overcoming the limitations of ICD, was sponsored by the College of American Pathologists. This resulted in the publication of the *Systemized Nomenclature of Pathology* (SNOP) in 1965. SNOP was the first system to use the concept of multiple axes. The axes represent non-intersecting concepts which a medical condition could be classified under, such as anatomical site, environment, history, or etiology [10].

The system was primarily intended for the coding of surgical and autopsy diagnoses. The same organization published a successor, the *Systemized Nomenclature of Medicine* (SNOMED) in 1969, the same year that WHO published ICD9 Version 9. Over the last three decades SNOMED has been adapted by many organizations as a controlled terminology for the indexing of the entire medical record. The international SNOMED version 3.4 (October 1996) contains more than 150,000 terms distributed in 11 axes as follows:

1. Anatomy
2. Morphology
3. Function
4. Disease/Diagnosis
5. Procedures
6. Occupations
7. Etiology – Living organisms
8. Etiology – Chemicals
9. Etiology – Physical agents, forces, activities
10. Etiology – Social context
11. Etiology – Other links

Each axis has an arbitrary number of hierarchically organized terms beneath it. Like ICD, SNOMED assigns an indexical numeric code to each term. The first digit expresses a general entity, while each succeeding digit specifies a more detailed location in the hierarchy. Every concept is built up from a term from one or more axes: for example, `acute rheumatic fever` is identified with term code D3-17112, where D indicates the code is for the Disease Axis, 3 indicates a disorder of the cardiovascular system and 17113 indicates acute rheumatic fever. The code can be associated with codes along other axes. For example, along the topographic axis – T-15000, indicates cardiac disorder, and along the morphology axis – M-41000, indicates that the morphology is acute inflammation.

Since SNOMED contains some overlap of terms in different axes, it is possible to form two different versions for the same concept. For example, `acute appendicitis` has a single code, but there are also terms and codes for `acute`, `acute inflammation`, `appendix`, and `in`. Thus, the concept could be expressed either as `appendicitis, acute`, as `acute inflammation, in, appendix`, or as `acute, inflammation NOS, in, appendix`. This makes it difficult to compare similar concepts that have been indexed in different ways, or to search for a term that exists in different forms within a medical record.

While SNOMED permits single terms to be combined to create complex terms, rules for the combination of terms have not been developed. Consequently such compositions, such as `nevus of left esophagus` may not be medically valid [12]. Since SNOMED also uses indexical numeric codes, the same problems occur as with ICD when a term is relocated in a hierarchy.

The introduction of SNOP sparked intense debate between the proponents of statistical classification systems and those of more complex multiaxial medical nomenclatures. The former held that the primary uses of these systems was for international comparisons of disease and for cost reimbursement. For these purposes, only broad statistics on general groups and classifications of disease were needed. The proponents of multiaxial nomenclature maintained that single-axis systems were designed for general statistics and would become too complicated for a detailed nomenclature. Also, when several diagnostic codes are of interest, it is much harder to formulate complex boolean queries in a single axis than in a multiaxis code. Finally, the resolution offered by multiaxis codes could always be mechanically collapsed into a general-purpose single axis code, while the reverse operation is not feasible [9].

At the very time that these debates intensified, events occurred which abruptly shifted interest to statistical classification. The drug thaliomide caused thousands of severe birth defects in Europe and Canada, where it had been recently approved, and in the United States, where it had been administered experimentally. As a result, sweeping regulations for pharmaceutical product development and adverse effect reporting were introduced [19]. The new regulations spawned a need for broader medical vocabularies for statistical classification of drug reactions and adverse events. Both WHO and the US Food and Drug Administration sponsored efforts to develop classifications systems specifically for this purpose.

2.3 COSTART and WHOART

The FDA published the *Dictionary of Adverse Reaction Terms* (DART) in 1967. DART was a hierarchical dictionary with the body-system category at the highest level. The FDA used DART for approximately 2 years, replacing it in 1969 with the *Coding Symbols for a Thesaurus of Adverse Reaction Terms* (COSTART). This new dictionary emphasized the specific adverse reactions over the system or organ affected, and allowed more than one search category to improve retrieval capabilities [45].

COSTART has a hierarchical arrangement of terms. At the top, there are 12 body-system categories. These are followed by subcategories and ultimately by coding symbols and preferred terms (a word or phrase of choice used to represent an adverse reaction). Instead of numbers, COSTART uses English language word symbols to record preferred terms. For example, the symbol ABDO PAIN codes `abdominal pain`. Coding symbols are also used for search categories (body-system classes) and their subcategories.

The same year, 1969, saw the publication of the *WHO Adverse Reaction Terminology* (WHOART) by the World Health Organization. WHOART is also hierarchical, with 30 system-organ classes followed in the hierarchy by high-level terms and preferred terms. WHOART assigns a numeric code to all system-organ classes, high-level terms, and preferred terms. Since WHOART was published in four languages (English, French, German, and Spanish), the use of non-language specific codes was an important feature which distinguished it from COSTART [45].

For the next quarter century, adverse event coding was dominated by WHOART (required by the European Union) and COSTART (mandated by the FDA). The FDA translated all adverse reactions reported to the agency into COSTART terminology [51]. To report to the World Health Organization, the FDA converted the COSTART codes into WHOART codes, using a fixed translation table [45]. In combination with an adverse reaction terminology, most organizations processing regulatory data also used a morbidity terminology - primarily ICD-9 in Europe and ICD9-CM in the United States. The Japanese developed their own versions of these international terminologies, *Japanese Adverse Reaction Terminology* (JART) and MEDIS [39, 47].

Over the years, organizations expressed dissatisfaction with these established terminologies. Deficiencies included a lack of specificity at lowest level of terms, limited data retrieval options (e.g., too few levels in the hierarchy, or capacity to retrieve data via one axis only) and ineffective handling of syndromes. Reassignment of codes in new versions forced existing data to be reevaluated. Use of one terminology for adverse reactions, and other for morbidity complicated data retrieval and analysis [39].

To address these deficiencies, some organizations modified the existing terminologies for company or regional use, or developed entirely new systems. Some of these, in particular the *Hoechst Adverse Reaction Terminology System* (HARTS), spread well beyond the parent company (Hoechst, now Aventis). This created additional problems. The use of different terminologies in separate geographic regions impaired international communication and necessitated the cumbersome conversion of data from one terminology to another. These problems most acutely hindered multinational pharmaceutical companies whose subsidiaries used multiple terminologies to fulfill the varying data submission requirements of various national regulatory agencies. As pharmaceutical companies merged and became global, it became increasingly difficult to manage the information required for product development and regulation.

2.4 MedDRA

In 1994, the International Conference on Harmonization (ICH) introduced multi-disciplinary regulatory communication initiatives to compliment their ongoing safety, quality, and efficacy harmonization efforts. This included the M1 initiative which sought to standardize international medical terminology in all phases of the regulatory process. The initiative resulted in the development of the *Medical Dictionary for Regulatory Activities* (MedDRA), based on the UK Medicines Control Agency's (MCA) medical terminology. Version 2, the first implementable version, was published in 1997. Since then, new releases have appeared regularly and MedDRA has rapidly been adopted an internationally accepted medical terminology for regulatory purposes. The FDA already recommends (though does not yet require) MedDRA, and the European Union will require MedDRA coding for submissions and safety reporting after January 2003 [39]. MedDRA is a multiple axial system that uses non-indexed codes, and includes a five level hierarchy of of terms.

Table 1. MedDRA terms in the 4.0 version.

Level	Name	# terms
SOC	System Organ class	26
HLGT	High Level Group Term	334
HLT	High Level Term	670
PT	Preferred Term	15000
LLT	Low Level Term	55000

The SOC level is roughly analogous to the highest levels of COSTART and WHOART. The HLGT and HLT levels group related terms beneath them by anatomy, pathology, etiology or function. The PT level contains distinct descriptors medical concepts including symptoms, signs, diseases, procedures and social characteristics. The LLT level contains terms which are synonyms or lexical variants of the PT to which they are linked. Since LLTs accommodate culturally unique terms, each LLT may not have a translation in all languages. An LLT can be linked to only one PT.

Let us trace the term `malignant neoplasm of the upper lobe` through the five levels of the MedDRA system. At the SOC level the term may be classified either on the the etiological axis provided by `Neoplasms benign and malignant`, or the anatomical axis provided by `Respiratory, thoracic and mediastinal disorders`. Selecting the former leads to the HLGT `Respiratory and mediastinal neoplasms malignant and unspecified`, the HLT `Respiratory tract and pleural neoplasms malignant cell type unspecified NEC`, the PT `Lung cancer stage unspecified` and finally the LLT `malignant neoplasm of upper lobe`.

If we select the anatomical axis, we begin with the SOC `Respiratory, thoracic and mediastinal disorders`, the HLGT `Respiratory tract neoplasms`, the HLT `Lower respiratory tract neoplasms` and the PT `Lung cancer stage unspecified`. Here we see that the PT was included under two HLT terms. While a PT can have more than one associated HLT (and HLGT, and SOC) there is at most one path down through the hierarchy (SOC-HLGT-HLT-PT) to any PT, so that counting errors are avoided when querying on the SOC, HLGT, and HLT levels. MedDRA also assigns a primary SOC (and therefore a primary path PT-HLT-HLGT-SOC) for consistency in standard reporting, based on most common usage.

2.5 Other Terminologies and Systems

While MedDRA looms large in the regulatory landscape, and can be said to subsume several of the earlier systems, including COSTART, WHOART, and HARTS, our survey would not be complete without describing at least the most widely used systems that currently fall outside its scope.

DSM

Perhaps the largest medical field with its own well-developed terminology is psychiatry. The American Psychiatric association first published its *Diagnostic and Statistical Manual of Mental Disorders* (DSM) in 1952. The most recent edition, DSM-IV was published in 1994, with minor modifications in 2000. This system, which has been adopted by many countries, has five axes:

1. Clinical Disorders
2. Personality Disorders and Mental Retardation
3. General Medical Conditions
4. Psychosocial and Environmental Problems
5. Global Assessment of Functioning Scale

Beneath the axes, diagnostic labels and their corresponding codes are grouped. Codes contain 3-5 digits. The leading 3-4 digits specify a specific entity, with the 5th digit providing additional specificity such as subtype or severity. To provide compatibility for reimbursement and other interested third parties, each DSM-IV code is associated with a ICD code equivalent [15].

ICPC

In 1972, the World Conference of Family Doctors initiated a project to develop a comprehensive system to classify the primary health care and patient-doctor interactions. A succession of efforts led to the publication of the International Classification of Primary Care (ICPC) in 1978. The second edition was published in 1998, primarily to coordinate the codes with the the 10th version of ICD [61]. Several European countries including the Netherlands, Denmark, and Norway have produced national ICPC implementations [26].

The ICPC system groups codes under a matrix of 17 chapters and 7 components. The chapters are:

A General and Unspecified
B Blood, Blood Forming Organs and Immune Mechanism
D Digestive

F Eye
H Ear
K Circulatory
L Musculoskeletal
N Neurological
P Psychological
R Respiratory
S Skin
T Endocrine, metabolic, and nutritional
U Urinary
W Pregnancy, child-bearing, family planning
X Female genital and breast
Y Male genital
Z Social problems

These are roughly analogous to the highest level in WHOART, COSTART, MedDRA, etc. The components, which offer a different organizational principle, are as follows:

1. Symptoms and Complaints
2. Diagnostic Screening and Preventative Measures
3. Medication and Treatment procedures
4. Test results
5. Administration
6. Referrals and other reasons for encounter
7. Diagnoses and Disease

Components 2-6 are referred to as *process* components. Codes are grouped under a Chapter and a Component. Codes consist of three characters and have a title of limited length. Each code is mapped to one or more corresponding ICD-10 codes. A patient-physician interaction is assigned a series of codes, for example `A03 - Fever, A30 - Full examination, A31 - Temperature Measurement, A63 - Plan for follow-up visit.`

Read

Originally created by Dr James Read to generate computer based morbidity records, the Read Classification Codes (RCC) were adopted by British National Health Service in 1994, and were expanded to to cover all fields of primary health care with the publication of Version 3.

This version includes a 5 level hierarchy and allows classification of terms along multiple axes. The non-indexical 5 character codes are cross-referenced to ICD10 codes. In Version 3.1, the current version, a set of qualifier terms such as anatomical site was added. Qualifier terms can be combined with existing terms to form composites which exist outside of the hierarchy. Terms and qualifiers are grouped into templates that describe the range of medically valid combinations. Many medical organizations, primarily in the United Kingdom, use the RCC for classification of medical records in clinical information systems [41].

UMLS

The proliferation of controlled vocabularies has long worried information retrieval specialists, who would like to cross-reference information coded according to different schemes. The National Library of Medicine is developing the *Unified Medical Language System* (UMLS) with the specific goal of addressing these issues [52]. However, the UMLS Metathesaurus is nowhere near completion.

Given the diverse constituencies of the various schemes, it is highly unlikely that a comprehensive uniform scheme will be adopted within a decade or two. This simple fact has major impact on the design of hybrid translation systems: since any such system needs to "have legs" (require no or little external effort to move it from one scheme to another), it makes little engineering sense for them to contain detail knowledge about the internal organization principles of any particular coding scheme, and even less sense to tie the behavior of the system to the semantics of such schemes. In other words, the fact that we are considering a narrow domain, medicine, does in no way imply that we can, or even should try to, utilize deeply domain-specific knowledge. To the contrary, in order to keep the system re-targetable for different controlled vocabularies, we should rely on as generic machine translation and machine learning techniques as we can. For this reason, in the following we put the emphasis on generally applicable techniques as opposed to systems that are tied very closely to any particular target vocabulary.

3 Hybrid Autocoding Systems

The ability to master natural language communication played a central role in the original Turing test [50], and for well over a decade Machine Translation (MT) was believed to be right around the corner. In spite of some very clear early warnings like [2], it was not until the appearance of the ALPAC report [21], that the actual difficulty of the problem became widely appreciated. But once fully automated high quality MT was recognized to be the extremely hard problem it is, attention turned to restricted/reformulated versions. As a first step, "high quality" was replaced by "with some effort understandable" and "fully automated" was replaced by "suitable for human post-processing". Thus the data flow followed the scheme outlined in Figure 1:

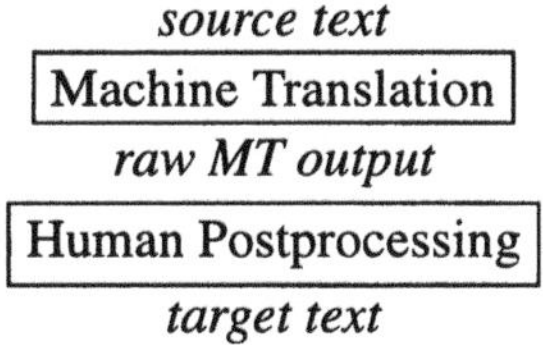

Figure 1. Machine-assisted translation (coding) workflow

Another important change was the move from unrestricted to domain-specific text. In fact, for many domains where the syntax is highly conventional and the content varies little, e.g., in weather reports [7], the labor-intensive post-processing step could be kept to a minimum. In general, interactive translator's workbench systems, where the results of the postprocessing step are continually fed back to the MT component, were found more effective than single-pass postprocessing.

It was soon realized that the success of such systems derives largely from the conventionalized, repetitive aspect of the task, and that conventionalization is best accomplished by controlling the source language. This is a very active research area to this day, but one that we can't do full justice to within the bounds of this chapter, and we refer the reader to the survey [60] and the proceedings of the biannual CLAW conferences published by the Language Technologies Institute at Carnegie Mellon University.

From the perspective of MT, the main distinguishing feature of the medical domain is the size of the vocabulary. The METEO system [7] of translating weather reports could be successful precisely because it knows less than 10^3 words, and it is possible to handcraft lexical entries reflecting the usage possibilities of each and every one of these. But even the tightly controlled vocabularies surveyed in the previous Section contain $10^4 - 10^6$ words and phrases, and our ability to handcraft meaningful lexical entries for each of them is limited.

Over the years, many researchers have noted that a great deal of the (controlled) vocabulary is *compositional* in the sense that the meaning of the whole expression can be recursively inferred from the meaning of its parts. To take advantage of this fact, one of course needs to define the meaning of the expressions in some knowledge representation scheme. A significant effort in this direction is GALEN, now OpenGALEN, which builds a whole conceptual model of the medical domain, with knowledge representation primitives such as ACTS_ON, IS_PART_OF, etc.

Since it is clearly the long-term goal of medical informatics to use such representations as the basis of inference, we include a discussion of the problems and main techniques here, but we caution the reader that the representation of medical knowledge is still in its infancy, with narrow, non-scalable prototypes dominating the field, while the use of autocoders is part of the mainstream production environment for pharmaceutical research and regulatory activities.

We distinguish three interrelated problems in the translation process, and discuss them in turn:

1. *Segmentation,* the problem of finding (the beginning and endpoints of) the words and phrases that are subject to translation
2. *Analysis,* the problem of identifying and recursively substituting smaller patterns that compositionally make up the larger patterns and
3. *Substitution,* the problem of providing the appropriate translation.

While separating the tasks at the conceptual level is clearly necessary, it should be emphasized that a working system need not contain separate modules for each, nor is it necessary to impose an execution order in which solving one of these problems precedes solving the other. In fact, lack of a rigid pipelined architecture, and the ability to delay decisions, is a key strength of the most important hybrid methods.

3.1 Segmentation

Finding the relevant words and phrases is a problem long familiar from machine translation. It is particularly acute in languages such as Chinese or Japanese where no whitespace separates the symbols, and in languages such as German where long compound words are common. Current segmentation and *entity extraction* techniques have their historical roots in string matching techniques – for a survey, see [46]. The key idea, built into Unix in the form of `grep` [49] is to describe string patterns in terms of regular expressions, and to speed up regular expression searches by means of compiling them into finite automata.

To check for an entity, the current generation of autocoders perform some combination of the following abstraction steps [22, 18, 40]

1. special character normalization – replacing # by `number`, @ by `at`, etc
2. spell correction – replacing `pian` by `pain`
3. acronym expansion – replacing ECG by `electrocardiogram`
4. undoing abbreviations – replacing `liv` by `liver`
5. capitalization – lowercase chars replaced by their uppercase counterpart
6. morphological analysis (stemming) – `fingers` replaced by `finger`
7. synonym-based data enrichment – replacing `ache` by `pain`
8. word deletion – omitting words low in information content

Notice that all these steps correspond to *generalized sequential mappings* that map input strings on output strings in an essentially memoryless fashion. The idea of "memoryless" computation may be somewhat counterintuitive for steps such as acronym expansion or data enrichment, but in fact dictionary entries can be thought of as containing two strings, an input and an output, and looking up a string in the dictionary is an operation that requires no writable memory. Such operations can be performed by a particularly simple kind of computing machinery, *finite state transducers* or FSTs. The techniques for compiling word lists into FSTs are well understood, and have been widely used in tokenization tasks since the introduction of `lex` [35].

In the eighties, the whole technology of string matching received a new impetus from the work of Koskenniemi [33], who showed how simple FSTs operating in parallel can be used to express complex relations between strings. Since both parallel and sequential combinations of FSTs can be compiled into single (albeit often much larger) FSTs, the road was open to express regularities by the kind of rules used by lexicographers and grammarians, compile the rules into FSTs, and combine the FSTs into a single, highly efficient program that will check for all the regularities at the same time. Crucially, the computational efficiency of the resulting FST is the same whether it is used in synthesis (generation from input to output) or analysis (recognition of input from output) mode. For more detailed discussion of these techniques see [42, 31]. In autocoding, and perhaps in all of machine translation, probabilistic FSTs are the single most important hybrid technique, since they enable the compilation of expert-written and machine-generated rules in a unified framework.

In spite of their mathematical similarity, not all the above steps are equally practical. Since special character normalization is the key to successfully matching certain entities, and the cost of implementing this step is negligible, it is best included in any system, but on the whole will affect only a small fraction, well below .1%, of the verbatims to be translated. Spell correction, on the other hand, affects a larger fraction, often as much as 1-2%, but general-purpose spellcheckers perform poorly in the medical domain, and implementing a domain-specific spellchecker is a major undertaking. Morphological analysis (stemming) yields a small but consistent improvement of 4-5% in autocoding, which is not nearly as good as that found in [34] for the most comparable situation in Information Retrieval (IR), short queries and short documents. The use of data enrichment is highly controversial, because the errors it can introduce, being semantically valid, are hard to spot. On the whole, we feel that this step should be performed indirectly, on the output knowledge representation, rather than directly, on the input dictionary words. Omitting *stopwords* that have low information content is very effective, but care must be taken in selecting the stopword list. For a summary of the effectiveness of these steps for IR see [14].

There is one widely used technique, word order normalization, which does not easily lend itself to reformulation in terms of FSTs. The idea, closely related to the "bag of words" model long familiar from information retrieval, is to simply list the words in some fixed (typically, alphabetical) order. By this method, many equivalent expressions such as `enlarged left kidney`, `left kidney, enlarged`, or `kidney, left enlarged`, can be mapped on a single normalized order, in this case `enlarged, kidney, left`. As a practical matter, word order normalization creates almost as many problems as it solves, since mechanical application of the procedure will map medically distinct conditions such as `pain without swelling` and `swelling without pain` on the same form `pain, swelling, without`.

Early entity extraction systems were entirely rule-based, with rules such as (paraphrase) "the word or phrase following `normal` or `abnormal` refers to a test result or body function". Such rules suffered from all the problems of classic expert systems: they were

- *expensive* – could only be generated by experts well versed both in the subject domain and in the intricacies of rule writing
- *brittle* – minor changes in the data often require radical revisions of the rule system
- *fragmentary* – relevant data may not be found even in large samples, and
- *monolithic* – changes cannot be effected in a modular fashion, every rule can interact with every other rule, to produce unexpected changes in overall behavior

With the introduction of statistical machine translation techniques [6], entity extraction became less of an art and more of a practicable technique. While such techniques address the cost, brittleness, and monolithicity issues, the problem of fragmentary data remains, and it has been the general experience that humans are far more capable of extracting the relevant patterns from fragmentary data than the currently known statistical techniques.

Therefore, the best systems still have a strong handcrafted/human supervised component, and the expertise of human coders still very much needs to be tapped into both at the stage of preparing the dictionaries and synonym tables and at the stage of assuring the quality of the machine output. How this "hybridization" itself can be performed automatically or semi-automatically will be discussed in Section 3.3.

3.2 Analysis

To a large extent, the task of finding the entities to be translated is intervowen with the parsing of the text: very often, it is not the first word or stem of the entity itself that provides the necessary clue, but rather the preceding word, and the same is true about detecting the end of an entity. That said, obtaining a full parse is seldom a necessary prerequisite to locating the phrases to be translated, and indeed the complexity of the full grammatical task is formidable, even in the medical domain, which itself is rather restricted compared to general-purpose English. Here we list some of the most salient properties that set adverse event coding apart from the bulk of the work on machine translation.

Table 2. Medical vs. general-purpose translation.

characteristic	adverse event coding	general MT
length	short (on average < 6 words)	long (multi-paragraph)
syntax	nonstandard	standard
	`acne facial recurrent`	`recurrent facial acne`
language	mixed	monolingual
	over 10% German, French, etc.	English
style	telegraphic `allergy arms and legs`	normal `skin allergy on the arms and legs`
vocabulary	specialized	general-purpose
	10^5 words and phrases added to	10^5 words in gen. vocabulary

Most of these differences extend well beyond adverse event coding: similar issues need to be faced in coding to the disease, symptom, and procedure classifications surveyed in Section 2, and in some cases, such as coding to drug names, specialized vocabulary may grow into the millions. To some extent, the above characteristics are correlated: the longer the text grows, e.g., in discharge reports, the more it sheds from its nonstandard properties and the more it becomes like general-purpose English. This would be an advantage if we had at our disposal high-quality parsers for general-purpose English just the way we have high-quality spellcheckers. Unfortunately, such parsers are beyond the state of the art, and we only see the downside of moving closer to ordinary English: more mistakes and ambiguities, a wider range of syntactic constructions, and an overall loosening of the implicit language control that is imposed by the rigid constraints of the genre.

Because of these factors, the analysis of medical text tends to follow, rather than lead, the state of the art in analyzing general-purpose texts. There the most important technique to have emerged in the past decade is *shallow parsing* or *chunking* [1], which abandons the goal of providing a full analysis, and views the extraction of grammatical categories such as Noun Phrase or Verb Phrase as an extension of the entity extraction task, relying on the same general finite-state techniques we discussed in Section 3.1 above. An important consequence of this trend is that the output of parsing (syntactic analysis) is no longer sufficiently detailed to provide input for knowledge representation (semantic analysis), and the words and phrases extracted from the text are substituted in more loosely defined semantic patterns [37]. This becomes particularly noticeable when we encounter medically meaningless expressions such as `removal of a renal cyst from the thyroid` [43]: while in principle a sophisticated knowledge representation scheme as found in OpenGALEN may be able to handle these correctly, exercising this capability over a wide range of constructions presupposes a level of parsing accuracy that is currently not available.

A relatively easy step is sentence- and clause-level parsing, i.e., to identify chunks separated by hyphens, parentheses, and other punctuation. For autocoding, as in the WebCoder system [32], this has the advantage that the chunks can be ranked according to their hypothesized hierarchical importance. Chunks shorter than four words that constitute the whole verbatim are considered *definitive* in the sense that they contain all the information necessary for substitution and receive rank 0, partial chunks have rank 1 and higher. As Table 3 shows, the number of less valuable (higher ranked) chunks emerging from a shallow parser falls of rapidly with rank. The quality of the information contained in lower-ranking chunks is also rapidly decreasing [32].

Table 3. Distribution of chunk ranks.

	WHOART tokens	*WHOART* types	*COSTART* tokens	*COSTART* types
trainset	64765	64515	50301	8038
testset	7195	7195	5589	1718
rank 0	62636	62366	50264	7946
rank 1	10759	8630	1796	556
rank 2	649	524	92	38
rank 3	22	22	2	2
rank 4	2	2	0	0

3.3 Substitution

Originally, the computational effort of autocoders was dominated by database lookup. These were little more than *translation memory* (TM) systems (for an overview, see [53]) that simply stored previously encountered entities with their proper encoding, as assigned by the human coder. The effectiveness of TM is almost entirely a function of the size of the translation memory or *synonym table* as it is called in autocoders. Since the tables are carefully constructed and debugged by human coders familiar with the coding schemes surveyed in Section 2 [17], their error rates are very low, but they miss a great deal. Today, in translating to a controlled medical vocabulary, the key problem lies not so much with providing the actual translation (which can be accomplished by standard hashing/database lookup methods), but with filling the synonym table with minimum human effort. Our interest is not so much with memorization as with Machine Learning (ML) techniques that can generalize better to data not seen in training, even at the cost of increased error.

Here we define two figures of merit standardly used in computational linguistics: by the *precision* of the system we mean the percentage of correct output relative to the total output, and by *recall* we mean the percentage of correct output relative to total input. How input and output is defined depends on the nature of the task, such as entity extraction, information retrieval, classification, or machine translation, but the intent is always the same: the precision measure is defined to be sensitive to false positives, while the recall measure is defined to be sensitive to false negatives. In practical applications, it is almost always possible to trade off precision for recall and conversely. TM systems are at one extreme of this tradeoff: precision is very high, but recall can be as low as 15%, and 40% is considered very good [5].

Rather than thinking of substitution as a mechanical database lookup step, it will be expedient to recast the whole problem as an instance of the general IR/text classification problem, or more specifically, as an instance of the message routing (MR) problem [24]. In supervised MR we are faced with the following problem: given some categories $C = \{C_1, C_2, \ldots, C_k\}$ and some "truthed" documents $F = \{F_1, F_2, \ldots, F_n\}$ with their true categories $t(F_i) \subset C$, for any new document F_{n+1}, find the most likely value of $t(F_{n+1})$, i.e., the category or categories associated to F_{n+1}. In a typical MR system files are first reduced to multisets or *bags* of words or word stems in the preprocessing stage, so that the task becomes classifying these bags of words, or stems, rather than the original files. Some systems, such as [32], go beyond words and employ word pairs and in general word n-grams, since these are more informative than word unigrams alone.

Given an arbitrary fixed ordering of words/stems, e.g., as created by a perfect hash function, bags are in one to one correspondence with vectors of a large dimensionality d (the number of words, usually several thousand to a few hundred thousand) having only nonnegative integer coefficients called in these applications the *counts*. Looked at this way, autocoding (and in general, supervised MR) is a classification problem in Euclidean space. Before turning to a discussion of the main machine learning approaches, however, we list the main aspects of the MR problem that set it apart from other important tasks such as Optical Character Recognition.

(1) The number of classes is large, often $10^4 - 10^5$
(2) The number of potential features (word n-grams) is astronomical
(3) Human judgments are available
(4) The data is repetitious

All machine learning algorithms solving supervised problems have two phases of operation: *training* and *classification*. In training, a mathematical model of the classes is created based on pairs consisting of verbatims and their translations. In classification, the model is applied to new data, and hypothesized translations (and confidences) are output. With some simplification, we can distinguish two broad types of training methods: those that create a separate model for each class, and those where knowledge about the individual classes is distributed over the whole system.

All handcrafted classifiers, based on (combinations of) keywords and rules, fall in the first type, where we also find automatically generated models such as Naive Bayes [38] and linear machines [25, 36]. The second type includes many variants of Artificial Neural Nets (ANNs) [44] and k-nearest neighbor classifiers (kNN). The string normalization techniques discussed in Section 3.1 are for the most part very conservative, assuring only that near-identical strings receive the same translation. We can therefore think of Translation Memory systems as nearest neighbor classifiers, but the abstract string-similarity neighborhoods around the input strings listed in the synonym table are very small.

Within the bounds of this survey we can't do full justice to all these techniques: for an overview of the main machine learning methods see [16], and for the state of the art in text classification see the annual SIGIR and TREC proceedings. But we emphasize here that much of what we know about Message Routing comes from relatively small collections of data, such as the original Reuters corpus, which has less than 30k documents and only 90 categories with both training and test examples. Almost all this knowledge needs to be critically reevaluated for autocoding. Because of (1), techniques that are not linear or at least near-linear in the number of classes are essentially useless even when they are provably superior on small data sets: this includes much of standard multivariate statistics, such as factor analysis [4] and regression methods [3].

One well-understood technique for dealing with this problem is divide and conquer: we can leverage the hierarchical structure of the classes by first classifying to the highest level, and building independent lower-level classifiers for each node. In [32] we built a bodysystem-level master, with bodysystem-specific slaves, at the cost of increasing reject rates by 1.3%. and error rates by 0.1%. It has been argued that in some cases no loss (or even accuracy gains) could be seen [30], but these results may depend on experiments being performed in a more noisy environment than provided by medical texts. Because of (3), we have much better training data than is customary in information retrieval: essentially, all previously coded material, and all preexisting synonym tables, are at our disposal. Also, these are very high quality both in terms of extremely low error rates (often coded twice by different coders and reconciled by a third coder if needed) and high consistency (mature coding practices). Laveraging this training material properly is the key to the success of hybrid systems [27]

Because of (2), feature selection can matter a great deal. A widely used technique is Singular Value Decomposition (SVD) [8, 20], which provides good dimension reduction, but the nonlinear computational cost is almost impossible to absorb for very large problems [48]. Since the same drug is likely to have the same adverse effect on different people, we also benefit from (4) – with the proper statistical techniques both (3) and (4) can be highly leveraged to mitigate the effects of (1) and (2).

4 Conclusions

Automatic translation to controlled medical vocabularies is not a solved problem. Because of the characteristics of the medical domain, many techniques that work well for less demanding data sets are not practical for autocoding, and properly leveraging the knowledge of human coders remains the key to developing and deploying successful systems.

Acknowledgment

The work on WebCoder was supported under SBIR grant # 2 R44 CA 65250 from the National Cancer Institute, National Institutes of Health. The authors would like to thank Michael Johnston, Jeremy Pool, and J. Michael Richards for their help at various stages of the project.

References

1. Steven Abney. 1991. Parsing by chunks. In Robert Berwick, Steven Abney, and Carol Tenny, editors, *Principle-based parsing*. Kluwer Academic Publishers.
2. Bar-Hillel, Y 1960 A demonstration of the nonfeasibility of fully automatic high quality translation. In FL Alt (ed): *The present status of automatic translation of languages* Academic Press, New York, 158–163
3. Peter Biebricher, Norbert Fuhr, Gerhard Lustig, Michael Schwanter, and Gerhard Knorz 1988 The automatic indexing system AIR/PHYS – from research to application 11th Int Conf on R&D in IR 333-342
4. Harold Borko and Myrna Bernick 1963 Automatic document classification. JACM 10 151–161
5. Sonja Brajovic 2002 Personal Communication
6. PF Brown, SA Della Pietra, VJ Della Pietra, and RL Mercer 1993. The mathematics of statistical machine translation: Parameter estimation. *Computational Linguistics* **19** 263–311.
7. M. Chevalier, J. Dansereau, and Poulin, G. 1978. TAUM-METEO: Description du Système. Technical report, Groupe de recherches pour la traduction automatique Université de Montréal
8. Christopher G. Chute and Yiming Yang. 1995. An overview of statistical methods for the classification and retrieval of patient events. *Methods of Information in Medicine*, 34:104–109.
9. Roger A. Cote, and Stanley Robboy. 1980. Progress in Medical Information Management, Systemized Nomenclature in Medicine (SNOMED). *Journal of the American Medical Association*, 243:756–762.

10. Roger A. Cote. 1978. The SNOP-SNOMED Concept; Evolution Towards a Common Medical Nomenclature and Classification. *Proceedings of the Seventh International Congress on Medical Records*
11. Cimino JJ., G Hripcsak, S Johnson and P Clayton 1989 Designing an introspective, controlled medical vocabulary. In: Kingsland L, ed. *Proceedings of the 13th annual symposium on computer Applications in Medical Care.* IEEE Computer Society Press, Washington, DC
12. Cimino JJ 1996 Review paper: coding systems in health care. *Methods Inf Med,* 35(4-5):273–284.
13. William Cullen. 1792 *Synopsis and Nosology* N. Palten Publishers Hartford, CT
14. Guy Divita, Allen C. Browne, and Thomas C. Rindflesch. 1998. Evaluating lexical variant generation to improve information retrieval. In *Proc. American Medical Informatics Association 1998 Annual Symposium,* Orlando, Florida.
15. American Psychiatric Society 2000. *Diagnostic and Statistical Manual of Mental Disorders DSM-IV, 4th Edition Text Revision* American Psychiatric Society, Washington, DC
16. Richard O. Duda, Peter E. Hart and David G. Stork 2000 Pattern Classification Wiley, New York, NY
17. Thérèse Dupin-Spriet and Alain Spriet. 1994. Coding errors: classification, detection, and prevention. *Drug Information Journal,* 28:787–790.
18. Christian Fizames. 1997. How to improve the medical quality of the coding reports based on WHOART and COSTART use. *Drug Information Journal,* 31:85–92.
19. Food, Drug, and Cosmetic Act 1962. Kefauver-Harris Amendment, 21 U.S.C. Section 355
20. Stephen I. Gallant. 1995. Exemplar-based medical text classification. *Belmont Research SBIR Proposal 1 R43 CA 65250-01.*
21. Paul Garvin 1966. Language and Machines. Computers in Translation and Linguistics. (ALPAC report, 1966). National Academy of Sciences.
22. Terry L. Gillum, Robert H. George, and Jack E. Leitmeyer. 1995. An autoencoder for clinical and regulatory data processing. *Drug Information Journal,* 29:107–113.
23. John Graunt 1662. *Natural and Political Observations Made Upon The Bills of Mortality London.*
24. Donna K. Harman, editor. 1994. *The Second Text REtrieval Conference (TREC-2).* National Institute of Standards and Technology, Gaithersburg, Maryland.
25. Wilbur H. Highleyman. 1962. Linear decision functions with application to pattern recognition. *Proceedings of the IRE,* 50:1501–1514.
26. Marc Jamoulle. 2001 ICPC use in the European Community *16th WONCA World Congress of Family Doctors* Durban, South Africa Available at http://www.ulb.ac.be/esp/wicc/icpc_2001.html, accessed March 12, 2002
27. Michael. C. Joseph, Kathy Schoeffler, Peggy A. Doi, Helen Yefko, Cindy Engle, and Erika F. Nissman. 1991. An automated COSTART coding scheme. *Drug Information Journal,* 25:97–108.

28. Lester S. King. 1966 Boissier de Sauvages and 18th century nosology. *Bulletin of Historical Medicine*, 40: 43–51
29. Lester S. King. 1958 In *Medical World of the Eighteenth Century* , Chapter 7: 193-226 University of Chicago Press, Chicago
30. Daphne Koller and Mehran Sahami 1997 Hierarchically classifying documents using very few words In *Proc. 14th Int Conf on Machine Learning* 170–178
31. András Kornai 1999. Extended finite state models of language Cambridge University Press, Cambridge, England
32. András Kornai and J. Michael Richards. 2002. In A. Abraham, M. Koeppen (Eds.) *Hybrid Information Systems*: 527–538 Physica Verlag, Heidelberg
33. K. Koskenniemi 1983. Two-level morphology: a general computational model for word-form recognition and production. Ph.D. thesis, University of Helsinki.
34. Robert Krovetz. 1993. Viewing morphology as an inference process. In *Proceedings of SIGIR93*, pages 191–202.
35. Michael E. Lesk 1975 Lex — a lexical analyzer generator. Technical Report 39, AT&T Bell Laboratories, Murray Hill, NJ,
36. David D. Lewis, Robert E. Schapire, James P. Callan, and Ron Papka. 1996. Training algorithms for linear text classifiers. In *Proceedings of SIGIR96*, pages 298–306.
37. David M. Magerman 1995. Statistical Decision-Tree Models for Parsing. In *Proceedings of ACL95*, pages 276–283
38. ME Maron 1961 Automatic Indexing: an experimental inquiry. JACM 8 404–417
39. MSSO-DI-6003-4.1.0 2001 *MedDRA Introductory Guide, version 4.1*
40. Keya Pitts, Kathleen Jelliffe, and and Lionel Benson 1999. Advances in Volume Encoding Clinical Data. *Drug Information Journal* 33/4: 1079–1092.
41. NHS Centre for Coding and Classification. 1993 Read codes and the terms projects: a brief guide. London: Department of Health
42. E. Roche and Y. Schabes 1997. Finite-State Devices for Natural Language Processing MIT Press, Cambridge, MA.
43. Rogers, JE and Rector, AL 1997 Terminological Systems: Bridging the Generation Gap. Annual Fall Symposium of American Medical Informatics Association. Nashville TN Hanley & Belfus Inc. Philadelphia PA: 610-614.
44. Rumelhart, David E., McClelland, James L., and the PDP Research Group. 1986 Parallel Distributed Processing: Explorations in the Microstructure of Cognition MIT Press, Cambridge MA
45. Alan Saltzman 1985. Adverse Reaction Terminology Standardization: A Report on Schering- Plough's Use of the WHO Dictionary and the Formation of the WHO Adverse Reaction Terminology Users Group (WUG) Consortium. *Drug Information Journal* 19:35–41
46. D. Sankoff and JB Kruskal Time Warps, String Edits and Macromolecules. Reading, Mass.: Addison-Wesley, 1983
47. Fred Schneiweiss Adverse Reaction Thesauri Used in the Pharmaceutical Industry 1987. *Drug Information Journal* **21** 299–302
48. Hinrich Schütze, David A. Hull, and Jan O. Pedersen. 1995. A comparison of

classifiers and document representations for the routing problem. In *Proceedings of SIGIR95*, pages 229–237.

49. Ken Thompson 1968. Regular expression search algorithm. *Communications of the ACM* **11**: 419–422
50. Alan Turing 1950 Computing machinery and intelligence. MIND (the Journal of the Mind Association), vol. 59, no. 236, pp. 433-60, 1950
51. Wayne M. Turner, Julie B. Milstien, Gerald A. Faich and George D. Armstrong 1986. The Processing of Adverse Reaction Reports at FDA. *Drug Information Journal* 20: 147-150.
52. National Library of Medicine ULS Knowledge Sources, Unified Medical Language System 13th Edition - Janary Release 202AA Documentation
53. Lynn E Webb 2000 Advantages and Disadvantages of Translation Memory: A Cost/Benefit Analysis. MA Thesis, Monterey Institute of International Studies.
54. http://www.who.int/library/historical/access/disease/index.en.shtml WHO Library Historical Collection Jacques Bertillion Nomenclatures des maladies (statistique de morbidité – statistique des causes de décès) arrêtées par la Commission internationale chargée de reviser les nomenclatures nosologiques (18-21 août 1900) pour etre en usage à partir du 1er janvier 1901
55. http://www.who.int/library/historical/access/disease/index.en.shtml WHO Library Historical Collection International Commission for the Decennial Revision of Nosological Nomenclature] Commission internationale pour la révision décennale des nomenclatures internationales des maladies, causes de décès, causes d'incapacité de travail, devant servir à l'établissement des statistiques nosologiques (classification Bertillon) quatrième session, 16-19 octobre 1929. 4e rév. Paris : Impr. nationale, 1930.
56. International Commission for the Decennial Revision of Nosological Nomenclature Nomenclatures 1940. Internationales des causes de décès 1938 (classification Bertillon): cinquième révision décennale effectuée par la Conférence internationale de Paris du 3 au 7 octobre 1938. La Haye, Institut international de statistique. Available at http://www.who.int/library/historical/access/disease /index.en.shtml, accessed March 12, 2002
57. World Health Organization 1965. *Manual for international Classification of diseases, 7th revision.* Geneva, Switzerland.
58. World Health Organization 1969. *Manual for international Classification of diseases, 8th revision.* Geneva, Switzerland.
59. World Health Organization 1992. *Manual for international Classification of diseases and health related Problems, 10th revision.* Geneva, Switzerland.
60. Wojcik, Richard, and Hoard, James 1995 Controlled Languages in Industry. Survey of the State of the Art in Human Language Technology, ed. Giovanni Battista Varile and Antonio Zampolli, 274–276.
61. WONCA International Classification Committee 1998 *International Classification of Primary Care, Second Edition* Oxford University Press, Oxford, England.

Chapter 18

A Genetic Programming for the Induction of Natural Language Parser

Olgierd Unold and Grzegorz Dulewicz

Summary. Genetic programming has been shown as a support for automated inference of automaton-driven parser of natural language. As an encoding scheme a modified form of edge encoding was used.

Keywords: natural language processing, genetic programming, edge encoding, grammatical inference

1 Introduction

When we try to deal with natural language processing (NLP) we have to start with a grammar of a natural language. But the grammars described in linguistic literature have an informal form and many exceptions. Thus, they are not useful to create final formal models of grammars, which make machine processing of sentences possible. These grammars can be a starting point for the attempts to create basic models of natural language grammar at the most. However, it requires expert knowledge. Machine learning based on a set of sample sentences can be the better way to find the grammar rules. This kind of learning (grammatical inference) allows to avoid the preparation of knowledge about the language for the NLP system. The examples of correct and incorrect sentences allow the NLP systems with the self-evolutionary parser to try to find the right grammar. This self-evolutionary parser can be improved on basis of new examples. Thus, the knowledge acquired in this way is flexible and easyly modifiable.

Unold [12] proposed theoretical bases for the use of two classes of evolutionary computation, that is evolutionary programming and genetic programming, that support automated inference of fuzzy automaton-driven parser of natural language. This chapter examines the use of edge encoding, one of the types of genetic programming, for induction of parser based on a fuzzy automaton.

2 Fuzzy Automaton-Driven Parser of NLP System

In [10] a fuzzy parser of natural language texts was proposed (so-called fPDAMS – fuzzy nondeterministic pushdown automaton with associative memory access) that works in the stratificational knowledge representation system (SKRS). Stratificational knowledge representation system is an attempt at formalizing the multi-layered structure of natural language. The SKRS is based on a multistratum semantic network whose nodes contain particular linguistic units (single words, the so-called structural-semantic components, sentences). The node-to-node links correspond to relations between particular linguistic units.

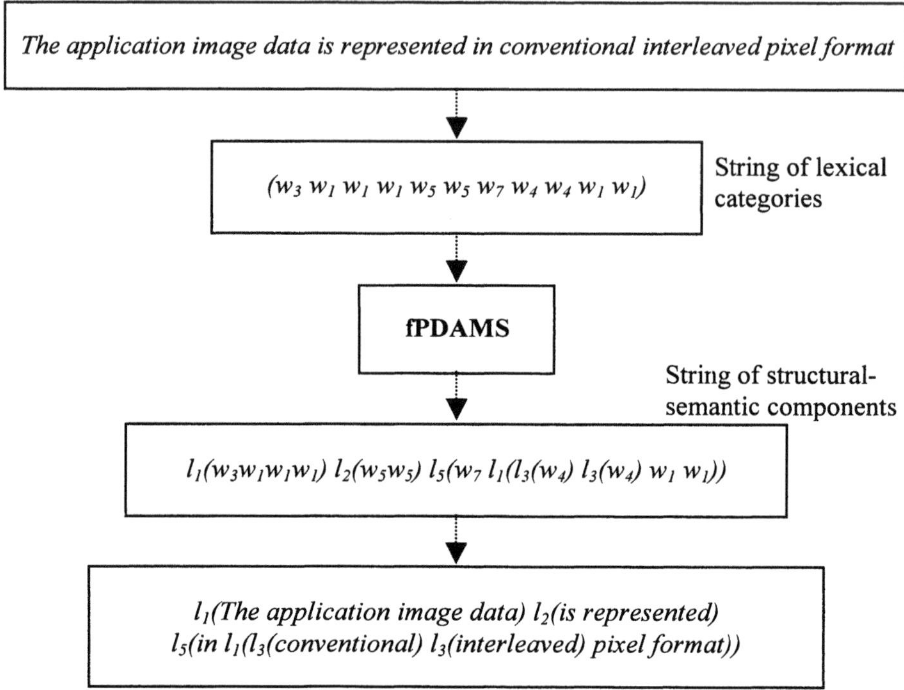

Figure 1. Sentence parsing into structural-semantic components by fPDAMS. The l_i symbols denote syntactic categories of substrings separated with the parentheses ( ) (l_1 – the nominal group, l_2 – the verbal group, l_3 – the adverbial group, l_5 – the prepositional group); the w_1 stands for a noun, w_3 for a determiner, w_4 for an adjective, w_5 for a verb, and w_7 for a preposition.

If a natural language text representing the acquired knowledge is to be mapped onto the stratificational semantic network, the particular linguistic units, such as word groups and sentences, must be isolated in the analyzed text. If we decide to limit our input texts to isolated clauses, we can limit the sentence decomposition task to two subtasks: one of word group decomposition into the particular words and another of sentence decomposition into word groups. In order to increase both knowledge integrity in the knowledge base and the efficiency of the search algorithm, an assumption has been made that it is not phrases, but the structural-semantic components (the components) that are the objective of the sentence decomposition, the components being word groups constituting semantic units which can no longer be decomposed, each of which has its own syntax characteristics.

Sentence decomposition into structural-semantic components in the SKRS is performed by a fuzzy automaton-driven parser fPDAMS. An automaton decomposition model in the SKRS can be used due to the reduction of the process of sentence decomposition into structural-semantic components to the process that isolates appropriate substrings of symbols in the string of symbols representing lexical categories.

The construction of the proposed analyzer is based on a fuzzy automaton [3]. The recent application of finite-state approach in NLP shows the usefulness of automata in this area of AI [8]. The basic advantage of applying a fuzzy automaton to process natural language lies in the existence of an established algorithm for the next state selection based on values of the characteristic functions, allowing the characteristic functions value to change with time. In that case fuzzy automaton-driven parser 'gets accustomed' to the input syntax constructions, imitating human behavior.

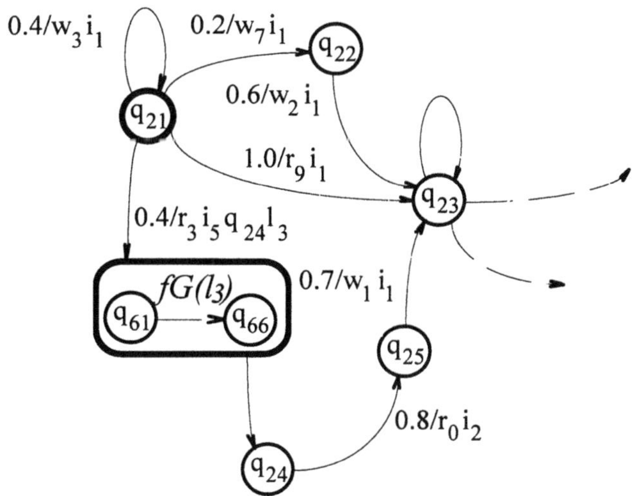

Figure 2. A fragment of the transition subgraph $fG(l_1)$ of the fuzzy subautomaton $fA(l_1)$.

Figure 2 shows a graph of a "hand-designed" subautomaton $fA(l_1)$ that isolates a substring that has a syntactical category of the nominal group from the string of the lexical categories of the words. There is another subautomaton, $fA(l_3)$, incorporated into the $fA(l_1)$ subautomaton structure, which accepts a substring that has a syntactical category of the adjectival group. The subautomaton $fA(l_3)$ is represented by the subgraph $fG(l_3)$ in the transition graph. The label $w_p i_s$ means that a transition along the edge is possible when w_p (i.e., lexical category of the word from the analyzed sentence) is the input symbol that is being read. During the transition the instruction i_s is performed. The label $w_p i_s q_t l_u$ means that during the transition, the instruction i_s is performed with the parameters q_t and l_u. The symbol r_k denotes a specific set of type w symbols; an edge marked with a label that contains the r_k

symbol represents a bundle of edges each of which is marked with one of the symbols belonging the r_k set.

The transition from the q_i to the q_j state at the input signal w_p has some degree of membership (usually denoted as $\mu_{ij} \in [0,1]$). The role for the characteristic function μ_{ij} which describes the transition relation of the fPDAMS automaton is as follows: for a given state q_i and input w_p of the automaton, the next state q_j of the automaton is determined so that to maximize the characteristic function value corresponding to the transition. Note that the fuzzy automaton is similar to the stochastic automaton. Both fuzzy and stochastic automata are examples of the acceptors of weighted languages. However, in stochastic automata the transitions are applied according to a probability distribution, there exists no uncertainty about the generated string. In fuzzy automata all applicable transitions are executed to some degree and transitions weights, in contrary to stochastic automaton, do not need to sum to 1, if there exist several alternative transitions at any state. The graph transition represents an analysis of a fragment of the input string of the lexical categories, the analysis result being stored on one of the six stacks of the fPDAMS automaton. The automaton operation is determined by a finite number of instructions. These instructions constitute the transition rules to allow the automaton to transit from one configuration to another. A limited number of instructions, straight rules of activity allow the use of fPDAMS as a kind of linguistic tool. The formal definition of the fPDMAS can be find in the Appendix.

An abridged form of fPDAMS is a PDAMS. The difference between both automata is that the PDAMS is a type of pushdown automaton rather than a fuzzy pushdown automaton. In PDAMS we simply skip the characteristic function μ_{ij}. The experiments that we have made base upon the PDAMS what doesn't decrease the general of our considerations.

3 Evolving fPDAMS Using Genetic Programming

Genetic programming [4], [5] breeds a population of rooted, point-labeled trees with ordered branches. From here also, to be able to use this class of evolutionary computation in evaluating the architecture of parser, one should first find a suitable method of mapping the transition graph of fPDMAS (or PDAMS) onto tree structure. There are two encoding techniques that we can apply, i.e., cellular encoding and edge encoding. Both methods rely in fact on operating on indirect structures, the so-called grammar-tree, which are subject to genetic programming process. Every such structure represents graph (for instance electric circuit [1], or of the transition graph of the fPDMAS). The grammar-tree is the genotype and the fPDMAS constructed in accordance with the tree's instruction is the phenotype.

The population of grammar-trees representing the fPDAMSs is subject to the standard scheme of evolutionary programming except that genetic programming allows sexual, genetic crossover operation. The recombination operator crossover is realized by exchanging the subtrees of two parent individuals. The mutation operator brings some variations to the genome and is realized by the inclusion of randomly initialized grammar-trees.

Genetic programming can operate on fPDMASs as follows:

1) Initially a population of parent fPDMASs is randomly or by hand constructed. The fPDMAS is represented by its grammar-tree.
2) The parents are tested in the environment, that is for each parent fPDMAS the collection of positive and negative examples of sentences is offered. The fitness of the automaton is then measured on the basis of the fraction of correctly analyzed sentences.
3) Offspring fPDAMSs are created by using the genetic operators: crossover and mutation. Operators can change both the topology and the connection weights. The topology-modifying operators can change the topology of the fPDMAS by: adding new state to the transition graph or removing the existing state. The operators changing the connection weigths operate on the terms $\mu w_p i_s q_t l_u$ modifing the individual elements.
4) The offsprings are evaluated over the existing environment in the same manner as their parents.
5) Those fPDMASs that provide the best fitness are retained to become parents of the next generation.
6) Steps 3-5 are repeated until the end condition is reached.

4 Cellular Encoding

Cellular encoding is a method for encoding families of similarly structured neural networks via genetic programming. Cellular encoding was previously proposed by Gruau [2] as a way of encoding a neural net on chromosomes that can be manipulated by the genetic algorithm. The cellular code is represented as a *grammar tree* with ordered branches whose nodes are labeled with name of *program symbols.* The reader must not confuse a grammar tree and tree grammar. A grammar tree means a grammar encoded as a tree, whereas tree grammar means a grammar that rewrites trees. This technique of encoding chromosomes is similar to the one used in genetic programming. A cell is a node of an oriented *network graph* with ordered connections. Each cell carries a duplicate copy of the cellular code (i.e., the grammar tree) and has an internal reading head that reads from the grammar tree. Typically, each cell reads from the grammar tree at a different position. The labels of the grammar tree represent instructions for cell development that act on the cell or on connections of the cell. During a step of the development process, a cell executes the instruction referenced by the symbol it reads, and moves its reading head down in the tree. We will refer to the grammar tree as a *program* and to each label as a *program-symbol.*

A cell also manages a set of internal registers, some of which are used during development, while others determine the weights and thresholds of the final neural net. The link register is used to refer to one of the several possibly fan-in connections (i.e., links) into a cell.

The order in which cells execute their genetic code is as follows. Once the cell executed operation denoted by symbol under its reading head it is entering at the end of execution queue. The next cell that executes its code is a cell from the head of the execution queue. If the cell divides then the new cell goes at the head of execution queue, and the original cell is placed at the tail of queue. New cell moves its reading head to the right subtree of grammar tree and the original cell moves its reading head to the left subtree. Development process stops when all cells from the queue reach the bottom of grammar tree. This order of execution implicates traversing grammar tree in breadth-fist, leftmost-children-first.

5 Edge Encoding for Induction of PDAMS Automaton

Another encoding scheme is edge encoding – a technique drawn up by Sean Luke and Lee Spector [7]. This method, just like cellular encoding, was originally developed to encode neural networks and is also capable of encoding a wide range of graphs. Similarly to cellular encoding, edge encoding is a tree-structured chromosome whose phenotype is a directed graph, optionally with labels. Edge encoding differs from cellular one in that the cellular encoding grows a graph by modifying the nodes in the graph while edge encoding operates especially on edges. Hence the leaf nodes in an edge-encoding chromosome represent unique edges in the resultant graph.

5.1 Program Symbols

Table 1 shows symbols specific to developing PDAMS automaton. They enable to modify topology of graph by adding edges, and semantic of the automaton by editing labels. To demonstrate the use of above-mentioned symbols a small automaton development will be introduced. The process of growth in sequenced steps is shown in Figure 3.

5.2 Fitness Evaluation

Each resultant automaton's graph obtained in the process of developing is represented a set of testing sentences. Automaton receives a note that depends on the quality of parsing. To calculate a fitness factor we apply a function consisting of several components (some of them based upon [6]).

Due to random generation of initial population most of individuals do not represent automaton graphs that would be able to parse correctly the entire input sentence. Fitness function must enable to pick up the automaton that could do anything about the input sentences.

Table 1. Program symbols for topology modification.

Symbol	Description
Double	Create edge F(a,b).[1]
DoubleRev	Create edge F(a,b) and reverse it.
Bud	Create node c. Create edge F(b,c).
BudRev	Create node c. Create edge F(b,c). Reverse edge F(b,c) to be F(c,b).
Split	Create node c. Change E to be E(a,c). Create edge F(c,b).
SplitRev	Create node c. Change E to be E(a,c). Create edge F(c,b). Reverse F to be F(b,c).
Loop	Create edge F(b,b).
Reverse	Reverse E to be E(b,a).
$\mathbf{w_1, w_2, ..., w_9}$	Label an edge with a "*w*" symbol, that is, define it to be an edge that can be traversed only on reading that w symbol.
$\mathbf{i_1, i_2, i_5, i_7}$	Label an edge with an "*i*" symbol, that is, define an automaton instruction that is executed while transition along this edge is performing.
SubA	Assign the head of E(a,b) (node b) to be an accepting state of a subautomaton.
A	Assign the head of E(a,b) (node b) to be an accepting state of a automaton.
S	Assign the head of E(a,b) to be the starting state.

[1] F(a,b) is an edge where a stands for node that is tail and b stands for node that is head of the edge.

The fitness function was decomposed down into components each of which respects a different aspect of PDAM's processing. First, there is the amount of input symbols consumed by automaton. The longer the part of the input sentence that it consumed, the higher the automaton's fitness should be. The relative length of consumed part of the sentence is used as a component of the fitness function. The following formulas define the component named *prefS*:

$$pref\_w = \frac{|w| - |v|}{|w|} \quad \text{and} \quad prefS = \sum_{w \in S} \frac{pref\_w}{|S|}, \tag{1}$$

where

$|w|$ - number of symbols in input sentence;
$|v|$ - length of unprocessed input;
$|S|$ - number of sentences in testing set.

The aim of the PDAMS is to break down the input sentence into structural-semantic components. The result of this process is written in one of PDAMS's stack. Therefore, another part of fitness function is a coefficient, name it *stackS*, that describes the content of that stack. The following formulas define coefficient *stackS*:

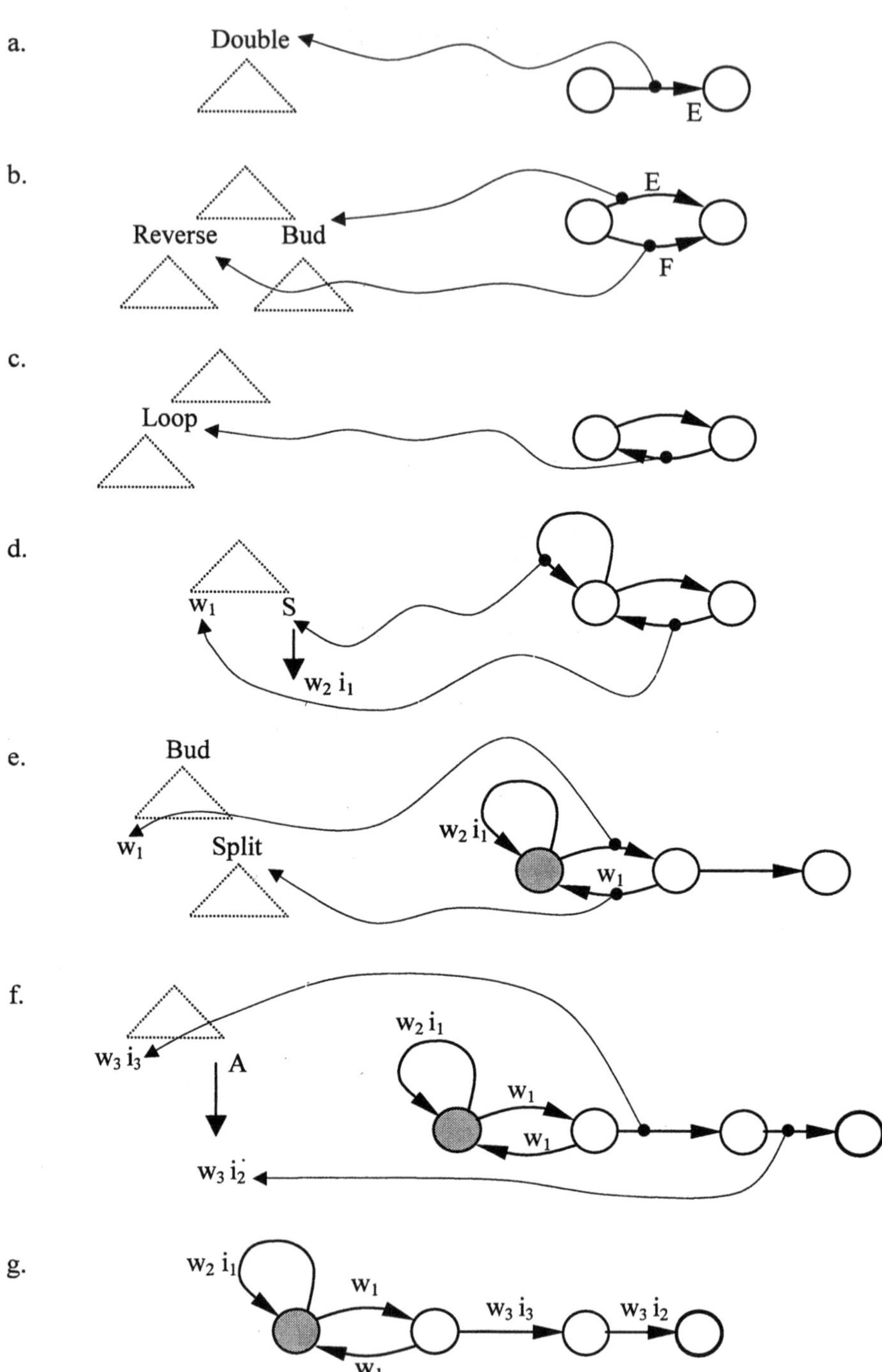

Figure 3. The growth of the example automaton.

$$stack\_w = \frac{|S_6|}{|w|} \quad \text{and} \quad stackS = \sum_{w \in S} \frac{stack\_w}{|S|} \tag{2}$$

The scheme encoding we used makes it possible to develop an automaton of any size. But during preliminary experiments the GA bred too huge automata, where relative number of used states (in relation to number of all states of automata) was about 0.01. We have added a mechanism for simply optimization of size of the automaton – two more components of fitness function that promote automata with higher usage of state and edges. The following formulas define these components – *su* and *eu*:

$$su = \frac{S_v}{S_t} \quad \text{and} \quad eu = \begin{cases} 0 \text{ if } V_t = 0 \\ \frac{V_v}{V_t} \text{ if } V_t > 0 \end{cases} \tag{3}$$

where
S_v – number of visited state;
S_t – total number of automaton states;
V_v – number of visited edges;
V_t – total number of automaton edges.

Finally we have defined total fitness as the following objective function:

$$fit = \frac{(a_1 prefS + a_2 stackS + a_3 eu + a_4 su)}{a_1 + a_2 + a_3 + a_4} \tag{4}$$

Coefficients a_1, a_2, a_3 and a_4 were estimated during the experiments.

We have used an unusual method to enrich the GA by specific knowledge. While automaton is working, its grammar tree (genotype that generated this automaton) is subject to assessment. Each node of the tree contains a counter, named *nScore*, that is modified each time the automaton makes a transition along the edge created by program symbol at that node. In simplest case counter is increased by one, if the analyzed sentence comes from set S_+ and decreased by one, if the sentence comes from set S_-. Furthermore, each node contains coefficient *pScore,* calculated after automaton ends its work, using the following formula.

$$pScore = \begin{cases} 0 \quad if \quad nScore = 0 \\ \frac{1}{nScore} \quad if \quad nScore > 0 \end{cases} \tag{5}$$

The coefficient pScore is a probability of choosing a node as a point for crossover or mutating operations. The more *nScore* value a node contains, the lower the probability of choosing the node is. As a result of using this method, some parts of genotype are weakened and other strengthened, which causes the building blocks to be created.

6 Results

6.1 Evolutionary Analyzer of Natural Language

For effective examining of the method presented above there was a prototype of evolutionary processing system (EPS) developed using C language. The core of the system is a collection of libraries. Each library encapsulates one encoding scheme and provides a set of functions to operate the scheme (creating and deleting objects such as genotype and phenotype, performing genetic operation, converting to/from user friendly form, etc.). The core of the system was supplied with the component to interact with the MATLAB computational environment. This additional part (called MEPS – Matlab Evolutionary Processing System) of the system is responsible for interacting with the core of the system from within MATLAB scripts and functions.

The MEPS system served as a base for an application working in the MATLAB environment – Evolutionary Analyzer of Natural Language (EAJN). The application was entirely developed in MATLAB language. It provides a simple user interface for setting up all parameters of experiments up and make it easy to generate reports of each aspect of the experiment in a nice form of HTML documents. All experiments were conducted using EANL.

6.2 Test Data and Parameters

To assess its performance the genetic algorithm was tested on three sets of sentences. The first set contained sample sentences generated by simple regular expressions. The second one was a subset of natural language sentences (adverb group). The last set contained full sentences of a natural language.

In all experiments the following parameters were used:

Initial population: Constant population size of 10. Randomly generated initial population. In spite of the use of constant population size the size of generated graphs varied from generation to generation. This was caused by the property of the encoding scheme that was used, which allowed generating phenotype of any size. The effect is similar to the use of variable population size - the amount of genetic material was changing during the evolution process.

Number of generation: It was different in each experiment. Each time the experiment was stopped after average fitness of the population achieved stable level. Due to different complexity in each case, the number of population the process had to run was different.

Selection: Tournament.

Mutation: Dynamic probability of mutation calculated according to the formula:

$$\text{pMutate} = (1 - \text{avgfit})/4 \tag{6}$$

where *avgfit* is an average fitness of the entire population in previous generation. Furthermore, mutation point was chosen with applying *pScore* coefficient.

Crossover: Standard crossover was used (exchanging subtrees). Cross point was chosen with applying pScore coefficient. In all experiments we use a crossover probability of 0.9.

6.3 Results

Conducted experiments were producing diverged results. In many probes the GA was losing for long time without finding any solution. Usually this led to reduction the size of the automata. But in some cases GA was hit the target. We present the results of experiments in which GA achieved the best scores.

Figure 4 shows variation of average fitness of entire population in experiments with sentences generated by regular expressions. Algorithm finds the optimal solution quite quickly. When using edge encoding, GA found automata optimal or at least near to optimal solution.

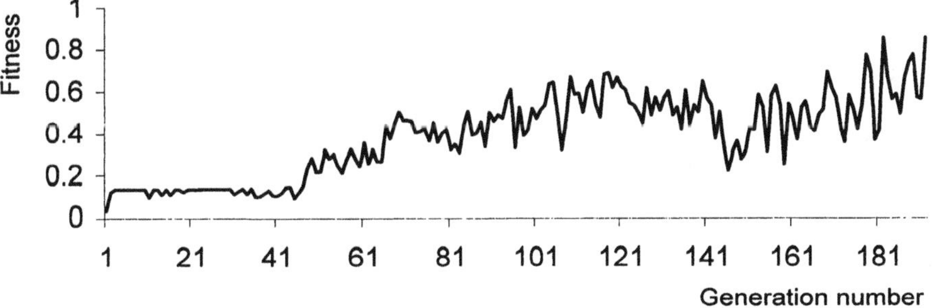

Figure 4. The average fitness experiment with regular expressions.

Figure 5 shows average fitness of population in experiment with English adverb group. As it can be seen, the maximal average fitness of solution found by GA is little more than 0.4. That indicates an increase of problem difficulty. After the fitness reached the level of 0.4 it kept this value for rest of the process. Apparently the GA was driven to a local optimum from which it could not escape.

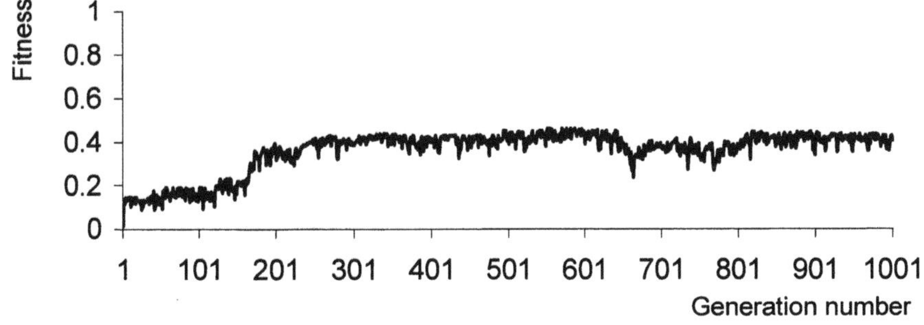

Figure 5. The average fitness in experiment with adverb group.

Figure 6 demonstrates average fitness of population in experiment with full natural language sentences. The more increase of difficulty causes even more reduction of population fitness. Also in this case the level of fitness stops increasing after it reached a certain value. Once again the GA was driven to a local optimum.

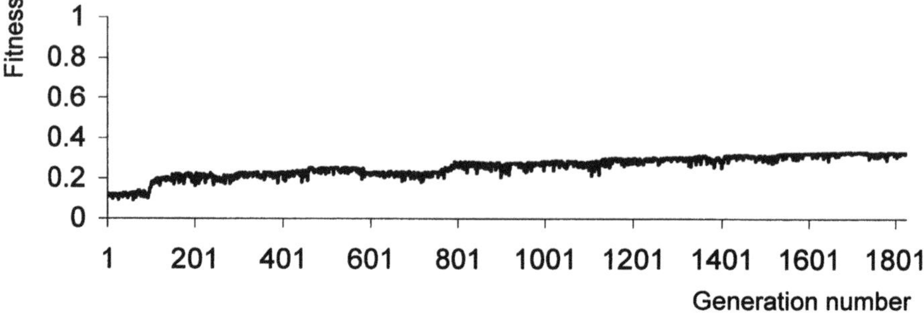

Figure 6. The average fitness in experiment with sentences of natural language.

The form of the fitness function apparently produces a difficult fitness landscape. This is the main reason for poor results of GA. Other aspects that influenced on the results are parameters of GA. There is a need to tune parameters such as probability coefficient of program symbols. They are significant for the structure of generating automata.

When analyzing graphs of generated automata, occurrence of similarly structured parts of graph can be clearly noticed. It seems to be the effect of appearing building blocks. Including specific knowledge by means of *pScore* coefficient of genotype was very helpful for that.

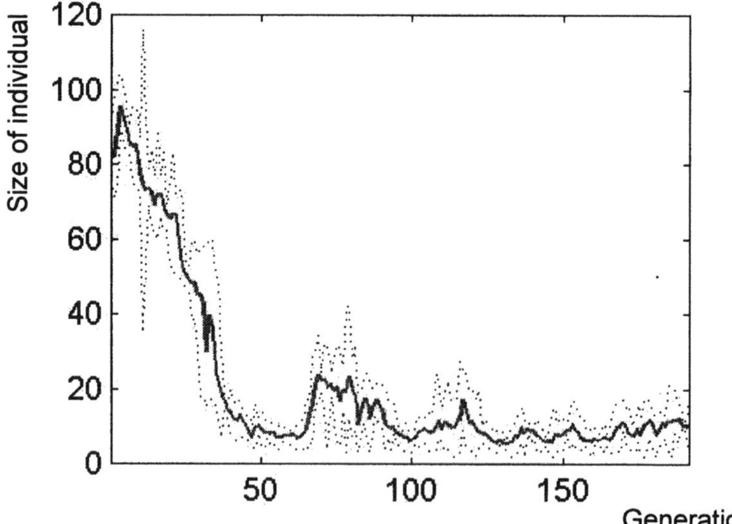

Figure 7. Average size of individual.

Another interesting aspect is the appearance of a variable amount of genetic material within a population. Figure 7 represents how the average size of individual was changing during one of the developing process. The further the GA from the solution is, the more genetic material population contains.

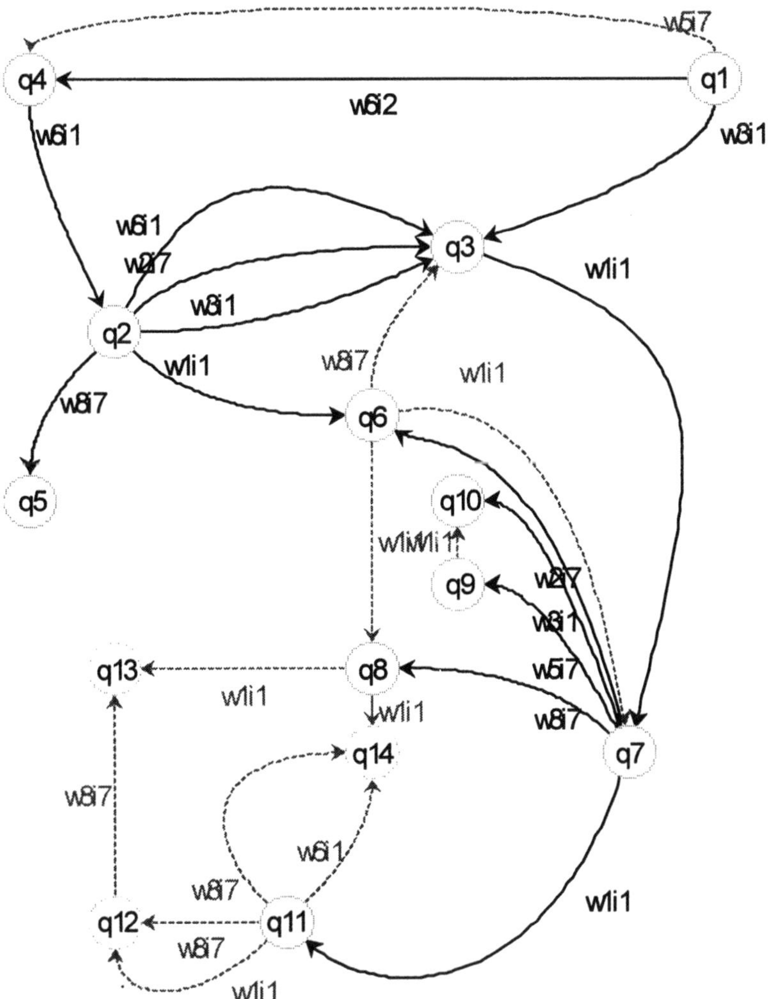

Figure 8. One of the automata achieved during the tests with sentences of natural language.

Although automata gained quite considerable complexity, actually none of generated automaton parsed an input sentence entirely properly. In most cases the content of stack was incorrect. This indicates that the objective function lacks the ability to pick out characteristics. Figure 8 shows one of automaton generated during the test with natural language sentences. Automaton doesn't contain instruction that writes to the stack moreover it lacks subautomaton structure, which is responsible for isolating substrings of sentence. The reason lies in selected probability for program symbols of edge encoding. Tuning those parameters up would improve the semantic of graph but requires many more experiments.

7 Conclusions

In this chapter, genetic programming has been shown as a support for automated inference of automaton-driven parser of natural language. As an encoding scheme a modified form of edge encoding was used. Basic edge encoding was enriched by a coefficient that is associated with each program symbol. This coefficient determines the probability of using a symbol for which it is defined. A set of such coefficients would be useful for constructing sophisticated methods for controlling of developing genotype by dynamic, or even evolutionary modification of these coefficients.

An interesting property of edge encoding is worth noticing. The size of the genotype and consequently, phenotype is not constant. During the developing process, the amount of genetic material varies just like in case of using variable size population does. Undoubtedly this property is very helpful in the process of solution finding. The further the GA is from solution is, the more genetic material population contains.

An unusual method of including specific knowledge to GA was provided – the assessment of genotype, which is helpful in arising building blocks.

The results of GA strongly depend on objective function. The one used in experiments was producing a very hard fitness landscape. It contained a lot of local optimums and rapid faults. The GA was often driven to a local optimum from which it did not escape.

Evolved graphs often lack stack instruction, which deprives them of the ability to parse correctly an input sentence. It is caused by two reasons. The first one is a set of probability of edge encoding program symbols. The second is the weakness of objective function – it cannot pick up characteristics responsible for creating appropriate structure of automaton.

8 Future Work

The method we propose shows great promise as a self-learning parser but still has few weak points. The results of GA strongly depend on the objective function. The one used in experiments was producing a very hard fitness landscape. It contained a lot of local optimums and rapid faults. The GA was often driven to a local optimum from which it did not escape.

Evolved graphs often lack stack instruction, which deprives them of the ability to parse correctly an input sentence. It is caused by two reasons. The first one is a set of probability of edge encoding operators. The second is the weakness of the objective function – it cannot pick up characteristics responsible for creating appropriate structure of the automaton.

Taking everything mentioned above into consideration we intend to follow several lines of research:

- experiment with 'stack-sensitive' operators,
- investigate alternative ways to compute objective function,
- consider automatically calculated probability of edge encoding operators.

References

1. Andre D, Bennet III F H, Koza J, Keane M, On the Theory of Designing Circuits using Genetic Programming and a Minimum of Domain Knowledge, Proc. of the 1998 IEEE Congress on Computional Intelligence WCCI'98, Anchorge, Alaska, 1998, pp. 130-135.
2. Gruau F, Cellular encoding of genetic neural networks, Technical report 92-91, Ecole Normale Superieure de Lyon, Institut IMAG, 1992.
3. Kandel A, Lee S C, Fuzzy Switching and Automata: Theory and Applications, Crane Russak, New York, 1979.
4. Koza J, Genetic Programming, MIT Press, Cambridge, MA, 1992.
5. Koza J, Genetic programming II, MIT Press, Cambridge, MA, 1994.
6. Lankhorst M M, A Genetic Algorithm for the induction of Nondeterministic Pushdown Automata. Computing Science Reports CS-R 9502, Department of Computing Science, University of Groningen, 1995.
7. Luke S, Spector L, Evolving Graphs and Networks with Edge Encoding: Preliminary Report. [in:] Koza J (eds.) Late-breaking Papers of Genetic Programming 96, Stanford Bookstore, 1996, pp. 117-124
8. Roche E, Schabes Y, Finite-State Language Processing, A Bradford Book, The MIT Press, Cambridge, Massachusetts, 1997.
9. Unold O, A Fuzzy Automaton Approach to Dialog Systems, Proc. of the IASTED International Conference-ASC'98, Cancun, Mexico, May 1998, pp. 215-218.
10. Unold O, Application of Fuzzy Sets in Natural Language Processing, Proc. of the 6th Congress on Intelligent Techniques and Soft Computing EUFIT'98, Aachen, Germany, September 1998, pp.1262-1266.
11. Unold O, Toward fuzziness in natural language processing, [in:] Roy R at al [eds.] Advances in Soft Computing – Engineering Design and Manufacturing, Springer Verlag, London, 1999, pp. 554-567.
12. Unold O, An evolutionary approach for the design of natural language parser. [in:] Suzuki Y at al [eds.] Soft Computing in Industrial Applications, Springer Verlag, London, 2000, pp. 293-297.

Appendix: Formal Definition of the fPDAMS

A.1 The fPDAMS Definition

Formally we consider the fPDAMS as a 10-tuple

$$\mathbf{fPDAMS} = (Q, \Gamma, F, \delta, z_0, q_0, q_f, L, T, p), \tag{7}$$

where

Q is a finite set of internal states ($q_i \in Q$),

$q_0 \in Q$ is a distinguished state, so-called an initial state,

$q_f \in Q$ is a distinguished state, so-called a terminal state,

$F \subset Q$ is a distinguished subset of the set Q,
so-called a set of terminal states of subautomata,

L is a finite set of l-markers ($L = \{l_1, l_2, l_3, l_4, l_5, l_6\}$), (8)

T is a finite set of t-markers ($T = \{t_1, t_2\}$), (9)

p is an element, so-called a p-marker,

$$\Gamma = W \cup Z_1 \cup Z_2 \cup Z_3 \cup Z_4 \cup Z_5 \cup Z_6, \tag{10}$$

where

W is an alphabet of an input tape:

$$W = \{w_0, w_1, w_2, w_3, w_4, w_5, w_6, w_7, w_8, w_9\}, \tag{11}$$

Z_1 is an alphabet of a stack S_1 ($Z_1 = Q$), (12)

Z_2 is an alphabet of a stack (Z_2 is a set of natural numbers),

Z_3 is an alphabet of a stack S_3 ($Z_3 = W \cup T \cup L$), (13)

Z_4 is an alphabet of a stack S_4 ($Z_4 = Q$), (14)

Z_5 is an alphabet of a stack S_5 ($Z_5 = W \cup T \cup L \cup p$), (15)

Z_6 is an alphabet of a stack S_6 ($Z_6 = W \cup T \cup L \cup p$), (16)

$$z_0 = (z_{00}, z_{01}, z_{02}, z_{03}, z_{04}, z_{05}, z_{06}) \in W \times Z, \tag{17}$$

where

$$Z = Z_1 \times Z_2 \times Z_3 \times Z_4 \times Z_5 \times Z_6, \tag{18}$$

is a 7-tuple of initial symbols of the input tape and the stacks respectively.

Let $M_i \in (d \times \{m_l, m_h, m_r\})$ for i = 1, ..., 6 be intended to represent the shift of an i-th head, where m_h denotes inertness of the head, m_l – move the head d-th tape square to the left, m_r – move the head d-th tape square to the right. M_0 represents the shift of the input tape.

Let
$$M = M_0 \times M_1 \times M_2 \times M_3 \times M_4 \times M_5 \times M_6 \tag{19}$$
and
$$\Phi = Q \times W \times Z. \tag{20}$$

Now the transition relation δ of fPDAMS is defined as follows:

$$\delta : \Phi \to 2^{\Phi x M} \tag{21}$$

Here, δ performs a mapping from the product space Φ to the set of the product space Φ × M. The product Φ is a 8-tuple consisting of a current state, an input symbol and six stack symbols. Each element of the set Φ × M consists of a next state and seven pairs, each of which defines a new symbol of the input tape or stacks and both direction and length of head motion.

δ is the fuzzy relation with a membership grade

$$\mu : \Phi \times \Phi \times M \to [0,1] \tag{22}$$

Thus $\mu(q_i, w_{j1}, z_k, q_j, w_{j2}, z_p, M_r) \in [0, 1]$, where $q_i, q_j \in Q$, $w_{j1}, w_{j2} \in W$, $z_k, z_p \in Z$, $M_r \in M$.

In the fPDAMS each transition has its own „strength" - a grade of membership in a fuzzy set.

The grade of transition for an input sequence of length *m* is defined by an *m*-ary fuzzy relation. The membership grade is as follows.

For an input word $w_{j1}w_{j2}...w_{jm} \in W^*$

$$\begin{aligned}\mu(q_{i1}, w_{j1}, z_{k1}, q_{im}, w_{jm}, z_{km}, M_{rm}) = \\ \max_{q_{i2}, q_{i3}, \dots, q_{im-1} \in Q} \min \quad & \mu(q_{i1}, w_{j1}, z_{k1}, q_{i2}, \sigma_1, z_{k2}, M_{r1}), \\ & \mu(q_{i2}, \sigma_2, z_{k2}, q_{i3}, \sigma_3, z_{k3}, M_{r2}), \dots, \\ & \mu(q_{im-1}, \sigma_j, z_{km-1}, q_{im}, w_{jm}, z_{km}, M_{rm}),\end{aligned} \tag{23}$$

where $z_{k1}, z_{k2}, \dots, z_{km} \in Z$, $M_{r1}, M_{r2}, \dots, M_{rm} \in M$, $\sigma_1, \sigma_2, \dots, \sigma_j \in W$.

Lets take notice of the fact that the fuzzy transition relation δ does not have to change in one move neither internal state of the fPDAMS nor input element being read.

Formally an activity of fPDAMS is described by a configuration. We consider the configuration of fPDAMS as a 7-tuple

$$K = (\alpha_0, \alpha_1, \alpha_2, \alpha_3, \alpha_4, \alpha_5, \alpha_6), \tag{24}$$

where α_0 is a triple (q_a, x, y), where q_a will stand for the current state of automaton, x represents the word written in input tape from the beginning to the current head position, y is the word written on the right of the head, while α_i (i = 1,..., 6) is a couple (q_a, v), where v denotes the word written in i-th stack (on the left of the head).

The binary relation

$$K_t \underset{\mu_{t,t+1}}{\Rightarrow} K_{t+1} \tag{25}$$

stands for the transition from the configuration K_t to the configuration K_{t+1} in one step with a membership grade $\mu_{t,t+1}$, where the subscript *t* denotes the configuration K at the time *t*.

The notation

$$K_t \overset{*}{\underset{\mu_{t,t+j}}{\Rightarrow}} K_{t+j} \tag{26}$$

stands for a configuration transition path

$$K_t \underset{\mu_{t,t+1}}{\Rightarrow} K_{t+1} \underset{\mu_{t+1,t+2}}{\Rightarrow} \dots \underset{\mu_{t+j-1,t+j}}{\Rightarrow} K_{t+j} \tag{27}$$

with a membership grade

$$\mu_{t,t+j} = \max_{t_{i+1}, t_{i+2}, \dots, t_{i+j-1}} \min \; \mu_{t,t+1}, \mu_{t+1,t+2}, \dots, \mu_{t+j-1,t+j} \tag{28}$$

The input word $w_{j1}w_{j2}\dots w_{jn}$ is accepted (splitted) by fPDAMS only if $K_t \overset{*}{\underset{\mu}{\Rightarrow}} K_f$ exists and $\mu > 0$, where

$$K_0 = ((q_0, z_{00}w_{j1}, w_{j2}w_{j3}\dots w_{jn}), (q_0, z_{01}),\dots,(q_0, z_{06})), \tag{29}$$

$$K_f = (\alpha_0, \alpha_1, \dots, \alpha_6) \text{ and } q_f \text{ in } \alpha_i \ (i=0, \dots,6). \tag{30}$$

The configuration K_0 is termed the initial configuration of fPDAMS, whereas the configuration K_f is called the terminal (final) configuration of fPDAMS. It should be noted that fPDAMS accepts the input word (divides it into subwords) as soon as it enters the terminal state.

A.2 The fPDAMS Instructions

The rules of the transitions from one configuration to the other one are clustered in the instructions below.

Let O, N, P, M, U denote the words over an alphabet of an input tape (O, N, P, M, U $\in W^*$),

A is a word over the stack alphabet S_1 ($A \in Q^*$),

B is a word over the stack alphabet S_2 ($B \in Z_2^*$),

C i J are words over the stack alphabet S_3 ($C,J \in (W \cup T \cup L)^*$),

D is a word over the stack alphabet S_4 ($D \in Q^*$),

E is a word over the stack alphabet S_5 ($E \in (W \cup T \cup L \cup p)^*$),

G is a word over the stack alphabet S_6 ($G \in (W \cup T \cup L \cup p)^*$).

Let K^* represent the set of all possible configurations of fPDAMS

$$K^* = \{K \mid K_0 \overset{*}{\underset{\mu}{\Rightarrow}} K \text{ and } \mu > 0 \text{ and}$$

$$K_0 = ((q_0, z_{00}w_{j1}, w_{j2}O), (q_0, z_{01}), \ldots, (q_0, z_{06})),$$
$$K_0, K \in q \times z_{00}W^* \times W^* \times z_{00}Z_1^* \times z_{02}Z_2^* \times \ldots \times z_{06}Z_6^*,$$
$$\text{where } q \in Q\}. \tag{31}$$

Each instruction of fPDAMS belongs to one of seven types. Now, we will show the formal notation of the instructions:

1) type i_1: $(\exists K_1, K_2 \in K^*)$

$$[\text{if}\,(K_1 \underset{\mu_{1,2}}{\Rightarrow} K_2 \text{ and } \mu_{1,2} > 0)$$
$$\text{then } (K_1 = ((q_i, z_{00}Nw_{jk}, w_{jk+1}P), (q_i, z_{01}A), (q_i, z_{02}B), (q_i, z_{03}C), (q_i, z_{04}D), (q_i, z_{05}E), (q_i, z_{06}G)),$$
$$K_2 = ((q_i, z_{00}Nw_{jk}w_{jk+1}, P), (q_i, z_{01}A), (q_i, z_{02}B), (q_i, z_{03}Cw_{jk}), (q_i, z_{04}D), (q_i, z_{05}E), (q_i, z_{06}G)))] \tag{32}$$

{execution of i_1 performs a pushing down onto the stack S_3 the symbol read by the input tape head and moves the input tape head one tape square to the right}

2) type i_2: $(\exists K_1, K_2 \in K^*)$

$$[\text{if}\,(K_1 \underset{\mu_{1,2}}{\Rightarrow} K_2 \text{ and } \mu_{1,2} > 0)$$
$$\text{then } (K_1 = ((q_i, z_{00}Nw_{jk}, w_{jk+1}P), (q_i, z_{01}A), (q_i, z_{02}B), (q_i, z_{03}C), (q_i, z_{04}D), (q_i, z_{05}E), (q_i, z_{06}GRSX)),$$
$$K_2 = ((q_i, z_{00}Nw_{jk}, w_{jk+1}P), (q_i, z_{01}A), (q_i, z_{02}B), (q_i, z_{03}CSX), (q_i, z_{04}D), (q_i, z_{05}E), (q_i, z_{06}GR)),$$
$$R, X \in (W \cup T \cup p)^*, S \in L^*)] \tag{33}$$

{execution of i_2 pops a group of symbols from stack S_6 and pushes it onto stack S_3. This group comprises the symbols from the set W, the l-marker and the string of succeeding l-markers}

3) type i_3: $(\exists K_1, K_2 \in K^*)$

$$[\text{if}\ (K_1 \underset{\mu_{1,2}}{\Rightarrow} K_2 \ \text{and}\ \mu_{1,2} > 0)$$

$$\text{then}\ (K_1 = ((q_i, z_{00}Nw_{jk}, w_{jk+1}P), (q_i, z_{01}A), (q_i, z_{02}B), (q_i, z_{03}C), (q_i, z_{04}D), (q_i, z_{05}E), (q_i, z_{06}G)),$$

$$K_2 = ((q_i, z_{00}Nw_{jk}, w_{jk+1}P), (q_i, z_{01}Aq_i), (q_i, z_{02}Bh), (q_i, z_{03}C), (q_i, z_{04}D), (q_i, z_{05}EpC), (q_i, z_{06}Gp)),$$

$$h \text{ is a natural number})] \quad (34)$$

{execution of i_3 stores the elements of current configuration in stack memory: the symbol of current state q_i is pushed down on top of stack S_1, the number of read input symbols h is pushed down on top of stack S_2, p-marker and next contents of stack S_3 are pushed on S_5, p-marker is pushed on S_6}

4) type i_4: $(\exists K_1, K_2 \in K^*)$

$$[\text{if}\ (K_1 \underset{\mu_{1,2}}{\Rightarrow} K_2 \ \text{and}\ \mu_{1,2} > 0)$$

$$\text{then}\ (K_1 = ((q_i, z_{00}Nw_{jk}, w_{jk+1}P), (q_i, z_{01}Aq_j), (q_i, z_{02}Bh), (q_i, z_{03}C), (q_i, z_{04}D), (q_i, z_{05}EpJ), (q_i, z_{06}GpY)),$$

$$K_2 = ((q_j, z_{00}Mw_{jh}, w_{jh+1}U), (q_j, z_{01}A), (q_j, z_{02}B), (q_j, z_{03}J), (q_j, z_{04}D), (q_j, z_{05}E), (q_j, z_{06}G)),$$

$$h \text{ is a natural number},\ Y \in (W \cup L \cup T)^*)] \quad (35)$$

{execution of i_4 performs a return of the automaton to the configuration (denoted as K_a) popped from stack memory. From the top of stack S_1 the symbol q_j is popped, q_j becomes a current state of automaton. The contents of the stack S_3 on the ground of the contents of the stack S_5 is reconstructed. The number h of read-in input symbols in configuration K_a is popped from the top of stack S_2, the input tape head is shifted to the h-th tape square. At last the string of symbols stretching from the top of the stack to the first p-marker is popped from the stack S_6}

5) type i_5 (with arguments q_k i l_t): $(\exists K_1, K_2 \in K^*)$

$$[\text{if}\ (K_1 \underset{\mu_{1,2}}{\Rightarrow} K_2 \ \text{and}\ \mu_{1,2} > 0)$$

$$\text{then}\ (K_1 = ((q_i, z_{00}Nw_{jk}, w_{jk+1}P), (q_i, z_{01}A), (q_i, z_{02}B), (q_i, z_{03}C), (q_i, z_{04}D), (q_i, z_{05}E), (q_i, z_{06}G)),$$

$$K_2 = ((q_i, z_{00}Nw_{jk}, w_{jk+1}P), (q_i, z_{01}A), (q_i, z_{02}B), (q_i, z_{03}Cl_t t_1), (q_i, z_{04}Dq_k), (q_i, z_{05}E), (q_i, z_{06}G)))] \quad (36)$$

{execution of i_5 performs a pushing down onto the stack S_4 the symbol q_k and onto the stack S_3 l_t and t_1-markers}

6) type i_6: $(\exists K_1, K_2 \in K^*)$

$$[\text{if} (K_1 \underset{\mu_{1,2}}{\Rightarrow} K_2 \text{ and } \mu_{1,2} > 0)$$

$$\text{then } (K_1 = ((q_i, z_{00}Nw_{jk}, w_{jk+1}P), (q_i, z_{01}A), q_i, z_{02}B), (q_i, z_{03}C), (q_i, z_{04}Dq_k), (q_i, z_{05}E), (q_i, z_{06}G)),$$
$$K_2 = ((q_k, z_{00}Nw_{jk}, w_{jk+1}P), (q_k, z_{01}A), (q_k, z_{02}B), (q_k, z_{03}C), (q_k, z_{04}D),(q_k, z_{05}E), (q_k, z_{06}G)))] \quad (37)$$

{execution of i_6 changes the current state of automaton to a state popped from stack S_4. The instruction stands for the return of control from a subautomaton}

7) type i_7: $(\exists K_1, K_2 \in K^*)$

$$[\text{if} (K_1 \underset{\mu_{1,2}}{\Rightarrow} K_2 \text{ and } \mu_{1,2} > 0)$$

$$\text{then } (K_1 = ((q_i, z_{00}Nw_{jk}, w_{jk+1}P), (q_i, z_{01}A), (q_i, z_{02}B), (q_i, z_{03}C), (q_i, z_{04}D), (q_i, z_{05}E), (q_i, z_{06}G)),$$
$$K_2 = ((q_i, z_{00}Nw_{jk}, w_{jk+1}P), (q_i, z_{01}A), (q_i, z_{02}B), (q_i, z_{03}), (q_i, z_{04}D),(q_i, z_{05}E), (q_i, z_{06}GCt_2)))] \quad (38)$$

{execution of i_7 pushes down the contents of stack S_3 and next t_2-marker onto the stack S_6. Afterwards the stack S_3 is cleaned}

A.3 The fPDAMS Algorithm

In the beginning, fPDAMS is in the initial configuration K_0. There is the string $w_{j1}w_{j2}...w_{jn}$ written in the input tape. The current state of the automaton is denoted by the symbol q_a. The algorithm of working fPDAMS proceeds as follows:

(1) if $q_a = q_f$ then success.

{if the automaton entered the terminal state, then it stops. There is the splitted string (w) in the stack S_6}.

(2) if $q_a \in F$ then execute i_6 and goto step 1.

{if the automaton entered one of the terminal states of the subautomaton, then top of the stack S_4 indicates the next state}

(3) $K(q_a) = \{K \mid K^a \underset{\mu}{\Rightarrow} K$ and
$\mu > 0$ and
$K^a = ((q_a, z_{00}N, P), (q_a, z_{01}A), (q_a, z_{02}B), (q_a, z_{03}C), (q_a, z_{04}D), (q_a, z_{05}E), (q_a, z_{06}G)),$
$K = ((q_x, z_{00}N', P'), (q_x, z_{01}A'), (q_x, z_{02}B'), (q_x, z_{03}C'), (q_x, z_{04}D'),(q_x, z_{05}E'), (q_x, z_{06}G'))$ and
$K^a, K \in K^*\}$ (39)

$\mu(q_a) = \{\mu \mid K^a \underset{\mu}{\Rightarrow} K$ and
$\mu > 0$ and
$K^a = ((q_a, z_{00}N, P), (q_a, z_{01}A), (q_a, z_{02}B), (q_a, z_{03}C), (q_a, z_{04}D), (q_a, z_{05}E), (q_a, z_{06}G)),$
$K = ((q_x, z_{00}N', P'), (q_x, z_{01}A'), (q_x, z_{02}B'), (q_x, z_{03}C'), (q_x, z_{04}D'),(q_x, z_{05}E'), (q_x, z_{06}G'))$ and
$K^a, K \in K^*\}$ (40)

{the set $K(q_a)$ represents the set of every possible configurations of fPDAMS in the current state. The set $\mu(q_a)$ represents the set of the membership functions of moves from the current state to the next ones}

(4) if $\|K(q_a)\| \geq 1$then

(4a) choose the next configuration $K_{next} \in K(q_a)$, where $\mu_{next} = \max_{\mu} \mu(q_a)$, and

$K(q_a)=K(q_a) \setminus K_{next}$.

{if for the current state there is the set consisting of more than one next state, then choose one with the maximum membership grade and decrease the set of next states}

(4b) if $\|K(q_a)\| > 0$ then execute i_3.

{if the set of next states is still non-empty then store the elements of current configuration in stack memory}

(4c) perform $K^a \underset{\mu}{\Rightarrow} K_{next}$ and goto step 1.

(5) if stack S_1 is empty then fail.

{there is no configuration of the automaton with the non-empty set of next states. fPDAMS does not accept ,nor does it split, the input word}

(6) execute i_4 and goto step 4.

{return of automaton to the configuration popped from stack memory}.

Index

- D -

- E -

List of Contributors

Ajith Abraham
Department of Computer Science
Oklahoma State University
700 N Greenwood Avenue, Tulsa, OK 74106-0700
USA
ajith.abraham@ieee.org

Rod Adams
Department of Computer Science,
University of Hertfordshire
England, UK
R.G.Adams@herts.ac.uk

S.J. Cavill
Department of Informatics
Cranfield University (RMCS)
Shrivenham, Swindon SN6 8LA
UK

Neil Davey
Department of Computer Science,
University of Hertfordshire
England, UK
N.Davey@herts.ac.uk

Joachim Denzler
Friedrich Alexander Universität Erlangen Nürnberg
91058 Erlangen
Germany

Rainer Deventer
Friedrich Alexander Universität Erlangen Nürnberg
91058 Erlangen
Germany

Maria do Carmo Nicoletti
UFSCar/DC
São Carlos - SP
Brazil
carmo@dc.ufscar.br

Grzegorz Dulewicz
VICTOR Ltd.
Poland
grzesiekd@visitorsoft.eu.org

D. Essam
School of Computer Science, University College
The University of New South Wales
Australian Defence Force Academy
Canberra, ACT2600
Australia
essam@cs.adafa.edu.au

Noriko Etani
Graduate School of Information Science
Nara Institute of Science and Technology
8916-5 Takayama-cho Ikoma-shi Nara 630-0101
Japan
netani@acm.org

A.J.B. Fogg
Department of Orthopaedics
Princess Margaret Hospital
Okus Road, Swindon SN1 4JU
UK

M.A. Foy
Department of Orthopaedics
Princess Margaret Hospital
Okus Road, Swindon SN1 4JU
UK

Luciano García
Facultad de Matemáticas y Computación
Universidad de La Habana
Colina Universitaria, Plaza, La Habana 10400
Cuba

Liqiang Geng
Department of Computer Science
University of Regina
Regina, Saskatchewan, Canada S4S 0A2
gengl@cs.uregina.ca

Stella George
Department of Computer Science,
University of Hertfordshire
England, UK
S.J.George@herts.ac.uk

Dušan Guller
Institute of Informatics
Comenius University
Mlynská dolina, 842 15 Bratislava
Slovakia
guller@fmph.uniba.sk

N.X. Hoai
School of Computer Science, University College
The University of New South Wales
Australian Defence Force Academy
Canberra, ACT2600
Australia
x.nguyen@student.adfa.edu.au

Jaakko Hollmén
Laboratory of Computer and Information Science
Helsinki University of Technology
P.O. Box 5400, FIN-02015 HUT
Finland
Jaakko.Hollmen@hut.fi

Howard J. Hamilton
Department of Computer Science
University of Regina
Regina, Saskatchewan, Canada S4S 0A2
hamilton@cs.uregina.ca

Muhammad Riaz Khan
AMEC Technologies, TTI,
400-111 Dunsmuir Street, Vancouver, BC V6B 5W3
Canada
riaz.khan@amec.com

András Kornai
Metacarta Inc
875 Massachusetts Ave, Cambridge MA 02139
USA
andras@kornai.com

Musa A. Mamedov
School of Information Technology and Mathematical Sciences
The University of Ballarat
Ballarat, Vic 3353
Australia

Luis Martí
Università degli Studi di Udine
Dipartimento di Matematica e Informatica
via delle Scienze 208, Udine 33100 (UD)
Italy
and
Facultad de Matemáticas y Computación
Universidad de La Habana
Colina Universitaria, Plaza, La Habana 10400
Cuba

R.I. McKay
School of Computer Science, University College
The University of New South Wales, Australian Defence Force Academy
Canberra, ACT2600
Australia
rim@cs.adafa.edu.au

Tomoko Murakami
Knowledge Media Laboratory
Corporate R&D Center, Toshiba Corp.
1 Komukai-Toshiba-cho, Saiwai-ku, Kawasaki 210-8582
Japan

M. Narasimha Murty
Department of Computer Science and Automation
Indian Institute of Science
Bangalore 560 012
India
mnm@csa.iisc.ernet.in

Kazumi Nakamatsu
Himeji Institute of Technology
Shinzaike, Himeji 670-0092
Japan

Heinrich Niemann
Friedrich Alexander Universität Erlangen Nürnberg
91058 Erlangen
Germany

Ryohei Orihara
Middleware Department
e-Solution Company, Toshiba Corp.
1 Toshiba-cho, Fuchu-shi, Tokyo 183-8511
Japan

Alberto Policriti
Università degli Studi di Udine
Dipartimento di Matematica e Informatica
via delle Scienze 208, Udine 33100 (UD)
Italy

Shigeaki Sakurai
Knowledge Media Laboratory
Corporate R&D Center, Toshiba Corp.
1 Komukai-Toshiba-cho, Saiwai-ku, Kawasaki 210-8582
Japan

Lisa Stone
PPD Informatics/Belmont Research
84 Sherman St, Cambridge MA 02140
USA

Naomichi Sueda
Department of Computer Science and Intelligent Systems
Faculty of Engineering, Oita University
700 Dannoharu, Oaza, Oita 870-1192
Japan

Ken Tabb
Department of Computer Science,
University of Hertfordshire
England, UK
K.J.Tabb@herts.ac.uk
http://www.health.herts.ac.uk/ken/vision/

S.J. Taylor
Department of Orthopaedics
Princess Margaret Hospital
Okus Road, Swindon SN1 4JU
UK

Goran Trajkovski
Computer and Information Sciences Department
Towson University
8000 York Road, Towson, MD 21252
USA
gtrajkovski@towson.edu

Joaquim Quinteiro Uchôa
UFLA/DCC
Lavras - MG
Brazil
joukim@comp.ufla.br

Olgierd Unold
Institute of Engineering Cybernetics
Wroclaw University of Technology
Wyb. Wyspianskiego 27, 50-370 Wroclaw
Poland
unold@ci.pwr.wroc.pl

M.L. Vaughn
Department of Informatics
Cranfield University (RMCS)
Shrivenham, Swindon SN6 8LA
UK

Juha Vesanto
Laboratory of Computer and Information Science
Helsinki University of Technology
P.O. Box 5400, FIN-02015 HUT
Finland
Juha.Vesanto@hut.fi

S.V.N. Vishwanathan
Department of Computer Science and Automation
Indian Institute of Science
Bangalore 560 012
India
vishy@csa.iisc.ernet.in

John Yearwood
School of Information Technology and Mathematical Sciences
The University of Ballarat
Ballarat, Vic 3353
Australia .

GPSR Compliance
The European Union's (EU) General Product Safety Regulation (GPSR) is a set of rules that requires consumer products to be safe and our obligations to ensure this.

If you have any concerns about our products, you can contact us on

ProductSafety@springernature.com

In case Publisher is established outside the EU, the EU authorized representative is:

Springer Nature Customer Service Center GmbH
Europaplatz 3
69115 Heidelberg, Germany

www.ingramcontent.com/pod-product-compliance
Ingram Content Group UK Ltd.
Pitfield, Milton Keynes, MK11 3LW, UK
UKHW021933200726
13853UKWH00010B/1356